Dirk Holtmann (Ed.)
Bioprocess Intensification

Also of interest

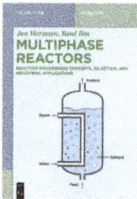

Multiphase Reactors
Reaction Engineering Concepts, Selection, and Industrial Applications
Harmsen, Bos, 2023
ISBN 978-3-11-071376-3, e-ISBN (PDF) 978-3-11-071377-0,
e-ISBN (EPUB): 978-3-11-071384-8

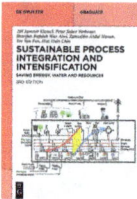

Sustainable Process Integration and Intensification
Saving Energy, Water and Resources
Klemeš, Varbanov, Wan Alwi, Manan, Fan, Chin, 2023
ISBN 978-3-11-078283-7, e-ISBN (PDF) 978-3-11-078298-1,
e-ISBN (EPUB) 978-3-11-078300-1

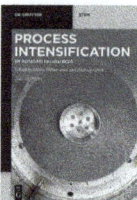

Process Intensification by Rotating Packed Beds
Skiborowski, Górak (Eds.), 2022
ISBN 978-3-11-072490-5, e-ISBN (PDF) 978-3-11-072499-8,
e-ISBN (EPUB) 978-3-11-072509-4

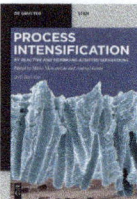

Process Intensification by Reactive and Membrane-assisted Separations
Skiborowski, Górak (Eds.), 2022
ISBN 978-3-11-072045-7, e-ISBN (PDF) 978-3-11-072046-4,
e-ISBN (EPUB) 978-3-11-072060-0

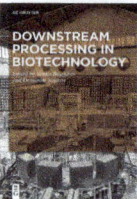

Downstream Processing in Biotechnology
Beschkov, Yankov (Eds.), 2021
ISBN 978-3-11-057395-4, e-ISBN (PDF) 978-3-11-057411-1,
e-ISBN (EPUB) 978-3-11-057400-5

Bioprocess Intensification

Edited by
Dirk Holtmann

DE GRUYTER

Editor
Prof. Dr.-Ing. Dirk Holtmann
Karlsruher Institut für Technologie (KIT)
Fritz-Haber-Weg 4
76131 Karlsruhe
dirk.holtmann@kit.edu

ISBN 978-3-11-076032-3
e-ISBN (PDF) 978-3-11-076033-0
e-ISBN (EPUB) 978-3-11-076037-8

Library of Congress Control Number: 2024934627

Bibliographic information published by the Deutsche Nationalbibliothek
The Deutsche Nationalbibliothek lists this publication in the Deutsche Nationalbibliografie;
detailed bibliographic data are available on the internet at http://dnb.dnb.de.

Contents

Axel Schmidt, Alina Hengelbrock and Jochen Strube

6   Continuous biomanufacturing in upstream and downstream processing — 117

Markus Stöckl, André Gemünde and Dirk Holtmann

7   Microbial electrotechnology – Intensification of bioprocesses through the combination of electrochemistry and biotechnology — 173

Michael E. Runda and Sandy Schmidt

Sera Bolat, Raphael Greifenstein, Matthias Franzreb and Dirk Holtmann

Jochen Schaub, Andreas Ankenbauer, Tobias Habicher,Michael Löffler, Nicolas Maguire, Dominique Monteil, Sebastian Püngel, Lisa Stepper, Fabian Stiefel, Judith Thoma, Andreas Unsöld, Julia Walther, Christopher Wayne and Thomas Wucherpfennig

List of contributing authors

Andreas Ankenbauer
DevOps Biologics DS Germany
HP BioP Launch&Innovation
Boehringer Ingelheim Pharma GmbH & Co. KG
Biberach a.d. Riß
Germany

Sera Bolat
Institute of Bioprocess Engineering and
Pharmaceutical Technology
University of Applied Sciences Mittelhessen
Wiesenstrasse 14
35390 Giessen
Germany

Eduardo J. S. Brás
NMI Natural and Medical Sciences Institute
University of Tübingen
Reutlingen
Germany
E-mail: eduardo.bras@nmi.de

André Delavault
Institute of Bio- and Food Technology Division II –
Technical Biology
KIT – Karlsruhe Institute of Technology
Karlsruhe
Germany
E-mail: andre.delavault@kit.edu
https://orcid.org/0000-0002-2232-1580

Anna Dinius
Institute of Biochemical Engineering
Technische Universität Braunschweig
Rebenring 56
38106 Braunschweig
Germany
And
Center of Pharmaceutical Engineering
Technische Universität Braunschweig
Franz-Liszt-Str. 35a
38106 Braunschweig
Germany

Pedro Carlos de Barros Fernandes
iBB—Institute for Bioengineering and Biosciences
Instituto Superior Técnico
Universidade de Lisboa
Av. Rovisco Pais
1049-001 Lisboa
Portugal
and
Associate Laboratory i4HB
Institute for Health and Bioeconomy at Instituto
Superior Técnico
Universidade de Lisboa
Av. Rovisco Pais
1049-001 Lisboa
Portugal
And
BioRG and DREAMS
Faculty of Engineering
Lusofona University (ULHT)
Campo Grande, 376
1749 024 Lisbon
Portugal
E-mail: pedro.fernandes@ist.utl.pt
https://orcid.org/0000-0003-0271-7796

Matthias Franzreb
Institute of Functional Interfaces
Karlsruhe Institute of Technology
Bld. 330
Hermann-von-Helmholtz-Platz 1
76344 Eggenstein-Leopoldshafen
Germany

Shiqi Gao
School of Chemical Engineering and Technology
Hebei University of Technology
Tianjin 300130
China

https://doi.org/10.1515/9783110760330-201

André Gemünde
Institute of Bioprocess Engineering and
Pharmaceutical Technology
University of Applied Sciences Mittelhessen
Wiesenstrasse 14
35390 Giessen
Germany
https://orcid.org/0000-0003-3825-7365

Raphael Greifenstein
Institute of Functional Interfaces
Karlsruhe Institute of Technology
Bld. 330
Hermann-von-Helmholtz-Platz 1
76344 Eggenstein-Leopoldshafen
Germany

Tobias Habicher
Bioprocess Development Biologicals
Boehringer Ingelheim Pharma GmbH & Co. KG
Biberach a.d. Riß
Germany

Lukas Hartmann
Institute of Process Engineering in Life Sciences
Karlsruhe Institute of Technology
Fritz-Haber-Weg 4
76131 Karlsruhe
Germany
E-mail: lukas.hartmann@kit.edu

Alina Hengelbrock
Clausthal University of Technology
Institute for Separation and Process Technology
Leibnizstr. 15
D-385678 Clausthal-Zellerfeld
Germany

Kevin P. Hoffmann
Institute of Biochemical Engineering
Technische Universität Braunschweig
Rebenring 56
38106 Braunschweig
Germany
And
Center of Pharmaceutical Engineering
Technische Universität Braunschweig
Franz-Liszt-Str. 35a
38106 Braunschweig
Germany

Dirk Holtmann
Institute of Bioprocess Engineering and
Pharmaceutical Technology
University of Applied Sciences Mittelhessen
Wiesenstrasse 14
35390 Giessen
Germany
and
Institute of Process Engineering in Life Sciences
Karlsruhe Institute of Technology
Karlsruhe
Fritz-Haber-Weg 4
76131 Karlsruhe
Germany
E-mail: dirk.holtmann@kit.edu

Yanjun Jiang
School of Chemical Engineering and Technology
Hebei University of Technology
Tianjin 300130
China
yanjunjiang@hebut.edu.cn

Weixi Kong
School of Chemical Engineering and Technology
Hebei University of Technology
Tianjin 300130
China

Zuzanna J. Kozanecka
Institute of Biochemical Engineering
Technische Universität Braunschweig
Rebenring 56
38106 Braunschweig
Germany
And
Center of Pharmaceutical Engineering
Technische Universität Braunschweig
Franz-Liszt-Str. 35a
38106 Braunschweig
Germany

Thomas Krieg
HP BioP Launch &Innovation
Boehringer Ingelheim PharmaGmbH &Co. KG
Biberach a.d. Riß
Germany
E-mail: thomas_2.krieg@boehringer-ingelheim.co

Rainer Krull
Institute of Biochemical Engineering
Technische Universität Braunschweig
Rebenring 56
38106 Braunschweig
Germany
And
Center of Pharmaceutical Engineering
Technische Universität Braunschweig
Franz-Liszt-Str. 35a
38106 Braunschweig, Germany
E-mail: r.krull@tubraunschweig.de
https://orcid.org/0000-0003-2821-8610

Kim B. Kuchemüller
Hamburg University of Technology
Institute of Bioprocess and Biosystems Engineering
Denickestr. 15
D-21073 Hamburg
Germany

Jianqiao Liu
School of Chemical Engineering and Technology
Hebei University of Technology
Tianjin 300130
China

Pengbo Liu
School of Chemical Engineering and Technology
Hebei University of Technology
Tianjin 300130
China

Yunting Liu
School of Chemical Engineering and Technology
Hebei University of Technology
Tianjin 300130
China
E-mail: ytliu@hebut.edu.cn

Michael Löffler
Bioprocess Development Biologicals
Boehringer Ingelheim Pharma GmbH & Co. KG
Biberach a.d. Riß
Germany

Nicolas Maguire
Bioprocess Development Biologicals
Boehringer Ingelheim Pharma GmbH & Co. KG
Biberach a.d. Riß
Germany

Dominique Monteil
Cell Culture
HP BioP Mammalian
Boehringer Ingelheim US Biopharma
Fremont, CA
USA

Johannes Möller
Hamburg University of Technology
Institute of Bioprocess and Biosystems Engineering
Denickestr. 15
D-21073 Hamburg
Germany
E-mail: johannes.moeller@tuhh.de

Katrin Ochsenreither
Biotechnologische Konversion
Technikum Laubholz GmbH
Göppingen
Germany
E-mail: katrin.ochsenreither@technikumlaubholz.de
https://orcid.org/0000-0002-5797-2789

Ralf Pörtner
Hamburg University of Technology
Institute of Bioprocess and Biosystems Engineering
Denickestr. 15
D-21073 Hamburg
Germany

Sebastian Püngel
Bioprocess Development Biologicals
Boehringer Ingelheim Pharma GmbH & Co. KG
Biberach a.d. Riß
Germany

Michael E. Runda
Department of Chemical and Pharmaceutical Biology
Groningen Research Institute of Pharmacy
University of Groningen
Antonius Deusinglaan 1
9713AV Groningen
The Netherlands

Jochen Schaub
Bioprocess Development Biologicals
Boehringer Ingelheim Pharma GmbH & Co. KG
Biberach a.d. Riß
Germany
E-mail: jochen.schaub@boehringer-ingelheim.com

Axel Schmidt
Clausthal University of Technology
Institute for Separation and Process Technology
Leibnizstr. 15
D-385678 Clausthal-Zellerfeld
Germany

Sandy Schmidt
Department of Chemical and Pharmaceutical Biology
Groningen Research Institute of Pharmacy
University of Groningen
Antonius Deusinglaan 1
9713AV Groningen
The Netherlands
E-mail: s.schmidt@rug.nl
https://orcid.org/0000-0002-8443-8805

Lisa Stepper
Bioprocess Development Biologicals
Boehringer Ingelheim Pharma GmbH & Co. KG
Biberach a.d. Riß
Germany

Fabian Stiefel
Bioprocess Development Biologicals
Boehringer Ingelheim Pharma GmbH & Co. KG
Biberach a.d. Riß
Germany

Markus Stöckl
Sustainable Electrochemistry
DECHEMA Research Institute
Theodor-Heuss-Allee 25
60486 Frankfurt am Main
Germany
E-mail: markus.stoeckl@dechema.de.
https://orcid.org/0000-0002-1372-7642

Jochen Strube
Clausthal University of Technology
Institute for Separation and Process Technology
Leibnizstr. 15
D-385678 Clausthal-Zellerfeld
Germany
E-mail: strube@itv.tu-clausthal.de

Christoph Syldatk
Institute of Bio- and Food Technology Division II –
Technical Biology
KIT – Karlsruhe Institute of Technology
Karlsruhe
Germany
E-mail: christoph.syldatk@kit.edu

Judith Thoma
Bioprocess Development Biologicals
Boehringer Ingelheim Pharma GmbH & Co. KG
Biberach a.d. Riß
Germany

Andreas Unsöld
Bioprocess Development Biologicals
Boehringer Ingelheim Pharma GmbH & Co. KG
Biberach a.d. Riß
Germany

Julia Walther
Bioprocess Development Biologicals
Boehringer Ingelheim Pharma GmbH & Co. KG
Biberach a.d. Riß
Germany

Christopher Wayne
DevOps Biologics DS Germany
HP BioP Launch&Innovation
Boehringer Ingelheim Pharma GmbH & Co. KG
Biberach a.d. Riß
Germany

Thomas Wucherpfennig
Bioprocess Development Biologicals
Boehringer Ingelheim Pharma GmbH & Co. KG
Biberach a.d. Riß
Germany

Lukas Hartmann, Thomas Krieg and Dirk Holtmann*

1 Intensification of bioprocesses – definition, examples, challenges and future directions

Abstract: Strategies to reduce cost and emission profiles are becoming increasingly important for the development of affordable and sustainable bio-based production. The overall objective of process intensification in different industries is to achieve substantial benefits in terms of cost, product concentration and quality, while eliminating waste and improving process safety. Intensification of bioprocesses could be a valuable tool for enhancing the efficiency and reducing resource consumption in bioproduction. In general, bioprocess intensification is defined as an increase in bioproduct output relative to cell concentration, time, reactor volume or cost. This brief overview provides a definition of process intensification in biotechnology, presents several general and specific examples, and addresses some of the current challenges.

Keywords: process intensification; biotechnology; bio-based production; resource efficiency; waste elimination; process safety

1.1 Defining of PI in biotechnology

In the 1970s, process intensification (PI) emerged as a strategy for developing completely new, breakthrough technologies in process engineering. The main motivation for the development of the first PI technologies was to reduce the size and investment costs of production plants in the chemical industry. In times of rising raw material costs, increased competition, shorter product lifecycles and changing consumer demands, industry is facing new challenges. Increasing the efficiency and flexibility of production processes is one strategy for meeting these challenges. PI can now be redefined as the key strategy for increasing the efficiency and flexibility of process operations in various industries. It leads to innovative equipment and technologies that open up new process windows. A valuable and comprehensive PI definition is given by Stankiewicz and Moulijn (2000): "Process intensification consists of the development of novel apparatuses and techniques that, compared to those commonly used today, are expected to bring dramatic improvements

*Corresponding author: Dirk Holtmann, Institute of Process Engineering in Life Sciences, Karlsruhe Institute of Technology, Fritz-Haber-Weg 4, 76131 Karlsruhe, Germany, E-mail: dirk.holtmann@kit.edu

Lukas Hartmann, Institute of Process Engineering in Life Sciences, Karlsruhe Institute of Technology, Fritz-Haber-Weg 4, 76131 Karlsruhe, Germany, E-mail: lukas.hartmann@kit.edu

Thomas Krieg, HP BioP Launch & Innovation, Boehringer Ingelheim Pharma GmbH & Co. KG, Biberach a.d. Riß, Germany, E-mail: thomas_2.krieg@boehringer-ingelheim.co

As per De Gruyter's policy this article has previously been published in the journal Physical Sciences Reviews. Please cite as:
L. Hartmann, T. Krieg and D. Holtmann "Intensification of bioprocesses – definition, examples, challenges and future directions" *Physical Sciences Reviews* [Online] 2023. DOI: 10.1515/psr-2022-0101 | https://doi.org/10.1515/9783110760330-001

in manufacturing and processing, substantially decreasing equipment-size to production-capacity ratio, energy consumption, or waste production, and ultimately resulting in cheaper, sustainable technologies" [1]. Unlike conventional process optimisation, PI can overcome process limitations resulting in more compact and efficient systems. Very often, more compact and cheaper systems can still be achieved by applying PI principles. Simultaneously, PI frequently leads to advanced processes in terms of flexibility, resource consumption, waste generation and operational safety. The more efficient use of raw materials and energy, as well as the reduction in costs, secures and enhances the competitiveness of companies. This results in both environmental and economic benefits. Commonly, PI is classified in terms of the underlying principles of (i) time – improving reaction kinetics, (ii) space – maximising homogeneity in the system, (iii) thermodynamics – maximising thermodynamic driving forces, and (iv) synergy – combining different unit operations in one reactor [2]. Ultimately, the PI definition can be simplified to "doing more with less resources". PI is enabling a transformation in chemical and biochemical engineering, enhancing its potential to address global issues such as climate change, limited resources and higher environmental standards. PI often resulted in processes with at least doubled performance. It can therefore be concluded that PI always involves quantitative statements. For example, a new reactor design will only qualify for PI if key parameters are substantially improved. One of the important questions, however, is the differentiation of PI and process optimisation (PO). Both PI and PO refer to performance improvements that lead to increased productivity. While PO usually adds incremental improvements in performance, the application of PI contributes not only to significant increases in productivity, but also to improvements in other environmental and economic indicators. PO can therefore be considered an approach of gradual improvement, whereas PI usually involves significant improvements through radically innovative principles. The field of PI in biotechnology can be divided into different main areas (see Table 1.1). Finally, PI can also be divided into two overarching sectors: (i) process-intensifying devices, such as novel reactors and intensive mixing, heat and mass transfer devices; and (ii) process-intensifying methods, such as new or hybrid separations, reaction and separation integration, multifunctional reactors and techniques using alternative energy sources (e.g. electrochemistry, ultrasonic) [1]. It must be mentioned, that the development of improved catalysts is not included in the general definition of PI. Similar applies to biotechnology, although significant improvements in metrics can also be achieved through enzyme and strain engineering.

1.2 Examples of PI in biotechnology

Biotechnology has a wide range of applications, from the removal of pollutants in environmental biotechnology to the production of vaccines. The catalysts used are equally diverse, ranging from isolated enzymes to complex human cells and even more complex biocenoses. The complexity of the processes also varies greatly, including

Table 1.1: General overview of PI technologies in biotechnology. In each case, a comparison is made between the existing technologies and the innovative processes of the intensified processes (adapted and extended from Burrek et al. (2022) [3]).

Topic	Established "classical" system	Challenges in application of the established systems	Intensified approaches to overcome the challenges
Energy input	Chemical stored energy in molecules e.g., glucose, formate, natural cofactors as mediators	–Costs –Changing composition of the reaction medium (viscosity, pH) –Interaction with downstream process	–Electrochemistry –Photochemistry –Ultrasound
Carbon source/ substrate	Carbohydrate-based processes	–Transition to circular economy/bioeconomy –Ethical issues ("food or fuel discussion") –Costs	–CO_2 conversion –"new" sustainable substrates, e.g., syngas, formate –Use of residual materials and by-product streams
Solvent	Water-based media/buffer with solid/dissolved substrates	–Limited solubility of substrates –Low product concentration –High reaction volumes in downstream processing	–Organic solvents –Ionic liquids –Supercritical fluids –Deep eutectic solvents –Solvent-free catalysis –Gas fermentation
Reactor	Stirred reactors, separation of biocatalysis and downstream processing	–Inadequate reaction performance (e.g., caused by in-sufficient mixing) –Limitations in heat- and mass-transfer	–Bubble columns –Rotating bed reactors –Micro reactors –Microfluidics –Enzyme membrane reactors
Process design	Unit operations are separated in different process steps	–Substrate and/or product inhibition –High number of process steps –Huge plant-related footprint	–*In situ* product removal –Integration of chemo- and biocatalysis
Process mode	Batch/fed-batch processes	–Fluctuating product quality –Insufficient utilisation of equipment	–Continuous bioprocessing –Repeated harvest/seeding iterations –Cell perfusion

decentralised facilities such as biogas processes, specialised fermentation plants and highly integrated biorefineries. Thus, an overview of the specific tasks in the different areas of application of PI in biotechnological processes is given in Table 1.2. In order to sub-categorise the PI technologies the table is divided in (i) alternative energy inputs, (ii) improved reaction conditions and (iii) reactors and process modes. This overview does

Table 1.2: Examples of PI technologies in biotechnology.

PI technology	Addressed main challenge	Literature examples (original literature and reviews)
Alternative energy input		
Electrochemistry combined with enzyme catalysis	Many of the industrially relevant enzymes are cofactor dependent, their supply and regeneration are often very complex	[3–17] a comprehensive overview is given in [18]
Electrochemistry combined with whole cells	In microbial electrosynthesis, metabolism is driven by electrons as redox equivalents or steered in electrofermentation	[19–28] see chapter "Microbial electrotechnology – intensification of bioprocesses through the combination of electrochemistry and biotechnology"
Photochemistry	Enzymatic or microbial reactions are driven by light energy	[3, 29–39] see chapter "Light-driven bioprocesses"
Ultrasonic	Increasing enzymatic activity, synthesising or regenerating cofactors, improving mixing or extraction	[3, 40–48]
Microwave	Applications are very often similar to those of ultrasonic	[45, 49–53]
PI to improve reaction conditions		
Immobilised enzymes	Enzyme stability and reusability, reduced downstream processing requirements	[54–66] see chapter "Process intensification using immobilized enzymes"
Solvents (in particular for enzymatic processes)	Stability of enzymes, solubility of educts and products	[67–78] see chapter "Intensification of biocatalytic processes by using alternative reaction media"
Influencing morphology (in particular filamentous fungi)	Various morphologies with different levels of productivity are found in many organisms	[79–85] see chapter "Intensification of bioprocesses with filamentous microorganisms"
Model based optimisation of processes	In biotechnological processes, the interactions are often extremely complex. They can only be optimised on the basis of models. Process optimisation based on modelling complex systems	[86–91] see chapter "Chapter "Bioprocess intensification with model-assisted DoE-strategies for the production of biopharmaceuticals"

Table 1.2: (continued)

PI technology	Addressed main challenge	Literature examples (original literature and reviews)
Reactors conditions and process modes		
"Unconventional" reaction conditions, e.g., pressurised reactors, rocking motion bioreactor, rotating bed reactors	Overcome process limitations due to limited substrate solubility and mass-transfer	[64, 92–100]
Miniaturised reactors and microfluidic devises	Increased heat- and mass-transfer, improved safety	[101–110] see chapter "Miniaturization and microfluidic Devices: an overview of basic concepts, fabrication techniques and applications"
Integration of microbial or enzymatic production and product removal in one step	Avoidance of process limitations due to product inhibition or volatility of the products	[111–122] see chapter "In-situ product removal"
Using advanced process modes (e.g., continuous bioprocessing, repeated harvest/seeding iterations, cell perfusion, intensified seed train)	Reduction of unproductive process times, better utilisation of substrate, biomass and reactor/equipment, improved product quality	[56, 112, 123–142] see chapters "Continuous Biomanufacturing in Upstream and Downstream Processing" and "Process intensification in biopharmaceutical process development and production"
Single-use technologies in up- and downstream	Increased productivity, reduced energy and media consumption, minimised risk of cross-contamination, reduced cleaning effort	[137–139, 143–147] see chapter "Process intensification in biopharmaceutical process development and production"
Combining chemical and enzymatic reaction steps	Takes advantage of both catalytic steps without need for intermediate purification	[39, 148–155] see chapter "Integration of chemo- and bio-catalysis to intensify bioprocesses"
Utilisation of waste and by-products from agriculture, municipality and various industries	Optimum use of resources through cascade utilisation	[156–167]

not claim to be comprehensive. For more detailed information and individual aspects, please refer to the cited and further literature.

1.3 Current challenges and future direction of PI implementation in biotechnology

Although many PI technologies have been successfully and promisingly tested at laboratory scale, only a limited number have been integrated into industrial bioprocesses.

This is often attributed to technological and economic barriers that hinder the transition from the development phase to actual industrial implementation. There are a number of different barriers to the implementation of the approaches in practice:

(i) The transition of PI technologies from initial laboratory application to full-scale production facilities poses significant technical and economic risks. The costs associated with new equipment and implementation are substantial. Development timelines often conflict with the need for rapid technological advancement to remain competitive in the global marketplace. In addition, significant manpower and time are required to ensure robust process control and safe operation of unproven technologies. Because PI must be considered as a holistic technology, it is always difficult to define a single key performance indicator. Therefore, the most important and devisice indicator is the potential of financial savings. In van der Wielen et al. (2021), the relationship between capital expenditure (CAPEX) and operational expenditure (OPEX) of intensified and non-intensified bio-based processes and their financial and climate impacts are exemplarily discussed [168].

(ii) There is a tendency in the bioprocessing industry to stick with established production technologies for as long as possible. Operators are generally sceptical about new technologies and processes, and prioritise the robustness, stability and safety of the existing process. New technologies are only integrated if they have several positive references from other (industrial) plants. This is also due to the lack of PI-specific metrics for benchmarking different technologies.

(iii) In addition, PI technologies are often optimised for a specific process and systems can sometimes only be operated within a narrow range of parameters. The resulting reduced flexibility in operating behaviour limits the range of applications and transferability. This is in contradiction with the requirements for flexible systems as described above.

It can be concluded that the challenges involved in implementing further PI technologies in industrial practice are considerable, but solvable. This will require proactive collaboration between stakeholders in research and technology development, plant engineering and construction, industry and regulators. As shown above, many of the technologies are highly interdisciplinary. Therefore, a need for more interaction between the disciplines is involved. The following actions and activities can help to overcome some of these barriers:

(i) Integration of process intensification into the academic training of process and biochemical engineers. Interdisciplinary training and the use of quantitative metrics should provide a solid foundation for future applications.

(ii) Accelerating technology development by screening methods and early-stage indicators for process intensification technologies. Digital biomanufacturing is accelerating the implementation of PI in the industrial biotechnology sector [169]. Digital twins from individual unit operations and holistic plant models should be used to help convince stakeholders.

(iii) Publication of successful developments and implementation examples of process-intensive technologies, highlighting the environmental and economic benefits by using established metrics.

From this, three short and concise guidelines can be derived for the continued successful implementation of PI in biotechnology: be creative, be quantitative and be economical.

References

1. Stankiewicz A, Moulijn JA. Process intensification: transforming chemical engineering. Chem Eng Prog 2000;96:22–33.
2. Van Gerven T, Stankiewicz A. Structure, energy, synergy, time – the fundamentals of process intensification. Ind Eng Chem Res 2009;48:2465–74.
3. Burek BO, Dawood AWH, Hollmann F, Liese A, Holtmann D. Process intensification as game changer in enzyme catalysis. Front Catal 2022;2:1–18.
4. García-Molina G, Natale P, Coito AM, Cava DG, A. C. Pereira I, López-Montero I, et al. Electro-enzymatic ATP regeneration coupled to biocatalytic phosphorylation reactions. Bioelectrochemistry 2023;152:108432.
5. Jack J, Fu H, Leininger A, Hyster TK, Ren ZJ. Cell-free CO_2 valorization to C6 pharmaceutical precursors via a novel electro-enzymatic process. ACS Sustain Chem Eng 2022;10:4114–21.
6. Varničić M, Zasheva IN, Haak E, Sundmacher K, Vidaković-Koch T. Selectivity and sustainability of electroenzymatic process for glucose conversion to gluconic acid. Catalysts 2020;10:269.
7. Zhu X, Fan X, Lin H, Li S, Zhai Q, Jiang Y, et al. Highly efficient electroenzymatic cascade reduction reaction for the conversion of nitrite to ammonia. Adv Energy Mater 2023;13:2300669.
8. Zhan P, Liu X, Zhang S, Zhu Q, Zhao H, Ren C, et al. Electroenzymatic reduction of furfural to furfuryl alcohol by an electron mediator and enzyme orderly assembled biocathode. ACS Appl Mater Interfaces 2023;15: 12855–63.
9. Sayoga GV, Bueschler VS, Beisch H, Utesch T, Holtmann D, Fiedler B, et al. Electrochemical H_2O_2 – stat mode as reaction concept to improve the process performance of an unspecific peroxygenase. New Biotechnol 2023;78:95–104.
10. Zhang Z, Li J, Ji M, Liu Y, Wang N, Zhang X, et al. Encapsulation of multiple enzymes in a metal–organic framework with enhanced electro-enzymatic reduction of CO_2 to methanol. Green Chem 2021;23: 2362–71.
11. Lee YS, Gerulskis R, Minteer SD. Advances in electrochemical cofactor regeneration: enzymatic and non-enzymatic approaches. Curr Opin Biotechnol 2022;73:14–21.
12. Yuan M, Abdellaoui S, Chen H, Kummer MJ, Malapit CA, You C, et al. Selective electroenzymatic oxyfunctionalization by alkane monooxygenase in a biofuel cell. Angew Chem Int Ed 2020;59:8969–73.
13. Siedentop R, Prenzel T, Waldvogel SR, Rosenthal K, Lütz S. Reaction engineering and comparison of electroenzymatic and enzymatic ATP regeneration systems. Chemelectrochem 2023;10:e202300332.
14. Zhang C, Zhang X, Fu Y, Zhang L, Kuhn A. Metal–organic framework functionalized bipolar electrodes for bulk electroenzymatic synthesis. J Catal 2023;421:95–100.
15. Abt M, Franzreb M, Jestädt M, Tschöpe A. Three-phase fluidized bed electrochemical reactor for the scalable generation of hydrogen peroxide at enzyme compatible conditions. Chem Eng J 2023;476:146465.
16. Tosstorff A, Dennig A, Ruff AJ, Schwaneberg U, Sieber V, Mangold KM, et al. Mediated electron transfer with monooxygenases – insight in interactions between reduced mediators and the co-substrate oxygen. J Mol Catal B Enzym 2014;108:51–8.

17. Bormann S, Hertweck D, Schneider S, Bloh JZ, Ulber R, Spiess AC, et al. Modeling and simulation-based design of electroenzymatic batch processes catalyzed by unspecific peroxygenase from A. aegerita. Biotechnol Bioeng 2021;118:7–16.
18. Sayoga G, Abt M, Teetz N, Bueschler V, Liese A, Franzreb M, et al. Quantitative and non-quantitative assessments of enzymatic electrosynthesis: a case study of parameter requirements. ChemElectroChem 2023;10:e202300226.
19. Stöckl M, Gemünde A, Holtmann D. Microbial electrotechnology – intensification of bioprocesses through the combination of electrochemistry and biotechnology. Phys Sci Rev 2023. https://doi.org/10.1515/psr-2022-0108.
20. Mateos R, Escapa A, San-Martín MI, De Wever H, Sotres A, Pant D. Long-term open circuit microbial electrosynthesis system promotes methanogenesis. J Energy Chem 2020;41:3–6.
21. Enzmann F, Holtmann D. Rational scale-up of a methane producing bioelectrochemical reactor to 50 L pilot scale. Chem Eng Sci 2019;207:1148–58.
22. Hengsbach JN, Sabel-Becker B, Ulber R, Holtmann D. Microbial electrosynthesis of methane and acetate-comparison of pure and mixed cultures. Appl Microbiol Biotechnol 2022;106:4427–43.
23. Ragab Aa, Shaw DR, Katuri KP, Saikaly PE. Effects of set cathode potentials on microbial electrosynthesis system performance and biocathode methanogen function at a metatranscriptional level. Sci Rep 2020; 10:19824.
24. Baek G, Rossi R, Saikaly PE, Logan BE. High-rate microbial electrosynthesis using a zero-gap flow cell and vapor-fed anode design. Water Res 2022;219:118597.
25. Prévoteau A, Carvajal-Arroyo JM, Ganigué R, Rabaey K. Microbial electrosynthesis from CO_2: forever a promise? Curr Opin Biotechnol 2020;62:48–57.
26. Jourdin L, Burdyny T. Microbial electrosynthesis: where do we go from here? Trends Biotechnol 2021;39: 359–69.
27. Cabau-Peinado O, Straathof AJJ, Jourdin L. A general model for biofilm-driven microbial electrosynthesis of carboxylates from CO_2. Front Microbiol 2021;12:1–17.
28. Abdollahi M, Al Sbei S, Rosenbaum MA, Harnisch F. The oxygen dilemma: the challenge of the anode reaction for microbial electrosynthesis from CO_2. Front Microbiol 2022;13:1–9.
29. Runda ME, Schmidt S. Light-driven bioprocesses. Phys Sci Rev 2023. https://doi.org/10.1515/psr-2022-0109.
30. De Santis P, Wegstein D, Burek BO, Patzsch J, Alcalde M, Kroutil W, et al. Robust light driven enzymatic oxyfunctionalization via immobilization of unspecific peroxygenase. ChemSusChem 2023;16:e202300613.
31. Burek BO, de Boer SR, Tieves F, Zhang W, van Schie M, Bormann S, et al. Photoenzymatic hydroxylation of ethylbenzene catalyzed by unspecific peroxygenase: origin of enzyme inactivation and the impact of light intensity and temperature. ChemCatChem 2019;11:3093–100.
32. Willot SJP, Fernández-Fueyo E, Tieves F, Pesic M, Alcalde M, Arends IW, et al. Expanding the spectrum of light-driven peroxygenase reactions. ACS Catal 2019;9:890–4.
33. Yun CH, Kim J, Hollmann F, Park CB. Light-driven biocatalytic oxidation. Chem Sci 2022;13:12260–79.
34. Meyer L-E, Eser BE, Kara S. Coupling light with biocatalysis for sustainable synthesis – very recent developments and future perspectives. Curr Opin Green Sustainable Chem 2021;31:100496.
35. Hoschek A, Toepel J, Hochkeppel A, Karande R, Bühler B, Schmid A. Light-dependent and aeration-independent gram-scale hydroxylation of cyclohexane to cyclohexanol by CYP450 harboring synechocystis sp. PCC 6803. Biotechnol J 2019;14:1800724.
36. Tüllinghoff A, Djaya-Mbissam H, Toepel J, Bühler B. Light-driven redox biocatalysis on gram-scale in Synechocystis sp. PCC 6803 via an in vivo cascade. Plant Biotechnol J 2023;21:2074–83.
37. Feyza Özgen F, Runda ME, Burek BO, Wied P, Bloh JZ, Kourist R, et al. Artificial light-harvesting complexes enable Rieske oxygenase catalyzed hydroxylations in non-photosynthetic cells. Angew Chem Int Ed 2020; 59:3982–7.

38. Assil-Companioni L, Büchsenschütz HC, Solymosi D, Dyczmons-Nowaczyk NG, Bauer KKF, Wallner S, et al. Engineering of NADPH supply boosts photosynthesis-driven biotransformations. ACS Catal 2020;10: 11864–77.
39. Özgen FF, Runda ME, Schmidt S. Photo-biocatalytic cascades: combining chemical and enzymatic transformations fueled by light. Chembiochem 2021;22:790–806.
40. Yoon J, Kim J, Tieves F, Zhang W, Alcalde M, Hollmann F, et al. Piezobiocatalysis: ultrasound-driven enzymatic oxyfunctionalization of C–H bonds. ACS Catal 2020;10:5236–42.
41. Thangarasu V, Siddharth R, Ramanathan A. Modeling of process intensification of biodiesel production from Aegle Marmelos Correa seed oil using microreactor assisted with ultrasonic mixing. Ultrason Sonochem 2020;60:104764.
42. Wu Z, Tagliapietra S, Giraudo A, Martina K, Cravotto G. Harnessing cavitational effects for green process intensification. Ultrason Sonochem 2019;52:530–46.
43. Panda D, Manickam S. Cavitation technology – the future of greener extraction method: a review on the extraction of natural products and process intensification mechanism and perspectives. Appl Sci 2019;9:766.
44. Liao J, Guo Z, Yu G. Process intensification and kinetic studies of ultrasound-assisted extraction of flavonoids from peanut shells. Ultrason Sonochem 2021;76:105661.
45. Călinescu I, Vinatoru M, Ghimpețeanu D, Lavric V, Mason TJ. A new reactor for process intensification involving the simultaneous application of adjustable ultrasound and microwave radiation. Ultrason Sonochem 2021;77:105701.
46. Pawar SV, Rathod VK. Role of ultrasound in assisted fermentation technologies for process enhancements. Prep Biochem Biotechnol 2020;50:627–34.
47. Kajarekar BR, Gogate PR. Ultrasound assisted intensification of streptomycin production based on fermentation. Chem Eng Process: Process Intensif 2022;171:108748.
48. Umego EC, He R, Huang G, Dai C, Ma H. Ultrasound-assisted fermentation: mechanisms, technologies, and challenges. J Food Process Preserv 2021;45:e15559.
49. Velasquez-Orta SB, Mohiuddin O, Orta Ledesma M-T, Harvey AP. 13 – process intensification of microalgal biofuel production. In: Jacob-Lopes E, et al., editor. 3rd generation biofuels. Sawston, UK: Woodhead Publishing; 2022:269–90 pp.
50. Aamir M, Afzaal M, Saeed F, Afzal A, Shah YA, Tariq I, et al. Chapter 19 – effect of synergism of sonication and microwave on fermentation and emulsification processes. In: Nayik GA, et al., editor. Ultrasound and microwave for food processing. Academic Press; 2023:497–535 pp.
51. Perino S, Chemat F. Green process intensification techniques for bio-refinery. Curr Opin Food Sci 2019;25: 8–13.
52. Kant Bhatia S, Kant Bhatia R, Jeon JM, Pugazhendhi A, Kumar Awasthi M, Kumar D, et al. An overview on advancements in biobased transesterification methods for biodiesel production: oil resources, extraction, biocatalysts, and process intensification technologies. Fuel 2021;285:119117.
53. Kulkarni RM, Sarkar B, Modhvadiya K, Patil S. Process intensification of microwave-assisted acetylation reaction of glycerol to synthesize fuel additives using porcine pancreas lipase catalyst. Biofuels 2023:1–8. https://doi.org/10.1080/17597269.2023.2244325.
54. Bolat, S, Greifenstein, R, Franzreb, M, Holtmann, D. Process intensification using immobilized enzymes. Phys Sci Rev 2023. https://doi.org/10.1515/psr-2022-0110.
55. Salvi HM, Yadav GD. Process intensification using immobilized enzymes for the development of white biotechnology. Catal Sci Technol 2021;11:1994–2020.
56. Meyer L-E, Hobisch M, Kara S. Process intensification in continuous flow biocatalysis by up and downstream processing strategies. Curr Opin Biotechnol 2022;78:102835.
57. Tamborini L, Fernandes P, Paradisi F, Molinari F. Flow bioreactors as complementary tools for biocatalytic process intensification. Trends Biotechnol 2018;36:73–88.

58. Žnidaršič-Plazl P. Biocatalytic process intensification via efficient biocatalyst immobilization, miniaturization, and process integration. Curr Opin Green Sustainable Chem 2021;32:100546.
59. Bormann S, Burek BO, Ulber R, Holtmann D. Immobilization of unspecific peroxygenase expressed in Pichia pastoris by metal affinity binding. Mol Catal 2020;492:110999.
60. Ströhle FW, Kranen E, Schrader J, Maas R, Holtmann D. A simplified process design for P450 driven hydroxylation based on surface displayed enzymes. Biotechnol Bioeng 2016;113:1225–33.
61. Mazzei R, Yihdego Gebreyohannes A, Papaioannou E, Nunes SP, Vankelecom IF, Giorno L. Enzyme catalysis coupled with artificial membranes towards process intensification in biorefinery – a review. Bioresour Technol 2021;335:125248.
62. Roura Padrosa D, Benítez-Mateos AI, Calvey L, Paradisi F. Cell-free biocatalytic syntheses of l-pipecolic acid: a dual strategy approach and process intensification in flow. Green Chem 2020;22:5310–6.
63. Buergler MB, Dennig A, Nidetzky B. Process intensification for cytochrome P450 BM3-catalyzed oxy-functionalization of dodecanoic acid. Biotechnol Bioeng 2020;117:2377–88.
64. Bolivar JM, Mannsberger A, Thomsen MS, Tekautz G, Nidetzky B. Process intensification for O_2-dependent enzymatic transformations in continuous single-phase pressurized flow. Biotechnol Bioeng 2019;116: 503–14.
65. Bolivar JM, López-Gallego F. Characterization and evaluation of immobilized enzymes for applications in flow reactors. Curr Opin Green Sustainable Chem 2020;25:100349.
66. Basso A, Serban S. Industrial applications of immobilized enzymes – a review. Mol Catal 2019;479:110607.
67. Delavault A, Ochsenreither K, Syldatk C. Intensification of biocatalytic processes by using alternative reaction media. Phys Sci Rev 2023. https://doi.org/10.1515/psr-2022-0104.
68. Hümmer M, Kara S, Liese A, Huth I, Schrader J, Holtmann D. Synthesis of (-)-menthol fatty acid esters in and from (−)-menthol and fatty acids – novel concept for lipase catalyzed esterification based on eutectic solvents. Mol Catal 2018;458:67–72.
69. Pätzold M, Weimer A, Liese A, Holtmann D. Optimization of solvent-free enzymatic esterification in eutectic substrate reaction mixture. Biotechnol Rep 2019;22:e00333.
70. Pätzold M, Siebenhaller S, Kara S, Liese A, Syldatk C, Holtmann D. Deep eutectic solvents as efficient solvents in biocatalysis. Trends Biotechnol 2019;37:943–59.
71. Pätzold M, Burek BO, Liese A, Bloh JZ, Holtmann D. Product recovery of an enzymatically synthesized (−)-menthol ester in a deep eutectic solvent. Bioproc Biosyst Eng 2019;42:1385–9.
72. Domínguez de María P. Green solvents and biocatalysis: a bigger picture. EFB Bioeconomy Journal 2023;3: 100056.
73. Domínguez de María P, Kara S, Gallou F. Biocatalysis in water or in non-conventional media? Adding the CO_2 production for the debate. Molecules 2023;28. https://doi.org/10.3390/molecules28186452.
74. Semproli R, Chanquia SN, Bittner JP, Müller S, Domínguez de María P, Kara S, et al. Deep eutectic solvents for the enzymatic synthesis of sugar esters: a generalizable strategy? ACS Sustainable Chem Eng 2023;11: 5926–36.
75. Holtmann D, Hollmann F. Is water the best solvent for biocatalysis? Mol Catal 2022;517:112035.
76. Hilberath T, van Troost A, Alcalde M, Hollmann F. Assessing peroxygenase-mediated oxidations in the presence of high concentrations of water-miscible co-solvents. Front Catal 2022;2:1–8.
77. van Schie MMCH, Spöring JD, Bocola M, Domínguez de María P, Rother D. Applied biocatalysis beyond just buffers – from aqueous to unconventional media. Options and guidelines. Green Chem 2021;23: 3191–206.
78. Li Z, Han Q, Wang K, Song S, Xue Y, Ji X, et al. Ionic liquids as a tunable solvent and modifier for biocatalysis. Catal Rev 2022:1–47. https://doi.org/10.1080/01614940.2022.2074359.
79. Dinius A, Kozanecka ZJ, Hoffmann KP, Krull R. Intensification of bioprocesses with filamentous microorganisms. 2023. https://doi.org/10.1515/psr-2022-0112.
80. Dinius A, Schrinner K, Schrader M, Kozanecka ZJ, Brauns H, Klose L, et al. Morphology engineering for novel antibiotics: effect of glass microparticles and soy lecithin on rebeccamycin production and cellular

morphology of filamentous actinomycete Lentzea aerocolonigenes. Front Bioeng Biotechnol 2023;11: 1–16.

81. Laible AR, Dinius A, Schrader M, Krull R, Kwade A, Briesen H, et al. Effects and interactions of metal oxides in microparticle-enhanced cultivation of filamentous microorganisms. Eng Life Sci 2022;22:725–43.

82. Meyer V, Cairns T, Barthel L, King R, Kunz P, Schmideder S, et al. Understanding and controlling filamentous growth of fungal cell factories: novel tools and opportunities for targeted morphology engineering. Fungal Biol Biotechnol 2021;8:8.

83. Holtmann D, Vernen F, Müller J, Kaden D, Risse J, Friehs K, et al. Effects of particle addition to Streptomyces cultivations to optimize the production of actinorhodin and streptavidin. Sustain Chem Pharm 2017;5: 67–71.

84. Walisko J, Vernen F, Pommerehne K, Richter G, Terfehr J, Kaden D, et al. Particle-based production of antibiotic rebeccamycin with Lechevalieria aerocolonigenes. Process Biochem 2017;53:1–9.

85. Etschmann MMW, Huth I, Walisko R, Schuster J, Krull R, Holtmann D, et al. Improving 2-phenylethanol and 6-pentyl-α-pyrone production with fungi by microparticle-enhanced cultivation (MPEC). Yeast 2015;32: 145–57.

86. Finnigan W, Citoler J, Cosgrove SC, Turner NJ. Rapid model-based optimization of a two-enzyme system for continuous reductive amination in flow. Org Process Res Dev 2020;24:1969–77.

87. Müller D, Klein L, Lemke J, Schulze M, Kruse T, Saballus M, et al. Process intensification in the biopharma industry: improving efficiency of protein manufacturing processes from development to production scale using synergistic approaches. Chem Eng Process: Process Intensif 2022;171:108727.

88. Sarkar S, Bhowmick TK, Gayen K. Process intensification for the enhancement of growth and chlorophyll molecules of isolated Chlorella thermophila: a systematic experimental and optimization approach. Prep Biochem Biotechnol 2023;53:634–52.

89. Sakiewicz P, Piotrowski K, Ober J, Karwot J. Innovative artificial neural network approach for integrated biogas – wastewater treatment system modelling: effect of plant operating parameters on process intensification. Renew Sustain Energy Rev 2020;124:109784.

90. Bayer B, Striedner G, Duerkop M. Hybrid modeling and intensified DoE: an approach to accelerate upstream process characterization. Biotechnol J 2020;15:2000121.

91. Bose SA, Rajulapati SB, Velmurugan S, Arockiasamy S, Jayaram K, Kola AK, et al. Process intensification of biopolymer polyhydroxybutyrate production by pseudomonas putida SS9: a statistical approach. Chemosphere 2023;313:137350.

92. Kheirkhah T, Neubauer P, Junne S. Controlling Aspergillus niger morphology in a low shear-force environment in a rocking-motion bioreactor. Biochem Eng J 2023;195:108905.

93. Bai Y, Moo-Young M, Anderson WA. Characterization of power input and its impact on mass transfer in a rocking disposable bioreactor. Chem Eng Sci 2019;209:115183.

94. Gómez-Ríos D, Junne S, Neubauer P, Ochoa S, Ríos-Estepa R, Ramírez-Malule H. Characterization of the metabolic response of Streptomyces clavuligerus to shear stress in stirred tanks and single-use 2D rocking motion bioreactors for clavulanic acid production. Antibiotics 2019;8. https://doi.org/10.3390/antibiotics8040168.

95. Ashok A, Devarai SK. l-Asparaginase production in rotating bed reactor from Rhizopus microsporus IBBL-2 using immobilized Ca-alginate beads. 3 Biotech 2019;9:349.

96. Hollenbach R, Muller D, Delavault A, Syldatk C. Continuous flow glycolipid synthesis using a packed bed reactor. Catalysts 2022;12. https://doi.org/10.3390/catal12050551.

97. Donzella S, Compagno C, Molinari F, Paradisi F, Contente ML. Boosting the catalytic performance of a marine yeast in a SpinChem® reactor for the synthesis of perillyl alcohol. React Chem Eng 2023;8:2963–6.

98. Knoll A, Maier B, Tscherrig H, Büchs J. The oxygen mass transfer, carbon dioxide inhibition, heat removal, and the energy and cost efficiencies of high pressure fermentation. In: Kragl U, editor. Technology transfer in biotechnology: from lab to industry to production. Berlin, Heidelberg: Springer; 2005:77–99 pp.

99. Knoll A, Bartsch S, Husemann B, Engel P, Schroer K, Ribeiro B, et al. High cell density cultivation of recombinant yeasts and bacteria under non-pressurized and pressurized conditions in stirred tank bioreactors. J Biotechnol 2007;132:167–79.
100. Ferreira P, Lopes M, Belo I. Use of pressurized and airlift bioreactors for citric acid production by Yarrowia lipolytica from crude glycerol. Fermentation 2022;8. https://doi.org/10.3390/fermentation8120700.
101. Heuer C, Preuß J, Habib T, Enders A, Bahnemann J. 3D printing in biotechnology – an insight into miniaturized and microfluidic systems for applications from cell culture to bioanalytics. Eng Life Sci 2022; 22:744–59.
102. Enders A, Grünberger A, Bahnemann J. Towards small scale: overview and applications of microfluidics in biotechnology. Mol Biotechnol 2022. https://doi.org/10.1007/s12033-022-00626-6.
103. Cardoso Marques MP, Lorente-Arevalo A, Bolivar JM. Biocatalysis in continuous-flow microfluidic reactors. In: Bahnemann J, Grünberger A, editors. Microfluidics in biotechnology. Cham: Springer International Publishing; 2022:211–46 pp.
104. Shi H, Nie K, Dong B, Long M, Xu H, Liu Z. Recent progress of microfluidic reactors for biomedical applications. Chem Eng J 2019;361:635–50.
105. Brás EJS, Fernandes PCdB. Miniaturization and microfluidic devices: an overview of basic concepts, fabrication techniques, and applications. Phys Sci Rev 2023. https://doi.org/10.1515/psr-2022-0102.
106. Ozdalgic B, Ustun M, Dabbagh SR, Haznedaroglu BZ, Kiraz A, Tasoglu S. Microfluidics for microalgal biotechnology. Biotechnol Bioeng 2021;118:1716–34.
107. Scheler O, Postek W, Garstecki P. Recent developments of microfluidics as a tool for biotechnology and microbiology. Curr Opin Biotechnol 2019;55:60–7.
108. Ortseifen V, Viefhues M, Wobbe L, Grünberger A. Microfluidics for biotechnology: bridging gaps to foster microfluidic applications. Front Bioeng Biotechnol 2020;8:1–12.
109. Burmeister A, Grünberger A. Microfluidic cultivation and analysis tools for interaction studies of microbial co-cultures. Curr Opin Biotechnol 2020;62:106–15.
110. Dusny C, Grünberger A. Microfluidic single-cell analysis in biotechnology: from monitoring towards understanding. Curr Opin Biotechnol 2020;63:26–33.
111. Teke GM, Tai SL, Pott RWM. Extractive fermentation processes: modes of operation and application. ChemBioEng Rev 2022;9:146–63.
112. Liu S-R, Yang X-J, Sun D-F. Enhanced production of ε-poly-l-lysine by immobilized Streptomyces ahygroscopicus through repeated-batch or fed-batch fermentation with in situ product removal. Bioproc Biosyst Eng 2021;44:2109–20.
113. Chen Y, Garg N, Luo H, Kontogeorgis GM, Woodley JM. Ionic liquid-based in situ product removal design exemplified for an acetone–butanol–ethanol fermentation. Biotechnol Prog 2021;37:e3183.
114. Tönjes S, Uitterhaegen E, De Brabander P, Verhoeven E, Delmulle T, De Winter K, et al. In situ product recovery as a powerful tool to improve the fermentative production of muconic acid in Saccharomyces cerevisiae. Biochem Eng J 2023;190:108746.
115. Kaur G, Garcia-Gonzalez L, Elst K, Truzzi F, Bertin L, Kaushik A, et al. Reactive extraction for in-situ carboxylate recovery from mixed culture fermentation. Biochem Eng J 2020;160:107641.
116. Teke GM, Pott RWM. Design and evaluation of a continuous semipartition bioreactor for in situ liquid-liquid extractive fermentation. Biotechnol Bioeng 2021;118:58–71.
117. Santos AG, de Albuquerque TL, Ribeiro BD, Coelho MAZ. In situ product recovery techniques aiming to obtain biotechnological products: a glance to current knowledge. Biotechnol Appl Biochem 2021;68: 1044–57.
118. Lukito BR, Basri N, Thong A, Hermansen C, Weingarten M, Peterson EC. Co-Culture of Kluyveromyces marxianus and Meyerozyma guilliermondii with in situ product recovery of 2-phenylethanol. J Agric Food Chem 2023;71:8991–7.
119. Salas-Villalobos UA, Gómez-Acata RV, Castillo-Reyna J, Aguilar O. In situ product recovery as a strategy for bioprocess integration and depletion of inhibitory products. J Chem Technol Biotechnol 2021;96:2735–43.

120. Hülsewede D, Meyer L-E, von Langermann J. Application of in situ product crystallization and related techniques in biocatalytic processes. Chem Eur J 2019;25:4871–84.
121. Červeňanský I, Mihaľ M, Markoš J. Pertraction-adsorption in situ product removal system: intensification of 2-phenylethanol bioproduction. Sep Purif Technol 2020;251:117283.
122. Selder L, Sabra W, Jürgensen N, Lakshmanan A, Zeng AP. Co-cultures with integrated in situ product removal for lactate-based propionic acid production. Bioproc Biosyst Eng 2020;43:1027–35.
123. Mahal H, Branton H, Farid SS. End-to-end continuous bioprocessing: impact on facility design, cost of goods, and cost of development for monoclonal antibodies. Biotechnol Bioeng 2021;118:3468–85.
124. Kornecki M, Schmidt, Lohmann, Huter, Mestmäcker, Klepzig, et al. Accelerating biomanufacturing by modeling of continuous bioprocessing – piloting case study of monoclonal antibody manufacturing. Processes 2019;7:495.
125. Vees CA, Neuendorf CS, Pflügl S. Towards continuous industrial bioprocessing with solventogenic and acetogenic clostridia: challenges, progress and perspectives. J Ind Microbiol Biotechnol 2020;47:753–87.
126. Coffman J, Brower M, Connell-Crowley L, Deldari S, Farid SS, Horowski B, et al. A common framework for integrated and continuous biomanufacturing. Biotechnol Bioeng 2021;118:1735–49.
127. Khanal O, Lenhoff AM. Developments and opportunities in continuous biopharmaceutical manufacturing. mAbs 2021;13:1903664.
128. Feidl F, Vogg S, Wolf M, Podobnik M, Ruggeri C, Ulmer N, et al. Process-wide control and automation of an integrated continuous manufacturing platform for antibodies. Biotechnol Bioeng 2020;117:1367–80.
129. Xu N, Liu S, Xu L, Zhou J, Xin F, Zhang W, et al. Enhanced rhamnolipids production using a novel bioreactor system based on integrated foam-control and repeated fed-batch fermentation strategy. Biotechnol Biofuels 2020;13:80.
130. Lapeña D, Olsen PM, Arntzen MØ, Kosa G, Passoth V, Eijsink VGH, et al. Spruce sugars and poultry hydrolysate as growth medium in repeated fed-batch fermentation processes for production of yeast biomass. Bioproc Biosyst Eng 2020;43:723–36.
131. Zhang F, Liu J, Han X, Gao C, Ma C, Tao F, et al. Kinetic characteristics of long-term repeated fed-batch (LtRFb) l-lactic acid fermentation by a Bacillus coagulans strain. Eng Life Sci 2020;20:562–70.
132. Biglari N, Orita I, Fukui T, Sudesh K. A study on the effects of increment and decrement repeated fed-batch feeding of glucose on the production of poly(3-hydroxybutyrate) [P(3HB)] by a newly engineered Cupriavidus necator NSDG-GG mutant in batch fill-and-draw fermentation. J Biotechnol 2020;307:77–86.
133. Stepper L, Filser FA, Fischer S, Schaub J, Gorr I, Voges R. Pre-stage perfusion and ultra-high seeding cell density in CHO fed-batch culture: a case study for process intensification guided by systems biotechnology. Bioproc Biosyst Eng 2020;43:1431–43.
134. Särnlund S, Jiang Y, Chotteau V. Process intensification to produce a difficult-to-express therapeutic enzyme by high cell density perfusion or enhanced fed-batch. Biotechnol Bioeng 2021;118:3533–44.
135. Göbel S, Jaén KE, Dorn M, Neumeyer V, Jordan I, Sandig V, et al. Process intensification strategies toward cell culture-based high-yield production of a fusogenic oncolytic virus. Biotechnol Bioeng 2023;120: 2639–57.
136. Wong HE, Chen C, Le H, Goudar CT. From chemostats to high-density perfusion: the progression of continuous mammalian cell cultivation. J Chem Technol Biotechnol 2022;97:2297–304.
137. Schulze M, Kues D, Gao W, Houser M, Scheibenbogen K, Husemann B, et al. Automation of integrated perfusion control simplifying process intensification of mammalian biomanufacturing in single-use bioreactors. Chem Ing Tech 2022;94:1968–76.
138. Müller J, Teale M, Steiner S,Junne S, Neubauer P, Eibl D, et al. Intensified and continuous mAb production with single-use systems. In: Pörtner R, editor. Cell culture engineering and technology: in appreciation to Professor Mohamed Al-Rubeai. Cham: Springer International Publishing; 2021:401–29 pp.
139. Schaub, J, Ankenbauer, A, Habicher, T, Löffler, M, Maguire, N, Monteil, D, et al. Process intensification in biopharmaceutical process development and production – an industrial perspective. Phys Sci Rev 2023. https://doi.org/10.1515/psr-2022-0113.

140. Müller J, Ott V, Eibl D, Eibl R. Seed train intensification using an ultra-high cell density cell banking process. Processes 2022;10:911.
141. Jordan M, Mac Kinnon N, Monchois V, Stettler M, Broly H. Intensification of large-scale cell culture processes. Curr Opin Chem Eng 2018;22:253–7.
142. Schulze M, Lemke J, Pollard D, Wijffels RH, Matuszczyk J, Martens DE. Automation of high CHO cell density seed intensification via online control of the cell specific perfusion rate and its impact on the N-stage inoculum quality. J Biotechnol 2021;335:65–75.
143. Schmidt SR. Process intensification based on disposable solutions as first step toward continuous processing. In: Process control, intensification, and digitalisation in continuous biomanufacturing. Hoboken, NJ: Wiley; 2022:137–78 pp.
144. Zijlstra G, Touw K, Koch M, Monge M, et al. Design considerations towards an intensified single-use facility. In: Single-use technology in biopharmaceutical manufacture. Hoboken, NJ: Wiley; 2019:181–92 pp.
145. Jossen V, Eibl R, Eibl D. Single-use bioreactors – an overview. In: Single-use technology in biopharmaceutical manufacture. Hoboken, NJ: Wiley; 2019:37–52 pp.
146. Eibl R, Löffelholz C, Eibl D. Single-use bioreactors – an overview. In: Single-use technology in biopharmaceutical manufacture. Hoboken, NJ: Wiley; 2010:33–51 pp.
147. Samaras JJ, Micheletti M, Ding W. Transformation of biopharmaceutical manufacturing through single-use technologies: current state, remaining challenges, and future development. Annu Rev Chem Biomol Eng 2022;13:73–97.
148. Muthmann N, Guez T, Vasseur J, Jaffrey SR, Debart F, Rentmeister A. Combining chemical synthesis and enzymatic methylation to access short RNAs with various 5′ caps. Chembiochem 2019;20:1693–700.
149. Bering L, Thompson J, Micklefield J. New reaction pathways by integrating chemo- and biocatalysis. Trends Chem 2022;4:392–408.
150. Liu Y, Liu P, Gao S, Wang Z, Luan P, González-Sabín J, et al. Construction of chemoenzymatic cascade reactions for bridging chemocatalysis and biocatalysis: principles, strategies and prospective. Chem Eng J 2021;420:127659.
151. Levin I, Liu M, Voigt CA, Coley CW. Merging enzymatic and synthetic chemistry with computational synthesis planning. Nat Commun 2022;13:7747.
152. Zhou Y, Wu S, Bornscheuer UT. Recent advances in (chemo)enzymatic cascades for upgrading bio-based resources. Chem Commun 2021;57:10661–74.
153. Kaspar F, Schallmey A. Chemo-enzymatic synthesis of natural products and their analogs. Curr Opin Biotechnol 2022;77:102759.
154. Wu S, Zhou Y, Gerngross D, Jeschek M, Ward TR. Chemo-enzymatic cascades to produce cycloalkenes from bio-based resources. Nat Commun 2019;10:5060.
155. Brehm J, Lewis RJ, Richards T, Qin T, Morgan DJ, Davies TE, et al. Enhancing the chemo-enzymatic one-pot oxidation of cyclohexane via in situ H_2O_2 production over supported Pd-based catalysts. ACS Catal 2022; 12:11776–89.
156. Langsdorf A, Drommershausen AL, Volkmar M, Ulber R, Holtmann D. Fermentative alpha-humulene production from homogenized grass clippings as a growth medium. Molecules 2022;27:8684.
157. Langsdorf A, Volkmar M, Holtmann D, Ulber R. Material utilization of green waste: a review on potential valorization methods. Bioresour Bioprocess 2021;8:19.
158. Langsdorf A, Volkmar M, Ulber R, Hollmann F, Holtmann D. Peroxidases from grass clippings for the removal of phenolic compounds from wastewater. Bioresour Technol Rep 2023;22:101471.
159. Volkmar M, Maus AL, Weisbrodt M, Bohlender J, Langsdorf A, Holtmann D, et al. Municipal green waste as substrate for the microbial production of platform chemicals. Bioresour Bioprocess 2023;10:43.
160. Schoeters F, Thoré ES, De Cuyper A, Noyens I, Goossens S, Lybaert S, et al. Microalgal cultivation on grass juice as a novel process for a green biorefinery. Algal Res 2023;69:102941.
161. Bilal M, Iqbal HMN. Sustainable bioconversion of food waste into high-value products by immobilized enzymes to meet bio-economy challenges and opportunities – a review. Food Res Int 2019;123:226–40.

162. Mahato N, Sharma K, Sinha M, Dhyani A, Pathak B, Jang H, et al. Biotransformation of citrus waste-I: production of biofuel and valuable compounds by fermentation. Processes 2021;9:220.
163. Hadj Saadoun J, Bertani G, Levante A, Vezzosi F, Ricci A, Bernini V, et al. Fermentation of agri-food waste: a promising Route for the production of aroma compounds. Foods 2021;10:707.
164. Beitel SM, Coelho LF, Contiero J. Efficient conversion of agroindustrial waste into D(−) lactic acid by Lactobacillus delbrueckii using fed-batch fermentation. BioMed Res Int 2020;2020:4194052.
165. Costa S, Summa D, Semeraro B, Zappaterra F, Rugiero I, Tamburini E. Fermentation as a strategy for bio-Transforming waste into resources: lactic acid production from agri-food Residues. Fermentation 2021;7:3.
166. Sabater C, Ruiz L, Delgado S, Ruas-Madiedo P, Margolles A. Valorization of vegetable food waste and by-products through fermentation processes. Front Microbiol 2020;11:1–11.
167. Alexandri M, Blanco-Catalá J, Schneider R, Turon X, Venus J. High L(+)-lactic acid productivity in continuous fermentations using bakery waste and lucerne green juice as renewable substrates. Bioresour Technol 2020;316:123949.
168. van der Wielen LAM, Mussatto SI, van Breugel J. Bioprocess intensification: cases that (don't) work. New Biotechnol 2021;61: 108–15.
169. Whitford WG. Bioprocess intensification: technologies and goals. In: Process control, intensification, and digitalisation in continuous biomanufacturing; Hoboken, NJ: Wiley; 2022:93–136 pp.

Eduardo J. S. Brás and Pedro Carlos de Barros Fernandes*

2 Miniaturization and microfluidic devices: an overview of basic concepts, fabrication techniques, and applications

Abstract: Miniaturization brings along significant advantages in the development, optimization, and implementation of chemical, biochemical, and related fields processes and assays. Given the reduced footprint, miniaturization allows a significant reduction in volumes to be processed while providing conditions for several conditions to be evaluated simultaneously. Accordingly, work can be performed timely in a space efficient manner, with significant costs savings. Microfluidics is the pinnacle of miniaturization, where the previous advantageous aspects are taken to the limit, with the added features of operation under well-defined and highly efficient mass and heat transfer environment. Additionally, microfluidic environment fosters process integration, monitoring, and control in a single framework. The present chapter aims to provide an overview of diverse applications of miniaturized devices, fabrication methods, and key issues in fluid dynamics that characterize a microfluidic environment.

Keywords: microfabrication; microfluidic devices; miniaturization.

2.1 Introduction

Miniaturization involves scaling down as to operate in set-ups with reduced footprints, which translates to significant savings in reagents costs. Those set-ups either mimic larger scale already existing counterparts or enable novel designs and experimental concepts. Additionally, miniaturization comes along with high parallelization, which enables the simultaneous assessment of several process variables, and often provides a unique opportunity to shift from batch to continuous operation, through, e.g., the use of continuous flow reactors, which overall translates in significant enhanced throughputs [1–3]. Moreover, when scaling of bio/chemical processes from laboratory

---

*Corresponding author: Pedro Carlos de Barros Fernandes, iBB—Institute for Bioengineering and Biosciences, Instituto Superior Técnico, Universidade de Lisboa, Av. Rovisco Pais, 1049-001 Lisboa, Portugal; Associate Laboratory i4HB, Institute for Health and Bioeconomy at Instituto Superior Técnico, Universidade de Lisboa, Av. Rovisco Pais, 1049-001 Lisboa, Portugal; and BioRG and DREAMS, Faculty of Engineering, Lusofona University (ULHT), Campo Grande, 376, 1749 024 Lisbon, Portugal, E-mail: pedro.fernandes@ist.utl.pt. https://orcid.org/0000-0003-0271-7796
Eduardo J. S. Brás, NMI Natural and Medical Sciences Institute at the University of Tübingen, Reutlingen, Germany, E-mail: eduardo.bras@nmi.de

As per De Gruyter's policy this article has previously been published in the journal Physical Sciences Reviews. Please cite as: E. J. S. Brás, P. C. de Barros Fernandes "Miniaturization and microfluidic devices: an overview of basic concepts, fabrication techniques, and applications" *Physical Sciences Reviews* [Online] 2023. DOI: 10.1515/psr-2022-0102 | https://doi.org/10.1515/9783110760330-002

to pilot scale/industrial scale is envisaged, miniaturization enables numbering-up (also known as scaling-out), e.g., linking several similar reactors together in parallel, like a computer blade server, to increase the overall working volume. Besides, should one unit break down, it can be easily interchanged with no impact on the remaining units, thus enabling uninterrupted operation. This provides a simple and straight forward alternative to the complex and time-consuming conventional scale-up approach. The latter consists of increasing the volume of the reactor used in the lab scale (scale-up) where dimensionless numbers, transport coefficients, and geometrical similarities are used to design a reactor in a larger scale, a time consuming and often far from a straightforward process [4–8]. Despite the simplicity of the concept, numbering-up brings along some fluid flow distribution and control issues [9]. Thus, when shifting from laboratory to production scale, numbering-up also relies on devices with increased dimensions but with a structure that preserves favorable mass and heat transfer features [10]. Accordingly, numbering-up has been shown to enable easy transfer from process design in laboratory scale to industrial scale production [11]. All these considered, miniaturization is clearly within the scope of process intensification. The latter can be defined as any chemical engineering (or related discipline) development that leads to a substantially cleaner, safer, smaller, and more energy efficient technology [12] and ultimately results in improved efficiency and safety of processes and products quality, while decreasing operating costs and waste [13, 14]. Miniaturization encompasses the use of vessels up to the mL range and at least one characteristic dimension of channels up to the mm scale. Should the latter be under 1 mm (and accordingly volumes in the µL scale should be considered) extremely high surface area to volume ratios (SVR), e.g., 50,000 m^{-1} are observed, as compared to under 100 m^{-1} up to 1000 m^{-1} found in conventional laboratory and production vessels. The set-up thus involves the use of microdevices, e.g., microreactors, where a microfluidic environment, the lower endpoint of miniaturization, emerges. Microfluidics is characterized by well-defined and controlled mass and thermal diffusion, with nonetheless efficient heat and mass transfer, therefore, enabling novel process windows of operation [15–17]. The high SVR observed in microfluidic devices results in changes in the relative relevance of external forces acting on the fluid, e.g., viscosity may become dominant over inertia, unlike what is typically observed in the macroscopic environment [18–20]. Still, and such as in macroscopic fluid flow, the relative impact of different forces in fluid flow in microchannels can be assessed though a set of dimensionless numbers [18, 19].

The production of miniaturized devices is rooted in either the manufacture of semiconductor devices or in precision machining [21]. However, it has evolved to address the requirements for handling fluids, which can involve diverse tasks such as mixing, separation, parameter monitoring and control, detection and quantification of analytes, and reaction conditions [22–25]. Microfluidic devices have thus to tackle different issues; operate under different physical–chemical conditions, e.g., mild or eventually extreme pH environments, aqueous or nonaqueous (involving, e.g., ionic liquids or organic solvents), mild and eventually relatively extreme temperatures, and a vast range of operating pressures; handle/manipulate particles, either biological, e.g., microbial and animal cells, or synthetic, e.g., magnetic particles; and perform occasionally conflicting tasks, e.g., effective mixing in the typical diffusion controlled microfluidic environment, to name a few of the hurdles that microfluidic devices may encounter [26–29]. To comply with such diversity, microfluidic devices have grown in complexity, from glass or polydimethylsiloxane (PDMS) single microchannel chip type to highly multiplexed (simultaneous assessment of a single sample for several targets) multichannel platforms with a huge level of integration for monitoring of analytes and control of fluid flow [30–32]. Given the diversity, different materials and accordingly diverse fabrication techniques have been introduced to provide tailor-made solutions for each intended application in a reproducible and cost-effective manner [22, 33]. Thus, the high level of parallelization of miniaturized devices, combined with fine spatial–temporal control of the microenvironment and integration of analytical sensors, results in high throughput and high-quality data per experiment. These features enable the faster development of chemical and bioprocesses, thus reducing time to market [34, 35] and ease the separation and purification of specific compounds from complex mixtures during drug development. Moreover, they also allow for drug screening, since modeling of physiological conditions and simulation of microenvironments at the single cell to organ/human level is feasible [36]. Further detailed examples of the impact of microfluidics in chemical, biological, and pharmaceutical fields can be found elsewhere [8, 37–40].

In this chapter, an overview on microfluidic devices is provided, which includes fabrication techniques, fundamentals of microfluidics and fluid dynamics, and basic aspects on the role of microfluidics in the development of bioprocesses.

2.2 Microfabrication techniques: advantages and limitations

As previously mentioned, the advent of microfabrication techniques, pioneered by the semiconductor industry, opened the way for a myriad of novel microfluidic-based applications. While this chapter does not intend to go into detail about microfabrication techniques, we do wish to provide a general overview of the possibilities and limitations of these techniques in the context of bioprocess intensification.

When developing of a microfluidic system, one needs to take into consideration the material that will comprise the system and within the context of this book several questions will immediately rise. Will the substrate/product of reaction be adsorbed by the material? Is optical transparency relevant? Does the process under study require extreme temperatures and pressures?

Once the aforementioned questions have been addressed, a new set of questions concerning the geometry and scale of the system arise. What scale is required for the system (nL range or hundreds of µL)? Will the system require integrated instrumentation such as valves and pumping modules? Is it necessary to increase the surface area of the system by introducing microfeatures such as pillars or columns?

Again, once all these questions have been addressed, one now can start to decide on what microfabrication techniques should be employed for the development of the system at hand. In the following sections, we will go over the most common techniques for patterning and their use in the field of bioprocess development. However, this is not intended as a deep guide on how to navigate the world of microfabrication and micro-engineering as that would be beyond the intended scope of the chapter, but rather a brief overview of current technologies for novel scientists to the field to further expand their knowledge.

2.2.1 Thermoplastic patterning techniques

Processing of thermoplastics has widespread use throughout all different fields of science, technology, and consumer goods. Most techniques take advantage of the glass transition range of these materials, which when reached turns the substrate somewhat liquid, allowing for its shape to be manipulated. Once the substrate temperature has been reduced, the material retains this shape. Furthermore, other techniques akin to those used for metal and glass processing are also available. The use of thermoplastic materials for biocatalysis brings some interesting possibilities to the system, such as optical transparency, low adsorption of chemical compounds, and in general are biocompatible substrates. The most widespread thermoplastics also benefit from a very low price point, with examples of commonly used thermoplastic materials including poly(methyl methacrylate) (PMMA), polystyrene (PS), and polycarbonate (PC). One drawback of the use of thermoplastic-based material can be the sterilization of the system. Traditional autoclaving approaches could lead to the melting of the material and UV irradiation can be ineffective due to the material itself absorbing the light.

2.2.1.1 Extrusion

Although extrusion techniques are not traditionally considered for microfabrication, it is the main means of production of microtubes and capillaries and thus we will make brief note of them here. The tubes themselves can then be functionalized, with enzymes for example, and directly used as a simple microreactor, as a standalone system [41], or coupled to more complex microfluidic systems [42]. This type of tubes often has internal diameters in the range of hundreds of micrometers and can be produced at lengths of several meters. However, there is little room for patterning or improving the system. The lack of patterning options makes them very simple but limited flow reactors.

2.2.1.2 (Micro)Injection molding

Injection molding is one of the most common production methods of thermoplastic and thermoplastic elastomer (TPE) components worldwide. In short, the raw polymer material is heated to its glass transition temperature, injected into a mold, and subsequently cooled down, taking the shape of the negative of the mold. It is widely used for the production of consumer goods in lots of several thousand due to the quick processing times and relatively good surface finish. Recent advances in microinjection molding refer to feature sizes of 100 µm [43, 44]. However, despite the low price per unit once a system is established, the initial cost of entry of this technology is often too high for academic applications. This high cost of entry relates not only to the price of the injection equipment themselves, which will heavily depend on the targeted throughput, but also to the cost of the necessary molds that can be in the thousands of euros depending on the complexity of the system [45].

2.2.1.3 Hot embossing

Alongside with injection molding, hot embossing is a widespread technique for the production of consumer goods. Here, instead of injecting the semi-liquid heated polymer into a closed mold, the polymer substrate is pressed with a mold using a controlled pressure and subsequently cooled down. Feature resolution is limited to the resolution of the mold used with dimensions being similar to the ones obtained through injection molding. However due to the lower cost of entry, this approach has more widespread use in academic circles, with applications in both cell culture [46] and enzyme-based applications [47]. One advantage of hot embossing approaches is the possibility of using manual clamps and ovens to achieve a patterned surface and not necessarily rely on the use of high-end hot presses. However, it is a rather time-consuming process.

2.2.1.4 CNC micro milling

Computer numerical control (CNC) systems can be used to guide motorized milling machines according to a given computer-assisted design (CAD). Advances in this field through the introduction of smaller end-mills and higher complexity machines, such as 5-axis stages, have allowed for the fabrication of very intricate and complex systems with dimensions in the tenths of micrometers [48, 49]. Often the limitation of CNC milling is related to the high production time when compared to previous techniques. Despite this, micro milled systems have seen widespread use across multiple applications within the field of bioprocess development [50–53].

2.2.1.5 Laser cutting

As opposed to the previous patterning techniques, where the process itself defined all three dimensions of the end product (width, length, and height/depth), laser cutting only allows for the patterning of two of these dimensions, typically in the X–Y plane. There is also the possibility of altering all three dimensions through laser engraving. However, this technique is not as common and at the time of writing it produces poor surface finishes when not followed up by careful surface treatment protocols [54]. When laser cutting is used, the height of the feature produced is entirely dependent on the original substrate thickness, as laser cutting only allows for through-layer cutting. Feature resolution is in the domain of hundreds of micrometers and processing speed is quite high. Despite these advantages, laser cutting frequently produces defects in the features of thermoplastics due either to heating in the area adjacent to the laser pulse or to relatively poor channel wall smoothness. Additionally, laser cutting does often imply a more complex bonding procedure in addition to the fact that if the final system requires multiple heights, several layers need to be processed separately. Furthermore, inner features such as columns are not possible by only using laser cutting. Nevertheless, laser cut systems have seen widespread use for several applications [55–57].

2.2.2 Glass patterning techniques

It is often the case in biochemical synthesis that the reaction under study cannot be performed in a polymer-based vessel, making glass an attractive alternative due to its chemical inertness. Glass also has the advantage of being reusable when compared to most polymer-based systems, as it can be easily autoclaved or even chemically etched for cleaning purposes. However, glass systems are harder to fabricate and in general much more expensive than thermoplastic-based systems. There are several options for patterning glass devices, such as micro milling, laser micromachining, or chemical etching, although these have seen limited use in the context of bioprocess development [58].

2.2.3 Laser ablation

Laser ablation consists of irradiating the material with short pulses of highly energetic light to vaporize solid material. Laser ablation differs from laser cutting due to the fact that there is much less heating of the nonirradiated areas and thus less lateral damage to the feature area. The depth of which the materials are removed is dependent on the material properties of the substrate, as not all types of glass can be processed through laser ablation, and on the wavelength of choice. In contrast to the laser cutting, where only 2D designs can be achieved, laser ablation allows for the production of 3D features. Recent reports at the time of writing mention the achievement of feature sizes below 10 μm [59–61].

2.2.4 PDMS patterning

Polydimethylsiloxane (PDMS) is the *de facto* most widely used material when it comes to microfluidic devices across all fields [47, 62–64]. This high usage stems from the fact that it is relatively cheap and allows for quick prototyping for design testing. It is chemically inert and biocompatible; however, it does tend to adsorb a wide variety of molecules, especially those that are hydrophobic in nature, which can be a hindrance for kinetic studies of reactions [65]. PDMS is also relatively simple to sterilize through the use of UV radiation. Despite this chemical inertness, it is commonplace to functionalize the surface of PDMS systems through prior surface activation with plasma treatments such as oxygen plasma. This allows the introduction of chemical groups for both surface passivation [66] and to improve enzyme [67, 68] or cell adhesion [69]. The main way of PDMS patterning is by casting a mixture of the monomer material mixed with a specific cross-linker on a previously fabricated mold, bake it and then remove the cross-linked PDMS structure. In terms of resolution, PDMS is dependent on the quality of the molds used, with features below 10 µm being easily achieved. However, if larger reaction areas are necessary, it may be cumbersome to produce (Figure 2.1).

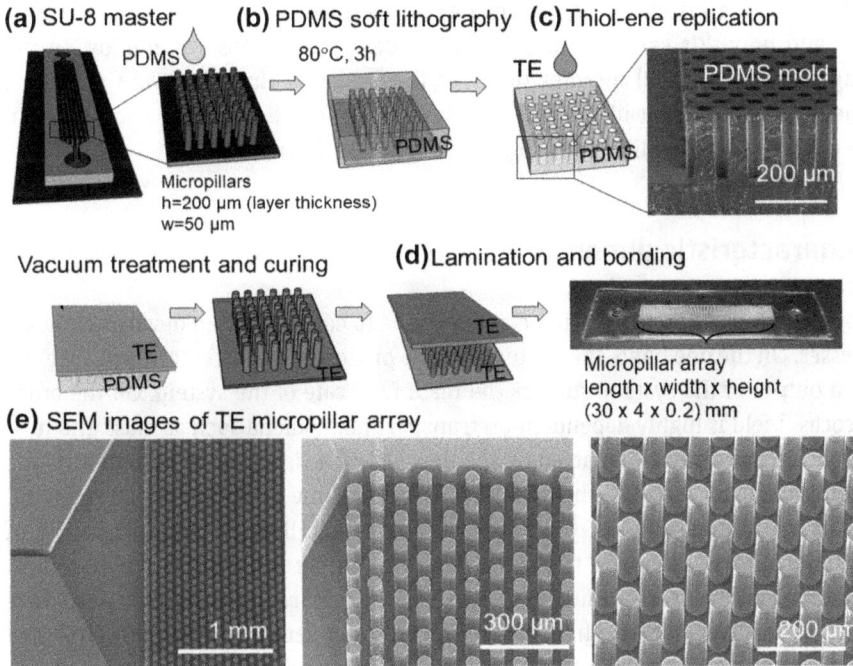

Figure 2.1: Typical workflow necessary for the production of PDMS-based microfluidic systems. Figure adapted from the work of Tähkä et al, under the Creative Common CC BY license [70].

2.2.5 Additive manufacturing

With the development of 3D printing technology and its rapid commercialization, several research groups took this opportunity and now use this technology to build cheaper microfluidic systems [71–73]. Even by using low to mid-end commercial stereo-lithography-based systems, it is possible to generate microfluidic devices with intricate and complicated designs for numerous applications [74]. Most of the resins used in these systems are not biocompatible; hence, they cannot be used directly in cell-based applications. However, they are still often used as cheaper alternatives for PDMS molding when compared to traditional silicon wafer–based molds. Recent developments in the field have led to a wide range of different resins with varying properties in terms of flexibility, hardness, and chemical inertness, some of which are also biocompatible. Current systems allow for feature resolution in the tenths of micrometers; however, this is not only dependent on the equipment being use but also on the resin itself.

2.3 Fluid dynamics and the microfluidic environment

Now that we covered the basics of the microfabrication techniques possible for the production of these systems, it is important to understand how to characterize their behavior and how this can be impacted by the design we choose to employ. In the following sections, we will attempt to cover the basics of fluid flow and transport phenomena in the context of microfluidic systems before delving into how we can actually employ microfluidics technology in bioprocess intensification.

2.3.1 Characteristic dimension

Fluid flow and fluid dynamics are important aspects to consider when developing novel bioprocesses. On the one hand, they determine the process outcome in terms of absolute input and output of the system, such as the mass flow rate of the system. On the other hand, process yield is highly dependent on transport phenomena such as mass and heat transfer, which are directly impacted by the flow characteristics of the system.

In order to characterize the behavior of the system, one needs to understand what the characteristic dimension of the system is. This will allow for the calculation of several dimensionless numbers that determine the fluidic behavior of the system in use. Depending on the process under consideration, the characteristic dimension of the system will typically be the length (L) or the hydraulic diameter of the channel (D_H). The latter is given by Eq. (2.1), where A is the area cross section of the channel, P is the wetted perimeter (part of the cross-sectional perimeter of a channel in contact with the fluid that it carries; in the present case, it is assumed the whole perimeter of the channel contacts the fluid), r is the channel radius, D is the channel diameter, H is the channel height, and a is the squared channel side.

$$D_H = \frac{4A}{P} \; ; D_{H\text{Circular}} = \frac{4\pi r^2}{2\pi r} = D \; ; D_{H\text{Rectangular}} = \frac{4W \times H}{2(W + H)} \; ; D_{H\text{Square}} = \frac{4a^2}{4a} = a \qquad (2.1)$$

2.3.2 Reynolds number

The Reynolds number (Re) is the ratio between the inertial forces and viscous forces in a fluid caused by different fluid velocities presented in the bulk of the fluid flow. An example of this phenomena is the lower fluid velocities closer to channel walls in a closed flow system compared to the center of the channel. The Re in a channel can be calculated according to Eq. (2.2), where v is the average fluid velocity, ρ is the fluid density, and μ is the dynamic viscosity of the fluid.

$$Re = \frac{vD_H\rho}{\mu} \qquad (2.2)$$

Depending on the value of the Re, the fluid flow regime can be characterized as being turbulent $Re > {\sim}4000$, while **laminar flow** is considered to be achieved at a $Re < {\sim}2000$. When working in microfluidic systems, the fluid regime will almost always be laminar in nature due to the very small D_H of the systems. This phenomenon has several advantages for the use of microfluidics in bioprocess development. Due to the lack of turbulence in the system, all mass transfer phenomena, perpendicular to the movement of the liquid, will be diffusion dependent instead of advection dependent as we will see in later sections of this chapter. The less chaotic nature of laminar flow makes it easier to predict and model with the help of the **Navier–Stokes** equations. This paves the way to very accurate process modeling without relying on high computational power.

The predictable nature of the fluid flow also allows for the generation of intricate systems, which allow a more detailed study in specific phenomena, for example, substrate uptake in biocatalysis or protein partition in a liquid–liquid extraction process. There are also limitations in the use of laminar flow; thus, by being dependent on diffusion for mass transport, large molecules require higher residence times to reach a catalyst, which can lead to the creation of depletion zones, subsequently lowering the efficiency of the system. Laminar flow can also be a hindrance when multiple substrates are required to be mixed together prior to the reaction chamber. To counteract this, there has been a lot of effort in the development of micromixers that can be easily integrated into the systems [75].

2.3.3 Péclet number

The Péclet numbers (*Pe*) are a class of dimensionless numbers that characterize the ratio between the impact of advection and diffusion in a transport event and can be applied to both mass transfer (*Pe$_M$*) and heat transfer (*Pe$_H$*). Both versions of the *Pe* number can be calculated according to Eq. (2.3), where *L* is the characteristic dimension of the system, *D* is the diffusivity, and *α* is the thermal diffusivity of the system.

$$Pe = \frac{\text{Advective Transport}}{\text{Diffusive Transport}} \; ; Pe_M = \frac{vL}{D} \; ; Pe_H = \frac{vL}{\alpha} \tag{2.3}$$

In most microfluidic systems, the longitudinal *Pe*, in other words, the *Pe* in the direction of the fluid flow, where the channel length is considered to be the characteristic dimension, tends to be very high. However, the radial *Pe* tends to be very low, which is a direct result of the laminar nature of the fluid flow and the lack of random streamlines in the fluid flow perpendicular to the direction of the channel that would be present in a turbulent system. The end result is a negligible fluid velocity in the radial direction of the microfluidic channel, which results in very small *Pe* numbers, demonstrating the dependence on diffusive transport for these systems.

2.3.4 Schmidt and Prandtl numbers

Another pair of dimensionless numbers relevant to transfer phenomena are the Schmidt (*Sc*) and the Prandtl (*Pr*) numbers, where the first relates to mass transfer, while the second to heat transfer. Both numbers refer to the ratio of transfer caused by momentum and by the existence of gradients of the referred physical quantity. In the case of the *Sc*, this would take the form of a concentration gradient of a specific molecule, while in the case of the *Pr*, this would refer to a temperature gradient in the system. Both numbers can be obtained by dividing the respective *Pe* by the *Re*.

2.3.5 Capillary number

The capillary number (Ca) gives us an understanding of the ratio of the viscous forces and the surface tension of the fluid in the system. It can be calculated according to Eq. (2.4) where *σ* is the surface tension of the fluid in the system or the interface tension between two adjacent fluids. It is of particular interest in systems that rely on flow through porous medium and multiphase flow systems, such as the case of a liquid–liquid extraction system or a droplet-based system.

$$Ca = \frac{\mu v}{\sigma} \tag{2.4}$$

2.3.6 Damköhler number

While the previous dimensionless numbers covered in this chapter all refer to fluid flow and transfer phenomena, there is a group of dimensionless numbers of particular interest in the context of bioprocess intensification – the Damköhler numbers (**Da**). The Da provide a ratio between the reaction rate in a system, i.e., enzyme conversion rate and the supply of the necessary substrate through advection, in the case of the First Damköhler number (Da_I) and through diffusion in the case of the Second Damköhler number (Da_{II}). The calculation of either Da number is dependent on the kinetics of the reaction being modeled.

There are other dimensionless numbers that can provide a more in-depth analysis of a fluidic system; however, it would be beyond the scope of this book to attempt to analyze them all. In the following sections of this chapter, we will proceed to show the direct application of microfluidics technology in bioprocess development and intensification by covering examples of both upstream and downstream processes.

2.4 Bioprocess development through microfluidics

(Bio)chemical processes overall can be split into two sections, the upstream, which concerns substrate preparation and the main conversion step(s), and the downstream, which are the product purification steps necessary to make the product usable and ready for sale. Microfluidic researchers have concerned themselves with both halves of the production line and worked on systems to improve different approaches to both ends of the production line. In the following sections, we will cover some of the most recent approaches used in both upstream and downstream processing developments for both fermentation-based processes and enzymatic-based approaches. Furthermore, we will make note of microfluidic platforms used not only to study process conditions but also those that can be used to improve catalyst efficiency, such as the case of evolutionary and screening systems.

2.4.1 Screening systems

2.4.1.1 Enzyme and cell screening

The discovery of new catalysts for biochemical conversion are of upmost importance, e.g., to withstand the harsh conditions encountered in several processes in the pharmaceutical, energy, and food sectors [76, 77] or to be active toward novel and non-natural substrates [78], whether these are novel enzymes from sea water samples [79] or genetically altered bacteria to enhance fermentation properties [80]. Traditional methods often employ the use of tedious well-plate assays that, in their own right, have seen an increase in throughput through the development of autonomous pipetting systems. However, they are still quite slow and labor intensive for high-throughput screening assays.

Digital and droplet microfluidics take advantage of the capillary number and low turbulence of these systems to become effective droplet generators, creating hundreds or even thousands of droplets per minute [81]. These systems comprise two immiscible phases, typically (but not necessarily) one aqueous and one organic phase and the typical mode of operation creates a droplet of the aqueous phase surrounded by the organic phase. Although the nomenclature is sometimes unclear, digital microfluidics also refer to the usage of electrowetting approaches, where the droplets are moved using an array of electrodes [82]. In these systems, the throughput of droplets under study is lower than the more standard droplet microfluidics. In both cases, these systems can be coupled with different types of sensors (photosensors, RAMAN, etc.) and active sorting systems (optical tweezers, electrophoresis, valves, etc.), which allow for high-throughput processing of the generated droplets [83–85].

If one considers each individual droplet as its own independent microreactor and capable of performing high-speed measurements of the reaction under study, then it becomes possible to sort out the most efficient individuals for the reaction under study [86]. This is of particular interest after performing genetic alterations to a known strain of bacteria or yeast [87], or when processing sea and soil samples in the pursuit of a new strain [88, 89]. By measuring the reaction outcome in each droplet, the active sorting mechanism can be signaled, resulting in the collection of the most efficient samples.

The same approach can be used for the directed evolution of enzymes of interest. In short, directed evolution consists of the generation of mutant libraries of known or synthetic enzymes and their subsequent sorting in an attempt to expedite the natural process [90]. These mutants can then be introduced into a droplet generation system and quickly sorted in accordance with the desired trait. Furthermore, the mutant DNA strands can be sorted into the droplets allowing for the enzyme to be produced and evaluated in the same isolated droplet [91]. Several recent reports have shown the efficiency of these systems, and some consider them to be the next big step in high-throughput enzyme screening [92–94]. Illustrative examples of the application of droplet microfluidics are given in Figure 2.2.

2.4.2 Upstream processing – biocatalysis

The use of microfluidics systems for biocatalysis can take many forms and serve different purposes. In this section, we will do a brief overview of the type of systems that can be created and the types of conversions that have been achieved. The reader should bear in mind that despite the chapter being focused on enzyme-based reactions, a lot of these approaches would be translatable to purely chemical reactions or to systems that would

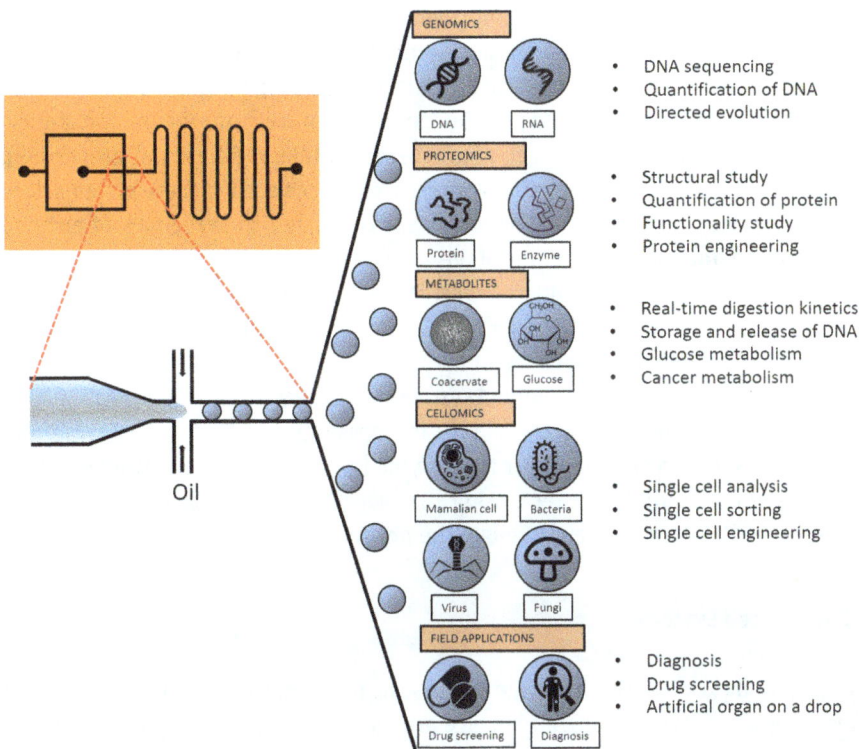

Figure 2.2: Examples of biotechnological applications of droplet microfluidics. Figure adapted from the work of Sanchéz Barea et al, under the Creative Common CC BY license [83].

rely on both types of reactions. Another important aspect for consideration is that most systems are traditionally used for optimization studies and subsequent scaling up of those process conditions as demonstrated by our own previous work [95]. However, it is reasonable to use these systems as means of production with the correct scaling out strategies [96].

2.4.2.1 Bare channels and patterned channels

Although traditionally stirred tank reactors (STRs) are the simplest type of reactor used at the macroscale, the easiest type to mimic are the plug flow or continuous tubular reactors. This can be achieved as simply as creating a straight or serpentine channel using the microfabrication techniques described before. In these types of systems, the enzyme needed is nonspecifically adsorbed to the surface of the microfluidic channel or covalently bonded to it, while the substrate is injected into the system. Although these systems have been widely used due to their simplicity, they often suffer from a major drawback, which is the lack of surface area to harbor a significant amount of the enzyme under study, despite being an improvement when compared with macroscale systems of this nature. Nevertheless, these systems have been successfully employed in the immobilization of lipases [41] and carboxylases [97], among others [98].

One solution to the low SVR is the introduction of features into the microchannels that would increase the system surface area. These can take the forms of groves [99], pillars [100], or columns [70]. However, the increased surface area of these systems does not counteract one other limiting factor of these systems, which is the lack of enzyme stability. In order to increase the lifetime of the enzymes being used, it is commonplace to encapsulate the enzyme in porous microbeads. These not only dramatically increase the surface area of the system, and thus the enzyme density, but also stabilize the enzymes for longer periods of time as we will see in the following section.

Despite this, these type of bare channel systems do present another advantage that is not seen in more complex systems, which is the possibility to flow the enzyme and substrate side-by-side and easily recover the enzyme for posterior analysis and reuse, or for the integration of liquid-based separation strategies such as the use of aqueous two-phase systems (ATPS) to immediately isolate the product of reaction [101, 102].

2.4.2.2 Packed bed systems

Packed bed systems are highly used at the macroscale for production purposes due to the aforementioned advantages of higher surface area and thus higher enzyme concentration in the system, in addition to the fact that the use of encapsulated and immobilized enzymes increase their lifetime. However, mass and heat transfer are hindered since hundreds of micrometers to mm-sized particles are used [9, 103].

When using microfluidic systems, it is possible to create systems that allow for the entrapment of the same type of microbeads that would be used at a larger scale; however, there are some considerations to be taken into account. The particles should have a diameter under 5% of the channel diameter to enable even flow, which typically leads to particle diameters under 50 nm [103]. The presence of these microbeads will increase the backpressure of the system, making it not only more difficult to perfuse but also if the system being employed is made of a soft compressible material such as PDMS, this can lead to channel deformation and abnormal flow patterns [8, 95, 103]. The packing strategy of the system has also to be taken into consideration, as well as the cross-section geometry of the channel being used. In both cases, if the channel is not properly packed, there will be areas of less fluidic resistance and once again abnormal flow lines may occur, which could impact the outcome of the process.

Despite these challenges, packed bed microreactors have an advantage over their simpler counterparts, that is, the scale-up process. In opposition to the bare channel systems, where the upscale process would be heavily dependent on the channel geometry due to differences in surface to volume ratios, which would impact transport phenomena, here mass and heat transfer will be highly dependent on the packing of the microbeads in the channel/reactor, thus making upscale processes more straightforward [95].

Overall, these type of microfluidic packed bed reactors are seen as an attractive tool for the optimization of both single step and multistep reactions.

2.4.2.3 Monolith reactors

To overcome the drawbacks of packed bed reactors, monolithic reactors were developed. In this case, instead of packing the microchannel with a suspension of microbeads or capsules, a single unit, porous filling is produced inside the microfluidic device. Monolithic reactors thus include a network of micro- or mesoporous materials, which may be either silica based or polymer based. Their porosity enables high enzyme concentration together with high mass transfer and flow rate at relatively low pressure, thus comparing favorably with packed-bed rectors [103–106]. Polymer-based monoliths are prone to low thermal stability and tend to swell; however, this can be overcome through extreme cross-linking [105, 106]. This type of system has seen use in transesterification [107], oxidation [108], and hydrolysis reactions [109].

Examples of different microfluidic set-ups are given in Figure 2.3.

2.4.3 Upstream processing – cell-based approaches

Whole cells have been used to produce a vast array of products, from simple molecules, e.g., ethanol, to complex molecules, e.g., proteins. Such processes rely on a myriad of different species ranging from bacteria and yeast to algae and mammalian cells. All these types of cell culture can also be tested using microfluidic systems. By confining the cells to smaller volume, it is possible to exert a higher degree of control on the culture conditions such as shear stress, chemical and temperature gradients, and residence

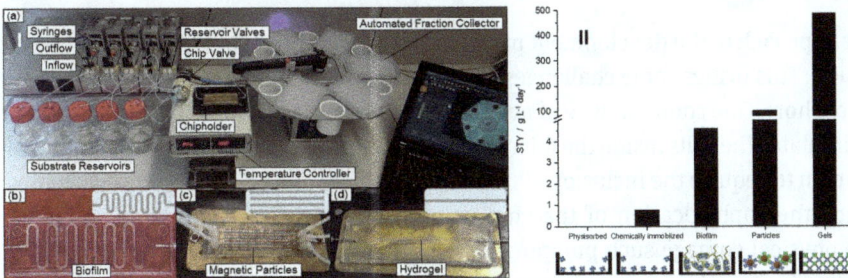

Figure 2.3: Demonstration of the required setup common to several types of microfluidic applications, as well as different approaches to a flow reactor used for the stereoselective reduction of 5-nitrononane-2,8-dione (I). Comparison of the reaction output from the different approaches used in this work (II). Figure adapted from the work of Peschke et al, under the Creative Common CC BY license [110].

times. However, due to the small dimensions of microfluidic systems, readouts to monitor the system can be made more difficult due to the low number of cells present per microfluidic device [86, 111].

In this section, we will cover the most common approaches used in the cultivation of both fixed cell cultures and suspended cell cultures.

2.4.3.1 Fixed cell culture

There is a wide range of methods for the culturing of cells; this may be performed in a typical suspension culture or colony based as in the case of solid-state fermentation, or there is the possibility of growing as a cohesive biofilm such as the case of some algae cultures. Microfluidic technology allows for in depth studies of all these types of culture approaches.

If the cells are adherent, the microfluidic channel itself can be used directly to seed the cells and allow them to grow as a film or even as a small colony [112]. Notable limitations of growing a biofilm in a microfluidic environment are mass transfer limitations and the risk of excessive biomass build-up that can result in channel clogging. Still, these can be overcome through droplet microfluidics, where forces, hence, generated enhance mass transfer and help to shape the biofilm [113]. The adhesion of the cells can be improved and patterned through specific surface functionalization protocols [114]; this has led to the development of a new method for quick screening assays referred to as cellular microarrays [115, 116], which nevertheless increases the complexity of the set-up. In some cases, when nonadherent cells are under study, these microarrays are achieved through microfluidic features that allow for the physical entrapment of the cells [62, 117]. These devices can be operated as to control growth in the device, so that overgrowth and cell proliferation outside the traps is prevented, thus enhancing the life span of the device.

2.4.3.2 Suspended cell culture

Another approach to the development of cell-based microbioreactors is the use of cells in suspension. This brings some challenges as residence times in microfluidic systems tend to be very short. One common way of avoiding this issue is to create a closed loop system that recirculates the cells inside the microfluidic device. When considering this approach, it is common to require the inclusion of microfluidic valves and on chip peristaltic pumps, increasing the sophistication of the chip being used. One also needs to consider the possible physical damage such pumping systems may cause to the cells [118, 119].

Another approach is relying on droplet microfluidics; here, the challenge is maintaining droplet stability long enough so cell replication has a chance to occur [120, 121]. An example of droplet microfluidic system for the screening of microbial cells is provided in Figure 2.4. Detailed information on recent developments involving suspended cell culturing in microfluidic environment can be found in a recent comprehensive review [30].

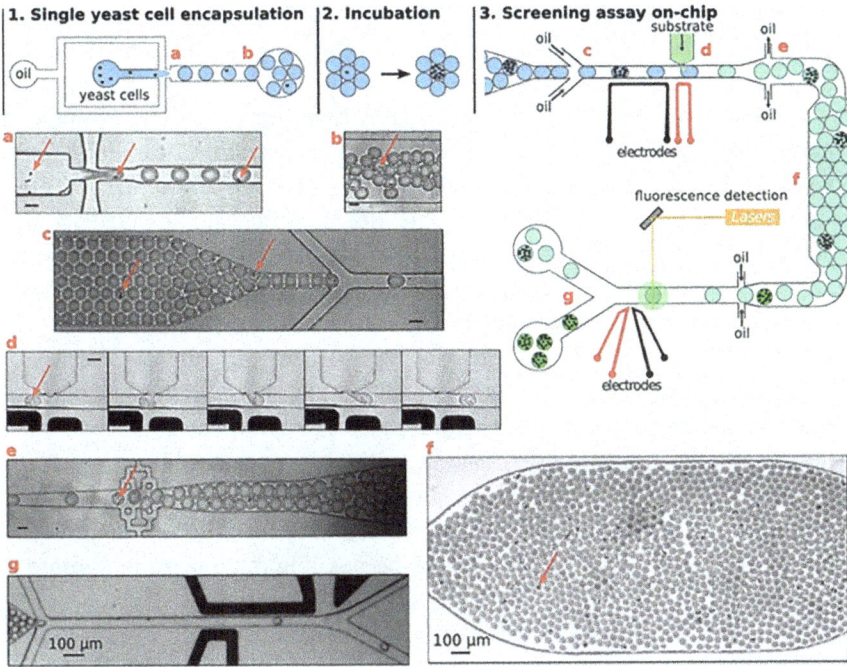

Figure 2.4: Workflow diagram and subsequent micrographs for a droplet microfluidic system for the screening of yeast cells for biotechnological applications. Figure adapted from the work of Beneyton et al, under the Creative Common CC BY license [121].

2.4.4 Downstream processing

As we know, in any biochemical production process, it is necessary to remove unnecessary elements from the final product. These might require rather straight forward processes, such as when leftover biomass is removed from the fermentation product, or a more refined approach, such as in the case of the removal of stereoisomers that could otherwise be toxic or counterproductive to the application of the final product. These steps fall under the downstream processing (DSP) portion of the production line.

While processes like density gradient centrifugation are not compatible with the notion of microfluidics, there are a wide array of DSP operations that can be not only studied and optimized using microfluidic devices but actually be employed at larger scales using appropriate scale-out strategies.

2.4.4.1 Cell lysis

When relying on the use of cells for the production of compounds of interest (CI), it is commonplace for the CI to be produced intracellularly, such as the case of certain proteins and enzymes. Hence, one of the first steps in the DSP sequence is the lysis of the produced biomass. There are two broad groups of lysis strategies, physical and (bio) chemical approaches.

When thinking of physical approaches to cell lysis, it is common to think of traditional methods like the French press or ultrasound sonication, which do not seem compatible with microfluidics technology. However, there are a vast array of approaches to mechanical cell lysis reported in the literature that mimic the effect produced in these traditional techniques [122]. The most common approaches rely on mechanical methods that destroy the integrity of the cell wall and membrane. These rely on the inclusion of features inside the microfluidic system, such as blades [123, 124], channel choke points [125], and narrow micro tunnels [126, 127] to damage the cell wall through the use of frictional and shear forces. Another approach is taking advantage of the instrumentation often coupled to microfluidic devices to allow for the integration of mechanical actuators such as pistons [128] and grinding setups [129, 130] that will apply force to a compressible microfluidic device, leading to the lysis of the cells inside the system. Other approaches, such as electrical disruption of cell walls [131] and the use of acoustic forces [132], have also been shown to be possible for integration within a microfluidic system.

It is common that mechanical lysis is not enough to disrupt the rigid cell wall presented by bacteria; here, (bio)chemical approaches are more suited. This approach consists in the mixing of chemical or enzymatic compounds that will disrupt the stability of the cell wall and membrane without compromising the structure of the CI. To study this effect, the microfluidic systems can take a more continuous approach, with long channels to study the kinetics of a given lysis agent [133] or a droplet approach [134]. An illustrative example of the use of a microfluidics to promote cell lysis is provided in Figure 2.5.

2.4.4.2 Membrane filtration

Filtration techniques are often one of the first separation methods employed as they are a cheap and efficient way of separating larger bodies from smaller molecules. Furthermore, advances in membrane material and functionalization have paved the way for affinity-based methods that also take advantage of this size-based separation. The integration of membranes in microfluidic systems is a very sought-after feature not only in the field of bioprocess development but also in other applications such as the development of organ-on-chip technologies and sample preparation [135].

Membrane integration in microfluidics typically consists of a device, which has two or more fluidic layers, separated by the membrane under study and can be used not only for studying ways to improve the separation through membrane screening [136, 137] but can also be used to study and prevent detrimental effects such as membrane fouling [138, 139] (Figure 2.6).

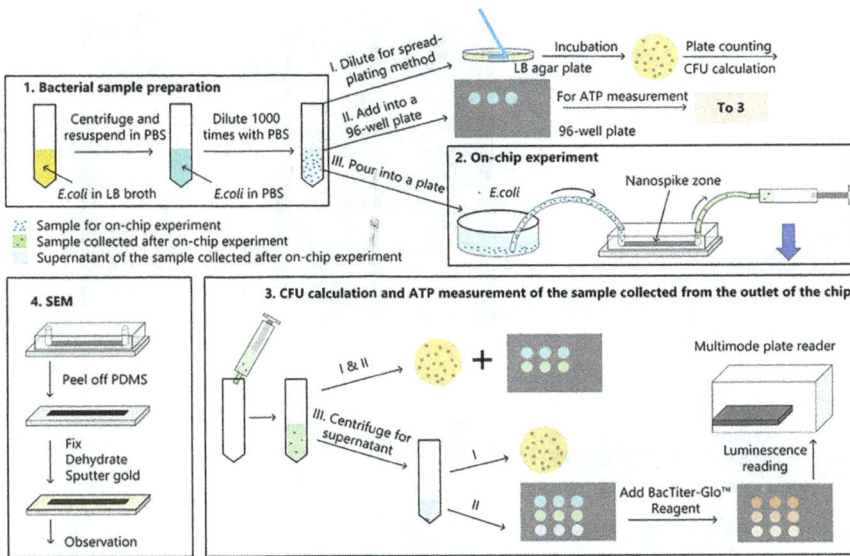

Figure 2.5: Use of a microfluidic system as a nonchemical approach to cell lysis. Cells are lysed as they pass through silicon nanospikes present inside the microchannel through which the cell suspension flows. To evaluate the efficiency of the set-up for cell lysis, colony forming units (CFU) samples of processed cell suspension were determined through the spread-plating method and compared to samples of cell suspensions without microfluidic device processing. Additionally, the supernatant of the processed cell suspension was monitored for ATP concentration, the presence of the latter further evidencing cell lysis. A scanning electron microscope (SEM) was also used to qualitatively evaluate cell lysis. Figure adapted from the work of Li et al, under the Creative Common CC BY license [123].

2.4.4.3 Liquid–liquid extraction

Once the CI is released, it is necessary to separate it from the other compounds present in the reaction mixture. One approach is the use of two immiscible liquids, which will form two phases, where the CI has different affinities to each, causing it to partition to one phase over the other. The first approaches relied on the use of an aqueous and an organic phase; however, this type of technology has evolved to a point where purely aqueous mixtures such as aqueous two-phase systems and ionic liquid–based systems can be used. This is of the upmost importance when considering the extraction of proteins and other biological compounds as they are often incompatible with organic solvents [140].

In terms of microfluidic applications, there is a wide selection of approaches for devices capable of screening assays to understand the best pairing of phases to extract a specific protein or compound of interest. These microfluidic devices, similar to some of the previous sections, can be traditional [141] or digital/droplet based [142, 143] and have been applied to a both small chemical molecules [141], as well as to antibodies [144] and

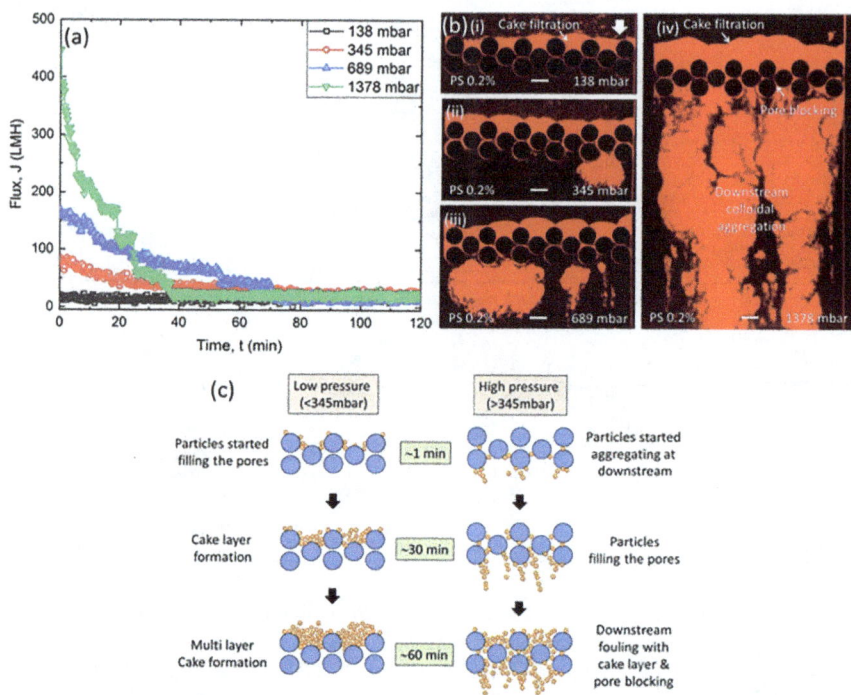

Figure 2.6: Study of membrane fouling mechanisms through the use of a microfluidic system. Figure adapted from the work of Debnath et al, under the Creative Common CC BY license [138].

enzymes [145, 146]. In addition to screening for the ideal phase pairing, microfluidic platforms can be used to screen actual process conditions including multiple extraction steps as demonstrated by our own previous work [147].

2.4.4.4 Chromatography

One of the cornerstones of molecular separation science is chromatography and whether the system under study is high performance liquid chromatography (HPLC), gas chromatography (GC) for analytical science, or size exclusion chromatography (SEC), they all rely on some interaction between the mobile and the stationary phases. This interaction can be of different natures such as ionic or hydrophobic or even be multimodal taking advantage of different molecular or affinity-based interactions in parallel [148].

Microfluidic technology can be a powerful tool in the optimization of extraction conditions, by both allowing for the screening of different stationary phases [149] but also, through proper instrumentation, for the screening of whole cycles of equilibration, capture, elution, and regeneration in an automated and efficient manner [150]. The typical structure used for these types of studies will consists of a large microfluidic channel to trap the stationary phase, connected to smaller channels to prevent the movement of the stationary phase, similar to the previously discussed packed-bed reactors [151]. Although most approaches are used for screening purposes, there are some reports of applying microfluidic systems for conducting actual HPLC studies at high pressures [152].

2.5 Conclusions

The concept of miniaturization encompasses the use of devices and/or platforms with small footprints, characterized by high surface to volume ratios. These culminate in microfluidic devices, which display channels with diameters in the range of tens to hundred micrometers, thus enabling precise control and modeling of experimental conditions. These features enable the simultaneous evaluation of multiple process variables with minute volumes required, with notable savings in time and consumables. Accordingly, miniaturization and particularly microfluidic devices have been paramount in the development of production processes, among them in the chemical, biochemical sectors and related fields, illustrative examples of which were presented in this chapter. This has been made possible as miniaturized devices have been produced to allow upstream (e.g., chemical/biochemical reactions and fermentations) and downstream (e.g., chromatography, liquid–liquid extraction, cell lysis) processing and to integrate analyte monitoring, to name some of the most relevant features. Given the diversity of materials available for microfluidic devices and intended goals, several fabrication techniques have been developed and implemented, some of which originate from electronics and semiconductors industries, which moreover illustrate the interdisciplinarity of microfluidics. The appealing features provided by microfluidic devices can be ascribed to the characteristic fluid dynamics that are created in this unique environment, which occur under given physical dimensions of the device and a set of conditions and physical properties defined by an array of dimensionless numbers. Although microfluidics has some limitations under specific operational conditions, e.g., handling of slurries or processes where heat and mass transfer are not a major issue, the attractive aspects of the concept suggest that the dissemination of microfluidic devices in process development and implementation to commercial scale will further increase, given the vast diversity of positive examples so far and the current observed trend. In particular, the use of microfluidic devices in cascade reactions (enzymatic and chemo-enzymatic) and set-ups where upstream and downstream are integrated in a single framework are, among others, expected to be of most relevance.

References

1. Aubin J, Commenge JM, Falk L, Prat L. Process intensification by miniaturization. In: Poux M, Cognet P, Gourdon C, editors. Green process engineering: from concepts to industrial applications. Boca Raton, FL, USA: CRC Press; 2015:77–108 pp.
2. Žnidaršič-Plazl P. Biocatalytic process intensification via efficient biocatalyst immobilization, miniaturization, and process integration. Curr Opin Green Sustain Chem 2021;32:100546.
3. Cussler EL. Non-selective membranes for separations. J Chem Technol Biotechnol 2003;78:98–102.
4. Dong Z, Wen Z, Zhao F, Kuhn S, Noël T. Scale-up of micro- and milli-reactors: an overview of strategies, design principles and applications. Chem Eng Sci X 2021;10:100097.
5. Togashi S, Miyamoto T, Sano T, Suzuki M. Microreactor system using the concept of numbering-up. In: Zhuang FG, Li JC, editors. New trends in fluid mechanics research. Berlin, Heidelberg, Germany: Springer Berlin Heidelberg; 2007:678–81 pp.
6. Enzmann F, Stöckl M, Zeng AP, Holtmann D. Same but different-Scale up and numbering up in electrobiotechnology and photobiotechnology. Eng Life Sci 2019;19:121–32.
7. Takors R. Scale-up of microbial processes: impacts, tools and open questions. J Biotechnol 2012;160:3–9.
8. Šalić A, Zelić B. Synergy of microtechnology and biotechnology: microreactors as an effective tool for biotransformation processes. Food Technol Biotechnol 2018;56:464–79.
9. Žnidaršič-Plazl P. Biotransformations in microflow systems: bridging the gap between academia and industry. J Flow Chem 2017:111–7. https://doi.org/10.1556/1846.2017.00021.
10. Jensen KF. Flow chemistry-Microreaction technology comes of age. AIChE J 2017;63:858–69.
11. Kockmann N, Gottsponer M, Zimmermann B, Roberge DM. Enabling continuous-flow chemistry in microstructured devices for pharmaceutical and fine-chemical production. Chem 2008;14:7470–7.
12. Reay D, Ramshaw C, Harvey A. Chapter 2 – process intensification – an overview. In: Reay D, Ramshaw C, Harvey A, editors. Process Intensification, 2nd ed. Oxford, UK: Butterworth-Heinemann; 2013:27–55 pp.
13. Haase S, Tolvanen P, Russo V. Process intensification in chemical reaction engineering. Processes 2022;10: 99.
14. Verdnik A, Pintaric ZN, Kravanja Z. Process intensification with microprocess engineering. Chem Eng Trans 2022;94:589–94.
15. Illg T, Löb P, Hessel V. Flow chemistry using milli- and microstructured reactors—from conventional to novel process windows. Bioorg Med Chem 2010;18:3707–19.
16. de Santis P, Meyer LE, Kara S. The rise of continuous flow biocatalysis – fundamentals, very recent developments and future perspectives. React Chem Eng 2020;5:2155–84.
17. Kiwi-Minsker L, Renken A. Microstructured reactors for catalytic reactions. Catal Today 2005;110:2–14.
18. Bruus H. Governing equations in microfluidics. In: Laurell T, Lenshof A, editors. Microscale acoustofluidics. Cambridge, UK: The Royal Society of Chemistry; 2014:1–28 pp.
19. Wibowo D, Zhao CX, He Y. Chapter 2 – fluid properties and hydrodynamics of microfluidic systems. In: Santos HA, Liu D, Zhang H, editors. Microfluidics for pharmaceutical applications. Oxford, UK: Elsevier; 2019:37–77 pp.
20. Gou Y, Jia Y, Wang P, Sun C. Progress of inertial microfluidics in principle and application. Sensors 2018;18: 1762.
21. Bojang AA, Wu HS. Design, fundamental principles of fabrication and applications of microreactors. Processes 2020;8:891.
22. Cardoso S, Silverio V. Chapter 2 – introduction to microfabrication techniques for microfluidics devices. In: Chappel E, editor. Drug delivery devices and therapeutic systems. London, UK: Academic Press; 2021: 19–30 pp.
23. Radadia AD. Microfluidics for biochemical and chemical reactions. In: Li D, editor. Encyclopedia of microfluidics and nanofluidics. Boston, MA, USA: Springer US; 2008:1195–207 pp.

24. Gharib G, Bütün İ, Muganlı Z, Kozalak G, Namlı İ, Sarraf SS, et al. Biomedical applications of microfluidic devices: a review. Biosensors 2022;12:1023.
25. Blesken C, Olfers T, Grimm A, Frische N. The microfluidic bioreactor for a new era of bioprocess development. Eng Life Sci 2016;16:190–3.
26. Darvas F, Dormán G, Hessel V, Ley SV, editors. Flow chemistry – applications. Berlin, Germany: De Gruyter; 2021.
27. Song Y, Cheng D, Zhao L. Microfluidics: fundamental, devices and applications. Weinheim, Germany: Wiley-VCH Verlag GmbH & Co. KGaA; 2018.
28. Zhang S, Wang Y, Onck P, den Toonder J. A concise review of microfluidic particle manipulation methods. Microfluid Nanofluidics 2020;24:24.
29. Afsaneh H, Mohammadi R. Microfluidic platforms for the manipulation of cells and particles. Talanta Open 2022;5:100092.
30. Kerk YJ, Jameel A, Xing X, Zhang C. Recent advances of integrated microfluidic suspension cell culture system. Eng Biol 2021;5:103–19.
31. Aralekallu S, Boddula R, Singh V. Development of glass-based microfluidic devices: a review on its fabrication and biologic applications. Mater Des 2023;225:111517.
32. Calero M, Fernández R, García P, García JV, García M, Gamero-Sandemetrio E, et al. A multichannel microfluidic sensing cartridge for bioanalytical applications of monolithic quartz crystal microbalance. Biosensors 2020;10:189.
33. Scott S, Ali Z. Fabrication methods for microfluidic devices: an overview. Micromachines 2021;12:319.
34. Marques MP, Szita N. Bioprocess microfluidics: applying microfluidic devices for bioprocessing. Curr Opin Chem Eng 2017;18:61–8.
35. Watts P, Wiles C. Micro reactors, flow reactors and continuous flow synthesis. J Chem Res 2012;36:181–93.
36. Li X, Fan X, Li Z, Shi L, Liu J, Luo H, et al. Application of microfluidics in drug development from traditional medicine. Biosensors 2022;12:870.
37. Boffito DC, Fernandez Rivas D. Process intensification connects scales and disciplines towards sustainability. Can J Chem Eng 2020;98:2489–506.
38. Luis SV, Garcia-Verdugo E, editors. Flow chemistry: integrated approaches for practical applications. Cambridge, UK: The Royal Society of Chemistry; 2019.
39. Trojanowicz M. Flow chemistry in contemporary chemical sciences: a real variety of its applications. Molecules 2020;25:1434.
40. Cardoso Marques MP, Lorente-Arevalo A, Bolivar JM. Biocatalysis in continuous-flow microfluidic reactors. Adv Biochem Eng Biotechnol 2022;179:211–46.
41. Bi Y, Zhou H, Jia H, Wei P. A flow-through enzymatic microreactor immobilizing lipase based on layer-by-layer method for biosynthetic process: catalyzing the transesterification of soybean oil for fatty acid methyl ester production. Process Biochem 2017;54:73–80.
42. Gruber P, Carvalho F, Marques MPC, O'Sullivan B, Subrizi F, Dobrijevic D, et al. Enzymatic synthesis of chiral amino-alcohols by coupling transketolase and transaminase-catalyzed reactions in a cascading continuous-flow microreactor system. Biotechnol Bioeng 2018;115:586–96.
43. Dempsey D, McDonald S, Masato D, Barry C. Characterization of stereolithography printed soft tooling for micro injection molding. Micromachines 2020;11:819.
44. Becker H, Gärtner C. Polymer microfabrication technologies for microfluidic systems. Anal Bioanal Chem 2008;390:89–111.
45. Bell MA, Becker KP, Wood RJ. Injection molding of soft robots. Adv Mater Technol 2022;7:2100605.
46. Puza S, Gencturk E, Odabasi IE, Iseri E, Mutlu S, Ulgen KO. Fabrication of cyclo olefin polymer microfluidic devices for trapping and culturing of yeast cells. Biomed Microdevices 2017;19:40.
47. Zhu Y, Chen Q, Shao L, Jia Y, Zhang X. Microfluidic immobilized enzyme reactors for continuous biocatalysis. React Chem Eng 2020;5:9–32.
48. O'Toole L, Kang CW, Fang FZ. Precision micro-milling process: state of the art. Adv Manuf 2021;9:173–205.

49. Kirsch B, Bohley M, Arrabiyeh P, Aurich J. Application of ultra-small micro grinding and micro milling tools: possibilities and limitations. Micromachines 2017;8:261.
50. Yusuf A, Garlisi C, Palmisano G. Overview on microfluidic reactors in photocatalysis: applications of graphene derivatives. Catal Today 2018;315:79–92.
51. Wouters B, Pirok BWJ, Soulis D, Garmendia Perticarini RC, Fokker S, van den Hurk RS, et al. On-line microfluidic immobilized-enzyme reactors: a new tool for characterizing synthetic polymers. Anal Chim Acta 2019;1053:62–9.
52. Guzzi F, Candeloro P, Coluccio ML, Cristiani CM, Parrotta EI, Scaramuzzino L, et al. A disposable passive microfluidic device for cell culturing. Biosensors 2020;10:18.
53. Mishra S, Liu YJ, Chen CS, Yao DJ. An easily accessible microfluidic chip for high-throughput microalgae screening for biofuel production. Energies 2021;14:1817.
54. Shaegh SAM, Pourmand A, Nabavinia M, Avci H, Tamayol A, Mostafalu P, et al. Rapid prototyping of whole-thermoplastic microfluidics with built-in microvalves using laser ablation and thermal fusion bonding. Sensor Actuator B Chem 2018;255:100–9.
55. Trantidou T, Friddin MS, Gan KB, Han L, Bolognesi G, Brooks NJ, et al. Mask-free laser lithography for rapid and low-cost microfluidic device fabrication. Anal Chem 2018;90:13915–21.
56. Huang JH, Harris JF, Nath P, Iyer R. Hollow fiber integrated microfluidic platforms for in vitro Co-culture of multiple cell types. Biomed Microdevices 2016;18:88.
57. Gruber P, Marques MPC, Sulzer P, Wohlgemuth R, Mayr T, Baganz F, et al. Real-time pH monitoring of industrially relevant enzymatic reactions in a microfluidic side-entry reactor (µSER) shows potential for pH control. Biotechnol J 2017;12:1600475.
58. Sugioka K, Xu J, Wu D, Hanada Y, Wang Z, Cheng Y, et al. Femtosecond laser 3D micromachining: a powerful tool for the fabrication of microfluidic, optofluidic, and electrofluidic devices based on glass. Lab Chip 2014;14:3447–58.
59. Soldera M, Alamri S, Sürmann PA, Kunze T, Lasagni AF. Microfabrication and surface functionalization of soda lime glass through direct laser interference patterning. Nanomaterials 2021;11:129.
60. Suryawanshi PL, Gumfekar SP, Bhanvase BA, Sonawane SH, Pimplapure MS. A review on microreactors: reactor fabrication, design, and cutting-edge applications. Chem Eng Sci 2018;189:431–48.
61. Domínguez MI, Centeno MA, Martínez TM, Bobadilla LF, Laguna ÓH, Odriozola JA. Current scenario and prospects in manufacture strategies for glass, quartz, polymers and metallic microreactors: a comprehensive review. Chem Eng Res Des 2021;171:13–35.
62. Brás EJ, Chu V, Aires-Barros MR, Conde JP, Fernandes P. A microfluidic platform for physical entrapment of yeast cells with continuous production of invertase. J Chem Technol Biotechnol 2017;92:334–41.
63. Yang T, Choo J, Stavrakis S, de Mello A. Fluoropolymer-coated PDMS microfluidic devices for application in organic synthesis. Chem 2018;24:12078–83.
64. Preetam S, Nahak BK, Patra S, Toncu DC, Park S, Syväjärvi M, et al. Emergence of microfluidics for next generation biomedical devices. Biosens Bioelectron X 2022;10:100106.
65. Grant J, Özkan A, Oh C, Mahajan G, Prantil-Baun R, Ingber DE. Simulating drug concentrations in PDMS microfluidic organ chips. Lab Chip 2021;21:3509–19.
66. Xia YM, Hua ZS, Srivannavit O, Ozel AB, Gulari E. Minimizing the surface effect of PDMS–glass microchip on polymerase chain reaction by dynamic polymer passivation. J Chem Technol Biotechnol 2007;82:33–8.
67. Kreider A, Richter K, Sell S, Fenske M, Tornow C, Stenzel V, et al. Functionalization of PDMS modified and plasma activated two-component polyurethane coatings by surface attachment of enzymes. Appl Surf Sci 2013;273:562–9.
68. Sugiura S, Edahiro J, Sumaru K, Kanamori T. Surface modification of polydimethylsiloxane with photo-grafted poly(ethylene glycol) for micropatterned protein adsorption and cell adhesion. Colloids Surf B Biointerfaces 2008;63:301–5.
69. Akther F, Yakob SB, Nguyen NT, Ta HT. Surface modification techniques for endothelial cell seeding in PDMS microfluidic devices. Biosensors 2020;10:182.

70. Tähkä S, Sarfraz J, Urvas L, Provenzani R, Wiedmer SK, Peltonen J, et al. Immobilization of proteolytic enzymes on replica-molded thiol-ene micropillar reactors via thiol-gold interaction. Anal Bioanal Chem 2019;411:2339–49.
71. Nielsen AV, Beauchamp MJ, Nordin GP, Woolley AT. 3D printed microfluidics. Annu Rev Anal Chem 2020; 13:45–65.
72. Ding L, Bazaz SR, Fardjahromi MA, McKinnirey F, Saputro B, Banerjee B, et al. A modular 3D printed microfluidic system: a potential solution for continuous cell harvesting in large-scale bioprocessing. Bioresour Bioprocess 2022;9:64.
73. Bellou MG, Gkantzou E, Skonta A, Moschovas D, Spyrou K, Avgeropoulos A, et al. Development of 3D printed enzymatic microreactors for lipase-catalyzed reactions in deep eutectic solvent-based media. Micromachines 2022;13:1954.
74. Heidt B, Rogosic R, Leoné N, Brás E, Cleij T, Harings J, et al. Topographical vacuum sealing of 3D-printed multiplanar microfluidic structures. Biosensors 2021;11:395.
75. Enders A, Siller IG, Urmann K, Hoffmann MR, Bahnemann J. 3D Printed microfluidic mixers—a comparative study on mixing unit performances. Small 2018;15:1804326.
76. Tsegaye B, Balomajumder C, Roy P. Microbial delignification and hydrolysis of lignocellulosic biomass to enhance biofuel production: an overview and future prospect. Bull Natl Res Cent 2019;43:51.
77. Mesbah NM. Industrial biotechnology based on enzymes from extreme environments. Front Bioeng Biotechnol 2022;10:870083.
78. Eggert T, Leggewie C, Puls M, Streit W, van Pouderoyen G, Dijkstra BW, et al. Novel biocatalysts by identification and design. Biocatal Biotransform 2004;22:141–6.
79. Ferrer M, Méndez-García C, Bargiela R, Chow J, Alonso S, García-Moyano A, et al. Decoding the ocean's microbiological secrets for marine enzyme biodiscovery. FEMS Microbiol Lett 2019;366:fny285.
80. Li N, Duan J, Gao D, Luo J, Zheng R, Bian Y, et al. Mutation and selection of *Oenococcus oeni* for controlling wine malolactic fermentation. Eur Food Res Tech 2015;240:93–100.
81. Ding Y, Howes PD, deMello AJ. Recent advances in droplet microfluidics. Anal Chem 2020;92:132–49.
82. Pang L, Ding J, Liu XX, Fan SK. Digital microfluidics for cell manipulation. TrAC, Trends Anal Chem 2019;117: 291–9.
83. Barea JS, Lee J, Kang DK. Recent advances in droplet-based microfluidic technologies for biochemistry and molecular biology. Micromachines 2019;10:412.
84. Weng L, Spoonamore JE. Droplet microfluidics-enabled high-throughput screening for protein engineering. Micromachines 2019;10:734.
85. Sohrabi S, Kassir N, Keshavarz Moraveji M. Droplet microfluidics: fundamentals and its advanced applications. RSC Adv 2020;10:27560–74.
86. Bjork SM, Joensson HN. Microfluidics for cell factory and bioprocess development. Curr Opin Biotechnol 2019;55:95–102.
87. Gach PC, Shih SCC, Sustarich J, Keasling JD, Hillson NJ, Adams PD, et al. A droplet microfluidic platform for automating genetic engineering. ACS Synth Biol 2016;5:426–33.
88. Yazdi SR, Agrawal P, Morales E, Stevens CA, Oropeza L, Davies PL, et al. Facile actuation of aqueous droplets on a superhydrophobic surface using magnetotactic bacteria for digital microfluidic applications. Anal Chim Acta 2019;1085:107–16.
89. Gorbatsova J, Jaanus M, Vaher M, Kaljurand M. Digital microfluidics platform for interfacing solid-liquid extraction column with portable capillary electropherograph for analysis of soil amino acids. Electrophoresis 2016;37:472–5.
90. Dalby PA. Strategy and success for the directed evolution of enzymes. Curr Opin Struct Biol 2011;21: 473–80.
91. Holstein JM, Gylstorff C, Hollfelder F. Cell-free directed evolution of a protease in microdroplets at ultrahigh throughput. ACS Synth Biol 2021;10:252–7.

92. Goto H, Kanai Y, Yotsui A, Shimokihara S, Shitara S, Oyobiki R, et al. Microfluidic screening system based on boron-doped diamond electrodes and dielectrophoretic sorting for directed evolution of NAD(P)-dependent oxidoreductases. Lab Chip 2020;20:852–61.
93. Chiu FWY, Stavrakis S. High-throughput droplet-based microfluidics for directed evolution of enzymes. Electrophoresis 2019;40:2860–72.
94. Fu X, Zhang Y, Xu Q, Sun X, Meng F. Recent advances on sorting methods of high-throughput droplet-based microfluidics in enzyme directed evolution. Front Chem 2021;9:666867.
95. Brás EJS, Domingues C, Chu V, Fernandes P, Conde JP. Microfluidic bioreactors for enzymatic synthesis in packed-bed reactors—multi-step reactions and upscaling. J Biotechnol 2020;323:24–32.
96. Brás EJS, Chu V, Conde JP, Fernandes P. Recent developments in microreactor technology for biocatalysis applications. React Chem Eng 2021;6:815–27.
97. Zhu Y, Huang Z, Chen Q, Wu Q, Huang X, So PK, et al. Continuous artificial synthesis of glucose precursor using enzyme-immobilized microfluidic reactors. Nat Commun 2019;10:4049.
98. Bolivar JM, Tribulato MA, Petrasek Z, Nidetzky B. Let the substrate flow, not the enzyme: practical immobilization of d-amino acid oxidase in a glass microreactor for effective biocatalytic conversions. Biotechnol Bioeng 2016;113:2342–9.
99. Zhu C, Gong A, Zhang F, Xu Y, Sheng S, Wu F, et al. Enzyme immobilized on the surface geometry pattern of groove-typed microchannel reactor enhances continuous flow catalysis. J Chem Technol Biotechnol 2019; 94:2569–79.
100. Nagy C, Kecskemeti A, Gaspar A. Fabrication of immobilized enzyme reactors with pillar arrays into polydimethylsiloxane microchip. Anal Chim Acta 2020;1108:70–8.
101. Vobecká L, Tichá L, Atanasova A, Slouka Z, Hasal P, Přibyl M. Enzyme synthesis of cephalexin in continuous-flow microfluidic device in ATPS environment. Chem Eng J 2020;396:125236.
102. Meng SX, Xue LH, Xie CY, Bai RX, Yang X, Qiu ZP, et al. Enhanced enzymatic reaction by aqueous two-phase systems using parallel-laminar flow in a double Y-branched microfluidic device. Chem Eng J 2018;335: 392–400.
103. Arshi S, Nozari-Asbemarz M, Magner E. Enzymatic bioreactors: an electrochemical perspective. Catalysts 2020;10:1232.
104. van der Helm MP, Bracco P, Busch H, Szymańska K, Jarzębski AB, Hanefeld U. Hydroxynitrile lyases covalently immobilized in continuous flow microreactors. Catal Sci Technol 2019;9:1189–200.
105. Ciemięga A, Maresz K, Malinowski J, Mrowiec-Białoń J. Continuous-flow monolithic silica microreactors with arenesulphonic acid groups: structure–catalytic activity relationships. Catalysts 2017;7:255.
106. Sachse A, Galarneau A, Coq B, Fajula F. Monolithic flow microreactors improve fine chemicals synthesis. New J Chem 2011;35:259.
107. Alotaibi M, Manayil JC, Greenway GM, Haswell SJ, Kelly SM, Lee AF, et al. Lipase immobilised on silica monoliths as continuous-flow microreactors for triglyceride transesterification. React Chem Eng 2018;3: 68–74.
108. Alotaibi MT, Taylor MJ, Liu D, Beaumont SK, Kyriakou G. Selective oxidation of cyclohexene through gold functionalized silica monolith microreactors. Surf Sci 2016;646:179–85.
109. Onbas R, Yesil-Celiktas O. Synthesis of alginate-silica hybrid hydrogel for biocatalytic conversion by β-glucosidase in microreactor. Eng Life Sci 2019;19:37–46.
110. Peschke T, Bitterwolf P, Hansen S, Gasmi J, Rabe K, Niemeyer C. Self-immobilizing biocatalysts maximize space-time yields in flow reactors. Catalysts 2019;9:164.
111. Lattermann C, Büchs J. Microscale and miniscale fermentation and screening. Curr Opin Biotechnol 2015; 35:1–6.
112. Kim J, Park HD, Chung S. Microfluidic approaches to bacterial biofilm formation. Molecules 2012;17: 9818–34.
113. David C, Heuschkel I, Bühler K, Karande R. Cultivation of productive biofilms in flow reactors and their characterization by CLSM. Methods Mol Biol 2020;2100:437–52.

114. Tong Z, Rajeev G, Guo K, Ivask A, McCormick S, Lombi E, et al. Microfluidic cell microarray platform for high throughput analysis of particle–cell interactions. Anal Chem 2018;90:4338–47.
115. Rothbauer M, Wartmann D, Charwat V, Ertl P. Recent advances and future applications of microfluidic live-cell microarrays. Biotechnol Adv 2015;33:948–61.
116. Willaert R, Goossens K. Microfluidic bioreactors for cellular microarrays. Fermentation 2015;1:38–78.
117. Puchberger-Enengl D, van den Driesche S, Krutzler C, Keplinger F, Vellekoop MJ. Hydrogel-based microfluidic incubator for microorganism cultivation and analyses. Biomicrofluidics 2015;9:014127.
118. Mozdzierz NJ, Love KR, Lee KS, Lee HLT, Shah KA, Ram RJ, et al. A perfusion-capable microfluidic bioreactor for assessing microbial heterologous protein production. Lab Chip 2015;15:2918–22.
119. Lee KS, Boccazzi P, Sinskey AJ, Ram RJ. Microfluidic chemostat and turbidostat with flow rate, oxygen, and temperature control for dynamic continuous culture. Lab Chip 2011;11:1730.
120. Periyannan Rajeswari PK, Joensson HN, Andersson-Svahn H. Droplet size influences division of mammalian cell factories in droplet microfluidic cultivation. Electrophoresis 2017;38:305–10.
121. Beneyton T, Thomas S, Griffiths AD, Nicaud JM, Drevelle A, Rossignol T. Droplet-based microfluidic high-throughput screening of heterologous enzymes secreted by the yeast *Yarrowia lipolytica*. Microb Cell Factories 2017;16:18.
122. Grigorov E, Kirov B, Marinov MB, Galabov V. Review of microfluidic methods for cellular lysis. Micromachines 2021;12:498.
123. Li L, Tian F, Chang H, Zhang J, Wang C, Rao W, et al. Interactions of bacteria with monolithic lateral silicon nanospikes inside a microfluidic channel. Front Chem 2019;7:483.
124. Yun SS, Yoon SY, Song MK, Im SH, Kim S, Lee JH, et al. Handheld mechanical cell lysis chip with ultra-sharp silicon nano-blade arrays for rapid intracellular protein extraction. Lab Chip 2010;10:1442.
125. Huang X, Xing X, Ng CN, Yobas L. Single-cell point constrictions for reagent-free high-throughput mechanical lysis and intact nuclei isolation. Micromachines 2019;10:488.
126. Dizaji AN, Ozturk Y, Ghorbanpoor H, Cetak A, Akcakoca I, Kocagoz T, et al. Investigation of the effect of channel structure and flow rate on on-chip bacterial lysis. IEEE Trans Nanobiosci 2021;20:86–91.
127. Hao WJ, Chen WJ, Chai MH, Yuan FF, Huang LM, Wei ZH, et al. Microfluidic platform based on site-specific post-imprinting modification of molecularly imprinted monolith with Connizzaro reaction to improve identification of N-myristoylated peptides. Sensor Actuator B Chem 2022;356:131338.
128. Kim YC, Kang JH, Park SJ, Yoon ES, Park JK. Microfluidic biomechanical device for compressive cell stimulation and lysis. Sensor Actuator B Chem 2007;128:108–16.
129. Flaender M, den Dulk R, Flegeau V, Ventosa J, Delapierre G, Berthier J, et al. Grinding Lysis (GL): a microfluidic device for sample enrichment and mechanical lysis in one. Sensor Actuator B Chem 2017;258: 148–55.
130. Berasaluce A, Matthys L, Mujika J, Antoñana-Díez M, Valero A, Agirregabiria M. Bead beating-based continuous flow cell lysis in a microfluidic device. RSC Adv 2015;5:22350–5.
131. Jeon H, Kim S, Lim G. Electrical force-based continuous cell lysis and sample separation techniques for development of integrated microfluidic cell analysis system: a review. Microelectron Eng 2018;198:55–72.
132. Lu H, Mutafopulos K, Heyman JA, Spink P, Shen L, Wang C, et al. Rapid additive-free bacteria lysis using traveling surface acoustic waves in microfluidic channels. Lab Chip 2019;19:4064–70.
133. Fradique R, Azevedo AM, Chu V, Conde JP, Aires-Barros MR. Microfluidic platform for rapid screening of bacterial cell lysis. J Chromatogr A 2020;1610:460539.
134. Shamloo A, Hassani-Gangaraj M. Investigating the effect of reagent parameters on the efficiency of cell lysis within droplets. Phys Fluids 2020;32:062002.
135. de Jong J, Lammertink RGH, Wessling M. Membranes and microfluidics: a review. Lab Chip 2006;6:1125.
136. Inci F. Benchmarking a microfluidic-based filtration for isolating biological particles. Langmuir 2022;38: 1897–909.
137. Chen X, Shen J, Hu Z, Huo X. Manufacturing methods and applications of membranes in microfluidics. Biomed Microdevices 2016;18:104.

138. Debnath N, Kumar A, Thundat T, Sadrzadeh M. Investigating fouling at the pore-scale using a microfluidic membrane mimic filtration system. Sci Rep 2019;9:10587.
139. Fung K, Li Y, Fan S, Fajrial AK, Ding Y, Ding X. Acoustically excited microstructure for on-demand fouling mitigation in a microfluidic membrane filtration device. J Membr Sci Lett 2022;2:100012.
140. Soares RRG, Azevedo AM, van Alstine JM, Aires-Barros MR. Partitioning in aqueous two-phase systems: analysis of strengths, weaknesses, opportunities and threats. Biotechnol J 2015;10:1158–69.
141. Alimuddin M, Grant D, Bulloch D, Lee N, Peacock M, Dahl R. Determination of log D via automated microfluidic liquid–liquid Extraction. J Med Chem 2008;51:5140–2.
142. Mary P, Studer V, Tabeling P. Microfluidic droplet-based liquid–liquid extraction. Anal Chem 2008;80: 2680–7.
143. Wells SS, Kennedy RT. High-throughput liquid–liquid extractions with nanoliter volumes. Anal Chem 2020; 92:3189–97.
144. Espitia-Saloma E, Vâzquez-Villegas P, Rito-Palomares M, Aguilar O. An integrated practical implementation of continuous aqueous two-phase systems for the recovery of human IgG: from the microdevice to a multistage bench-scale mixer-settler device. Biotechnol J 2016;11:708–16.
145. Meagher RJ, Light YK, Singh AK. Rapid, continuous purification of proteins in a microfluidic device using genetically-engineered partition tags. Lab Chip 2008;8:527.
146. Silva DFC, Azevedo AM, Fernandes P, Chu V, Conde JP, Aires-Barros MR. Determination of partition coefficients of biomolecules in a microfluidic aqueous two phase system platform using fluorescence microscopy. J Chromatogr A 2017;1487:242–7.
147. Brás EJS, Soares RRG, Azevedo AM, Fernandes P, Arévalo-Rodríguez M, Chu V, et al. A multiplexed microfluidic toolbox for the rapid optimization of affinity-driven partition in aqueous two phase systems. J Chromatogr A 2017;1515:252–9.
148. Bao B, Wang Z, Thushara D, Liyanage A, Gunawardena S, Yang Z, et al. Recent advances in microfluidics-based chromatography—a mini review. Separations 2020;8:3.
149. Nascimento A, Pedro MNS, Pinto IF, Aires-Barros MR, Azevedo AM. Microfluidics as a high-throughput solution for chromatographic process development – the complexity of multimodal chromatography used as a proof of concept. J Chromatogr A 2021;1658:462618.
150. Pinto IF, Santos DR, Soares RRG, Aires-Barros MR, Chu V, Azevedo AM, et al. A regenerable microfluidic device with integrated valves and thin-film photodiodes for rapid optimization of chromatography conditions. Sensor Actuator B Chem 2018;255:3636–46.
151. Kecskemeti A, Gaspar A. Particle-based liquid chromatographic separations in microfluidic devices – a review. Anal Chim Acta 2018;1021:1–19.
152. Lazar IM, Trisiripisal P, Sarvaiya HA. Microfluidic liquid chromatography system for proteomic applications and biomarker screening. Anal Chem 2006;78:5513–24.

Yunting Liu*, Shiqi Gao, Pengbo Liu, Weixi Kong, Jianqiao Liu and
Yanjun Jiang*

3 Integration of chemo- and bio-catalysis to intensify bioprocesses

Abstract: Nature has evolved highly efficient and complex systems to perform cascade reactions by the elegant combination of desired enzymes, offering a strategy for achieving efficient bioprocess intensification. Chemoenzymatic cascade reactions (CECRs) merge the complementary strengths of chemo-catalysis and bio-catalysis, such as the wide reactivity of chemo-catalysts and the exquisite selective properties of biocatalysts, representing an important step toward emulating nature to construct artificial systems for achieving bioprocess intensification. However, the incompatibilities between the two catalytic disciplines make CECRs highly challenging. In recent years, great advances have been made to develop strategies for constructing CECRs. In this regard, this chapter introduces the general concepts and representative strategies, including temporal compartmentalization, spatial compartmentalization and chemo-bio nanoreactors. Particularly, we focus on what platform methods and technologies can be used, and how to implement these strategies. The future challenges and strategies in this burgeoning research area are also discussed.

Keywords: asymmetric synthesis; chemoenzymatic cascades; compartmentalization; co-immobilization; nanoreactor.

3.1 Introduction

Bioprocess intensification (BPI) emphasizes the efficiency, sustainability, cost effectiveness, and yield of biomanufacturing, resulting in economically improved processes. The state-of-the-art strategies for BPI are divided into sections according to upstream (biocatalyst engineering and immobilization techniques), bioreactor (continuous processing) and downstream processing steps in a conventional bioprocess. Apart from that, cascade reactions have also emerged as a revolutionary tool to intensify bioprocesses. Cascade reactions are a series of reactions whereupon the products of a prior step are directly used as the substrate/reactant for the following reaction step.

*Corresponding authors: Yunting Liu and Yanjun Jiang, School of Chemical Engineering and Technology, Hebei University of Technology, Tianjin 300130, China, E-mail: ytliu@hebut.edu.cn (Y. Liu), yanjunjiang@hebut.edu.cn (Y. Jiang)
Shiqi Gao, Pengbo Liu, Weixi Kong and Jianqiao Liu, School of Chemical Engineering and Technology, Hebei University of Technology, Tianjin 300130, China

As per De Gruyter's policy this article has previously been published in the journal Physical Sciences Reviews. Please cite as: Y. Liu, S. Gao, P. Liu, W. Kong, J. Liu and Y. Jiang "Integration of chemo- and bio-catalysis to intensify bioprocesses" *Physical Sciences Reviews* [Online] 2023. DOI: 10.1515/psr-2022-0103 | https://doi.org/10.1515/9783110760330-003

All this occurs inside the same reactant tank without isolation of intermediates. This type of reactions is beneficial and attractive because it avoids the need of separation and purification of reaction intermediates as preparation for subsequent reaction steps [1]. Nature has evolved highly efficient and complex systems to perform cascade reactions by the elegant combination of desired enzymes, offering a strategy for achieving efficient bioprocess intensification. Nature's enzymatic cascade reactions have been one of the biggest inspirations for the current manufacturing of fine chemicals and pharmaceuticals. However, this strategy is far from meeting the unlimited need of synthesizing new compounds due to the enzyme's intrinsic low productivity and the narrow substrate scope. The multistep one-pot chemoenzymatic cascade reactions (CECRs) have become the focus of current efforts in green chemistry and synthetic biology [2], which exhibit the following advantages: (1) acquiring synergistic catalytic abilities that can widen substrate scope, improve reactivity and enhance stereochemical control of chemical reactions, (2) increasing yields and synthetic efficiency, and minimizing the time- and resource-consuming, tedious purification steps and waste production, (3) efficiently managing unstable intermediates, and avoiding the use of toxic reagents. In fact, these are one of the goals to be achieved by BPI. Therefore, CECRs themselves can be seen as a major strategy for BPI due to the improvements in reaction process and results, although the improvements were rarely described in terms of 'intensification'.

Despite being quite elegant, developing chemoenzymatic cascade catalysis is not a trivial task due to mutual inactivation, incompatible reaction conditions, and disfavoured kinetics [3]. For example, chemical catalysts operate smoothly at harsh conditions (e.g., organic solvents, high temperature), but they are inactivated in aqueous medium. Whereas biocatalysts perform efficiently in an aqueous solution, at room temperature and neutral pH, but they are unstable in organic solvents and high temperatures. Additionally, for such cascade reactions to occur, the process must be energetically favorable. If this is not the case, the cascade reactions should be redesigned to ensure that the reaction can reach high conversion. Therefore, looking for catalysts that are compatible with each other while retaining their respective activity, and conditions that permit all steps of the reaction sequence to proceed successfully are two main issues that restrict the development of this field. In recent years, much advancement in this field has been achieved with various strategies and techniques developed for constructing chemoenzymatic systems [4]. Among them, this chapter highlight temporal compartmentalization, spatial compartmentalization and nanoreactor strategies.

3.2 Temporal compartmentalization strategy

The term temporal compartmentalization refers to a design methodology where the different cascading reactions are separated by time. That is, the second step of the reaction is not allowed to proceed before the current reaction step is finished. Briefly, strategies to employ temporal compartmentalization entails the addition of reactants and changes of reaction conditions such as temperature, pH at selected times. Cascade reactions can be performed in concurrent mode where all catalysts and reagents are added at the initial step without changing the reaction conditions in subsequent reaction steps, or in sequential mode with the execution of the subsequent step (e.g., addition of catalyst, reagent, cofactor, co-solvent, etc.) only after the completion of the prior step. The sequential process not only can meet all requirements of the temporal compartmentalization, but also allow to design and control appropriate reaction sequence, which helps the search and optimization of compatible environment for the various catalytic steps. Accordingly, this strategy has become very popular and important towards developing one-pot CECRs.

Metal-catalyzed C-C cross-coupling has been recognized as a robust and versatile synthetic tool for C-C bond formation in organic chemistry, such as Suzuki, Heck and Negishi reactions, etc. Moreover, a key advantage of such reactions is that they can be effectively performed in water, the preferred solvent for enzymes, which allows their combination with biotransformation [5]. The one-pot CECRs integrating a metal mediated cross-coupling and an enantioselective biotransformation has been hailed as a major breakthrough and will be the topic for discussion.

Scheme 3.1 shows the first reported (by Gröger et al.) instance of utilizing the Suzuki cross-coupling reaction in aqueous medium with a subsequent bio-reduction [6]. These two steps required quite different conditions: the reaction between phenylboronic acid and p-bromo-acetophenone (Suzuki reaction) was conducted at 70 °C in basic conditions. The following bioreduction process is catalyzed by alcohol dehydrogenase from *Rhodoccocus* sp. and requires a neutural pH and room temperature conditions. A hurdle to conduct this one-pot process was that the free phosphine and boronic acid tend to inhibit the enzyme activity. This can be easily solved by simply removing the phosphine and adding 1-equivalent of boronic acid ensuring that no excess boronic acid remained to inhibit the enzyme. Ultimately, the CECR was achieved by simple adjustments of temperature and pH prior to enzymatic step and provided a series of chiral biaryl alcohols.

From this pioneering work of Gröger, successive improvements in both catalysts and medium have enabled to develop this sequential *mo*-CMCR towards wide scope, high robustness and applicability. Several years later, the overall sequential process was conducted by using water-soluble palladium catalysts, which essentially allowed the use of room temperature [7]. The groups of Schmitzer, Gröger and González–Sabín leveraged

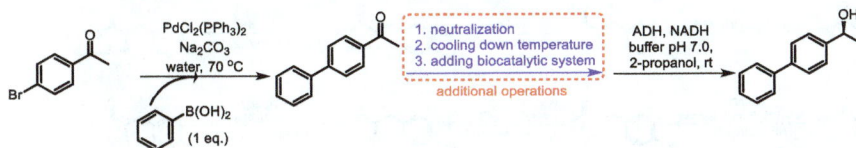

Scheme 3.1: Temporal compartmentalization strategy for chemoenzymatic synthesis of chiral biaryl alcohols through Pd-catalyzed Suzuki cross-coupling and enzymatic reduction.

the benefits of biphasic-systems to mitigate the effects of the solubility problem of the reactants, leading to significant increase of the substrate concentration [8]. Vicente et al. achieved the enantioselective synthesis of an odanacatib precursor through this CECR [9]. Cacchi et al. utilized protein-stabilized palladium nanoparticles to catalyze Suzuki reaction to avoid using sensitive phosphines [10].

Garg et al. developed a sequential cascade combining a Ni-catalyzed Suzuki coupling and a ketoreductase (KRED)-mediated bioreduction, producing enantioenriched diarylmethanol derivatives [11]. The Suzuki reaction could take place successfully by using Ni(cod)$_2$ (15 mol%) as the catalyst, SIPr (30 mol%) as the ligand and water (0.5 M) as the reaction medium at 60 °C. After completing this step, cooling down reaction mixture, neutralizing pH to 7, and enzyme and cofactor additions were required to start the enzymatic transformation. The KRED can achieve the desired asymmetric reduction employing an i-PrOH–NADPH cofactor recycling system, producing chiral alcohols in high yields and enantiomeric excesses (ee).

The Suzuki reaction and an enzymatic (hydro)amination can be used to produce chiral biaryl amino acid and amines. Turner et al. reported 3-step one-pot reaction cascades entailing Suzuki reaction along with: (i) phenylalanine ammonia lyases (PALs) catalyzed hydroamination of arylpropenoic acids (Scheme 3.2A); or (ii) D-amino acid dehydrogenase catalyzed reductive amination of α-keto acids (Scheme 3.2B) [12]. Boc-protection of free amines was required for a successful Suzuki reaction. Furthermore, the reaction temperature for these two steps were too high for the relatively low temperature required for the enzymatic process. To solve this problem, a sequential approach was taken. First, the enzymatic synthesis of 4-bromophenylalanine was performed at mild temperature (25 or 37 °C). Then a mixed solvent of deionised water and THF, and protection reagent of Boc$_2$O were added, and the solution was microwave heated (200 W, 90 °C) to protect the primary amino group to provide the desired Boc-protected 4-bromophenylalanine derivatives. Finally, the phenylboronic acid, the Pd-catalyst and the base were injected to the reaction, initiating the Pd-catalyzed Suzuki reaction with the aid of microwave irradiation at 120 °C. A series of chiral biaryl amino acid derivatives could thus be obtained. For obtaining chiral biaryl amines, Bornscheuer and co-workers reported an application of the Suzuki reaction followed by a trans-amination reaction which was catalyzed by a modified amine transaminases (ATAs) [13]. One of the important points for the success of this process was that the modification/

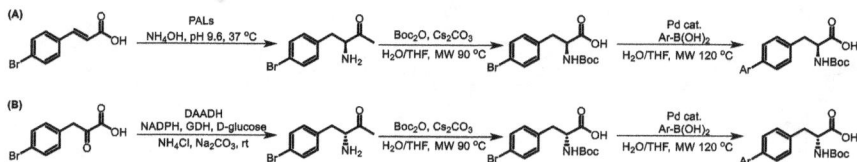

Scheme 3.2: Temporal compartmentalization strategy for chemoenzymatic synthesis of chiral biaryl amino acid.

engineering of the ATAs from *Asperguillus fumigatus* resulted in broader substrate scope and improved stability. The main benefit of using this biaryl cross-coupling in the transamination is that it can avoid the additional amine group protection.

Hartwig and Zhao combined a Rh-catalyzed diazocoupling and an ene-reductase (ER)-catalyzed reduction for synthesizing 2-aryl 1,4-dicarbonyl compounds (Scheme 3.3A) [14]. The diazocoupling was conducted using methylene chloride (DCM) as slovent at −78 °C. Subsequently, DCM was evaporated and all components required for the reduction step were added. The enzymatic reduction was performed efficiently in buffer containing 2.5% DMSO at room temperature, where glucose dehydrogenase was utilized for cofactor recycling, producing the final products with >99% ee. In addition, Gröger and co-workers synthesized bulk chemical nitriles from readily available alkenes through an HCN-free three-step sequence, involving Rh-catalyzed hydroformylation, spontaneous aldoxime formation, and enzymatic aldoxime dehydration (Scheme 3.3B) [15]. In the first step, a phase separation was employed to achieve a reuse of the expensive metal catalyst. To start the enzymatic step, the heat-treatment of the reaction mixture resulting from aldoxime formation was required, which is important to help decompose hydroxylamine that could reduce the enzyme stability.

Apart from the coupling reaction, there are many other types of metal-catalyzed reactions have been integrated with biotransformations. González–Sabín and co-workers reported a sequential process connecting a Ru(IV)-catalyzed isomerization of allylic alcohols with an ω-transaminases (ω-TA)-catalyzed asymmetric bioamination of the *in situ*-formed ketone intermediates to produce the chiral primary amines [16]. Assembling the two steps in a one-pot simultaneous process is unattainable due to the strong inhibitory effect on the enzyme system from metal catalyst, which could be easily

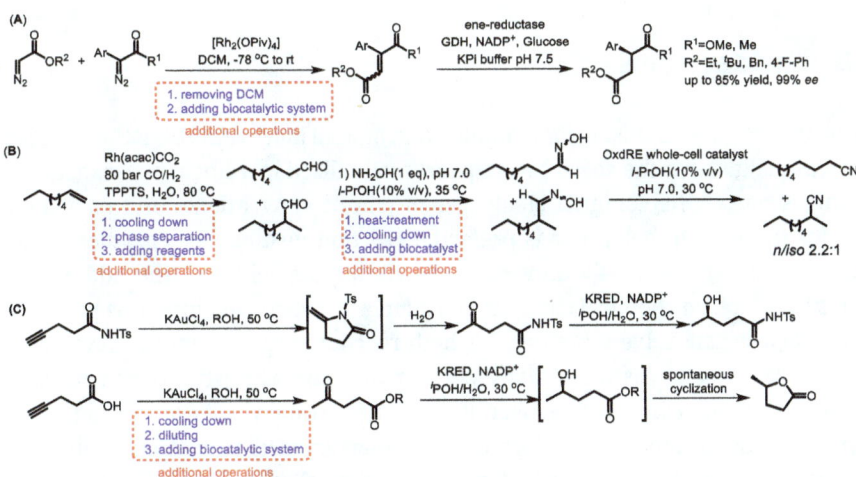

Scheme 3.3: Temporal compartmentalization strategy for chemoenzymatic cascades combining. (A) Rh-catalyzed diazocoupling and ER-mediated reduction; (B) Rh-catalyzed hydroformylation, spontaneous aldoxime formation, and enzymatic aldoxime dehydration; (C) Au-catalyzed cycloisomerization and KRED catalyzed bioreduction.

addressed just by diluting the reaction concentration suitable for the enzymatic step. Recently, they achieved the combination of the Pd/Au-catalyzed cycloisomerization of alkynes that contained a tethered nucleophile and an asymmetric bioreduction in aqueous media relying on the operation of dilution (Scheme 3.3C) [17]. This process involved the initial Pd or Au-catalyzed cycloisomerization of the substrates, concomitant hydrolysis of intermediate five-membered heterocycles, and the keto group of the latter undergoing a bioreduction as the final step. This provides a variety of enantiopure valuable molecules, such as 1,4-diols, lactones, and γ-hydroxy-carbonyl compounds. In the same year, they developed an aqueous one-pot sequential CECR that combines an organocatalyst (AZADO) catalyzed oxidation reaction with the subsequent ω-TA-catalyzed transamination, where the additional dilution operation between the two steps was still required to realize the enzymatic step [18].

3.3 Spatial compartmentalization strategy

Intracellular spatial organization is critical to eliminating the negative effects of toxic intermediates, side reactions and sluggish turnover rates. This technique creates separated micro-sized environments that are designed for cellular processes, thus allowing for multi-enzymatic reactions. Several approaches have been developed to circumvent incompatibility and mutual inactivation by physically separating the catalytic centers by shielding [19]. In this section, the most pivotal and impactful methods for spatial compartmentalization will be discussed, including biphasic systems, water-based micellar solution, membrane filters, and natural whole cells.

3.3.1 Biphasic systems

Apart from the aforementioned incompatibility and mutual inactivation, another issue to consider was the fact that the species were poorly solubilized in a single-phase medium. In fact, this is one of the common challenges of joining conventional and enzymatic catalysts, for the reason that water is preferred reaction medium for enzymes, while chemical catalysts are generally only dissolved in organic solvents. The addition of co-solvents to solubilize chemical catalysts is a simple strategy to address this problem. Identifying appropriate solvent systems for both reaction steps is vital for developing one-pot chemoenzymatic cascades. Many performance characteristics of the reaction (such as catalyst stability and activity, overall reaction rates and yield) are affected by the selection of the reaction media. Therefore, different reaction media are usually employed to solve specific cases. A Friedel-Crafts-type asymmetric alkylation of α,β-unsaturated aldehydes (catalyzed by peptide) was combined with laccase-catalyzed oxyamination to furnish indole-based derivatives functionalized by oxygen in a mixture H_2O/THF (2/1 v/v) [20]. A concurrent cascade in a system of pH 6.0 buffer/MeCN (1:1) for the conversion of

1,4-dihydroxybenzene to α-arylated aldehydes was designed by marrying oxidation catalyzed by laccase with subsequent the organocatalyzed α-arylation [21]. An efficient method for direct C-H hydroxylation in H_2O/CH_3CN (9:1) was developed based on sequential photoredox/enzymatic [22]. A cooperative chemoenzymatic reaction was developed by combining photocatalyzed alkene isomerization with ene-reductases enabled reduction of C=C, which was conducted in pH 7.5 Tris buffer containing DMSO (10% v/v) to generate valuable enantioenriched products [23]. Combining an organo-catalyst (AZADO) and an ω-TA, an one-pot oxidation-transamination sequential process was implemented for the asymmetric conversion of racemic alcohols into amines, where a biphasic solvents system of $H_2O/PhCF_3$ (4:1) was used for the organocatalytic step while pH 7.5 KPi buffer containing DMSO (15% v/v) was used for the enzymatic step.

In the systems above, the co-solvents are miscible with water, which sometimes still suffered from the incompatibility and mutual inactivation issues. Biphasic systems with immiscible liquid phases that act as two separate (macro)compartments *i.e.* no mixing and contacting between the different types of catalysts. This can provide compatible conditions and improve solubility, and have been proved to be an effective way to perform CECRs, especially for concurrent ones. Haak and co-workers reported biphasic system with a dynamic kinetic resolution (DKR) of racemic β-haloalcohols that included haloalcohol dehalogenase (HheC)-catalyzed cyclization in buffer and iridacycle-catalyzed racemization in toluene (Scheme 3.4A) [24]. The biphasic system reduced the interaction between metal catalyst and enzyme, thus enabling this simultaneous one-pot process. Kroutil et al. developed a toluene/buffer biphasic system that entailed deracemization by oxidation/reduction sequence using the same iridacycle along with an ADH [25]. The nonstereoselective oxidation of racemic chlorohydrins to α-chloro ketones (which was catalyzed by Ir) occurred using toluene as a solvent. The obtained ketones were then reduced stereoselectively (catalyzed by ADH) in buffer. The compatibility of metal

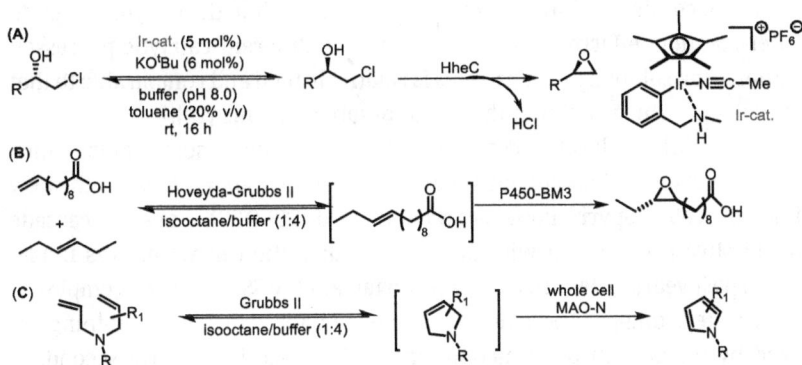

Scheme 3.4: Biphasic system (buffer/toluene) for: (A) chemoenzymatic synthesis of enantioenriched epoxides through racemization catalyzed by Ir and asymmetric cyclization catalyzed by HheC; (B) chemoenzymatic synthesis of enantioenriched epoxides through Ru-catalyzed RCM and P450-BM3-catalyzed epoxidation; (C) chemoenzymatic synthesis of furans through RCM catalyzed by ruthenium and laccase/TEMPO-catalyzed aromatization.

and enzyme was attributed to the identification of orthogonal reagents for each step. However, only moderate enantioselectivities were obtained due to the secondary alcohols could be racemized by the iridacycle.

A cooperative, biphasic tandem catalysis composed of a cross-metathesis (CM) reaction catalyzed by ruthenium and a stereoselective enzymatic epoxidation in buffer has been developed to synthesize enantioenriched epoxides (Scheme 3.4B) [26]. A 2nd-generation Hoveyda-Grubbs catalyst was employed due to their air stability along with its activity in protic media. Isooctane was employed as the non-aqueous phase because it is suitable for CM as well as biocompatible with the enzyme (P450-BM3 from *Bacillus megaterium*). Compared with the stepwise process, the reaction in tandem resulted in a higher yield. This was likely due to the irreversible enzymatic reaction that result in the metathesis equilibrium shifting towards the right. Using a reaction media that contained isooctane and buffer, another two one-pot processes were developed by the group of Castagnolo: a sequential application of CM catalyzed by ruthenium and aromatization catalyzed by monoamine oxidase (MAO-N) (Scheme 3.4C) [27], and a combination of CM catalyzed by ruthenium with aromatization catalyzed by laccase/TEMPO [28]. Key to the success of these one-pot cascades is the use of biphasic systems where the organic substrates do not react with the enzyme, avoiding their deactivation.

3.3.2 Aqueous micellar solutions

The notion of conducting organic reactions in water is often abandoned since most uncharged organic compounds are simply insoluble, especially at room temperature. Surfactants with solubilizing ability are usually utilized to tackle this drawback by providing a micellar environment, in which substrates and catalysts might readily interact. These micellar nanoreactors are considered dual-phase systems where the chemical catalysts are confined within the hydrophobic core while the enzymes remain in the bulk water solution, effectively separating the two different catalytic processes. In micellar solutions, the solubility of substrates is relatively high and comparable to that of even organic solvent, which can omit the use of organic solvents [29].

Lipshutz and co-workers developed a series of novel 'tailored' surfactants, which proved to be very useful for various aqueous chemical reactions [30]. With one of the new surfactants (1-octadecyl-5-oxopyrrolidine-3-carboxylic acid, C18-OPC), a one-pot cascade coupling a Suzuki-Miyaura coupling with an aldol condensation in water was implemented [31]. In the past years, a designer/tailored surfactant (TPGS-750-M) was employed in a couple of two-step, one-pot sequential chemoenzymatic cascades combining reactions catalyzed by metal with bioreduction, enabling synthesis of chiral secondary alcohols (Scheme 3.5) [32]. This tailor-made surfactant contains vitamin E as hydrophobic moiety, forming micelles that act as reservoirs for housing lipophilic substrates and metal catalysts. The metal-catalyzed reaction was performed in the core of micelles and the enzymatic reaction was performed outside of the micelles, *i.e.* in the bulk aqueous solution. This ultimately enabled the enzymes and metals to be compatible, thus avoiding mutual inhibition.

The beneficial impact of surfactants was also demonstrated in a one-pot combination of laccase/TEMPO-catalyzed deoximation with subsequent bioreduction or bioamination for synthesizing enantiopure chiral alcohols/amines [33]. In this process, the application of Cremophor® (polyethoxylated castor oil) as a surfactant that was tolerated by all catalytic species increased the degree of substrate solvation, and also enhanced the enzymatic performance, thereby enabling substrate concentrations up to 100 mM in the deoximation-bioreduction cascade.

Recently, Wu et al. reported a polymer material that is amphiphilic and also offered catalytical properties. This polymer can be simply synthesized by grafting hydrophobic alkyl chains onto hydrophilic hyperbranched polyglycerol (hPG) emulsification agent. The resulting emulsion allowed chemical reactions to occur at the emulsion interface and additionally encapsulated the active biocatalysts. The chemoenzymatic reaction was performed in a solvent-free emulsion stabilized by an hPG-based amphiphile [34]. The first step was the hydrolysis of ethyl diacetate to ethylene glycol by CALB, while the second step is the acetalization of benzaldehyde and ethylene glycol.

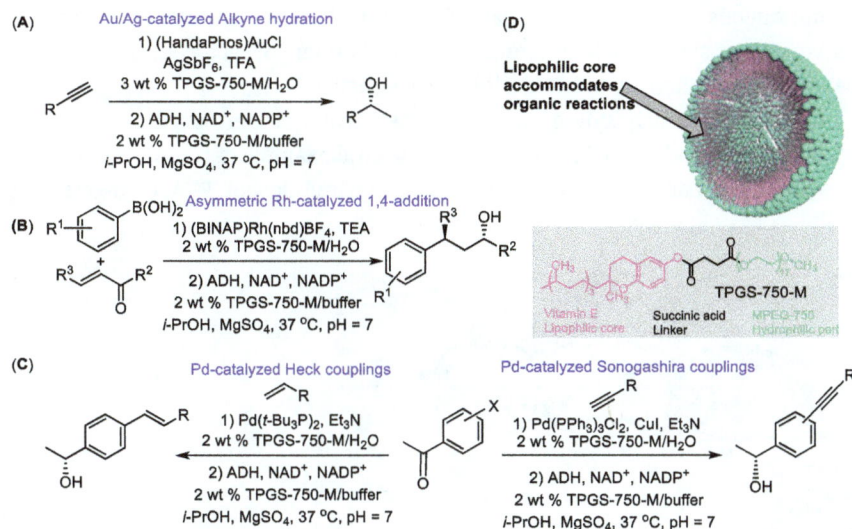

Scheme 3.5: CECRs by aqueous micellar solutions: (A) Au/Ag catalyzed alkyne hydration and bioreduction; (B) asymmetric Rh-catalyzed 1,4-addition and bioreduction; (C) Pd-catalyzed Heck and Sonogashira coupling and bioreduction; (D) micelles formed by TPGS-750-M in water. Reproduced from Ref. [32] with permission from Springer Nature, 2019.

3.3.3 Encapsulation

Another method for achieving a separation of chemo-and bio-catalysis is the encapsulation of a metal catalyst into an environment that is akin to an enzyme active site. Ward et al. created a novel artificial transfer-hydrogenase (ATHase) by encapsulating a biotinylated [Cp*Ir(Biot-p-L)Cl] pianostool complex within the framework of the protein streptavidin (Sav) [35]. Sav effectively created a neutral shield that protects the metal complex against interacting with other catalytic species outside the protein shell, thus avoiding the inactivation of the iridium complex and the enzymes. A few simultaneous cascade reactions were accomplished by combining ATHase with a variety of redox enzymes, including a three-enzyme cascade reaction for stereoselective deracemization of cyclic amines and the four-enzyme cascade reaction for the transformation of L-lysine to its acid form. Recently, they also developed an NAD(P)H-dependent ATHase that was used for a four-enzyme cascade to synthesize chiral secondary amines (Scheme 3.6A) [36]. Moreover, they tailored the ATHase for the regeneration of NADH derivatives, which was combined with a range of ene-reductases for the synthesis of chiral maleimides [37].

Toste and co-workers created a metal compartment by encapsulating Au (I) or Ru (II) specie in a Ga4L6 (L = *N,N'-bis*(2,3-dihydroxy-benzoyl)-1,5-diaminonapthalene) supramolecular cluster (Scheme 3.6B) [38]. Encapsulation by the supramolecules stabilizes the organo-metallic catalysts. This increased their aqueous solubility and also prevented the direct contact and reaction of metal complexes with other components. They achieved various simultaneous chemoenzymatic cascades employing the supramolecular assemblies with various enzymes such as esterases or alcohol dehydrogenases, in which the enantioselectivity and the rate of biocatalysis was enhanced.

The encapsulated-decarboxylase and a metathesis catalyst were used to synthesize 4,4'-dihydroxystilbene and proved to be another impactful example of compartmentalized cascade reactions [39]. Large size (~3 mm diameter) polyvinyl alcohol (PVA) cryo-gels were

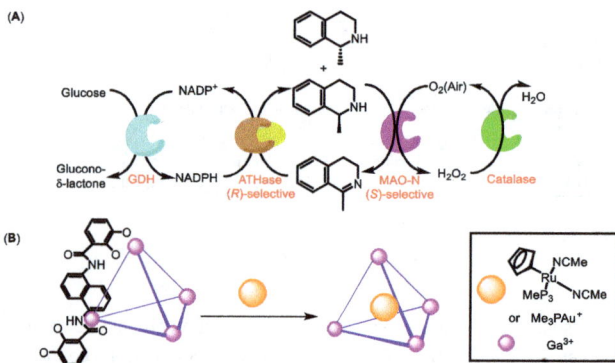

Scheme 3.6: CECRs by encapsulation: (A) CECRs of four-enzyme cascade based on ATHase; (B) Systematic illustration of encapsulating Au(I) or Ru(II) within Ga4L6. Reproduced from Ref. [38] with permission from Springer Nature, 2012.

employed as a matrix to separate enzymes into stable aqueous compartments, enabling the biotransformations by strictly water-dependent biocatalysts in the aqueous phase with the nonaqueous phase outside the capsules. With this concept, they realized enzymatic decarboxylation in the non-aqueous solvent by enclosing the cofactor-free phenolic acid decarboxylase [derived from *Bacillus subtilis* (*bs*PAD)] with (PVA/PEG) cryogels. The combination of enzymatic decarboxylation and ruthenium-catalyzed metathesis produced substituted stilbene-derived compounds with diverse biological activities.

3.3.4 Semi-permeable membrane separation

The size-sieving and selection effects of semi-permeable membranes make them effective spatial compartmentalization tools. They can compartmentalize different parts of the catalytic system by the filtering of specific species. As a natural semi-permeable membrane, cell membrane not only can maintain the stability of the internal reaction system, but also can prevent the external toxic substances from entering the cell interior. In whole-cell catalytic systems, the cell membrane can act as a natural barrier between enzymes and metals, preventing their mutual inactivation and maintaining their activity. Organism (bacteria, fungi and yeast) have been successfully used for the biosynthesis of metal nanoparticles (MNPs). It should be mentioned that in most cases the whole cells are fragile and will inevitably be destroyed under agitation conditions. Thus, the technologies regarding whole-cell immobilization and reactor engineering have been developed to enhance the stability of whole cells.

Lloyd and co-workers created aerobic *E. coli* cultures with palladium nanoparticles (Pd NPs) coated on the cell membrane via a bioreduction reaction. These cultures excessively produced recombinant MAO-N (Scheme 3.7A) [40]. It should be noted that any interaction between the enzymes and palladium was mostly avoided due to the

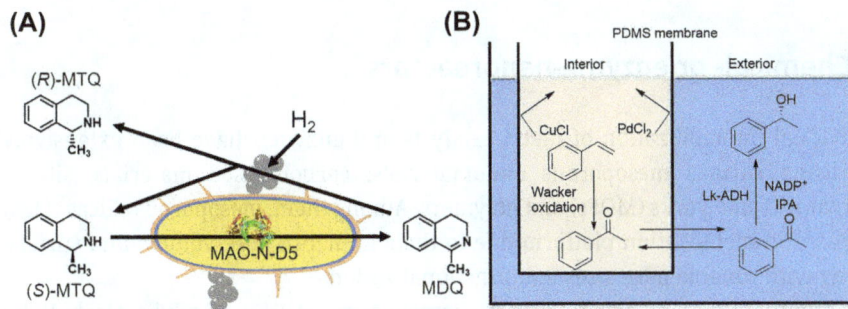

Scheme 3.7: CECRs by semi-permeable membrane separation: (A) The cyclic deracemization of MTQ via the imine MDQ using palladized biocatalyst with the MAO-N insert. Reproduced from Ref. [40] with permission from American Chemical Society, 2011. (B) PDMS thimble for *mo*-CECR combining Pd/Cu-catalyzed Wacker oxidation and ADH-catalyzed reduction. Reproduced from Ref. [42] with permission from Wiley, 2015.

separating effect of the cell membrane, resulting in a near full retention of catalytic activity over five oxidation/reduction cycles in the enantioselective deracemization reactions. Qu et al. reported an efficient method for nanocatalyst synthesis through the *in situ* growth of ultrasmall MNPs on the bacterial surface [41]. The bacteria-MNPs hybrids were used as catalysts to achieve a highly efficient hydrogenation reaction.

Beyond cell membranes, artificial and hydrophobic polydimethylsiloxane (PDMS) thimbles were also used in chemoenzymatic cascades. Gröger's group achieved the combination of Wacker oxidation (catalyzed by palladium/copper) with an enzymatic asymmetric reduction using a PDMS membrane (Scheme 3.7B) [42]. The Wacker oxidation proceeded inside of the PDMS thimble while the stereoselective reduction proceeded at the outside of the PDMS material. The size-sieving and selection effects of the PDMS thimble enabled the diffusion of intermediate from the interior to the outside while limited the transport of the metallic salts and biocatalytic system, and thus avoided the mutual inactivation. Latham et al. also developed a one-pot cascade combining halogenation enhanced by flavin-dependent halogenase (Fl-Hal) and palladium-catalyzed Suzuki coupling by applying the PDMS thimble [43]. The compartmentalization approach simplified the experimental operations and significantly improved the synthetic efficiency.

3.4 Nanoreactors

The fabrication of nanoreactors by assembly of insoluble materials and catalysts has become an effective way to circumvent incompatibility. Nanoreactors are composed of catalysts immobilized into nanosized compartments, while the substrates are allowed to diffuse between different "compartments" to react with the catalysts. Catalytic nanoreactors are divided into two categories: chemo- or bio-nanoreactors containing a single catalytic species (chemical catalysts or enzymes), and the chemo-bio nanoreactors with multiple catalytic species (metal-enzyme integrated catalysts).

3.4.1 Chemical- or enzyme-nanoreactors

The individual immobilization of metal catalysts and enzymes have been extensively studied using various mesoporous nanomaterials: carbon-based materials, silicas, metal-organic frameworks (MOFs) and polymers. Among them, mesoporous silicas (MSs) are excellent immobilization platform due to their immense pore volume and surface area along with tunable pore sizes and functionalization.

The group of Akai reported a novel chemoenzymatic DKR of allylic alcohols by the coupling of CALB and homogeneous catalyst O=V(OSiPh$_3$)$_3$, and obtained the corresponding chiral esters [44]. However, the interactions of metal and biocatalyst led to mutual inactivation in the reaction process. To tackle this problem, they immobilized the oxovanadium catalyst on mesoporous silica (V-MPS), which had a rigid and ordered hexagonal pore structure with uniform mesopores, approximately 3 nm in diameter (Scheme 3.8A) [45]. The surface of the mesoporous silica was modified by covalently attaching vanadium species. This ultimately increased the material's catalytic activity, minimized the interactions between both the catalysts while enabled substrates easily accessible to the metal center. The marrying of V-MPS catalyzed racemization with KR promoted by commercially available immobilized CALB produced esters in excellent yield and enantioselectivity. In this system, the reaction sites were divided and thus achieved excellent compatibility. Moreover, the immobilized catalysts were recovered simply by centrifugation, and the material was shown to be an excellent catalyst that remained active over 6 cycles. Very recently, they achieved the DRK of a tertiary alcohol which is a particularly difficult substrates for applying the DKR process. This was performed by combining *Candida Antarctica* lipase A (CALA) and V-MPS4 (pore diameter of the MPS is 4 nm), although the compatibility of both catalysts was not well resolved [46].

The deconstruction of biomass into commodity chemicals such as 5-hydroxymethyl furfural (HMF) has gradually become a research hotspot since the increasingly problems of fossil fuels and environment. Zhao et al. synthesized HMF from glucose through a CECR coupling an enzymatic isomerization to an acid promoted dehydration [47]. To carry out the one-pot process, the tolerance of enzyme for organic solvents that are essential for dehydration step is another challenge apart from the compatibility of the both catalysts. These problems were solved by immobilization. A thermophilic glucose isomerase was immobilized in an amino-functionalized MS. This particular combination increased the enzyme activity and organic solvent tolerance. The immobilized enzyme was combined with an acid (-SO$_3$H) functionalized MS in a blend of THF/H$_2$O (4:1 v/v), producing HMF with up to 30% yield. Later on, cellulase was added to this system by Lee et al. which allowed the HMF to be furnished directly from cellulose [48]. In this work, iron oxide-encapsulated mesoporous silica nanoparticles (MSNs) were employed in the fabrication of immobilized enzymes (*i.e.* cellulase-Fe$_3$O$_4$@MSN and isomerase-Fe$_3$O$_4$@MSN).

MOFs with advantages of thermal stability, structural tunability, high porosity and easy functional modification, allow judicious allocation of catalytic species in compartments, thus becoming a fascinating platform for immobilization of metal or/and enzyme,

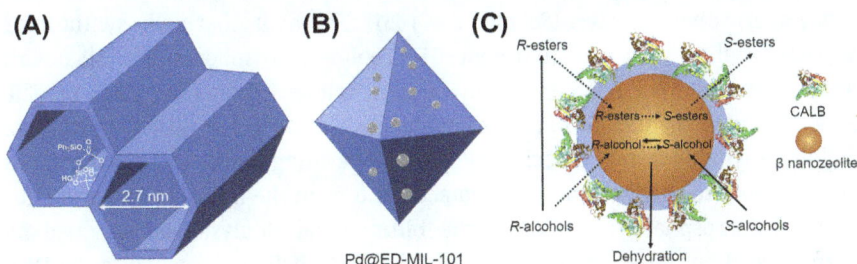

Scheme 3.8: CECRs by nanoreactors: (A) Deduced structure of V-MPS. Reproduced from Ref. [45] with permission from Wiley, 2013. (B) Schematic illustration of the preparation of Pd@ED-MIL-101. Reproduced from Ref. [49] with permission from Elsevier Science, 2019. (C) The core–shell nanozeolite@enzyme bifunctional microsphere.

which have captured increasingly attention of researchers. Li et al. combined a MOF-based metal catalyst with CALB in microwave-assisted one-pot DKR of aliphatic amines [49]. In the DKR process, alkalic additives are usually used to suppress an acid-promoted condensation side reaction, which may produce an undesired byproduct (ethylbenzene). To avoid the addition of additives, they synthesized ED-MIL-101 with basic property by grafting ethylenediamine (ED) onto the matrix of MIL-101, which acted as a host for the anchoring of colloidal Pd NPs (~3.5 nm) that were generated by employing oleylamine and borane *tert*-butylamine complex to reduce Pd(acac)$_2$ (Scheme 3.8B). In addition, enzyme aggregates of CALB (CALB-CLEAs) were crosslinked among enzyme molecules were applied as biocatalysts, which could reduce the additional supports. The combination of these two heterogeneous catalysts improved the selectivity in racemization process, thus minimizing the side reaction.

Gröger et al. demonstrated a CECR for synthesizing 1,3-diols in organic media, which combines an organocatalytic asymmetric aldol reaction with a ADH-catalyzed reduction based on a immobilization approach, which benefits from the use of an acrylate-based superabsorber as supports [50]. The co-immobilized enzyme and cofactors in the superabsorber retained in an "aqueous phase", which permit all necessary additives (such as NAD$^+$, NADH and buffer) to be in close to the enzyme, provided a suitable environment for enzymatic reaction inside the matrix and protected the enzyme from the influence of external organic solvents. First, the proline-derivative-catalyzed aldol reaction of benzaldehyde with acetone in lipophilic organic solvents (cyclohexane, chloroform, ethyl acetate) afforded the desired 1,3-hydroxyketone in up to 95% yield and 95% ee. In the second step, the bioreduction promoted by a coimmobilized ADH from *Rhodococcus* sp. produced the final (1*R*,3*S*)-diols in 89% yield and with a high selectivity of >99% ee and >35:1 dr. In addition, this chemoenzymatic cascade was also achieved by utilizing immobilized organo- and biocatalyst in different compartments of fixed-bed reactors, leading to the two steps being fully compatible. This co-immobilization strategy offered very efficient organic-solvent-compatible biocatalysts and could provide a basis for further CECRs.

Inorganic catalysts also can be used as carriers to immobilize enzymes, resulting in bifunctional catalysts. Zhang et al. constructed a core–shell nanozeolite@enzyme bifunctional microsphere catalyst (Scheme 3.8C) [51]. The hydrothermally synthesized colloidal β-nanozeolites were used to prepare H-β zeolite microspheres (Hβ-ZMSs) with secondary mesopores by a polymerization-induced colloid aggregation method. The Hβ-ZMSs, as the catalyst cores (5–7 μm in diameter), were covered by a shell of cationic polydiallyldimethylammonium chloride (PDDA), generating a core–shell-structured (Hβ-PDDA) MSs whose shell was used to protect CALB from the inhibition by acid sites of H-β zeolite microspheres. Ultimately, the bifunctional catalyst was obtained by immobilizing CALB on the external surface of the PDDA shell, and applied to the DKR of aromatic secondary alcohols, in which acid (H-β zeolite) catalytic racemization and enzymatic (CALB) KR were combined. Very recently, zeolite nanocrystals were fabricated by aerosol-based techniques [52]. These hollow spheres possessed large pores that accesses for enzymes. These enzymes are then anchored and cross-linked to form CLEAs, creating the bifunctional catalysts, which were applied in a chemoenzymatic cascade reaction: glucose oxidase produced H_2O_2 *in situ* and the latter was subsequently utilized by zeolite to promote the epoxidation of olefin.

3.4.2 Metal-enzyme co-immobilization

One of the initial reports of anchoring both palladium nanoparticles and CALB onto aminopropyl-functionalized siliceous mesocellular foams (AmP-MCF) was demonstrated by Bäckvall [53]. First, Pd^{2+} were bonded (coordination) to the amino groups of AmP-MCF and reduced with sodium borohydride, producing palladium nanoparticles (~2 nm). Subsequently, CALB was immobilized covalently on the Pd-AmP-MCF via Schiff-base forming reaction (Scheme 3.9A). The proximity of Pd NPs and CALB in the same compartment reduced the distance of intermediate diffusion and exhibited notable efficiency as a catalyst when used towards the chemoenzymatic dynamic-kinetic resolution of amines. Yadav et al. reported a similar example of co-immobilization. Palladium nanoparticles NPs and CALB were immobilized on mesocellular foam sequentially, and the synthesized catalyst proved to be effective for the one-pot synthesis of (R)-phenyl ethyl acetate [54].

The spatial separation of two types of catalysts through co-immobilization can effectively avoid the incompatibility between metal catalysts and enzymes. Prasad prepared a catalyst that was composed of a surface-anchored enzyme on a metal/silica (core/shell) nanoparticle structure [55]. In this work, gold NPs were infused into the mesopores of the silica gel by a sol-gel synthesis procedure. Epoxy groups were introduced to the surface of the mesoporous silica, which were then used for immobilization of glucosidase. Spatially separated co-immobilization and the micropore structures are responsible for the exceptional ability of the catalyst to expedite the cascade reaction that converts 4-nitrophenyl-β-glucopyranoside to 4-aminophenol.

Yang et al. designed a yolk–shell@shell nanosized reactor in which the Pd NPs and CALB were anchored in separated domains (Scheme 3.9B) [56]. At first, Pd^{2+} was linked to the aminopropyl group modified mesoporous silica nanosphere via coordination and reduced to palladium nanoparticles through sodium borohydride, generating inner core Pd/NH_2-MSN. During the process, amino groups served as coordination sites, stabilizing the palladium nanoparticles. Then, a sacrificial silica layer coating on Pd/NH_2-MSN was etched with the assistance of organosilane BTME (1,2-bis(trimethoxysilyl)ethane), establishing the yolk–shell-structured Pd/NH_2-MSN@BTME with minimal to no damage of inner Pd/NH_2-MSN core. After additional coating of a large-pore mesoporous silica shell by a biphasic-stratification approach, CALB was immobilized into the pore of mesoporous silica shell via physical adsorption to prepare a yolk–shell@shell nanoreactor. The yolk–shell@shell structure of the catalyst provided a complete spatial separation

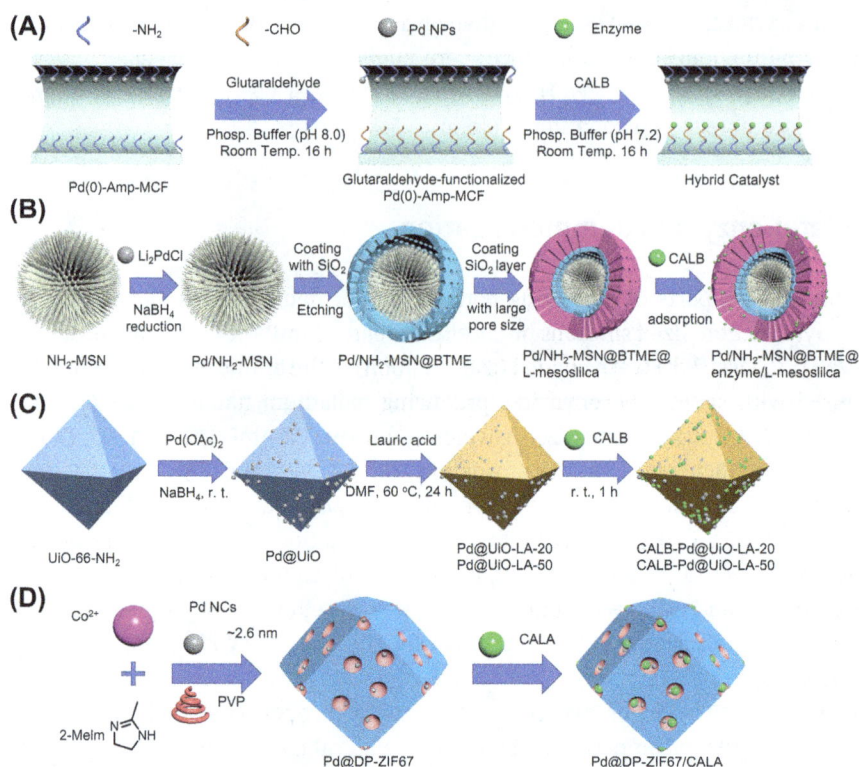

Scheme 3.9: Co-immobilization of metal nanoparticless and enzymes into, (A) aminopropyl-functionalized siliceous mesocellular foams (AmP-MCF). Reproduced from Ref. [53] with permission from Wiley, 2013. (B) mesoporous silica nanoparticles (MSNs) Reproduced from Ref. [56] with permission from The Royal Society of Chemistry, 2017. Co-immobilization of metal NPs and enzymes into, (C) zirconium-based MOF scaffold, Reproduced from Ref. [58] with permission from Wiley, 2019. UiO-66. (D) cobalt-based zeolitic imidazolate frameworks. Reproduced from Ref. [59] with permission from Wiley, 2019.

of metals catalysts and enzymes, which solved the incompatibility problem and thus lead to an excellent catalytic performance in chemoenzymatic DKR of 1-phenylethylamine.

Wu et al. published a stepwise technique to anchor Pd NPs along with enzymes onto mesoporous silica nanoparticles (MSN) separately [57]. Pd NPs were first loaded into MSN by *in situ* reduction of Pd(OAc)$_2$ with NaBH$_4$ to obtain a hybrid Pd@MSN, and then long-chain alkanes were introduced to form hydrophobic Pd@mMSN. Finally, CALB was anchored on the Pd@mMSN via hydrophobic interaction obtaining a bifunctional CALB@Pd@mMSN. Furthermore, it is possible to disperse the catalyst into multiple organic solvents with a wide range of polarity, and exhibited exceptional catalytic performance for cascade reaction synthesizing benzyl hexanoate in toluene.

Qi et al. prepared a biohybrid catalyst by stepwise co-immobilization of Pd NPs and CALB on UiO-66-NH$_2$ for cascade reactions (Scheme 3.9C) [58]. In this work, the palladium nanoparticles and CALB were first immobilized (*via* reduction and adsorption) on the pore and the surface of UiO-66-NH$_2$. Significantly, the hydrophobicity of the catalyst can be changed by swapping the ligands of the MOF with lauric acid, which improved its catalytic performance in different organic solvents. The resulting material demonstrated exceptional catalytic performance in toluene solvent. Lee and co-workers prepared large size mesoporous cobalt-based ZIF (DP-ZIF67) for the immobilization of both palladium nanoparticles and CALA, in which unsaturated metal cations, MNPs and enzyme were used as catalytic species for cascade reactions (Scheme 3.9D) [59]. Jiang et al. built a nanoreactor by positioning platinum nanoparticles and L-amino acid oxidase inside and outside of UiO-66 separately to avoid the deactivation of L-amino acid oxidase by platinum nanoparticles. The small micropore size ensures channeling of the byproduct, H$_2$O$_2$, from L-amino acid oxidase to the platinum nanoparticles, resulting in quick elimination [60]. Li et al. constructed a MOF-based fluorescent sensor to detect lactose by co-immobilizing Au nanoclusters (AuNCs), β-galactosidase (β-Gal), and glucose oxidase (GOx) [61].

More and more materials have been applied to construct chemo-bio nanoreactors. Wu et al. reported the first work on immobilizing both palladium nanoparticles and CALB on carbon nitride (C$_3$N$_4$) [62]. First, C$_3$N$_4$ was synthesized by subjecting dicyandiamide to a heat treatment at 550 °C for 4 h. Then, Pd NPs were immobilized on C$_3$N$_4$ by an impregnation-reduction method. Finally, CALB was covalently anchored onto C$_3$N$_4$ using glutardialdehyde as a crosslinker (Scheme 3.10A). The nanoreactors proved to be exceptional catalyst for the transformation of benzaldehyde to benzyl hexanoate. Gao et al. designed a core–shell structured nanocatalyst with a mesoporous metals (MMs) core and enzyme-immobilized polydopamine (PDA) shell [63]. In this work, PdPt metal core was synthesized and placed inside the PDA shell by reduction of metal ions with dopamine, and the mesoporous structures of PdPt metal core and PDA shell were created simultaneously with the assistance of surfactant. Finally, CALB was bioadhered to the surface of the polydopamine shell (Scheme 3.10B). The high accessibility (pore volume and surface area) of the mesopores ensured relatively rapid mass transfer, leading to an excellent catalytic performance in chemoenzymatic cascades.

Like MOFs, covalent organic frameworks (COFs) are a class of ordered and crystalline microporous material. The main difference between the two stems from the lack of metal centers in the case of COFs. More recently, COFs have been applied to build chemo-bio nanoreactors by Jiang and co-workers [64]. In this report, metal nanoparticles and enzymes were immobilized inside and outside the pore of the COFs (Tz-Da) separately. Pd or Pt NPs were dispersed in the Tz-Da by reducing the precursor with triazinyl. Additionally, surface covalently-anchored organophosphorus hydrolase (OPH) improved the enzyme stability (Scheme 3.10C). The obtained chemo-bio nanoreactors demonstrated exceptional catalytic properties and reversibility in the cascade degradation of organophosphate nerve agents.

Jiang and co-workers fabricated a semi-heterogeneous metal–enzyme-integrated catalyst using soluble porous imine molecule cages (CC3). This material was soluble in

(A)

(B)

(C)

(D)

(E)

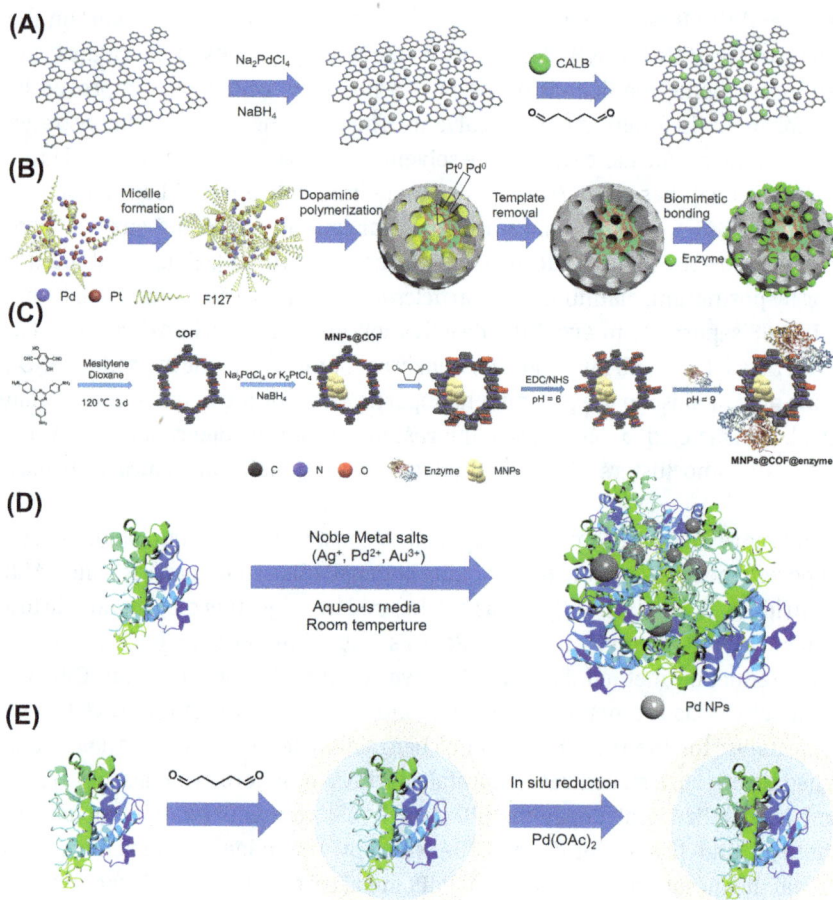

Scheme 3.10: Immobilization of metal nanoparticles and enzymes into, (A) carbon nitride (C_3N_4). (Reproduced from Ref. [62] with permission from Wiley, 2019) (B) polydopamine shell. Reproduced from Ref. [63] with permission from American Chemical Society, 2019. (C) covalent organic frameworks (Tz-Da). (Reproduced from Ref. [64] with permission from American Chemical Society, 2022.) Schematic illustration the strategy used to synthesize. (D) CALB-Pd NPs nanohybrids. Reproduced with permission from Ref. [66]. Copyright (2013) The Royal Society of Chemistry. (E) Pd/CALB-Pluronic nanohybrids. Reproduced with permission from Ref. [68]. Copyright (2019) Springer Nature.

organic solvents under the right temperature conditions, which conveniently fell under the temperature range for reactions of interest. After the catalytic reaction, the catalyst can be precipitated out of solution by decreasing the temperature to room temperature, effectively recovering the catalyst [65]. Pd ions were first stabilized by CC3 and reduced by $NaBH_4$ to prepare Pd@CC3, and then CALB was covalently-anchored to Pd@CC3 via a 3-component isocyanide-based Ugi reaction. The CC3 play multirole, such as support, stabilizer, solubilizer, homogenizer and heterogenizer. The semi-heterogeneous catalyst exhibited high performance in chemoenzymatic DKR of amines.

3.4.3 Bioconjugation

It has become more popular in the recent years to use protein for binding/stabilizing and reducing metal cations to metal nanoparticles. This is mostly due to the large amount of reducing functional groups that exist in protein molecules. Palomo and co-workers achieved the synthesis of enzyme-metal nanohybrids by this strategy [66]. Metal ions and CALB were mixed in an aqueous solution containing a co-solvent, producing the nanobiohybrids as precipitates (Scheme 3.10D). In this process, the reducing agent for nanoparticle formation was the enzyme, avoiding the aggregation of NPs. The enzyme also acted as a stabilizing agent.

The group of Bäckvall designed a palladium nanoparticle-based biohybrid catalyst, in which a cross-linked enzyme aggregate (CLEA) of CALB functioned as active support for the palladium nanoparticles [67]. CALB was first aggregated with glutaraldehyde with the assistant of $Na(CN)BH_3$ in isopropanol, and subsequently, Pd ions were entrapped and reduced onto the cross-linked network of aggregated CALB, generating amorphous metallic palladium-CALB CLEA. One key benefit of the CLEA method is that support materials are not needed. Ge et al. synthesized protein-polymer conjugates by the covalent bonding between of CALB and an aldehyde-functionalized polymer (Pluronic F-127) (Scheme 3.10E) [68]. In this process, due to the confinement of the protein–polymer, the size of Pd NPs could be precisely controlled down to 0.8 nm by adjusting the concentration of metal salts, which increased the catalytic activity of catalysts in DKR of primary amines in toluene at 55 °C.

The application of immobilized catalysts in flow reactors is a powerful technique for greener, more sustainable manufacturing. Flow chemistry allows reaction sequences to be conducted through sequentially separated reactors with different conditions, thus circumventing the incompatibilities between both the catalysts and the reaction conditions. Flow processing also has the potential to accelerate biotransformations due to enhanced mass transfer, making large-scale production more economically feasible in significantly smaller equipment with a substantial decrease in reaction time, from hours to a few minutes, and improvement in space-time yield, with increases of up to 650-fold as compared to batch processes [69]. Therefore, the continuous-flow chemoenzymatic cascade reactions stand among the most important protocols to achieve bioprocess intensification. Since several excellent reviews have exhaustively discussed this subject, only a brief introduction was given in this chapter.

3.5 Discussion and conclusions

The CECRs bridging chemo-and bio-catalysis and merging the advantages of both catalytic fields represents major advancement in technology towards the goal of mimicking biology to develop green and efficient methods for the production of value-added organic molecules. The CECRs also can achieve an efficient bioprocess intensification, but the combined issues of side reactions and deactivation of catalysts make it highly challenging. Mimicking but simplifying the metabolic process, wherein a complex and well-developed multi-enzyme cascade catalytic systems works efficiently, several strategies and technologies have been developed to tackle these hurdles, including temporal compartmentalization, spatial compartmentalization and catalytic nanoreactors, which stimulate the development of this active research area. In the past decades, it is clear that technological achievements in nanomaterials have led to a shift of the state-of-the-art strategies from a macroscopic to a microscopic level. On the both levels, searching for compatible catalysts and reaction conditions is still the major challenges in the future. Therefore, the researches to develop corresponding strategies will continue.

The incompatibility between chemo-and bio-catalysis was mainly due to the conflicting reaction conditions associated with the two steps or the mutual inhibition of reaction components (e.g., catalysts, reagents, solvents, cofactors, etc.). Temporal compartmentalization, *i.e.*, implementing chemoenzymatic cascade reactions via a step-wise fashion, can avoid the coexistence of all incompatible factors by consecutive addition of reaction components and supplementary operations, and therefore is more technically effective. Although numerous examples have demonstrated the applicability of this strategy, the benefit of it lies mostly in avoiding the isolation of reaction intermediates. Truly concurrent cascades, however, would expand the synthetic possibilities and enable reactions that would not be possible otherwise, for instance, chemoenzymatic DKR representing one of the most successful examples. Spatial compartmentalization and nanoreactor strategies are often used to circumvent the inherent incompatibility between the different catalysts, allowing for concurrent cascades. Spatial compartmentalization can be easily achieved by biphasic systems, aqueous micellar solution, artificial metalloenzymes, membrane filtration, and whole cells. Albeit effective, this strategy suffers from the catalyst recovery and reuse. On the contrary, constructing nanoreactors is advantageous in providing heterogeneous benefits including improved stability and reusability. Particularly, rational co-immobilization of metal catalysts and enzymes on the same supports, fabricating metal-enzyme hybrid catalysts, also gains additional advantages, such as the compartmentation, confinement, and proximity effects. The drawback of nanoreactor strategy is that it cannot be applied to the cascade reactions with incompatible reaction conditions. Therefore, the three strategies have their own advantages and disadvantages. The choice of the preferred strategy depends heavily on the requirements and characteristics of the given catalytic tasks.

A straightforward solution is to fabricate reliable and robust catalysts, specifically, to endow enzymes with the capability to accept harsh reaction conditions and to develop chemical catalysts meeting the 12 principles of green chemistry. It is exciting that the progress in protein engineering, directed evolution, catalyst technology, and material science have offered numerous opportunities to develop these novel catalysts. In addition, the advances in these fields prompt innovations in biocatalysis, especially the identification of novel enzyme classes and catalytic capabilities for efficient biotransformations, which may change the negative fact that the catalytic library of enzymes is still greatly limited compared with synthetic methods. It is expected that future enzymes with artificial catalytic activities will become extremely important complements to traditional catalysts. As a result, the types of biocatalytic reactions will significantly increase, which will offer more choices to design compatible reaction sequences and CECRs.

Living cell, as a perfect 'chemical factory', offers valuable guidance to construct efficient nanoreactors. It is expected that more efforts will be made to fabricate nanoreactors with higher structural and compositional complexities, in which different catalysts can be anchored at different sites. Utilizing multicompartment nanoreactors, a larger number of reaction steps can be integrated in a CECR, which has already been demonstrated with hierarchical mesoporous silica, MOFs and polymers. In addition, the advances on material science offer more choices to obtain appropriate matrix for constructing nanoreactors that possess tailored catalytic capacity. However, substrates tend to diffuse randomly and can lead to undesired side reactions. Consequently, the next step would be to navigate reaction species for the catalytic centers, even to follow desired diffusion pathway of the reaction sequence. Creating nanoreactors with selection channels would achieve this goal. Last but not the least, the integration of organometallic complexes into nanoreactors should be fully estimated, which will substantially expand the design space opportunities due to the versatile reactivity of the organometallic catalysts.

Mutual inactivation between chemocatalysis and biocatalysis has been identified in many studies, which has been a major challenge for constructing CECRs. In Gröger's works, triphenylphosphane and boronic acid in Suzuki cross-coupling had strong negative influence on activity of ADH enzyme; copper additives in Wacker-Tsuji oxidation had a moderate to strong suppression of the biotransformation [6]. In Inagaki's work, the addition of BSA to the reaction solution containing [RhCp*(bpy)Cl]Cl drastically reduced the catalytic activity due to the deactivation of the Rh complex via nonspecific interactions with the BSA protein [70]. In Ward's work, the direct contact between [Cp*Ir(biot-p-L)Cl] complex and monoamine oxidases led to the decrease in activity of the both catalysts, *i.e.*, mutual inactivation [35]. Understanding molecular reasons for these inactivation effects can open the way for the development of other novel strategies to achieve the chemoenzymatic reactions. However, the systematic studies investigating these interactions remain rare due to the lack of crystal structure of enzymes or difficulty in the detection of the dynamic change of enzymes' conformation. Generally, the inactivation mechanism was attributed to the coordination between metal ions and enzymes, commonly via cysteine, histidine, and tryptophan residues. Hollmann and co-workers revealed the mutual inactivation mechanism between $[Cp^*Rh(bpy)(H_2O)]^{2+}$ and the alcohol dehydrogenase from *Thermus sp.* ATN1 (TADH). TADH binds 4 equiv. of $[Cp^*Rh(bpy)(H_2O)]^{2+}$ without detectable decrease in catalytic activity and stability. Higher molar ratios lead to enzyme inactivation, suggesting that the metal complex functions as an 'unfolding catalyst'. This detrimental activity can be circumvented using strongly coordinating buffers (e.g. $(NH_4)_2SO_4$) while preserving its activity as NAD(P)H regeneration catalyst under electrochemical reaction conditions [71]. Consequently, more efforts should be dedicated to explore the mutual inactivation mechanisms, especially at the molecular level, which will put the basis for various future developments of chemoenzymatic cascades.

References

1. Schrittwieser JH, Velikogne S, Hall M, Kroutil W. Artificial biocatalytic linear cascades for preparation of organic molecules. Chem Rev 2017;118:270–348.
2. Rudroff F, Mihovilovic MD, Gröger H, Snajdrova R, Iding H, Bornscheuer UT. Opportunities and challenges for combining chemo-and biocatalysis. Nat Catal 2018;1:12–22.
3. Schmidt S, Castiglione K, Kourist R. Overcoming the incompatibility challenge in chemoenzymatic and multi-catalytic cascade reactions. Chem Eur J 2018;24:1755–68.
4. Liu Y, Liu P, Gao S, Wang Z, Luan P, Gonzalez-Sabin J, et al. Construction of chemoenzymatic cascade reactions for bridging chemocatalysis and Biocatalysis: principles, strategies and prospective. Chem Eng J 2021;420:127659.
5. Ríos-Lombardía N, García-Álvarez J, González-Sabín J. One-pot combination of metal-and bio-catalysis in water for the synthesis of chiral molecules. Catalysts 2018;8:75.
6. Burda E, Hummel W, Gröger H. Modular chemoenzymatic one-pot syntheses in aqueous media: combination of a palladium-catalyzed cross-coupling with an asymmetric biotransformation. Angew Chem Int Ed 2008;47:9551–4.
7. Borchert S, Burda E, Schatz J, Hummel W, Gröger H. Combination of a Suzuki cross-coupling reaction using a water-soluble palladium catalyst with an asymmetric enzymatic reduction towards a one-pot process in aqueous medium at room temperature. J Mol Catal B 2012;84:89–93.
8. Gauchot V, Kroutil W, Schmitzer AR. Highly recyclable chemo/biocatalyzed cascade reactions with ionic liquids: one-pot synthesis of chiral biaryl alcohols. Chem Eur J 2010;16:6748–51.
9. González-Martínez D, Gotor V, Gotor-Fernández V. Chemoenzymatic synthesis of an odanacatib precursor through a Suzuki-miyaura cross-coupling and bioreduction sequence. ChemCatChem 2019;11:5800–7.
10. Prastaro A, Ceci P, Chiancone E, Boffi A, Cirilli R, Colone M, et al. Suzuki-Miyaura cross-coupling catalyzed by protein-stabilized palladium nanoparticles under aerobic conditions in water: application to a one-pot chemoenzymatic enantioselective synthesis of chiral biaryl alcohols. Green Chem 2009;11:1929–32.
11. Dander JE, Giroud M, Racine S, Darzi ER, Alvizo O, Entwistle D, et al. Chemoenzymatic conversion of amides to enantioenriched alcohols in aqueous medium. Commun Chem 2019;2:1–9.
12. Ahmed ST, Parmeggiani F, Weise NJ, Flitsch SL, Turner NJ. Chemoenzymatic synthesis of optically pure L-and D-biarylalanines through biocatalytic asymmetric amination and palladium-catalyzed arylation. ACS Catal 2015;5:5410–3.
13. Dawood AW, Bassut J, de Souza RO, Bornscheuer UT. Combination of the Suzuki–Miyaura cross-coupling reaction with engineered transaminases. Chem Eur J 2018;24:16009–13.

14. Wang Y, Bartlett MJ, Denard CA, Hartwig JF, Zhao H. Combining Rh-catalyzed diazocoupling and enzymatic reduction to efficiently synthesize enantioenriched 2-substituted succinate derivatives. ACS Catal 2017;7: 2548–52.
15. Plass C, Hinzmann A, Terhorst M, Brauer W, Oike K, Yavuzer H, et al. Approaching bulk chemical nitriles from alkenes: a hydrogen cyanide-free approach through a combination of hydroformylation and biocatalysis. ACS Catal 2019;9:5198–203.
16. Ríos-Lombardía N, Vidal C, Cocina M, Morís F, García-Álvarez J, González-Sabín J. Chemoenzymatic one-pot synthesis in an aqueous medium: combination of metal-catalysed allylic alcohol isomerisation–asymmetric bioamination. Chem Commun 2015;51:10937–40.
17. Rodriguez-Alvarez MJ, Rios-Lombardia N, Schumacher S, Perez-Iglesias D, Moris F, Cadierno V, et al. Combination of metal-catalyzed cycloisomerizations and biocatalysis in aqueous media: asymmetric construction of chiral alcohols, lactones, and γ-hydroxy-carbonyl compounds. ACS Catal 2017;7:7753–9.
18. Liardo E, Ríos-Lombardía N, Morís F, Rebolledo F, González-Sabín J. Hybrid organo-and biocatalytic process for the asymmetric transformation of alcohols into amines in aqueous medium. ACS Catal 2017;7:4768–74.
19. Rabe KS, Mgller J, Skoupi M, Niemeyer CM. Cascades in compartments: en route to machine-assisted biotechnology. Angew Chem Int Ed 2017;56:13574–89.
20. Akagawa K, Umezawa R, Kudo K. Asymmetric one-pot sequential friedel–crafts-type alkylation and α-oxyamination catalyzed by a peptide and an enzyme. Beilstein J Org Chem 2012;8:1333–7.
21. Suljić S, Pietruszka J, Worgull D. Asymmetric bio-and organocatalytic cascade reaction-laccase and secondary amine-catalyzed α-arylation of aldehydes. Adv Synth Catal 2015;357:1822–30.
22. Betori RC, May CM, Scheidt KA. Combined photoredox/enzymatic C–H benzylic hydroxylations. Angew Chem Int Ed 2019;58:16490–4.
23. Litman ZC, Wang Y, Zhao H, Hartwig JF. Cooperative asymmetric reactions combining photocatalysis and enzymatic catalysis. Nature 2018;560:355–9.
24. Haak RM, Berthiol F, Jerphagnon T, Gayet AJA, Tarabiono C, Postema CP, et al. Dynamic kinetic resolution of racemic beta-haloalcohols: direct access to enantioenriched epoxides. J Am Chem Soc 2008;130:13508–9.
25. Mutti FG, Orthaber A, Schrittwieser JH, de Vries JG, Pietschnig R, Kroutil W. Simultaneous iridium catalysed oxidation and enzymatic reduction employing orthogonal reagents. Chem Commun 2010;46:8046–8.
26. Denard CA, Huang H, Bartlett MJ, Lu L, Tan Y, Zhao H, et al. Cooperative tandem catalysis by an organometallic complex and a metalloenzyme. Angew Chem Int Ed 2014;53:465–9.
27. Risi C, Zhao F, Castagnolo D. Chemo-enzymatic metathesis/aromatization cascades for the synthesis of furans: disclosing the aromatizing activity of laccase/TEMPO in oxygen-containing heterocycles. ACS Catal 2019;9:7264–9.
28. Scalacci N, Black GW, Mattedi G, Brown NL, Turner NJ, Castagnolo D. Unveiling the biocatalytic aromatizing activity of monoamine oxidases MAO-N and 6-HDNO: development of chemoenzymatic cascades for the synthesis of pyrroles. ACS Catal 2017;7:1295–300.
29. Lipshutz BH, Ghorai S, Cortes-Clerget M. The hydrophobic effect applied to organic synthesis: recent synthetic chemistry "in water". Chem Eur J 2018;24:6672–95.
30. Gabriel CM, Lee NR, Bigorne F, Klumphu P, Parmentier M, Gallou F, et al. Effects of co-solvents on reactions run under micellar catalysis conditions. Org Lett 2017;19:194–7.
31. Armenise N, Malferrari D, Ricciardulli S, Galletti P, Tagliavini E. Multicomponent cascade synthesis of biaryl-based chalcones in pure water and in an aqueous micellar environment. Eur J Org Chem 2016;2016: 3177–85.
32. Cortes-Clerget M, Akporji N, Zhou J, Gao F, Guo P, Parmentier M, et al. Bridging the gap between transition metal-and bio-catalysis via aqueous micellar catalysis. Nat Commun 2019;10:2169.
33. Cordeiro RSC, Ríos-Lombardía N, Morís F, Kourist R, González-Sabín J. One-pot transformation of ketoximes into optically active alcohols and amines by sequential action of laccases and ketoreductases or ω-transaminases. ChemCatChem 2019;11:1272–7.

34. Sun Z, Zhao Q, Haag R, Wu C. Chemoenzymatic cascades enabled by combining catalytically active emulsions and biocatalysts. ChemCatChem 2022;14:e202101556.
35. Köhler V, Wilson YM, Dürrenberger M, Ghislieri D, Churakova E, Quinto T, et al. Synthetic cascades are enabled by combining biocatalysts with artificial metalloenzymes. Nat Chem 2013;5:93–9.
36. Okamoto Y, Köhler V, Ward TR. An NAD (P) H-dependent artificial transfer hydrogenase for multienzymatic cascades. J Am Chem Soc 2016;138:5781–4.
37. Okamoto Y, Kohler V, Paul CE, Hollmann F, Ward TR. Efficient in situ regeneration of NADH mimics by an artificial metalloenzyme. ACS Catal 2016;6:3553–7.
38. Wang ZJ, Clary KN, Bergman RG, Raymond KN, Toste FD. A supramolecular approach to combining enzymatic and transition metal catalysis. Nat Chem 2013;5:100–3.
39. Gómez Baraibar Á, Reichert D, Mügge C, Seger S, Gröger H, Kourist R. A one-pot cascade reaction combining an encapsulated decarboxylase with a metathesis catalyst for the synthesis of bio-based antioxidants. Angew Chem Int Ed 2016;55:14823–7.
40. Foulkes JM, Malone KJ, Coker VS, Turner NJ, Lloyd JR. Engineering a biometallic whole cell catalyst for enantioselective deracemization reactions. ACS Catal 2011;1:1589–94.
41. Bing W, Wang F, Sun Y, Ren J, Qu X. Catalytic asymmetric hydrogenation reaction by in situ formed ultra-fine metal nanoparticles in live thermophilic hydrogen-producing bacteria. Nanoscale 2021;13:8024–9.
42. Sato H, Hummel W, Groeger H. Cooperative catalysis of noncompatible catalysts through compartmentalization: wacker oxidation and enzymatic reduction in a one-pot process in aqueous media. Angew Chem Int Ed 2015;54:4488–92.
43. Latham J, Henry JM, Sharif HH, Menon BRK, Shepherd SA, Greaney MF, et al. Integrated catalysis opens new arylation pathways via regiodivergent enzymatic C–H activation. Nat Commun 2016;7:1–8.
44. Akai S, Tanimoto K, Kanao Y, Egi M, Yamamoto T, Kita Y. A dynamic kinetic resolution of allyl alcohols by the combined use of lipases and [VO (OSiPh3)3]. Angew Chem Int Ed 2006;45:2592–5.
45. Egi M, Sugiyama K, Saneto M, Hanada R, Kato K, Akai S. A mesoporous-silica-immobilized oxovanadium cocatalyst for the lipase-catalyzed dynamic kinetic resolution of racemic alcohols. Angew Chem Int Ed 2013; 52:3654–8.
46. Kühn F, Katsuragi S, Oki Y, Scholz C, Akai S, Gröger H. Dynamic kinetic resolution of a tertiary alcohol. Chem Commun 2020;56:2885–8.
47. Huang H, Denard CA, Alamillo R, Crisci AJ, Miao Y, Dumesic JA, et al. Tandem catalytic conversion of glucose to 5-hydroxymethylfurfural with an immobilized enzyme and a solid acid. ACS Catal 2014;4:2165–8.
48. Lee YC, Dutta S, Wu KCW. Integrated, cascading enzyme-/chemocatalytic cellulose conversion using catalysts based on mesoporous silica nanoparticles. ChemSusChem 2014;7:3241–6.
49. Wang M, Wang X, Feng B, Li Y, Han X, Lan Z, et al. Combining Pd nanoparticles on MOFs with cross-linked enzyme aggregates of lipase as powerful chemoenzymatic platform for one-pot dynamic kinetic resolution of amines. J Catal 2019;378:153–63.
50. Heidlindemann M, Rulli G, Berkessel A, Hummel W, Gröger H. Combination of asymmetric organo-and biocatalytic reactions in organic media using immobilized catalysts in different compartments. ACS Catal 2014;4:1099–103.
51. Wang W, Li X, Wang Z, Tang Y, Zhang Y. Enhancement of (stereo) selectivity in dynamic kinetic resolution using a core-shell nanozeolite@enzyme as a bi-functional catalyst. Chem Commun 2014;50:9501–4.
52. Smeets V, Baaziz W, Ersen O, Gaigneaux EM, Boissiere C, Sanchez C, et al. Hollow zeolite microspheres as a nest for enzymes: a new route to hybrid heterogeneous catalysts. Chem Sci 2020;11:954–61.
53. Engström K, Johnston EV, Verho O, Gustafson KPJ, Shakeri M, Tai CW, et al. Co-immobilization of an enzyme and a metal into the compartments of mesoporous silica for cooperative tandem catalysis: an artificial metalloenzyme. Angew Chem Int Ed 2013;52:14006–10.
54. Magadum DB, Yadav GD. Design of tandem catalyst by co-immobilization of metal and enzyme on mesoporous foam for cascaded synthesis of (R)-phenyl ethyl acetate. Biochem Eng J 2018;129:96–105.

55. Ganai AK, Shinde P, Dhar BB, Gupta SS, Prasad BLV. Development of a multifunctional catalyst for a "relay" reaction. RSC Adv 2013;3:2186–91.
56. Zhang X, Jing L, Chang F, Chen S, Yang H, Yang Q. Positional immobilization of Pd nanoparticles and enzymes in hierarchical yolk–shell@ shell nanoreactors for tandem catalysis. Chem Commun 2017;53: 7780–3.
57. Zhang N, Hübner R, Wang Y, Zhang E, Zhou Y, Dong S, et al. Surface-functionalized mesoporous nanoparticles as heterogeneous supports to transfer bifunctional catalysts into organic solvents for tandem catalysis. ACS Appl Nano Mater 2018;1:6378–86.
58. Wang Y, Zhang N, Zhang E, Han Y, Qi Z, Ansorge-Schumacher MB, et al. Heterogeneous metal-organic-framework-based biohybrid catalysts for cascade reactions in organic solvent. Chem Eur J 2019;25:1716–21.
59. Dutta S, Kumari N, Dubbu S, Jang S, Kumar A, Ohtsu H, et al. Highly mesoporous metal-organic frameworks as synergistic multimodal catalytic platforms for divergent cascade reactions. Angew Chem Int Ed 2020; 132:3444–50.
60. Wu Y, Shi J, Mei S, Katimba HA, Sun Y, Wang X, et al. Concerted chemoenzymatic synthesis of α-keto acid through compartmentalizing and channeling of metal–organic frameworks. ACS Catal 2020;10:9664–73.
61. Guo M, Chi J, Zhang C, Wang M, Liang H, Hou J, et al. A simple and sensitive sensor for lactose based on cascade reactions in au nanoclusters and enzymes co-encapsulated metal-organic frameworks. Food Chem 2021;339:127863.
62. Wang Y, Zhang N, Hübner R, Tan D, Loffler M, Facsko S, et al. Enzymes immobilized on carbon nitride (C3N4) cooperating with metal nanoparticles for cascade catalysis. Adv Mater Interfac 2019;6:1801664.
63. Gao S, Wang Z, Ma L, Liu Y, Gao J, Jiang Y. Mesoporous core–shell nanostructures bridging metal and biocatalyst for highly efficient cascade reactions. ACS Catal 2019;10:1375–80.
64. Zhao H, Liu G, Liu Y, Liu X, Wang H, Chen H, et al. Metal nanoparticles@ covalent organic framework@ enzymes: a universal platform for fabricating a metal–enzyme integrated nanocatalyst. ACS Appl Mater Interfaces 2022;14:2881–92.
65. Gao S, Liu Y, Wang L, Wang Z, Liu P, Gao J, et al. Incorporation of metals and enzymes with porous imine molecule cages for highly efficient semiheterogeneous chemoenzymatic catalysis. ACS Catal 2021;11: 5544–53.
66. Filice M, Marciello M, del Puerto Morales M, Palomo JM. Synthesis of heterogeneous enzyme–metal nanoparticle biohybrids in aqueous media and their applications in C–C bond formation and tandem catalysis. Chem Commun 2013;49:6876–8.
67. Görbe T, Gustafson KP, Verho O, Kervefors G, Zheng H, Zou X, et al. Design of a Pd (0)-CalB CLEA biohybrid catalyst and its application in a one-pot cascade reaction. ACS Catal 2017;7:1601–5.
68. Li X, Cao Y, Luo K, Sun Y, Xiong J, Wang L, et al. Highly active enzyme–metal nanohybrids synthesized in protein–polymer conjugates. Nature Catal 2019;2:718–25.
69. Tamborini L, Fernandes P, Paradisi F, Molinari F. Flow bioreactors as complementary tools for biocatalytic process intensification. Trends Biotechnol 2018;36:73–88.
70. Himiyama T, Waki M, Maegawa Y, Inagaki S. Cooperative catalysis of an alcohol dehydrogenase and rhodium-modified periodic mesoporous organosilica. Angew Chem Int Ed 2019;131:9248–52.
71. Poizat M, Arends IW, Hollmann F. On the nature of mutual inactivation between [Cp* Rh (bpy)(H$_2$O)]$^{2+}$ and enzymes-analysis and potential remedies. J Mol Catal B Enzym 2010;63:149–56.

André Delavault*, Katrin Ochsenreither and Christoph Syldatk

4 Intensification of biocatalytic processes by using alternative reaction media

Abstract: Performances of biocatalytic processes in industry are often limited by productivity, product concentration and biocatalyst stability. Reasons can be such as unfavourable reaction thermodynamics, low water solubility of the substrates or inhibition caused by high substrate or product concentrations. A way to overcome these limitations and to enhance economic competitiveness of the process can be process intensification (PI) using an alternative reaction medium. Very early in industrial biotransformation processes, it was shown that many interesting target products of organic synthesis are much more soluble and sometimes even more stable in non-conventional reaction media than in buffered aqueous solutions. Moreover, the absence of water is also generally desired to prevent side and degradation reactions as well as microbial contamination, which in turn eliminates the need to work under sterile conditions thereby reducing energy expenditure. In addition, it was also discovered early on that solvents can influence the activity and stability of enzymes quite differently depending on their water affinity and thus if they form rather monophasic or biphasic systems with the latter.

Keywords: optimisation; unconventional media; biocatalysis; green solvents; productivity; biotransformation

4.1 Introduction

According to the definition, solvents are substances in which gases, liquids or solids can be dissolved without undergoing a chemical reaction. Enzyme reactions usually take place in water as a solvent, in which, in addition to the enzymes themselves, salts and organic compounds such as sugars, organic acids, amino acids, lipids and many other structurally complex compounds are usually homogeneously dissolved. In contrast to the above definition, water is not only a solvent in many enzyme reactions, but also a reaction partner and substrate at the same time.

*Corresponding author: André Delavault**, Institute of Bio- and Food Technology Division II – Technical Biology, KIT – Karlsruhe Institute of Technology, Karlsruhe, Germany, E-mail: andre.delavault@kit.edu. https://orcid.org/0000-0002-2232-1580
Katrin Ochsenreither, Biotechnologische Konversion, Technikum Laubholz GmbH, Göppingen, Germany, E-mail: katrin.ochsenreither@technikumlaubholz.de. https://orcid.org/0000-0002-5797-2789
Christoph Syldatk, Institute of Bio- and Food Technology Division II – Technical Biology, KIT – Karlsruhe Institute of Technology, Karlsruhe, Germany, E-mail: christoph.syldatk@kit.edu

As per De Gruyter's policy this article has previously been published in the journal Physical Sciences Reviews. Please cite as:
A. Delavault, K. Ochsenreither and C. Syldatk "Intensification of biocatalytic processes by using alternative reaction media" *Physical Sciences Reviews* [Online] 2023. DOI: 10.1515/psr-2022-0104 | https://doi.org/10.1515/9783110760330-004

Water is characterized by its pronounced dipole character due to its positive and negative partial charges. It can dissociate and, with the possibility of forming hydrogen bonds, contributes to the dissolution of salts and organic compounds up to the formation of micelle and membrane structures in the cell. Water molecules interact with the enzyme surface forming the so-called hydration shell around the protein. This hydration shell is of critical importance for the structure, stability and function of the enzyme. Complete withdrawal ("stripping") of the hydration shell by solvents or salts result usually in inactivity or degradation of the protein.

Intuitively, one might think that water should be used as default media "wherever possible", e.g., in reactions with non-water-depending equilibriums. Despite its apparent "greenness", water is not necessarily a solvent choice for biocatalysis, and the reason is multifactorial. First off, poor solubility of hydrophobic substrate in water will lead to low reagent loading and thus inferior productivities preventing process intensification. Second off, the relatively high boiling point of water can be problematic towards parameters specific to the conversion process (e.g., heating and mixing) and thus, in addition, energy expenditure for its recovery is a limitation as well when generating wastewaters. Third off, water can potentially complexify or simply prevent product recovery due to discrepancy between solvent(s), substrate(s) and product(s) [1]. These factors should then be considered when designing a bioprocess and its down-stream processing (DSP) which we will address later in this chapter. However, the current cutting-edge research that aims at increasing the relevance of water as a media in biocatalysis consists in micellar transformations using surfactants to bridge chemical and enzymatic catalysis [2, 3]. Water as a media for biocatalysis should thus not be systemically dismissed as it still as a key role to play in that framework.

In industry the performance of biocatalytic processes is often limited regarding productivity, achieved product concentrations and the stability of the biocatalysts. Reasons can be unfavourable reaction thermodynamics, low water solubility, i.e., availability, of the substrates or inhibitions e.g., caused by high substrate or product concentrations. A way to overcome these limitations and to enhance economic competitiveness of the process can be process intensification (PI) by choosing an alternative reaction medium.

It has been shown very early in industrial biotransformation processes that many interesting target products of organic synthesis are much more soluble and sometimes even more stable in non-conventional reaction media than in buffered aqueous solutions. Moreover, it has been shown that the absence of water also generally prevents side and degradation reactions as well as microbial contamination, which in turn eliminates the need to work under sterile conditions and reduces energy expenditure. However, it was also discovered early on that solvents can influence the activity and stability of enzymes quite differently depending on water solubility and whether they are present in monophasic or biphasic systems.

Furthermore, since the 1980s it is known that many enzymes exhibit properties in non-conventional reaction media that differ from their native aqueous buffer. It was demonstrated that many enzymes of the EC 3 enzyme class (hydrolases) can catalyze the reverse synthesis reaction in almost anhydrous nonpolar solvents instead of the hydrolysis reaction [4, 5]. It was shown first that lipases could catalyze the synthesis of ester bonds instead of hydrolysis in almost anhydrous organic solvents although with much lower activity (Equation (4.1)).

Hydrolysis: Ester + Water → Carboxylic Acid + Alcohol
Reverse Hydrolysis: Ester + Water ← Carboxylic Acid + Alcohol

$$K = \frac{[\text{Carboxylic Acid}] \times [\text{Alcohol}]}{[\text{Ester}] \times [\text{Water}]} \tag{4.1}$$

To favorize the reaction in the direct sense, meaning toward the target product, it is crucial that converted water molecules should either be removed/sequestrated from the reaction, or its formation alleviated by selecting suitable reaction conditions or substrates. This can be achieved *in situ* by continuous distillation or use of molecular sieves [6, 7].

Regardless, using free, immobilized enzymes or whole living cell biocatalysts in non-conventional reaction media has been of great importance for a long time for both academia and industry. In addition to the long-established use of polar and non-polar organic solvents in biocatalysis, newer alternative media have seen light such as supercritical fluids (SFs), ionic liquids (ILs) and deep eutectic solutions or systems (DESs).

4.2 Organic solvents as media

Organic solvents contain carbon atoms and, in contrast to the water molecule, have molecular skeletons of different sizes to which many different functional groups can be bonded that ultimately determines the solvent properties such as hydrophobicity, hydrophilicity and polarity. Well over a hundred different organic solvents are described in the literature and are commercially available [8].

Besides classification into hydrophilic (water-miscible) and hydrophobic (non-water-miscible), polar and non-polar solvents or a classification according to their chemical composition, e.g., chlorinated and non-chlorinated solvents, there are many other possibilities of classification, which are important for the safe handling of solvents, such as flammability, toxicity, water hazard or biodegradability. A number of these criteria already indicate possible interactions of solvents with biological systems. An important way of classifying solvents for practical applications is the eluotropic series, which groups the most common organic solvents according to their elution effect or elution selectivity in the separation or elution of substances in chromatography, whereby this depends strongly on the stationary phase used in each case.

The denaturing effect of many solvents, such as acetone or ethanol, on proteins in the presence of higher concentrations has been known for a long time and is used specifically for the precipitation of proteins. It was not until the 1980s that first systematic investigations were carried out in order to identify guidelines for the handling of enzymes in combination with organic solvents. In this context, selected enzyme reactions were experimentally investigated in a large number of organic solvents in order to determine the influence of parameters such as polarity, dielectric constant and $\log P$ value of the respective solvents on the enzyme activity [8].

The $\log P$ value is a parameter used in physical chemistry and is defined as the logarithm of the partition coefficient of the respective substance X at 25 °C in a two-phase system of n-octanol and water (Equation (4.2)).

$$\log P = \log \frac{[X]\text{Octanol}}{[X]\text{Water}} \tag{4.2}$$

According to this parameter, solvents can be categorized into four large groups (Table 4.1).

– completely water-miscible solvents ($\log P$ –2.5 to 0)
– partially water-miscible solvents ($\log P$ 0 to 2)
– low water miscible solvents ($\log P$ 2 to 4)
– non-water miscible solvents ($\log P$ > 4)

Table 4.1 is in line with the following case study: Although for polarity and the dielectric constant of the respective solvents, no systematic correlation with the transesterification activity of lipase was apparent, a clear influence was observed in the case of the $\log P$ value (Figure 4.1). While the addition of completely or partially water-miscible solvents ($\log P$ –2.5 to 2) to aqueous reaction media up to a certain concentration was possible without loss of enzyme activity, application of the enzymes in pure solvents led to their

Table 4.1: The $\log P$ value and water miscibility of organic solvents and their effect on protein structure and enzyme activity. (Modified after Faber [9])

Log P value	Examples	Water miscibility	Effect on protein structure and enzyme activity
–2.5 to 0	Acetone, ethanol, DMFA, DMSO	Completely water-miscible	Depending on the solvent and up to a certain concentration solubility of lipophilic compounds is increased without inactivating the enzymes solubilized in them.
0 to 2	Propanol, ethyl acetate, butanol, hexanol	Partially water-miscible	Usually, rapid enzyme inactivation occurs in the presence of these solvents.
2 to 4	Toluene, octanol, styrene, n-hexane	Slightly water miscible	Effects on protein structure and enzyme activity are possible depending on the enzyme but cannot be predicted. Empirical laws based on screening needs to be found on a case-specific approach.
>4	Diphenyl ether, octane, dodecane, butyl oleate	Not water miscible	Usually, no effect on protein structure and preservation of enzyme activity.

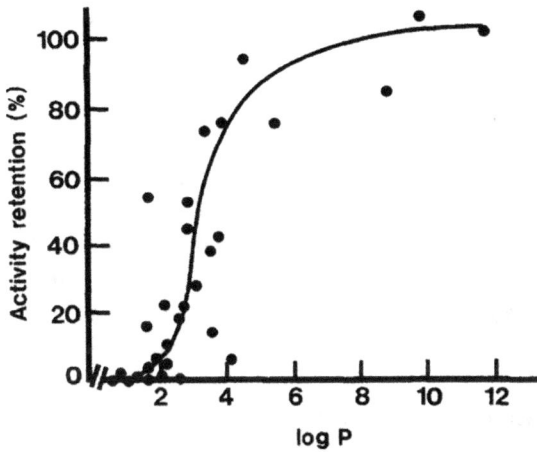

Figure 4.1: Initial activity of a lipase-catalyzed transesterification between tributyrin and heptanol in various nearly anhydrous organic solvents as a function of log P value. (After Halling et al. [10])

inactivation due to stripping of the hydration shell. The use of enzymes in solvents of log P from 2 to 4 and non-water-miscible solvents (log P > 4) was possible with preservation of activity in many cases and even at high temperatures.

Memo box 4.1

Classification for use of organic solvents in biocatalysis in relation to their log P-value

The log P value is a parameter used in physical chemistry, which describes the solubility of a compound X at 25 °C in a mixture of water and n-octanol. It is defined as the logarithm of the partition coefficient of the respective compound X in the two-phase system of n-octanol and water.

With the aid of this parameter, solvents can be categorized into four groups: 1. completely water-miscible solvents (log P −2.5 to 0) 2. partially water-miscible solvents (log P 0 to 2) 3. low water miscible solvents (log P 2 to 4) and 4. non-water miscible solvents (log P > 4).

Completely water-miscible solvents (log P −2.5 to 0) can lead to a withdrawal of the hydration shell of the biocatalyst since the affinity of these solvents for water is higher than that of proteins. In contrast, non-water-miscible solvents (log P > 4) will often even stabilize the hydration shell of enzymes and thus prevent their denaturation. For the other two groups, the effects are not necessarily predictable.

Additionally, water activity a_W of these solvents was identified as an important factor influencing the enzyme activity [11], a parameter which also plays an important role, for example, in the preservation of foodstuffs and other sensitive or perishable products. The water activity a_W is defined as the ratio of the water vapor partial pressure in the solvent (p) to the saturation vapor pressure of pure water (p_0) at a certain temperature (Equation (4.3)).

$$a_{W=}\frac{p}{p_0}$$ (4.3)

The a_W value is therefore dimensionless and ranges between 0 (no water available) and 1 (formation of condensation water). In biochemical reactions, the a_W value indicates the proportion of available water in the reaction system. It is of decisive importance for enzyme reactions and for microbial growth. Indeed, bacteria generally require an a_W value of at least 0.98 for growth while most fungi only require 0.7. The a_W value can be influenced by high salt or sugar concentrations which is used, for example, in the preservation and storage of food.

The water content (w), also known as natural water content or natural moisture content, is the ratio of the weight of water to the weight of the solids or liquids. This ratio is usually expressed as a weight percentage (wt%). When voids are filled with air, water content is equal to 0 (complete dryness). Water activity and water content do not necessarily correlate linearly in solvent systems. This is especially relevant when using ILs or DESs in order to modulate viscosity and optimize mass and heat transfers in view of process optimization. These points will be developed later.

If we now consider the interaction of enzymes or, more generally, of proteins with the various groups of solvents mentioned above, completely water-miscible solvents (log P –2.5 to 0) can lead to a withdrawal of the hydration shell. Indeed, the precipitation of proteins can be caused when higher concentrations of acetone or ethanol are used in aqueous media since the affinity of these solvents for water is higher than that of the proteins. A similar effect is also observed, for example, in the so-called salting out of proteins in which ammonium sulfate can be used as a precipitating agent of proteins [12].

Even if the course of the activity of enzymes in different solvents seems to be very different depending on their water content, it is almost identical after normalization to the essential water content of the respective enzyme that is necessary for activity [11]. However, different enzyme groups differ significantly regarding the essential water content required for activity.

In contrast, non-water-miscible solvents (log P > 4) can stabilize the hydration shell of enzymes, which is necessary for activity – even at high temperatures – and thus prevent their denaturation [13]. The influence on the activity of enzymes in partially water-miscible (log P 0 to 2) and slightly water-miscible solvents (log P 2 to 4) cannot be clearly predicted. If enzymes are used in two-phase systems consisting of an aqueous and an organic solvent phase there is, however, often an inactivation of the enzymes at the interface, while in the case of monophasic presence of these solvents with the addition of water until saturation and the formation of a second phase, like the non-water-miscible solvents, no loss of activity generally occurs.

Denaturation of enzymes in or by solvents can be avoided by various measures [14]. Very often, the denaturation of enzymes can be successfully prevented by immobilization and "freezing" in their active conformation, as demonstrated in fundamental work in the 1990s [15]. However, when using anhydrous enzymes in a non-polar, non-water-miscible solvent, it is important to note that minimum water content in the reaction system is essential for their activity. As a rule of thumb, water should be present in the reaction system so that each enzyme molecule is fully hydrated [12]. Moreover, within the solvent enzymes will be present in the structure in which they were previously precipitated or dried from the aqueous reaction mixture at a certain pH (*pH memory effect*) [11]. Before using enzymes in organic solvents or any other unconventional media, they should therefore be buffered and dried again if necessary. The log P value, the water activity a_W and water content (w) are crucial parameters that are essential to consider when working with enzymes not only in organic solvents, but also in the other unconventional reaction media mentioned in the following subparts.

Memo box 4.2

The importance of water activity for enzyme reactions in unusual reaction media

The water activity a_W is defined as the ratio of the water vapor partial pressure in a solvent (p) to the saturation vapor pressure of pure water (p_0) at a certain temperature. The a_W value is dimensionless and ranges between 0 (no water available) and 1 (formation of condensation water).

In biochemical reactions the a_W value indicates the proportion of available water in the reaction system. Completely water-miscible solvents (log P –2.5 to 0) can lead to a withdrawal of the hydration shell, an effect that can be used to precipitate proteins, since the affinity of these solvents for water is higher than that of proteins, while low- or non-water-miscible solvents (log P > 4) can even stabilize the essential hydrate shell of enzymes and thus prevent denaturation and maintain their activity even in extreme conditions.

4.2.1 Anhydrous or nearly anhydrous neat organic solvent systems

Enzyme activity in almost anhydrous apolar organic solvents was demonstrated for the first time in the mid-1980s, thus disproving a dogma that had been valid until then. It was shown first that lipases could catalyze the synthesis of ester bonds instead of hydrolysis in

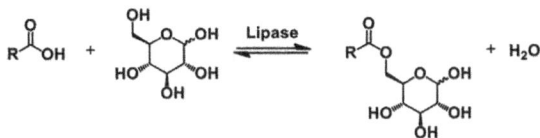

Figure 4.2: Lipase-catalysed synthesis of fatty acid glucose esters with R ≥ C12. (After Arcens et al. [16])

almost anhydrous organic solvents although with much lower activity. Figure 4.2 is a practical application of that principle for the biocatalyzed production of glycolipids (Equation (4.4)).

$$K = \frac{[\text{Fatty Acid}] \times [\text{Glucose}]}{[\text{Sugar ester}] \times [\text{Water}]} \tag{4.4}$$

Subsequently, it was not only demonstrated that other enzymes from the EC class 3 (hydrolases) could be used the same way but also that essential water content is decisive for the respective enzyme activities, which differ significantly for the different enzyme groups [17, 18].

Since condensation reactions produce water, equilibrium is reached above a certain concentration. In order to avoid this, water should either be removed from the reaction, or its formation should be avoided altogether by selecting suitable reaction conditions or substrates. The enantioselectivity of the corresponding enzyme reaction may strongly depend on the type of solvent used.

Examples from the literature of enzymatic reactions in nearly anhydrous organic solvents are:
- Enantioselective synthesis of chiral esters starting from fatty acids, fatty acid methyl esters or vinyl derivatives of fatty acids and alcohols, catalyzed by lipases and esterases.
- Transesterification reactions catalyzed by lipases and esterases.
- Enantioselective syntheses of amides and peptides with peptidases
- Synthesis of oligosaccharides and glycolipids with glycosidases.

By immobilizing the enzymes using various immobilization techniques that can be divided in carrier-bound or carrier-free, a stabilization of the respective active conformations can be achieved, which allows direct use in completely water-miscible organic solvents such as ethanol without denaturation even at high temperatures. Use of cross-linked enzyme crystals (CLECs) or cross-linked enzyme aggregates (CLEAs) are well-known representative of carrier-free immobilization techniques. In contrast, immobilization on carriers can be achieved via covalent bonding, chemi- or physisorption and thus a "freezing" of the enzyme conformation can be achieved. This was impressively demonstrated in the case of the immobilization of chymotrypsin [19]. However, the dry CLECs and the gel-like CLEAs differ significantly in their water activity, which can be of great importance for corresponding applications of such immobilized enzymes. Future strategies for carrying out enzymatic reactions in almost anhydrous organic solvents could be to make enzymes soluble in organic solvents by appropriate modification, such as "PEGylation" [20].

If one wants to use enzymes in almost anhydrous organic solvents the following basic rules should be considered [9]:

- Hydrophobic (little or non-) water miscible solvents (log P > 3) are more compatible for enzymes than hydrophilic (fully or partly) water miscible solvents (log P value < 3).
- The essential water layer around the enzyme molecules should be preserved in order to not hinder the enzyme activity. In the case of hydrophobic solvents, this can be ensured, for example, by water saturation by simply shaking out the solvents with water in the separating funnel beforehand. If CLEAs are used, their water content must also be considered.
- Before using the respective enzyme in an almost water-free solvent, "micro pH" corresponding to the pH optimum of the enzyme should be ensured. If necessary, the enzyme must be "re-buffered" and dried again beforehand.
- Since the enzyme is no longer dissolved in the solvent, diffusion of the reactants and products to and away from the enzyme surface should be ensured by efficient stirring, shaking or ultra-sonication to ensure mass and heat transfer with the reaction bulk.
- Stabilization of the enzymes can be achieved either by addition of stabilizing agents such as non-active proteins, sugars, polyalcohols, polymers as well as salts or by immobilization.

4.2.2 Organic solvents mixed with aqueous buffers: one-phase and two-phase systems

Meanwhile, the use of enzymes not only in almost anhydrous nonpolar organic solvents but also in partially or even fully water-miscible solvents is an established technique in the chemical and pharmaceutical industries. Stability-enhanced immobilized enzymes usually have been designed and used as chemical heterogeneous catalyst [21]. Industrial processes for the biotransformation of poorly water-soluble steroids with the addition of completely water-miscible solvents such as methanol, ethanol, dimethyl sulfoxide (DMSO) or dimethyl formamide (DMFA) have been described for steroid biotransformations with living cells in the literature since the 1940s. As a rule, growing or dormant living microorganism cells are used in combination to the poorly or non-water-soluble starting substrate dissolved in the corresponding solvent to improve the bioavailability or to enable easier dosing [22]. The subsequent processing of the products from the reaction mixtures is usually carried out by liquid extraction. A practical application of water-miscible solvents is for example their addition to influence the reaction equilibrium in protease-catalyzed reactions such as the trypsin-catalyzed exchange of amino acids in modified insulin molecules.

As alternative method to the described procedure for improving the availability of poorly water-soluble substrates, by the addition of water-miscible solvents, is the use of solvents that are not or only slightly water miscible, such as cyclohexane, methyl isobutyl ketone (MIBK), toluene or ethyl acetate in a two-phase systems [10]. This has the advantage that substrates and products are dissolved in the organic phase and thus significantly higher concentrations can be achieved than with the addition of water-miscible solvents. The enzymatic reaction takes place at the phase interface. Compounds that are sensitive in the aqueous phase can thus be stabilized, and possible product inhibition can be limited. The separation of the product from the unreacted residual substrate can as well be carried out either at the end of the reaction after phase separation or even continuously during the reaction, forming thus an *in-situ* product removal (ISPR) from the organic phase, which can also be used to avoid product inhibition.

However, frequently observed possible disadvantages can be a denaturation of the biocatalysts at the phase interface or inhibitory effects due to the solvent used. To avoid direct contact of the biocatalysts with the solvent phase, it is therefore advisable to use either special membrane systems or special immobilization techniques [21]. The aqueous phase of a two-phase system can be reduced to an absolute minimum called a micro-aqueous system. This can be of interest, for example, if water-soluble coenzymes are involved in the reaction. Avoiding direct contact of the enzyme(s) with the organic solvent can be essential for maintaining enzyme activity. One possibility is the use of reversed micelles as demonstrated for the enzymatic synthesis of L-tryptophane from indole and L-serine [23]. In addition to the aqueous micro- and organic phase, this reaction system also contains surfactants and co-surfactants to separate the aqueous and organic phase from each other, which, however, makes the subsequent work-up of reactants and products as well as the reuse of the enzymes present in the aqueous phase difficult. In addition, the transport of reactant from the organic phase to the enzyme in the aqueous phase and of the product from the reverse micelle back into the organic phase must be ensured during the reaction.

Another possibility is the use of Pickering emulsions. These are named after the British chemist Percival Spencer Umfreville Pickering, who in 1907 described the phenomenon that emulsions can be stabilized by the addition of solid particles which adsorb onto the interface between the aqueous and organic phases. A natural example is homogenized milk, in which the emulsion is stabilized by casein molecules at the interface between the fatty and aqueous phases. In biocatalysis, such reaction systems may be suitable for microencapsulation of enzymes [24]. Thus, basic criteria for the selection of suitable solvents for enzyme reactions in two-phase systems are [9]:
–   Capacity of the solvent for reactant and product solubilization
–   Partition coefficients in the aqueous and organic phases
–   Absence or presence of denaturing effects at the phase interface
–   Scalability, flammability, toxicity and reuse of the solvents

Reactions in two-phase systems can be of interest for biocatalysis especially when reactant and product are each preferentially present in different phases. A natural reaction system for enzymatic reactions in two-phase systems is lipase-catalyzed reactions with fats or oils in water, whereby the reaction of highly viscous triglycerides or fats can be promoted by the addition of nonpolar solvents. In contrast to esterases, many lipases are only fully active when an interface is present in the reaction system and thereby do not exhibit typical Michaelis–Menten kinetics. Further examples of reactions in two-phase systems described in the literature concern, for example, the recovery of water-soluble and enantiomerically pure amino acids when starting from racemic amino acid esters mixtures present in the solvent phase, the recovery of L-menthol from racemic D,L-menthyl acetate or the biotransformation reactions with poorly water-soluble steroids where both reactant and product are present in the solvent phase [4].

Memo box 4.3

Basic criteria for the selection of suitable solvents for enzyme reactions in two-phase systems

– Capacity of the solvent for reactant and product solubilization
– Partition coefficients in the aqueous and organic phases
– Absence or presence of denaturing effects at the phase interface
– Scalability, flammability, toxicity and reuse of the solvents

4.3 Low-transition-temperature mixtures (LTTMs)

4.3.1 General principles and definitions

LTTMs are a new generation of designer solvents emerged in the last decade as promising green media for multiple applications such as biocatalysis [25]. They are generally prepared by mixing natural high-melting-point starting materials that result in a stable liquid at considerably lower temperatures formed due to molecular interactions that can be of ionic nature as well. Among this family of mixtures, deep-eutectic solvents (DESs) were presented as promising alternatives to ionic liquids (ILs) and even more recently natural deep eutectic solvents (NaDESs) were introduced to describe specific LTTMs obtained from the combination of ubiquitous molecules, *in fine*, naturally occurring primary metabolites [26] (Figure 4.3).

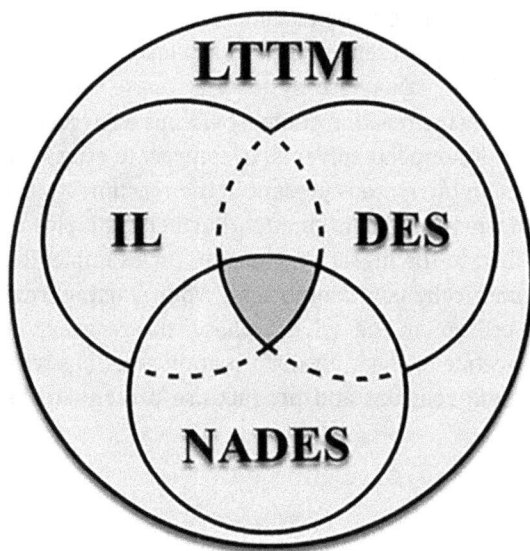

Figure 4.3: Classification of low transition temperature mixtures. (After Durand et al. [27])

4.3.2 Ionic liquids (ILs)

ILs are salts that are liquid at temperatures below 100 °C without the addition of water. They are usually composed of an organic cationic compound such as imidazolium, pyridinium, pyrrolidinium, guanidinium, which may be alkylated, and an anionic compound such as a halide, tetrafluoroborates, trifluoroacetates, imides or amides (Figure 4.4). When these compounds are combined with each other, charge delocalization and steric effects hinder the formation of a stable crystal lattice, in a way that the solid crystal structure in the individual compounds is broken down resulting in the formation of an IL.

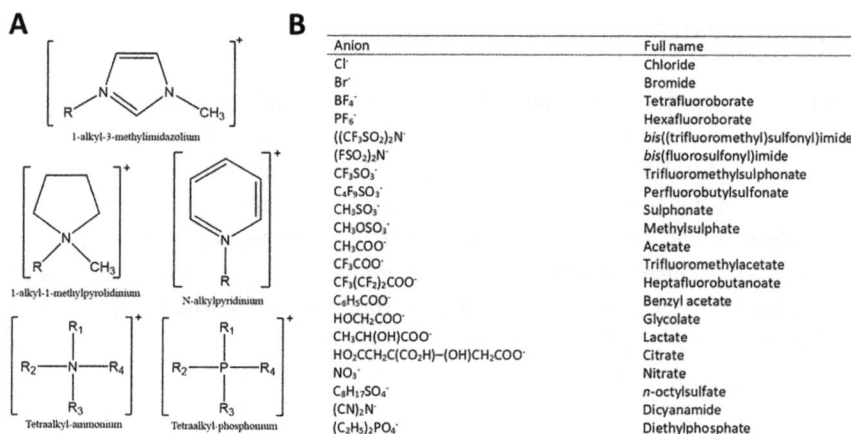

Anion	Full name
Cl⁻	Chloride
Br⁻	Bromide
BF_4^-	Tetrafluoroborate
PF_6^-	Hexafluoroborate
$((CF_3SO_2)_2N^-$	bis((trifluoromethyl)sulfonyl)imide
$(FSO_2)_2N^-$	bis(fluorosulfonyl)imide
$CF_3SO_3^-$	Trifluoromethylsulphonate
$C_4F_9SO_3^-$	Perfluorobutylsulfonate
$CH_3SO_3^-$	Sulphonate
$CH_3OSO_3^-$	Methylsulphate
CH_3COO^-	Acetate
CF_3COO^-	Trifluoromethylacetate
$CF_3(CF_2)_2COO^-$	Heptafluorobutanoate
$C_6H_5COO^-$	Benzyl acetate
$HOCH_2COO^-$	Glycolate
$CH_3CH(OH)COO^-$	Lactate
$HO_2CCH_2C(CO_2H)-(OH)CH_2COO^-$	Citrate
NO_3^-	Nitrate
$C_8H_{17}SO_4^-$	n-octylsulfate
$(CN)_2N^-$	Dicyanamide
$(C_2H_5)_2PO_4^-$	Diethylphosphate

Figure 4.4: Commonly used cations (A) and anions (B) of ionic liquids. (Reproduced after Moniruzzamana et al. [28])

There are many possible combinations to create ILs enabling the adjustment of their respective desired physicochemical properties to adapt them to technical requirements, e.g., to influence the solubility of substances that are difficult to dissolve in water, which is of great interest when carrying out enzymatic reactions [29]. Compared to organic solvents, ILs are also characterized by thermal stability and being difficult to ignite, as well as having a very low, hardly measurable vapor pressure. Even if there is no danger of explosion or poisoning by inhalation due to their non-volatility, as it is the case with many organic solvents, waste water could possibly be problematic due to the unresolved biodegradability, which is why their recyclability can be an important criterion for applications. Nevertheless, due to the large number of possible combinations, ILs are highly interesting reaction media for biocatalysis [30], and the aim is to achieve desired physicochemical properties with the lowest possible toxicity by combination of suitable components.

4.3.3 Deep eutectic solutions (DES) and naturally deep eutectic solutions (NADES)

Closely related to ILs, and also referred to by some authors as a "special class of biodegradable ILs", are DESs and NaDESs [31]. By definition, a solution or alloy of different substances is called "eutectic" if its components are present in a certain ratio that it becomes liquid or solid at a defined temperature. The corresponding point in the phase diagram is called a eutectic (Figure 4.5). The prerequisite for this effect is that the melting temperature of the mixing system is below the melting temperatures of the individual components and that a miscibility gap exists in the solid which remains until the melting temperature of the eutectic is reached. The "eutectic point" is the point in the phase diagram of a multicomponent system which is characterized by the concentration ratio of the eutectic and by its melting temperature (the "eutectic temperature").

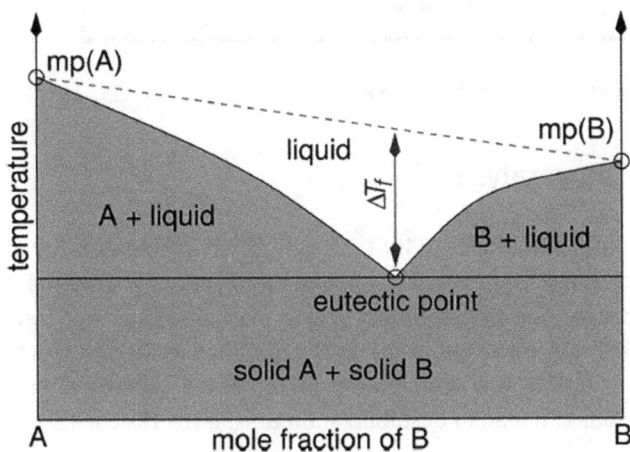

Figure 4.5: Phase diagram of a deep eutectic solution. (After Smith et al. [32])

In contrast to ionic liquids, eutectic solutions usually consist of an ammonium compound, such as choline or urea, in combination with organic compounds such as sugars, sugar alcohols or carboxylic acids, which act as donors to form hydrogen bonds (Figure 4.6). All building blocks can be obtained from renewable raw materials and, as individual components, having melting points above 100 °C. If mixtures of these are prepared in certain proportions and heated, solutions of varying viscosity are formed, which remain liquid even after cooling to room temperature.

Deep eutectic solutions (or systems) are inexpensive reaction media and, like ionic liquids, non-flammable and liquid. Unlike ionic liquids, however, they are usually harmless from a toxic point of view and are generally readily biodegradable while offering a large variety of combination and application possibilities as ILs.

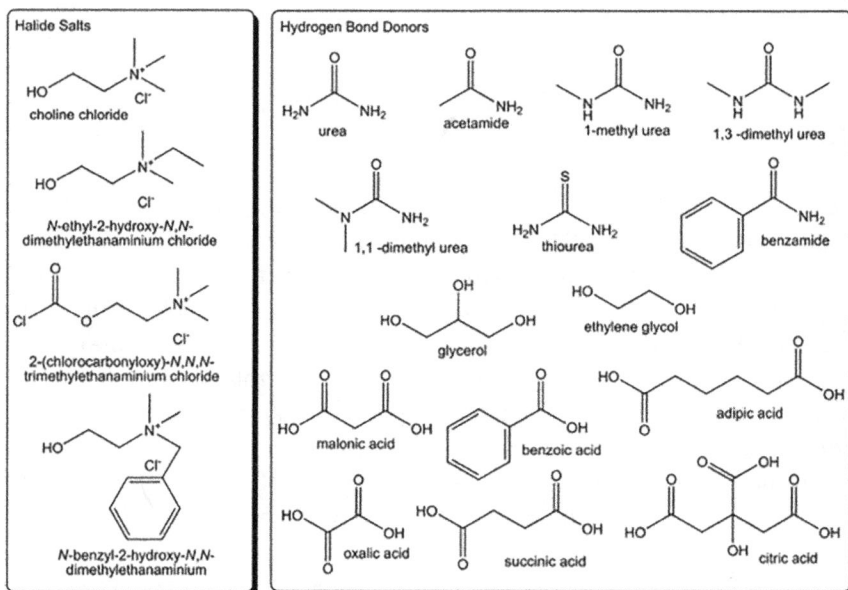

Figure 4.6: Components of deep eutectic solutions. (After Smith et al. [32])

4.3.4 Use of LTTMs for biocatalysis

The interest in the use of enzymes in ILs dates to 1984, when the activity and stability of alkaline phosphatase in the cleavage of the model compound *p*-nitrophenyl phosphate was investigated for the first time [33]. Subsequently, it was discussed that the interaction of enzymes with ILs containing kosmotropic ions can enhance the structural behavior of water molecules and thus also contribute to an increased stability of proteins. Oppositely, chaotropic ions can lead to destabilization thus, if the ILs consists of two rather chaotropic ions, there is a high probability that enzymes will lose activity in the latter. To summarize, hydrophilic ILs increase the tendency to form hydrogen bonds, making it possible for enzymes to dissolve in it and even maintain their activity with a high probability. The first examples of targeted biocatalytic applications of enzymes in ionic liquids date back to 2000 with the description of the peptidase-catalyzed synthesis of Z-aspartame. Since then, there has been an increasing body of work addressing the effect of ionic liquids on the structure, activity and stability of enzymes.

Regarding DESs or NaDESs, possible applications of deep eutectic solutions in enzyme catalysis are their direct use as a non-volatile solvent, considering the essential water concentration for the respective enzyme [31]. In addition to the sometimes-high viscosity, a disadvantage can be that, in contrast to organic solvents and ILs, the addition or formation of water during the reaction can lead to a breakdown of the electrostatic interactions between the components of the deep eutectic solution. However, this effect can in turn be used specifically in the work-up of the reaction mixtures. In any case, denaturing effects on the enzymes used are expected to be less than in the case of many organic solvents and ILs.

The use of ILs and DESs as alternatives to organic solvents has been described for lipase-catalyzed transesterifications, amidations and epoxidations, among others, using both single-phase and two-phase systems [29]. Indeed, since several ILs do not mix with ethers it enables the possibility for working up the products by simple liquid extraction. Furthermore, the use of ILs for the encapsulation (also known as coating) of enzyme when used in organic solvents is described in order to achieve, for example, an improvement in substrate transport from the reaction bulk to the enzyme in addition to prevent enzyme inactivation at the interface. Furthermore, ILs can be used to dissolve lignocellulose and its individual components improving thus their availability for enzymatic reactions. Likewise, there are already successful applications for whole cell biocatalysis in ILs e.g., to produce chiral alcohols starting from prochiral ketones. Since ILs have no vapor pressure, volatile compounds can be more easily separated from them by evaporation than from organic solvents [29]. In summary, it can be said that ILs for enzyme catalysis offer a variety of interesting application possibilities as an alternative reaction medium to organic solvents. However, in addition to the still high costs, the main challenges still to be faced are their recyclability and the still unclear biodegradability and environmental compatibility [34].

Despite their low cost and accessibility of DESs, there are so far only a manageable number of examples in the literature of enzyme reactions therein, mainly with lipases [35]. One of the first published papers dates to 2008 and describes lipase-catalyzed transesterification using an ethyl fatty acid and 1-butanol as substrates. Other papers describe the use of these reaction systems in the lipase-catalyzed production of biodiesel from highly viscous triacyl glycerides and methanol, in *n*-alkylation of primary aromatic amines, and in epoxide hydrolysis. Interest in DESs as new reaction media for enzymes is growing as they can be used as "2-in-1" systems in which a partner of the eutectic mixture can act as a substrate of a biocatalyzed reaction simultaneously [36]. This has been proven to inherently remove solubility limitations of certain substrates and even drive reactions in the direct sense as a virtually unlimited amount of substrate is a practical application of Le Chatelier's thermodynamic principle [37, 38]. In addition, the non-volatility, non-flammability, non-toxicity, good environmental compatibility and the possibility of producing DESs based on renewable raw materials are making them promising and attractive for future industrial applications as green solvents.

Despite encouraging prospects wastewater and organic waste management must be taken in consideration when designing biocatalytic processes especially when employing alternative solvent systems. Unlike organic solvents and the wastewaters they generate, DESs do not have to undergo incineration or distillation; thus, their release in nature could be considered, once the compound(s) of interest have been recovered, as their biodegradability and biocompatibility have been demonstrated in more recent studies [39, 40]. In contrast, some ILs seem to represent a real threat to soils and aquatic environments [34]. Recyclability, in other words, re-use of either neat ILs or DES mixtures could then be a viable solution to a straightforward wastewater management as well as organic waste and water pollution reduction in addition to tremendous atom economy as it has been reported in several studies [41–43]. Moreover, it could be foreseeable that choline chloride-rich wastes generated by such DES-mediated process could be re-valorized in feed additives or as agrochemical active ingredients [44, 45].

Future challenges for enzyme catalysis are not only to exploit any pre-existing potential, but also to expand the application and combination possibilities of DESs by using further salts and hydrogen bond donors in a similar way as it has already been done for ILs. It should also be added that these innovative approaches, which primary use is to reach sustainability while remaining efficient and competitive to traditional methods should undergo a thorough Life Cycle Assessment (LCA) and its iterations for a holistic approach [46]. These tools and metrics such as E-factor, total carbon dioxide release (TCR), space-time yield, biocatalyst yield… amongst other values for an atomistic approach, are needed to go beyond the simple claims of "greenness" and assert objectively that not only a process but also its product(s) are (1) safe for the environment, operators and consumers as well as (2) equally or ideally more efficient than alternative/previously used synthesis pathways [47–50].

As a final follow-up subpart, it is logical to mention DSP strategies adapted to biocatalysis. Meyer and coworkers rightfully reported that "converting a substrate enzymatically to a product is only half the story", indeed, the product and biocatalyst must be efficiently recovered in order to reduce costs and environmental impact as well as intensifying bioprocesses [51]. Flow biocatalysis arouse as a viable tool for not only the conversion process but also for product isolation. The type of reactor employed for a continuous flow set up will be highly dependent on how the biocatalyst is formulated (carrier-bound or carrier free), this applies likewise for product recovery and its physicochemical properties. Packed-bed reactors (PBR), continuous stirred tank reactors (CSTR), coil reactors, microreactors and tube-in-tube reactors are mostly encountered to carry continuous flow biotransformations and usually lead to improved reaction performances [52–54]. The product(s) can also be continuously recovered, allowing thus massive gains in productivity in comparison to classical "stop-and-go" setups, by methods such as continuous (1) liquid–liquid extraction, (2) adsorptive downstream processing and (3) crystallization and precipitation [51]. Continuous flow strategy shows moreover good synergy with alternative media like DESs, thus highlighting a technological match. Several reports mention the use of notably cross-linked enzymes for the production monobenzoate glycerol (α-MBG) accumulating thereby up to $10\,g\,L^{-1}$ of product while recycling the biocatalyst up to 6 times [55–57].

Memo box 4.4

Ionic liquids (ILs) and deep eutectic solutions (DES)

Ionic liquids are salts that are liquid at temperatures below 100 °C without the addition of water. They are usually composed of an organic cationic compound such as imidazolium, pyridinium, pyrrolidinium, guanidinium, which may be alkylated, and an anionic compound such as a halide, tetrafluoroborates, trifluoroacetates, imides or amides. When these compounds are combined with each other, charge delocalization and steric effects hinder the formation of a stable crystal lattice, the solid crystal structure in the individual compounds is broken down, and a so-called "ionic liquid" is formed.

A solution or alloy of different substances is called "eutectic", if its components are in such a ratio to each other that it becomes liquid or solid at a certain temperature. The melting temperature of the mixing system is below the melting temperatures of the individual components. The corresponding point in the phase diagram is called a eutectic. Natural eutectic solutions (NADES) usually consist of a mostly inexpensive organic ammonium compound, such as choline or urea, in combination with organic compounds such as sugars, sugar alcohols or carboxylic acids, which act as donors to form hydrogen bonds. In case of NADES, the building blocks are obtained from renewable raw materials.

Metrics such as LCA, E-factor, TCR, productivities, biocatalyst yield … to cite a few, should be applied not only to confirm sustainability of biocatalyzed processes in unconventional media but also, in tandem, enable factually the qualification of a (bio-)process as "intensified". Refer to the works of Prof. R. A. Sheldon and Prof. J. M. Woodley who both extensively covered Green Chemistry principles [58] and bioprocess metrics [59] applied to biocatalysis to learn more about the key parameters of sustainability and performance in the context of bioprocess intensification.

Continuous flow is a topical strategy to achieve process intensification and sustainability. It can benefit for both bioconversion and product isolation in order to increase yields and productivities. A wide range of highly customizable set up exist and are mostly tuned on a "case-specific" basis which can help making use of alternative media relevant for industrial applications.

4.4 Diverse reaction systems for biocatalysis

4.4.1 Supercritical fluids

Substances can occur in solid, liquid and gaseous form depending on temperature and pressure (Figure 4.7). At the so-called triple point, all three phases are in equilibrium and supercritical fluids properties lie between those of gases and liquids. The critical point is the state of a substance which is characterized by the equalization of the densities of the liquid phase and the gas phase. The differences between both physical states cease to exist under these conditions. Above this point, a fluid is said to be "supercritical" or "in a supercritical state". This occurs when a gas is subjected to an ever-increasing pressure and the distances between the gas molecules decrease until, on reaching the critical pressure, they become as large as those of molecules in the liquid phase.

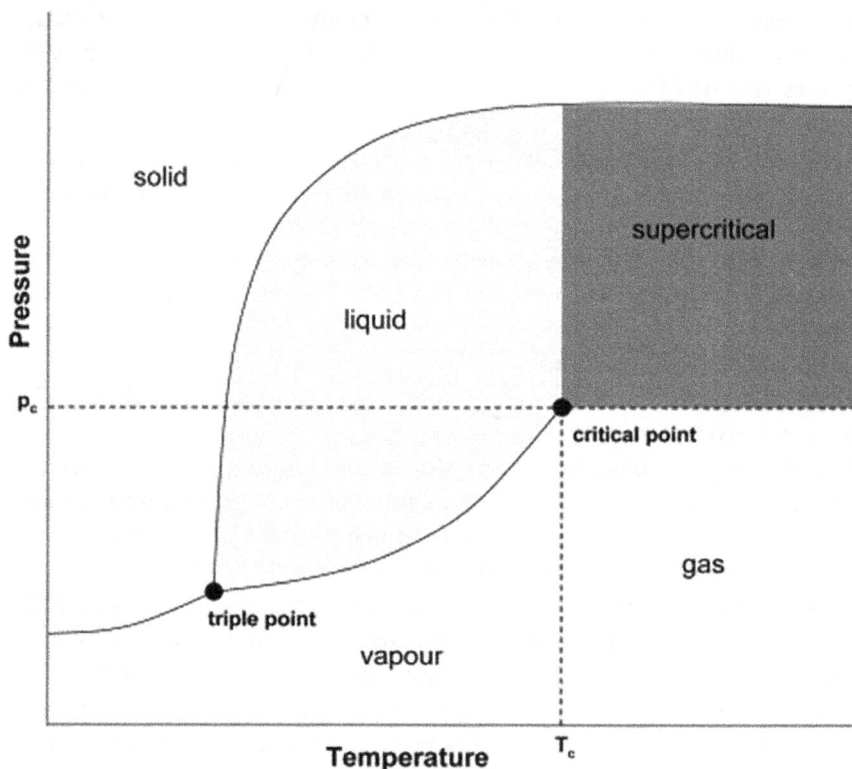

Figure 4.7: Phase diagram for a supercritical fluid. (After Hobbs and Thomas [60])

An advantage of these reaction systems is that the solvent properties of a super-critical fluid depend strongly on its respective density, which can be adjusted over a relatively wide range [61]. In this context, a higher density generally also increases the solubility of most substances, while a lower density lowers it. Supercritical fluids can thus combine the solubility of liquids with the low viscosity of gases, which is a great advantage because a substance dissolved in the supercritical fluid can be separated from the reaction medium again very easily at normal pressure. In contrast to working with organic solvents, ILs and DESs, carrying out reactions in supercritical fluids requires however, in all cases, significantly more complex equipment and process technology, since work must be carried out at alternating pressure. Carbon dioxide, ethane, propane, ethylene and fluoroform are substances that are commonly used as supercritical fluids [60]. Especially supercritical carbon dioxide is a much-used reaction medium due to its affordability and availability, its non-toxicity when handled appropriately. In addition, its separation from dissolved substances is relatively straightforward and has long been used, for example, in food technology as a non-toxic alternative for extraction purposes to decaffeinate coffee and tea. It is formed as soon as the pressure and temperature exceed 304.13 K (30.980 °C) and 7.375 MPa (73.75 bar), respectively.

When working with enzymes in nonpolar organic solvents, it was mentioned that an essential water concentration on the surface of the enzyme molecules is necessary to remain activity and the same principle applies to supercritical fluids. This can be achieved by water saturation, which is doable in the case of supercritical carbon dioxide [60]. As one of the first enzyme reactions in supercritical carbon dioxide, the cleavage of the model compound p-nitrophenyl phosphate by alkaline phosphatase was described in 1985 [62]. Other enzymes were successfully used in various supercritical fluids such as subtilisin by the company Carlsberg as well as various immobilized and commercially available lipases when used for transesterification reactions. Among other things, it was shown that for kinetic resolutions therein, the E value is affected by pressure and temperature. Furthermore, asymmetric reduction reactions with alcohol dehydrogenases as well as carboxylation reactions with various decarboxylases were successfully carried out in supercritical carbon dioxide [62].

Despite the positive results and the advantages of the adjustability of the solvent properties by modification of pressure and temperature as well as the possibility of simple product separation, the use of supercritical fluids in enzyme catalysis has not yet become widely established. The reasons for this are presumably to be seen in the limited solvent spectrum and the necessary high equipment costs and expertise to run such setup.

4.4.2 Gas, liquid and solid solvent free approaches

As an alternative to the use of enzymes in aqueous reaction systems, with or in organic solvents, ILs, DESs or supercritical fluids, there is also the possibility of almost completely dispensing the use of solvents. In solvent-free biocatalysis, substrates and products can be present either in the gas, liquid or solid phase. Even if such an approach initially appears unusual, there are a number of examples in the literature of reactions with soluble and immobilized enzymes or whole cells that are used directly in the substrate without the addition of solvents [60].

The approach is relatively simple and involves bringing the biocatalyst directly into contact with the substrate or substrates, which are then also the reaction medium. In all cases, the water activity must again be considered. When carrying out reactions in the gas phase this can be done, for example, simply by passing the substrate over the respective biocatalyst. This can be, for example, living whole cells of microorganisms immobilized on a carrier. Subsequent separation of substrate and product can then be accomplished by cold traps. Examples of this are cis-epoxidation reactions on gaseous alkenes.

When carrying out reactions in the liquid phase, in which the latter is both substrate and solvent, the biocatalyst is added directly to the liquid substrate phase. Here, in addition to the essential water concentration for the enzyme, it should be noted that mixing can become the rate-determining step.

Solid-to-solid biocatalysis poses a particular challenge, as the reaction takes only place at the interface between enzyme and substrate. In some cases, minimal amounts of water or solvent must be added as an adjuvant. In this case, one also speaks of a "heterogeneous eutectic reaction". Examples of such reactions from the literature are various peptide syntheses with the immobilized proteases subtilisin or thermolysin, including the successful synthesis of the dipeptide sweetener precursor aspartame (Z-AspPheOMe) in 3 M concentration, or lipase-catalyzed syntheses of esters from sugar alcohols and fatty acids [63]. With immobilized penicillin acylase, the successful synthesis of ampicillin from equimolar amounts of solid lyophilized 6-aminopenicillanic acid and D-phenylglycine methyl ester was achieved, whereby the essentially required water concentration was made available in the form of salt hydrates. The scale-up of such solid-phase biocatalysis reactions pose a challenge in terms of equipment and process technology although so-called "mechanoenzymatic" reactions is a strategy on the rise and despite these challenges it is a promising tool to support biocatalysis in achieving industry-satisfying performances [64].

4.5 Conclusions

The performance of enzymatic reactions in unconventional reaction media offers a wide range of innovative possibilities. The main motivations originating from these are to significantly increase the solubility of substrates and products compared to aqueous reaction systems, to simplify subsequent work-up and to avoid microbial contamination as well as side and degradation reactions. Many enzymes exhibit different properties in unconventional reaction media and catalyze different reactions than in aqueous systems. For example, hydrolytic enzymes can be successfully used for synthesis reactions, making specific use of their properties of exquisite regioselectivity, stereoselectivity and substrate specificity. When using enzymes in alternative reaction media, consideration of the water activity (a_W) is always an essential prerequisite, in addition, inactivation of enzymes in unconventional reaction media can be avoided by immobilization.

Acknowledgments: The authors would like to thank the editor Prof. Dirk Holtmann for his guidance and review of this article before its publication.

References

1. Holtmann D, Hollmann F. Is water the best solvent for biocatalysis? Mol Catal 2022;517:112035.
2. Dussart-Gautheret J, Yu J, Ganesh K, Rajendra G, Gallou F, Lipshutz BH. Impact of aqueous micellar media on biocatalytic transformations involving transaminase (ATA); applications to chemoenzymatic catalysis. Green Chem 2022;24:6172–8.
3. Gröger H, Gallou F, Lipshutz BH. Where chemocatalysis meets biocatalysis: in water. Chem Rev 2023;123: 5262–96.
4. Ueda M, Tanaka A, Fukui S. Enhancement of carnitine acetyltransferase synthesis in alkane-grown cells and propionate-grown cells of Candida tropicalis. Arch Microbiol 1985;141:29–31.
5. Zaks A, Russell AJ. Enzymes in organic solvents: properties and applications. J Biotechnol 1988;8:259–69.
6. Duan Y, Du Z, Yao Y, Li R, Wu D. Effect of molecular sieves on lipase-catalyzed esterification of rutin with stearic acid. J Agric Food Chem 2006;54:6219–25.
7. Yan Y, Bornscheuer UT, Schmid RD. Efficient water removal in lipase-catalyzed esterifications using a low-boiling-point azeotrope. Biotechnol Bioeng 2002;78:31–4.
8. Laane C, Boeren S, Vos K, Veeger C. Rules for optimization of biocatalysis in organic solvents. Biotechnol Bioeng 1987;30:81–7.
9. Faber K. Biotransformations in organic chemistry. Berlin, Heidelberg: Springer; 2011.
10. Halling P. Biocatalysis in multi-phase reaction mixtures containing organic liquids. Oxford: Pergamon Journals Ltd; 1987:47–84 pp.
11. Zaks A, Klibanov AM. Enzyme-catalyzed processes in organic solvents. Proc Natl Acad Sci USA 1985;82: 3192–6.
12. Rupley JA, Gratton E, Careri G. Water and globular proteins. Trends Biochem Sci 1983;8:18–22.
13. Zaks A, Klibanov AM. Enzymatic catalysis in organic media at 100 degrees C. Science 1984;224:1249–51.
14. Iyer PV, Ananthanarayan L. Enzyme stability and stabilization—aqueous and non-aqueous environment. Process Biochem 2008;43:1019–32.
15. Mozhaev VV, Melik-nubarov NS, Sergeeva MV, Šikšnis V, Martinek K. Strategy for stabilizing enzymes part one: increasing stability of enzymes via their multi-point interaction with a support. Biocatalysis 1990;3: 179–87.
16. Arcens D, Grau E, Grelier S, Cramail H, Peruch F. Impact of fatty acid structure on CALB-catalyzed esterification of glucose. Eur J Lipid Sci Technol 2020;122:1900294.
17. Zaks A, Klibanov AM. The effect of water on enzyme action in organic media. J Biol Chem 1988;263:8017–21.
18. Zaks A, Klibanov AM. Enzymatic catalysis in nonaqueous solvents. J Biol Chem 1988;263:3194–201.
19. Mozhaev VV, Melik-nubarov NS, Šikšnis V, Martinek K. Strategy for stabilizing enzymes Part Two: increasing enzyme stability by selective chemical modification. Biocatalysis 1990;3:189–96.
20. Castro GR, Knubovets T. Homogeneous biocatalysis in organic solvents and water-organic mixtures. Crit Rev Biotechnol 2003;23:195–231.

21. Hudson JM, Heffron K, Kotlyar V, Sher Y, Maklashina E, Cecchini G, et al. Electron transfer and catalytic control by the iron-sulfur clusters in a respiratory enzyme, E. coli fumarate reductase. J Am Chem Soc 2005; 127:6977–89.

22. Mahato SB, Garai S. Advances in microbial steroid biotransformation. Steroids 1997;62:332–45.

23. Eggers DK, Lim DJ, Blanch HW. Enzymatic production of L-tryptophan in liquid membrane systems. Bioprocess Eng 1988;3:23–30.

24. Rother C, Nidetzky B. Enzyme immobilization by microencapsulation: methods, materials, and technological applications. In: Encyclopedia of industrial biotechnology. Hoboken: John Wiley & Sons, Ltd; 2014:1–21 pp.

25. Francisco M, van den Bruinhorst A, Kroon MC. Low-transition-temperature mixtures (LTTMs): a new generation of designer solvents. Angew Chem Int Ed 2013;52:3074–85.

26. Choi YH, van Spronsen J, Dai Y, Verberne M, Hollmann F, Arends IW, et al. Are natural deep eutectic solvents the missing link in understanding cellular metabolism and physiology? Plant Physiol 2011;156:1701–5.

27. Durand E, Lecomte J, Villeneuve P. From green chemistry to nature: the versatile role of low transition temperature mixtures. Biochimie 2016;120:119–23.

28. Moniruzzaman M, Nakashima K, Kamiya N, Goto M. Recent advances of enzymatic reactions in ionic liquids. Biochem Eng J 2010;48:295–314.

29. van Rantwijk F, Sheldon RA. Biocatalysis in ionic liquids. Chem Rev 2007;107:2757–85.

30. Potdar MK, Kelso GF, Schwarz L, Zhang C, Hearn MTW. Recent developments in chemical synthesis with biocatalysts in ionic liquids. Molecules 2015;20:16788–816.

31. Paiva A, Craveiro R, Aroso I, Martins M, Reis RL, Duarte ARC. Natural deep eutectic solvents – solvents for the 21st century. ACS Sustainable Chem Eng 2014;2:1063–71.

32. Smith EL, Abbott AP, Ryder KS. Deep eutectic solvents (DESs) and their applications. Chem Rev 2014;114: 11060–82.

33. Magnuson DK, Bodley JW, Evans DF. The activity and stability of alkaline phosphatase in solutions of water and the fused salt ethylammonium nitrate. J Solut Chem 1984;13:583–7.

34. Thuy Pham TP, Cho C-W, Yun Y-S. Environmental fate and toxicity of ionic liquids: a review. Water Res 2010; 44:352–72.

35. Durand E. Solvants de type eutectiques profonds: nouveaux milieux réactionnels aux réactions de lipophilisation biocatalysées par les lipases? 2013.

36. Pätzold M, Siebenhaller S, Kara S, Liese A, Syldatk C, Holtmann D. Deep eutectic solvents as efficient solvents in biocatalysis. Trends Biotechnol 2019;37:943–59.

37. Delavault A, Opochenska O, Laneque L, Soergel H, Muhle-Goll C, Ochsenreither K, et al. Lipase-catalyzed production of sorbitol laurate in a "2-in-1" deep eutectic system: factors affecting the synthesis and scalability. Molecules 2021;26:2759.

38. Delavault A, Grüninger J, Kapp D, Hollenbach R, Rudat J, Ochsenreither K, et al. Enzymatic synthesis of alkyl glucosides by β-glucosidases in a 2-in-1 deep eutectic solvent system. Chem Ing Tech 2022;94:417–26.

39. Modla G, Lang P. Removal and recovery of organic solvents from wastewater by distillation. In: Bogle IDL, Fairweather M, editors. Computer aided chemical engineering. Amsterdam: Elsevier; 2012, 30:637–41 pp.

40. Zhao B-Y, Xu P, Yang FX, Wu H, Zong MH, Lou WY. Biocompatible deep eutectic solvents based on choline chloride: characterization and application to the extraction of rutin from *Sophora japonica*. ACS Sustainable Chem Eng 2015;3:2746–55.

41. Cui Y, Li C, Bao M. Deep eutectic solvents (DESs) as powerful and recyclable catalysts and solvents for the synthesis of 3,4-dihydropyrimidin-2(1H)-ones/thiones. Green Process Synth 2019;8:568–76.

42. Liang X, Fu Y, Chang. J. Effective separation, recovery and recycling of deep eutectic solvent after biomass fractionation with membrane-based methodology. Sep Purif Technol 2019;210:409–16.

43. Jeong KM, Lee MS, Nam MW, Zhao J, Jin Y, Lee DK, et al. Tailoring and recycling of deep eutectic solvents as sustainable and efficient extraction media. J Chromatogr A 2015;1424:10–17.

44. Fouladi P, Salamat R, Ahmadzade A, Aghdam Shahryar H, Noshadi A. Effect of choline chloride supplement on the internal organs and carcass weight of broilers chickens. J Anim Vet Adv 2008;7:1164–7.
45. Sclapari T, Bramati V. New uses of choline chloride in agrochemical formulations; 2012.
46. Wehner D, Prenzel T, Betten T, Briem AK, Hong SH, Ilg R. The Sustainability Data Science Life Cycle for automating multi-purpose LCA workflows for the analysis of large product portfolios. E3S Web Conf 2022; 349:11003.
47. Domínguez de María P. Biocatalysis, sustainability, and industrial applications: show me the metrics. Curr. Opin. Green Sustainable Chem. 2021;31:100514.
48. Lima-Ramos J, Tufvesson P, Woodley JM. Application of environmental and economic metrics to guide the development of biocatalytic processes. Green Process Synth 2014;3:195–213.
49. Sheldon RA. Metrics of green chemistry and sustainability: past, present, and future. ACS Sustainable Chem Eng 2018;6:32–48.
50. de María PD. On the need for gate-to-gate environmental metrics in biocatalysis: fatty acid hydration catalyzed by oleate hydratases as a case study. Green Chem 2022;24:9620–8.
51. Meyer L-E, Hobisch M, Kara S. Process intensification in continuous flow biocatalysis by up and downstream processing strategies. Curr Opin Biotechnol 2022;78:102835.
52. Benítez-Mateos AI, Contente ML, Padrosa DR, Paradisi F. Flow biocatalysis 101: design, development and applications. React Chem Eng 2021;6:599–611.
53. Santi M, Sancineto L, Nascimento V, Braun Azeredo J, Orozco EVM, Andrade LH, et al. Flow biocatalysis: a challenging alternative for the synthesis of APIs and natural compounds. Int J Mol Sci 2021;22:990.
54. Cosgrove SC, Mattey AP. Reaching new biocatalytic reactivity using continuous flow reactors. Chem Eur J 2022;28:e202103607.
55. Guajardo N, Schrebler RA, Domínguez de María P. From batch to fed-batch and to continuous packed-bed reactors: lipase-catalyzed esterifications in low viscous deep-eutectic-solvents with buffer as cosolvent. Bioresour Technol 2019;273:320–5.
56. Guajardo N, Ahumada K, Domínguez de María P. Immobilization of Pseudomonas stutzeri lipase through cross-linking aggregates (CLEA) for reactions in deep eutectic solvents. J Biotechnol 2021;337:18–23.
57. Guajardo N, Ahumada K, Domínguez de María P. Immobilized lipase-CLEA aggregates encapsulated in lentikats® as robust biocatalysts for continuous processes in deep eutectic solvents. J Biotechnol 2020;310: 97–102.
58. Sheldon RA. Biocatalysis and green chemistry. In: Green biocatalysis. Hoboken: John Wiley & Sons, Ltd; 2016:1–15 pp.
59. Woodley JM. Benchmarking the sustainability of biocatalytic processes. In: Handbook of green chemistry. Hoboken: John Wiley & Sons, Ltd; 2018:207–30 pp.
60. Hobbs HR, Thomas NR. Biocatalysis in supercritical fluids, in fluorous solvents, and under solvent-free conditions. Chem Rev 2007;107:2786–820.
61. Mesiano AJ, Beckman EJ, Russell AJ. Supercritical biocatalysis. Chem Rev 1999;99:623–34.
62. Matsuda T. Recent progress in biocatalysis using supercritical carbon dioxide. J Biosci Bioeng 2013;115: 233–41.
63. Hollenbach R, Delavault A, Gebhardt L, Ochsenreither K, Syldatk C. Lipase-mediated mechanoenzymatic synthesis of sugar esters in dissolved unconventional and neat reaction systems. ACS Sustainable Chem Eng 2022;10:10192–202.
64. Kaabel S, Friščić T, Auclair K. Mechanoenzymatic transformations in the absence of bulk water: a more natural way of using enzymes. Chembiochem 2020;21:742–58.

Johannes Möller*, Kim B. Kuchemüller and Ralf Pörtner

5 Bioprocess intensification with model-assisted DoE-strategies for the production of biopharmaceuticals

Abstract: The demand for highly effective biopharmaceuticals and the need to reduce manufacturing costs are increasing the pressure to develop productive and efficient bioprocesses. For this purpose, model-based process design concepts have been developed. Although first approaches were proposed, model-based process designs are still not state-of-the-art for cell culture processes during development or manufacturing. This highlights a need for improved methods and tools for optimal experimental design, optimal and robust process design and process optimization for the purposes of monitoring and control during manufacturing. In this review, an overview of the state of the art of model-based methods, their applications, further challenges, possible solutions and specific case studies for intensification of process development for production of biopharmaceuticals is presented. As a special focus, problems related to data generation (culture systems, process mode, specifically designed experiments) will be addressed.

Keywords: cell culture; intensification; model-assisted DoE; process design and optimization; Quality by Design

5.1 Introduction

The demand for highly effective biopharmaceuticals and the need to reduce manufacturing costs are increasing the pressure to develop productive and efficient bioprocesses [1–3]. As a consequence, the need for more complex and intensified bioprocesses rise and novel methods for process development have been introduced, e.g., changing an upstream manufacturing process from simple *batch* to more complex *fed-batch* or perfusion processes. Biopharmaceutical and pharmaceutical manufacturing are strongly influenced by the process analytical technology initiative (PAT) and quality-by-design (QbD) methodologies, which have been designed to enhance the understanding of more integrated processes [4, 5]. The major aim of this effort can be summarized as

Johannes Möller and Kim B. Kuchemüller have equally contributed to the manuscript (shared first authorship).

*Corresponding author: Johannes Möller, Hamburg University of Technology, Institute of Bioprocess and Biosystems Engineering, Denickestr. 15, D-21073 Hamburg, Germany, E-mail: johannes.moeller@tuhh.de
Kim B. Kuchemüller and Ralf Pörtner, Hamburg University of Technology, Institute of Bioprocess and Biosystems Engineering, Denickestr. 15, D-21073 Hamburg, Germany

As per De Gruyter's policy this article has previously been published in the journal Physical Sciences Reviews. Please cite as: J. Möller, K. B. Kuchemüller and R. Pörtner "Bioprocess intensification with model-assisted DoE-strategies for the production of biopharmaceuticals" *Physical Sciences Reviews* [Online] 2023. DOI: 10.1515/psr-2022-0105 | https://doi.org/10.1515/9783110760330-005

developing an improved understanding of the entire manufacturing process, including the development of technologies to perform online measurements and real-time control and optimization [6]. Additionally, the integration of the process knowledge into mathematical process models is emphasized. As detailed by Kroll et al. [7], model-based methods are increasingly used in all areas of biopharmaceutical process technology, e.g., for experimental design, process characterization, process design, monitoring and control. Benefits of these methods are lower experimental effort, process transparency, clear rationality beyond decisions and increased process robustness. Furthermore, model-based methods can help to implement regulatory requirements as suggested by recent QbD and qualification and validation initiatives.

Currently, the basis of bioprocess characterization and optimization are mostly experience-guided or based on statistical methods such as design of experiment (DoE) methods that require a large number of time-consuming experiments [8]. Cultivation systems are now available for this purpose, which allow for a large number of parallel experiments under controlled conditions, mostly based on single-use technology (Figure 5.1) [9]. To some extend theses culture systems are designed as scale-down models for large-scale reactor systems [10–13].

Linking mathematical models with experimental methods have been implemented along the process development cycle (Figure 5.1) to significantly reduce development time and costs and to finally establish a Digital Twin, a virtual representations of the whole manufacturing process. Furthermore, if combined with advanced PAT, higher automation and more efficient workflows can be established during routine manufacturing [4, 14, 15].

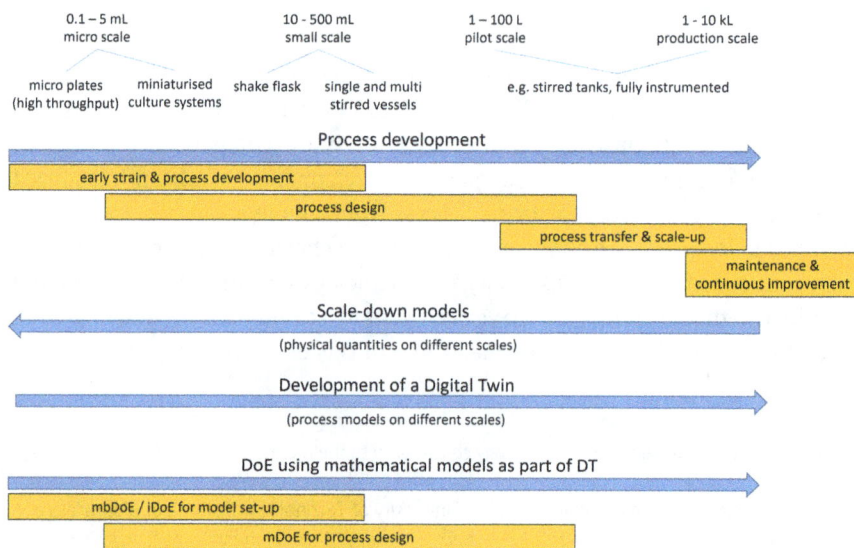

Figure 5.1: Model-assisted tools for process development (DT, Digital Twin; mbDoE, model-based Design of Experiment; iDoE, intensified Design of Experiment; mDoE, model-assisted Design of Experiment).

Such methods are still not state-of-the-art for cell culture processes during development or manufacturing, although first approaches have been proposed. This highlights a need for improved methods and tools for optimal experimental design, enhanced and robust process design, and process optimization for the purposes of monitoring and control during manufacturing. In this review, an overview of model-based methods, their applications, further challenges, possible solutions and specific case studies for intensification of process development is presented for the upstream manufacturing of biopharmaceuticals. The challenges in data generation (culture systems, process mode, specifically designed experiments) will be addressed additionally.

5.2 Model assisted process development

5.2.1 Mathematical models for model-assisted process design

In the following, mathematical process models as the main component in model assisted process design will be introduced (based on [16]). The type and the degree of complexity of mathematical process models are key for their function and usability [17]. The model should describe the real phenomena as accurately as needed with a clear focus on simplicity and adaptability. Thereby, the model evolves over time, e.g., starting with a simple structure and well-known effects and adapting stepwise to newly available data and observations [18, 19]. In general, different models can describe the same biological process with a comparable quality [8].

In the following, different model classes are introduced and their potential application in DTs is highlighted in a general manner. An overview on most prominent model classes is given in Figure 5.2. Problems related to data generation used for model set-up and validation as well as with the parameter identification will be discussed later.

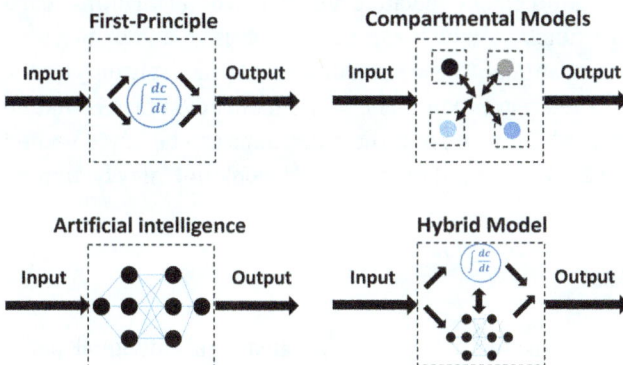

Figure 5.2: Overview about established model classes.

5.2.1.1 First-principle or mechanistic models

These models are based on fundamental physical/chemical laws and a proven understanding of the relationships and interactions [20, 21]. Therefore, they are based on well-known material or energy balances and derived from this proposition, i.e. first principles. These balances can then be mathematically transferred and adapted to the modelled system (e.g., mass and energy balances) and different other material characteristics can be added. In the field of biopharmaceutical manufacturing processes, first principle-derived models are used to describe the biological functions of the grown organisms. Hence, the complexity of biological systems is very large with thousands of reactions involved, which are not understood in every detail and first principle/ mechanistic models can then be used to describe known growth characteristics (e.g., cell growth based on consumed glucose). Most published models consider the "bio"-phase using heuristically derived linear/non-linear mathematical expressions, which are distinguished according to the considered complexity of the model (engineering approach, reviewed in [22]). Novel efforts were made to explain segregated biological classes, such as different cell cycle phases, their interaction, and impact on cell growth and cell culture quality in systems biology [7, 23].

5.2.1.2 Artificial intelligence

Artificial intelligence (AI) is broadly seen as theoretical and practical computer systems, which perform with human-like intelligent behaviour without human intervention [24]. Mathematical algorithms are used for data analysis and computational learning, known as machine learning. Common classes of machine learning are supervised, unsupervised and reinforcement learning [25]. Supervised learning use previous examples and a predefined output value/function to train the computer system. Unsupervised machine learning targets the identification of hidden structures/clusters in data without a reward function commonly used in cluster analysis and pattern recognition applications. In reinforcement machine learning, the algorithms learn based on a cumulative reward function, which can include positive and/or negative feedback, like training tricks to an animal. These modelling techniques rely only on the used data and are sometimes called regression models. The main advantage is that no prior knowledge is needed for modelling and that the algorithms can be implemented efficiently and fast. But, the capacities of an AI model for extrapolation may be limited due to the missing causality.

5.2.1.3 Compartmental models

In compartment models, the investigated system is separated into individual parts, i.e., compartments, with defined sub-models, their shape, and their interactions. The most prominent example is computational fluid dynamic (CFD) simulations predicting fluid flow behaviour in 3D shapes and constructions. Using CFD, the following phenomena could be investigated and used to design and understand cell culture processes [26–28]:

- Heterogeneous distribution of components, such as nutrients and oxygen.
- Gradients in the applied system, e.g., pH gradients.
- Whole system hydrodynamic characteristics, such as average fluid flow.
- Local-distributed hydrodynamic characteristics, e.g., shear stress.
- Stationary and time-resolved fluid flow characteristics.
- Optimization of bioreactor geometry for scale-down and scale-up.

Disadvantages are the high computational power required to solve complex geometries and the strong dependency of the quality of prediction on the mesh and boundary conditions. Exemplary, Freiberger et al. [26] investigated the link between hydrodynamics obtained using CFD and biological process behaviour of mammalian cell culture with different impeller set-up.

5.2.1.4 Hybrid approaches

In hybrid modelling different model sub-classes (e.g., mentioned above) are summarized into one combined model. Mostly, the structured understanding obtained in first-principle or mechanistic models are the backbone of a hybrid model describing the main investigated metabolic regulations [29]. Effects, which cannot be expressed by the mechanistic equations or which are not understood so far are then incorporated by data-driven machine learning tools or artificial intelligence [30]. An improved usage of the available knowledge and data is enabled if the empirical process understanding is coupled with AI models.

5.2.2 Strategies for model-assisted process development

In industrial applications, an optimization of several conflicting objectives at a time is typically targeted, leading to compromises. Multi-objective optimization is challenging, and its application is still not state-of-the-art in the context of biopharmaceutical manufacturing processes. Furthermore, minimization of the number of empirical experiments and the model-assisted exploration of the process design space are targeted. Even if tremendous progress has been achieved so far, there is still work to be carried out in order to realize the full potential of the process systems engineering toolbox [6].

Model-assisted design of experiments (mDoE) is the combination of statistical DoE with mathematical process models to enable a knowledge-driven bioprocess development in the context of QbD. Using this method, the obvious limitations of pure statistical DoE methods (e.g. scattering a design space with experiments only) can be avoided. The design as well as the optimization of bioprocesses can be improved, as it accelerates the design space exploration and thereby the time needed to identify the optimum combination of critical process parameters (CPP) for the variables of interest decreases. This concept is further discussed in Section 5.4.

Bayer et al. [31] suggested an intensified DoE (iDoE) coupled with hybrid modelling to generate process knowledge and simultaneously accelerate process characterization. To reduce the experimental workload, CPPs are changed during the cultivation to address for the dynamic changes in the process. A time-resolved hybrid model can be built on iDoE data to describe the occurring process dynamics, because it captures the whole process.

Hernández Rodríguez et al. [32–34] developed a coupled workflow with uncertainty-based upstream simulation and Bayes optimization using Gaussian processes. Within the workflow, multiple optimization goals and boundaries were considered. The power of the method was shown based on the optimization of an industrial seed train and the total process variability and duration was significantly reduced. Furthermore, they illustrated the application of Bayesian parameter estimation and Bayesian updating for seed train prediction to an industrial Chinese hamster ovarian (CHO) cell culture process, coupled with a mechanistic model. It was shown that through integration of new data by the Bayesian updating method, process variability (i.e., batch-to-batch) could be considered.

During process design and optimisation, cultivation conditions on different scales have a significant impact on cell performance. This can be facilitated by model-assisted prediction methods, whereby the performance depends on the prediction accuracy, which can be improved by inclusion of prior process knowledge, especially when only few high-quality data is available, and description of inference uncertainty, providing, apart from a "best fit"-prediction, information about the probable deviation in form of a prediction interval.

5.3 Critical aspects of data generation for setting up mechanistic mathematical models

The following discussion is dedicated in the first place to mechanistic models (see Section 5.2) as these are often applied in model-based strategies such as mDoE, however, could certainly also be led in a similar way for other modelling approaches. As explained above, mechanistic models rely on proven interactions, which can be expressed as cause-effect relationships. For the cultivation of biological organisms, common examples are the dependency of cell specific growth rate on substrate or metabolite concentration, of changing metabolic activity of the organisms due to temperature deviations etc. [7]. The performance of cells, e.g., metabolic values such as cell growth, cell specific uptake of nutrients or production of metabolites or the product of interest depends to a large extent on culture conditions, e.g. culture systems and process strategies (e.g. *batch/fed batch* or continuous) as well as on the investigated scale [35, 36]. The experimental design has to be selected accordingly to reflect such influencing factors. In the authors opinion these questions have hardly been addressed so far, as outlined in the following. First, problems related to the determination of metabolic values (e.g., growth rates, cell specific uptake and production rates) from raw data on cell density (viable and total), concentration of substrates and metabolites are discussed. A literature review evaluates culture systems for data generation and the impact of culture conditions and scale on kinetic model parameters, respectively. Finally advanced experimental strategies for data generation and evaluation with respect to model-set up will be discussed. A deeper discussion of "data mining"-strategies can be found in the relevant literature [7, 37, 38].

5.3.1 Problems related to the determination of metabolic values

Typically, a mechanistic kinetic model consists of differential algebraic equations that can be derived from the time-dependent mass balances of participating components with appropriate mechanistic or empirical kinetic equations [18, 39], a specific example is given in [40]. To describe the changing metabolism, mathematical equations are derived including specific numerical values, known as model parameters. The important variables to be defined in these equations are the specific rates for e.g. cell growth, cell death, and substrate consumption/metabolite production that dynamically change during the culture process [23, 41, 42]. To express these effects within the model, the underlying cause-effect-relationships have to be expressed in kinetic equations and the numerical values of the kinetic model parameters have to be identified. As only a small proportion of these model parameters are experimentally accessible, they have to be determined based on experimental data. Thus, it is required to have accurate and broad time-resolved measurements of respective variables (e.g. cell density (viable and total), concentration of substrates and metabolites) [8].

5.3.2 Impact of culture conditions and scale on model parameters

As was shown by Kyriakopoulos et al. [43], model parameters are system and process dependent and must be validated accordingly. They reviewed kinetic modelling of mammalian cell culture bioprocessing, and "found that the simple unstructured-unsegregated approach utilizing empirical Monod-type kinetics based on limiting substrates and inhibitory metabolites is used quite often due to its traceability and simple formalism". Using the cited references for this type of kinetic models (see [43], SUPPORTING INFORMATION, Table 5.S1. Unstructured-Unsegregated bioprocess kinetic models), we analysed the culture conditions applied for model set-up (total: 30 references, including those without detailed information or those using literature data for model set-up). The following culture systems were used (number of studies in brackets): flasks (3), spinner (5), shake flasks (3), bench top stirred tank reactors (culture volume <1–5 L)

(9), pilot scale stirred tank reactors (culture volume 10–20 L) (4), production scale stirred tank reactors (culture volume 500–5000 L) (1), culture systems compared on different scales (4). The following strategies were used as process modes: *batch* (9), *batch* or/and *fed-batch* (13), continuous (4), multiple (*batch/fed-batch* and continuous) (2). Based on these data and a deeper evaluation of the respective references, the following conclusions can be drawn: The focus of most studies was on model set-up, less on specific cause-effect-relations ships or problems related to parameter estimation. Mostly experiments have not especially been designed for evaluation of specific cause-effect-relationships (exception: [44–46]). The impact of culture conditions was considered only to some extent. Especially specific details for process parameters used for characterisation of bioreactors such as power-to-volume-ratio (P/V), $k_L a$, mixing time etc. were not available.

Of particular interest in this context are publications that have considered different cultivation systems on different scales and different process modes. Shirsat et al. [18] analysed *batch* and *fed-batch* cultivations on different scales with culture volumes between 50 mL and 15 L. The identified kinetic parameters varied between scales and process modes. Craven et al. [47] used data from *batch, fed-batch* and continuous cultivations on different scales (3 L and 15 L) for set-up of models based on a Monod kinetic structure and parameter estimation. The experimentally determined parameters had the greatest influence on model performance. They changed with scale and mode of operation. The remaining parameters, which were estimated using a differential evolutionary algorithm, were not as crucial. Further studies confirmed the impact of scale and operation mode on model parameters as well [48, 49]. Arndt et al. [35] and Möller et al. [36] introduced a model-based workflow to quantify differences in the process dynamics between bioreactor scales and thus to enable a more knowledge-driven scale-up. This workflow was tested on different CHO cell lines and applications from laboratory to production scale. In contrast, Pörtner et al. [41, 50] found that irrespective of the cultivation mode, i.e., *batch, fed-batch* or chemostat, the growth of a hybridoma cell line followed the same kinetics and, therefore, it was possible that data gained from *batch* experiments could be extended to *fed-batch* and continuous cultures.

In summary, most studies confirmed an impact of culture conditions and scales on kinetic and model parameters. To some extent this reflects the different performance of the cells, which might depend on culture conditions (medium, pH, DO, DCO_2 – e.g., in flasks and spinner cultures these parameters cannot be controlled), or differences in the hydrodynamics between scales and operation modes. It is still not well established to characterize the cultivation system with respect to engineering parameters such as P/V, $k_L a$ or mixing time. The relevance of a detailed bioreactor characterisation for model parameter estimation on different scales was demonstrated by [26, 36]. Furthermore, little attention has been paid to the fact whether the experiments are suitable for determining the model parameters used within the model. E.g., for meaningful identification of Monod-constants within limiting model terms the limiting substrate has to be identified. This can be either glucose or glutamine or other substrates such as amino acids. Similarly, relevant inhibiting effects from metabolic waste products or by-products must be present in the experiments, which is hardly the case for typical *batch* cultures. Some suggestions with respect to proper selection of a culture system will be discussed in the following section.

5.3.3 Advanced experimental strategies for data generation and evaluation used for model-set up

An overview on culture systems generally suitable in this respect is given in Table 5.1. The above discussion has shown that mostly conventional cultivation systems such as flasks, spinner, shake flasks or stand-alone stirred reactors have been used so far for data generation for model set-up. Surprisingly, multiple culture systems, which are state-of-the-art in process development [51, 52] have hardly been applied in this context. Over the last decade, several attempts have been made to develop miniaturized high throughput culture systems (for review see [53]). All these systems are aimed at an increased speed for cell culture process development mimicking the performance in larger scale. Most of these systems are designed to perform cultures under well controlled conditions, to some extend new sensor techniques can be integrated for continuous monitoring of quality attributes that could previously only be estimated offline [54]. As stated above, culture systems/bioreactors used for kinetic studies and model-set up should be characterised extensively, especially if data on different scales is to be compared. Recommendations for process engineering characterisation are given in [55]. An example for a successful model-based scale-up up to pilot scale in well characterised single-use reactors has been given by [36]. As high throughput and miniaturized culture systems allow for a larger number of multiple experiments to be performed in parallel, they call for efficient experimental design, data handling and data analysis [13].

5.4 Case study: application of model-assisted DoE within a knowledge-driven workflow

The mDoE as part of a knowledge-driven workflow for process design (Figure 5.3) serves as a tool for the qualitative design, assessment and evaluation of processes. Within the workflow, a process-related target (i.e., product titer) is optimized, and the model supports in the recommendation and evaluation of DoE designs. In the mDoE, experimental designs are created and evaluated using the mathematical model. The task can be on the optimization of process parameters, such as medium components, or the comparison of different test plans or boundary conditions (e.g., quality functionals, influence of prior knowledge). Depending on the respective task and/or the existing

Table 5.1: Properties of culture systems appropriate for data generation for model set-up, relationship between process capabilities (control, modes, information output) and number of parallel experiments in culture systems on different scale (adapted from [49]).

Type of system	Multiplexing per unit (depending on configuration)	Common scales	Agitation	Integrated sensors for DO and pH	Control of DO and pH	Process mode (*batch*, *fed-batch*, continuous/perfusion)	Multiple sampling
Microplates	Depending on plate (e.g. 96 wells per plate)	0.1–5 mL	Orbital shaker	No	No	*batch*	No
Miniaturized culture systems	Up to 48	10–15 mL	Orbital shaker, stirrer	Partly	Partly	*batch, fed-batch*, partly perfusion	No
Static flasks	1	Up to 75 mL	No	No	No	*batch*	No
Spinner reactors	Up to 16	Up to 500 mL	Stirrer	No	No	*batch, fed-batch* (bolus feeding)	Yes
Shake flasks	Up to 16	Up to 500 mL	Shaker	Partly	Partly	*batch, fed-batch* (bolus feeding)	Yes
Multi-reactors on lab scale	Up to 24	Up to 250 mL	Stirrer	Yes	Yes	*batch, fed-batch*, perfusion	Yes
Stand alone bioreactors (stirred, wave-type)	1	1–100 L	Stirrer/rocker	Yes	Yes	*batch, fed-batch*, perfusion	Yes

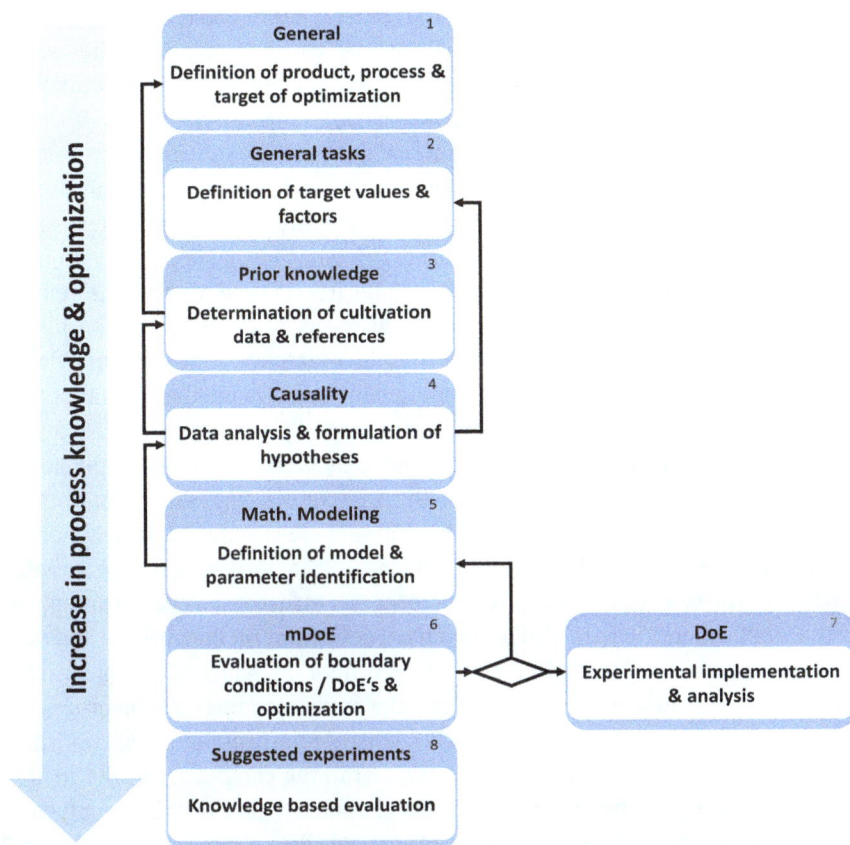

Figure 5.3: Structure of the knowledge-driven workflow (adapted from [56]).

knowledge, individual steps of the workflow can be left out. Findings/assumptions can also be added iteratively since new knowledge is generated with the implementation of the workflow. It should be emphasized that the comparatively intensive evaluation of the process data using the mDoE methods, partly in iterative processes, promotes understanding of the process significantly. Compared to DoE, the mathematical modelling ensures an in-depth evaluation of the process understanding and the process dynamics. This leads to knowledge-driven decisions and thus to process optimization. In the following the individual steps of the mDoE-workflow will be explained in detail and examples will be given.

At the beginning of the mDoE-workflow, the specific task, mainly the target(s) (e.g., max. titer, QA's) have to be defined and the general cultivation and process conditions are formulated (Figure 5.3, Box 5.1). It must be defined which product (e.g. antibodies, number of cells) is formed, which process (e.g., strain, medium, process control) is to be examined and which optimization goal (e.g., productivity, process time, statistical boundary conditions) is desired. The general tasks must then be defined (Figure 5.3, Box 5.2), whereby it must be precisely defined which factors and target values are to be examined. When defining the questions, it should be noted that in addition to process optimization, boundary conditions or various DoE's can also be examined. Due to the increased application and rapid evaluation using the mDoE, an increase in process understanding can also be defined as a possible goal. In the following data analysis step the available data and prior knowledge is used to identify the relevant cause-effect relationships for model set-up. After disclosure of prior knowledge (e.g. based on literature and/or expert knowledge and/or cultivation data), it is checked whether the amount of data is suitable for answering the questions in terms of plausibility and quality (Figure 5.3, Box 5.3). If necessary, the questions must be adapted or additional knowledge generated. Although prior knowledge has to be available at this point, the amount of data to be generated is still small compared to the experimental implementation of one and sometimes several test plans. The advantage is that only data that increases the added value/level of knowledge is generated. Based on the prior knowledge, assumptions for the description of the cell behaviour are formulated iteratively (Figure 5.3, Box 5.4) and a mathematical model (Figure 5.3, Box 5.5) is developed. It is important to select a suitable model structure. The use of simple model structures has turned out to be advantageous for the mDoE application. This can also be expanded depending on the purpose and over time. The special benefit of the mDoE can be seen here, as knowledge is generated iteratively. With each new run of the mDoE, the understanding of the process is increased. In general, however, it must be checked whether the cause-effect relationships are suitable for answering the specific task. In general, the complexity of the mathematical model is a key parameter for the successful application of mDoE. During early stages of development with only a low number of available data, the mathematical model may include mostly rather simple equations. These are based on the formulation of mathematical links (i.e., cause-effect-relationships) between cell growth, metabolism, and corresponding product formation. The selected data should be used to cover typical known effects, e.g., inhibitions or limitations. Certainly, the number of experiments that can be performed at this stage, preferably in small scale is usually limited and the number of incoming data sets has a significant impact on the quality of the fit. Therefore, an in-depth evaluation of the influence of prior knowledge on parameter identification is suggested, as discussed in Section 5.3 and shown in Figure 5.4 for a different amount of input data on the goodness of fit criteria R^2 [56]. In the case shown here, the complexity of the mathematical process model increases from *batch* to *fed batch*. In the mathematical model for the *fed batch*, for example, an inhibition by ammonia might be taken into account, so that this effect is only included in the parameter adjustment after an increased number of data sets [8]. With increasing complexity of the mathematical process model, a strong focus must be placed on the quality of the data. With high quality of the data sets, however, the amount of data can be kept low. This is where the advantage of the mDoE methods comes into play, since the iterative, model-based evaluation only generates data with a high level of knowledge gain.

A : *batch* process   **B : *fed-batch* process**

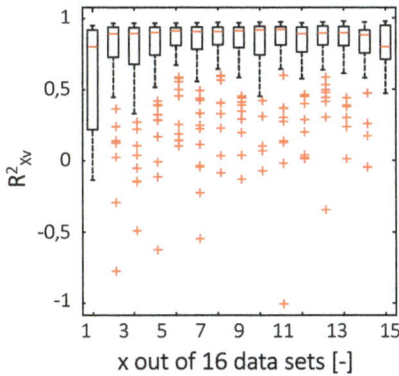

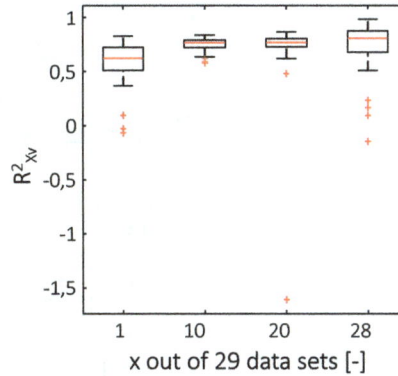

Figure 5.4: Evaluation of the influence of prior knowledge on parameter identification (Quality functional G3 as shown in Figure 5.5, n = 100 runs, median as well as the first and third quartile, minimum and maximum were determined, outliers are marked by +.) A: *batch* process, B: *fed-batch* process (adapted from [56]).

To evaluate the quality of the mathematical model, statistical criteria (e.g., coefficient of determination) are used for assessment. "Weighted non-linear least square" methods are widely used estimation methods for static optimization problems (for review see [56]). The aim is to minimize the squared deviation between n measured $y_{k,i}$ and simulated data points $_{k,i}$ for l variables multiplied by a weighting factor $w_{k,i}$ for k = 1, ..., l and i = 1, ..., n for a fixed time period t = $(t_1, ..., t_n)$. The weighting factor allows quantities to be given more importance, for example because some readings or data points are more reliable than others. At the same time, measurement deviations can be taken into account or the comparability of quantities of different dimensions can be guaranteed through standardization, with the difference here being in the following quality functionals. When used in mDoE methods, the merit functionals are used to determine the specific model parameters.

Using the example of data from CHO cultivations, Figure 5.5 shows that a critical evaluation of various quality functionals is urgently required. In this case 4 different quality functionals (G1 – G4) have been tested. All examined quality functionals (Figure 5.5A) reach a stationary value after a certain number of iterations, whereby the number of iterations is taken into account as a decisive factor when choosing a quality functional. However, simply considering the progression of the quality functionals as a function of the number of iterations cannot be sufficient, since there is no guarantee that the adjustment will be implemented with a high level of quality. Therefore, the coefficient of determination for X_v (R^2_{Xv}) and an overall coefficient of determination of all data series (R^2_{all}) are shown in Figure 5.5B as a function of the number of iterations. It can be seen that just the quality functional G3 performed satisfactorily in both cases. Therefore, we selected quality functional G3 as appropriate in this case.

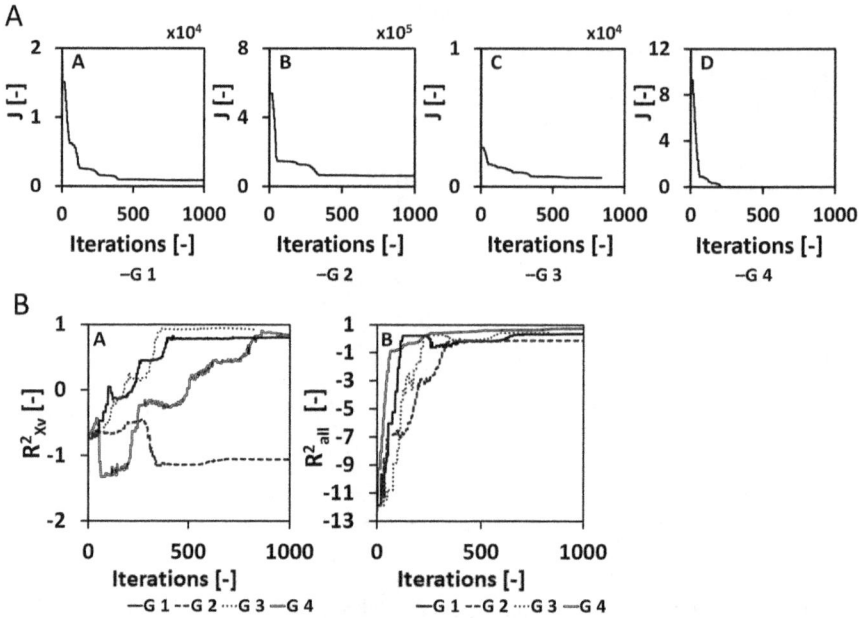

A

B

G1: $\quad J = \sum\limits_{k=1}^{l} \sum\limits_{i=1}^{n} w_{k,i} \dfrac{(y_{k,i} - \hat{y}_{k,i})^2}{y_{k,i,max}^2}$

G2: $\quad J = \sum\limits_{k=1}^{l} \sum\limits_{i=1}^{n} w_{k,i} \dfrac{(y_{k,i} - \hat{y}_{k,i})^2}{0,05 \cdot y_{k,i,max}^2}$

G3: $\quad J = \sum\limits_{k=1}^{l} \sum\limits_{i=1}^{n} w_{k,i} \sqrt{\dfrac{1}{n} \sum\limits_{i=1}^{n} (y_{k,i} - \hat{y}_{k,i})^2} \cdot \dfrac{1}{y_{k,i,max} - y_{k,i,min}}$

G4: $\quad J = -\dfrac{1}{l} \sum\limits_{k=1}^{l} R^2$

$R^2 = 1 - \dfrac{\sum_{i=1}^{n}(y_i - \hat{y}_i)^2}{\sum_{i=1}^{n}(y_i - \bar{y}_i)^2}$

Figure 5.5: Comparison of different quality functionals (G1 – G4) (adapted from [56]). (A) Function values of the four quality functionals, represented by the number of iterations. (B) Comparison of the calculated coefficients of determination for X_v and the overall coefficient of determination of all data series depending on the number of iterations of the various quality functionals.

Afterwards, a statistical DoE design is chosen (Figure 5.3, Box 5.6). The choice of an experimental design significantly influences the implementation of DoE and mDoE. In the author's understanding, the appropriate and rational selection of an experimental design is often neglected. DoEs are mostly selected based on heuristics within a given scientific field, and there is no guided decision-making workflow yet. In order to enable the selection of a suitable experimental design for the respective objective, different criteria have to be considered. The selection should be based on basic settings (number of factors, number of factor levels, the regression model and the number of test runs) as well as design-specific properties (block formation, orthogonality and rotatability). A scheme to select a design has been discussed by Kuchemüller et al. [22], an example is given in Figure 5.6.

The performance of a mDoE (Figure 5.3, Box 5.7 and 5.8) has been shown successfully for several applications [35, 36, 40, 57, 58] Figure 5.7.

5.5 Application potential of mDoE methods

mDoE methods can be seen as novel tools to drastically reduce development time and costs. With the mDoE, optimal process designs are detected with high significance, which lead to knowledge-based experimental designs. In general, two approaches are possible: (1) If there is little data, the experimental design space is initially set-up as broadly as possible in order to verify hypotheses. (2) If enough data is available, the design space is

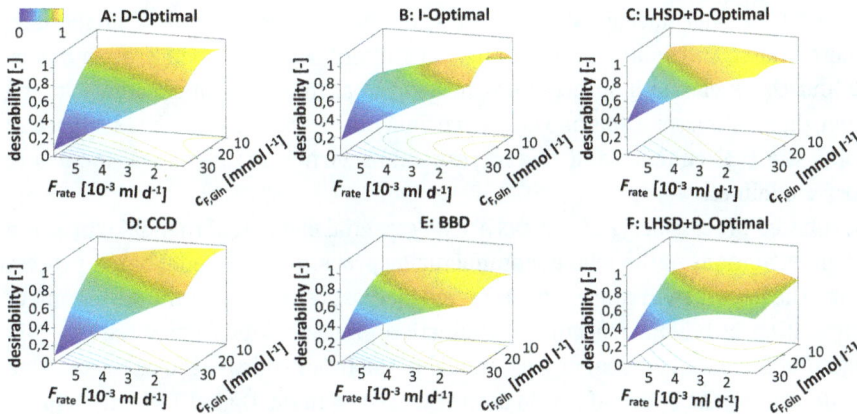

Figure 5.6: Comparison of D-optimal, I-optimal, LHSD + D-optimal design, CCD and BBD in a CHO-fed-*batch* process (30 factor combinations at maximum desired antibody concentration). The start time and the D-glucose concentration of the inflow were also varied, but fixed here as an example. The start time of the inflow is set at 96 h and the D-glucose concentration in the inflow at 222 mmol l^{-1}. The desirability is calculated depending on the maximum cell number and antibody concentration as well as the minimum ammonium concentration. (adapted from [56]).

reduced stepwise. This increases the knowledge gained and reveals gaps in knowledge. It predicts properties that could endanger the safety of the end product and dynamically makes changes to mitigate these risks [16].

However, mDoE can only be applied if there is an understanding of the mechanistic relationships and this is regarded as meaningful decision-making for process development and optimization using DoE in QbD. Overall, a dynamic analysis of the data sets provides a deepened understanding compared to a pure endpoint analysis [58]. Effects are detected more appropriately. The mDoE concept enables knowledge about the investigated experimental system to be summarized in a computer-aided system. At the same time, physical laws are linked to metabolic understanding. This offers the opportunity to design and evaluate cultural systems and to implement this on different scales. In addition, there are other advantages of the mDoE [16]:

- Constant increase in the level of knowledge about the process.
- Increasing understanding of the process and its influence on cell growth, phenotyping, epigenetic criteria, prognostic markers etc. in many process steps.
- Reduced development costs for experimental designs to define rapid and efficient cell expansion, accelerating time to clinical.
- Evaluation, screening and virtual testing of new configurations/settings before the experiments.

Nevertheless, conventional DoE can still be used for initial screening studies and can also lead to process optimization in several rounds. However, there are still some challenges to be overcome. Especially in the manufacture of advanced therapy medicinal products (ATMPs), case-specific, personalized processing may be required compared to the production of biopharmaceuticals [16]. In addition to these technical challenges, researchers and scientists are needed who have interdisciplinary and cross-industry training and cover classic bioprocess engineering, real industrial applications in cell cultivation, digitization solutions and advanced analytics. Further research is required to closely link the real cultures and the corresponding virtual mathematical models to regulatory guidelines.

Model-based process design concepts are an essential element of Digital Twin's, as a virtual representations of the whole manufacturing process. Both, model-based process designs and the Digital Twin concept have gained increasing interest for the development and optimization of biopharmaceutical production processes (for review see [14, 15]). With regard to the mathematical process model in the mDoE, this is seen as the starting point for describing the entire life cycle of the bioprocess in the Digital Twin. It includes the understanding of the process, whereby the degree of model complexity can be gradually increased in the course of the studies carried out.

Even if the potential of model-based tools has widely been shown, the implementation in industrial practice has to be improved. As stated by Kroll et al. [7]

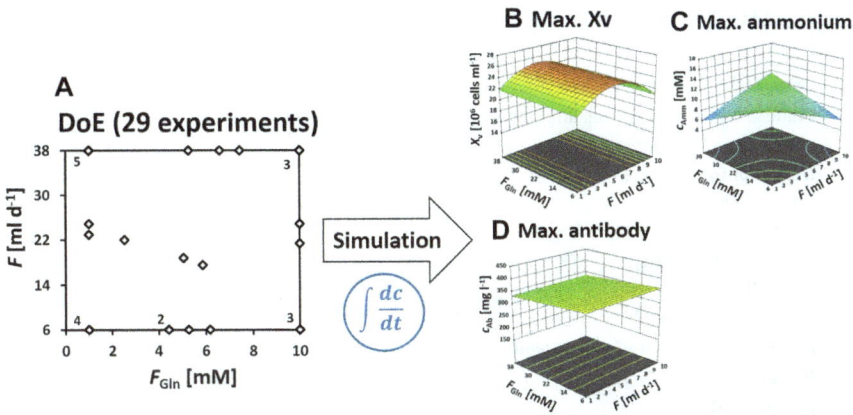

Figure 5.7: Application of the mDoE approach for the optimization of a *fed-batch* process, (A) initially planned experiments in a DoE (*n* = 29 experiments), Response surfaces are based on the model-assisted simulations of each experiment and the following response surfaces were predicted: (B) maximal viable cell density, (D) maximal ammonium concentration, (D) maximal antibody concentration. See [40] for further details. *F*, feeding rate; F_{Gln}, glutamine concentration in feed; F_{Glc}, glucose concentration in feed.

"today, despite these advantages, the potential of model-based methods is still not fully exhausted in bioprocess technology. This is due to a lack of (i) acceptance of the users, (ii) user-friendly tools provided by existing methods, (iii) implementation in existing process control systems and (iv) clear workflows to set up specific process models." To further boost the application of model-based tools in industry, a realistic discussion of pro's and con's is needed.

Symbols used

DCO_2	dissolved carbon dioxide concentration [mmol L^{-1}]
DO	dissolved oxygen concentration [mmol L^{-1}]
F	feed rate [ml d^{-1}]
F_{Gln}	feed rate glutamine [ml d^{-1}]
J	quality functional [–]
k_La	volumetric mass transfer coefficient [s^{-1}]
P/V	power-to-volume-ratio [–]
R^2	regression coefficient [–]
X_v	viable cell density [cells mL^{-1}]
$Y_{k,i}$	measured data points
$_{k,i}$	simulated data points
$w_{k,i}$	weighting factor

Abbreviations

CFC	Computational Fluid Dynamic
CHO	Chinese hamster ovary cell line
CPP	critical process parameters
DoE	Design of Experiments
DT	Digital Twin
LHSD	Latin Hypercube Sample Design
mAb	monoclonal antibody
iDoE	intensified DoE
mDoE	model assisted DoE
mbDoE	model-based Design of Experiment
PAT	Process Analytical Technology
QbD	Quality-by-Design

References

1. Nelson AL, Dhimolea E, Reichert JM. Development trends for human monoclonal antibody therapeutics. Nat Rev Drug Discov 2010;9:767–74.
2. Walsh G. Biopharmaceutical benchmarks 2014. Nat Biotechnol 2014;32:992–1000.
3. DiMasi JA, Grabowski HG, Hansen RW. Innovation in the pharmaceutical industry: new estimates of R&D costs. J Health Econ 2016;47:20–33.
4. Abt V, Barz T, Cruz-Bournazou MN, Herwig C, Kroll P, Möller J, et al. Model-based tools for optimal experiments in bioprocess engineering. Curr Opin Chem Eng 2018;22:244–52.
5. Möller J, Pörtner R. Model-based design of process strategies for cell culture bioprocesses: state of the art and new perspectives. In: Gowder SJT, editor. New insights into cell culture technology. Zagreb: InTech; 2017.
6. Bayer B, Dalmau Diaz R, Melcher M, Striedner G, Duerkop M. Digital Twin application for model-based DoE to rapidly identify ideal process conditions for space-time yield optimization. Processes 2021;9:1109.
7. Kroll P, Hofer A, Ulonska S, Kager J, Herwig C. Model-based methods in the biopharmaceutical process lifecycle. Pharmaceut Res 2017;34:2596–613.
8. Bayer B, Duerkop M, Pörtner R, Möller J. Comparison of mechanistic and hybrid modeling approaches for characterization of a CHO cultivation process: requirements, pitfalls and solution paths. Biotechnol J 2023; 18:e2200381. e2200381.
9. Eibl R, editor. Single-use technology in biopharmaceutical manufacture, 2nd ed. Newark: John Wiley & Sons Incorporated; 2019.
10. Delvigne F, Takors R, Mudde R, van Gulik W, Noorman H. Bioprocess scale-up/down as integrative enabling technology: from fluid mechanics to systems biology and beyond. Microb Biotechnol 2017;10:1267–74.
11. Neubauer P, Anane E, Junne S, Cruz Bournazou MN. Potential of integrating model-based design of experiments approaches and process analytical technologies for bioprocess scale-down. Adv Biochem Eng Biotechnol 2021;177:1–28.
12. Neubauer P, Junne S. Scale-down simulators for metabolic analysis of large-scale bioprocesses. Curr Opin Biotechnol 2010;21:114–21.
13. Sandner V, Pybus LP, McCreath G, Glassey J. Scale-down model development in ambr systems: an industrial perspective. Biotechnol J 2019;14:e1700766.
14. Herwig C, Pörtner R, Möller J, editors. Digital Twins: Applications to the design and optimization of bioprocesses. Advances in biochemical engineering, biotechnology. Cham: Springer; 2021, vol 177.

15. Herwig C, Pörtner R, Möller J. Digital Twins: Tools and concepts for smart biomanufacturing. Advances in biochemical engineering/biotechnology ser. Cham: Springer International Publishing AG; 2021, vol 176.
16. Möller J, Pörtner R. Digital Twins for tissue culture techniques—concepts, expectations, and state of the art. Processes 2021;9:447.
17. Narayanan H, Luna MF, Stosch MV, Cruz Bournazou MN, Polotti G, Morbidelli M, et al. Bioprocessing in the digital age: the role of process models. Biotechnol J 2020;15:e1900172.
18. Shirsat N, Mohd A, Whelan J, English NJ, Glennon B, Al-Rubeai M. Revisiting Verhulst and Monod models: analysis of batch and fed-batch cultures. Cytotechnology 2015;67:515–30.
19. Sanderson C, Barford J, Barton G. A structured, dynamic model for animal cell culture systems. Biochem Eng J 1999;3:203–11.
20. Moser A, Appl C, Brüning S, Hass VC. Mechanistic mathematical models as a basis for Digital Twins. Adv Biochem Eng Biotechnol 2021;176:133–80.
21. Gargalo CL, de las Heras SC, Jones MN, Udugama I, Mansouri SS, Krühne U, et al. Towards the development of Digital Twins for the bio-manufacturing industry. In: Digital Twins. Cham: Springer; 2020:1–34 pp.
22. Kuchemüller KB, Pörtner R, Möller J. Digital Twins and their role in model-assisted design of experiments. Adv Biochem Eng Biotechnol 2021;177:29–61.
23. Möller J, Korte K, Pörtner R, Zeng A-P, Jandt U. Model-based identification of cell-cycle-dependent metabolism and putative autocrine effects in antibody producing CHO cell culture. Biotechnol Bioeng 2018; 115:2996–3008.
24. Hamet P, Tremblay J. Artificial intelligence in medicine. Metab Clin Exp 2017;69S:S36–40.
25. Alanazi HO, Abdullah AH, Qureshi KN. A critical review for developing accurate and dynamic predictive models using machine learning methods in medicine and health care. J Med Syst 2017;41:69.
26. Freiberger F, Budde J, Ateş E, Schlüter M, Pörtner R, Möller J. New insights from locally resolved hydrodynamics in stirred cell culture reactors. Processes 2022;10:107.
27. Rosseburg A, Fitschen J, Wutz J, Wucherpfennig T, Schlüter M. Hydrodynamic inhomogeneities in large scale stirred tanks – influence on mixing time. Chem Eng Sci 2018;188:208–20.
28. Haringa C, Vandewijer R, Mudde RF. Inter-compartment interaction in multi-impeller mixing: Part I. Experiments and multiple reference frame CFD. Chem Eng Res Des 2018;136:870–85.
29. Stosch MV, Davy S, Francois K, Galvanauskas V, Hamelink JM, Luebbert A, et al. Hybrid modeling for quality by design and PAT-benefits and challenges of applications in biopharmaceutical industry. Biotechnol J 2014;9:719–26.
30. Stosch MV, Oliveira R, Peres J, Feyo de Azevedo S. Hybrid semi-parametric modeling in process systems engineering: past, present and future. Comput Chem Eng 2014;60:86–101.
31. Bayer B, Striedner G, Duerkop M. Hybrid modeling and intensified DoE: an approach to accelerate upstream process characterization. Biotechnol J 2020;15:e2000121.
32. Hernández Rodríguez T, Sekulic A, Lange-Hegermann M, Frahm B. Designing robust biotechnological processes regarding variabilities using multi-objective optimization applied to a biopharmaceutical seed train design. Processes 2022;10:883.
33. Hernández Rodríguez T, Posch C, Schmutzhard J, Stettner J, Weihs C, Pörtner R, et al. Predicting industrial-scale cell culture seed trains-A Bayesian framework for model fitting and parameter estimation, dealing with uncertainty in measurements and model parameters, applied to a nonlinear kinetic cell culture model, using an MCMC method. Biotechnol Bioeng 2019;116:2944–59.
34. Hernández Rodríguez T, Posch C, Pörtner R, Frahm B. Dynamic parameter estimation and prediction over consecutive scales, based on moving horizon estimation: applied to an industrial cell culture seed train. Bioproc Biosyst Eng 2021;44:793–808.
35. Arndt L, Wiegmann V, Kuchemüller KB, Baganz F, Pörtner R, Möller J. Model-based workflow for scale-up of process strategies developed in miniaturized bioreactor systems. Biotechnol Prog 2021;37:e3122.

36. Möller J, Hernández Rodríguez T, Müller J, Arndt L, Kuchemüller KB, Frahm B, et al. Model uncertainty-based evaluation of process strategies during scale-up of biopharmaceutical processes. Comput Chem Eng 2020; 134:106693.

37. Herwig C, Garcia-Aponte OF, Golabgir A, Rathore AS. Knowledge management in the QbD paradigm: manufacturing of biotech therapeutics. Trends Biotechnol 2015;33:381–7.

38. Posch AE, Koch C, Helmel M, Marchetti-Deschmann M, Macfelda K, Lendl B, et al. Combining light microscopy, dielectric spectroscopy, MALDI intact cell mass spectrometry, FTIR spectromicroscopy and multivariate data mining for morphological and physiological bioprocess characterization of filamentous organisms. Fungal Genet Biol: FG & B 2013;51:1–11.

39. Chotteau V, Hagrot E, Zhang L, Mäkinen MEL. Mathematical modelling of cell culture processes. In: Pörtner R, editor. Cell culture engineering and technology. Cell engineering. Cham: Springer International Publishing; 2021:431–66 pp.

40. Möller J, Kuchemüller KB, Steinmetz T, Koopmann KS, Pörtner R. Model-assisted design of experiments as a concept for knowledge-based bioprocess development. Bioproc Biosyst Eng 2019;42:867–82.

41. Pörtner R, Schäfer T. Modelling hybridoma cell growth and metabolism—a comparison of selected models and data. J Biotechnol 1996;49:119–35.

42. Zeng AP, Deckwer WD. Model simulation and analysis of perfusion culture of mammalian cells at high cell density. Biotechnol Prog 1999;15:373–82.

43. Kyriakopoulos S, Ang KS, Lakshmanan M, Huang Z, Yoon S, Gunawan R, et al. Kinetic modeling of mammalian cell culture bioprocessing: the quest to advance biomanufacturing. Biotechnol J 2018;13: e1700229.

44. Frame KK, Hu WS. Kinetic study of hybridoma cell growth in continuous culture. I. A model for non-producing cells. Biotechnol Bioeng 1991;37:55–64.

45. Frame KK, Hu WS. Kinetic study of hybridoma cell growth in continuous culture: II. Behavior of producers and comparison to nonproducers. Biotechnol Bioeng 1991;38:1020–8.

46. Miller WM, Blanch HW, Wilke CR. A kinetic analysis of hybridoma growth and metabolism in batch and continuous suspension culture: effect of nutrient concentration, dilution rate, and pH. Biotechnol Bioeng 1988;32:947–65.

47. Craven S, Shirsat N, Whelan J, Glennon B. Process model comparison and transferability across bioreactor scales and modes of operation for a mammalian cell bioprocess. Biotechnol Prog 2013;29:186–96.

48. Xing Z, Bishop N, Leister K, Li ZJ. Modeling kinetics of a large-scale fed-batch CHO cell culture by Markov chain Monte Carlo method. Biotechnol Prog 2010;26:208–19.

49. Teixeira AP, Alves C, Alves PM, Carrondo MJT, Oliveira R. Hybrid elementary flux analysis/nonparametric modeling: application for bioprocess control. BMC Bioinf 2007;8:30.

50. Pörtner R, Schilling A, Lüdemann I, Märkl H. High density fed-batch cultures for hybridoma cells performed with the aid of a kinetic model. Bioprocess Eng 1996;15:117–24.

51. Betts JI, Baganz F. Miniature bioreactors: current practices and future opportunities. Microb Cell Fact 2006; 5:21.

52. Kim BJ, Diao J, Shuler ML. Mini-scale bioprocessing systems for highly parallel animal cell cultures. Biotechnol Prog 2012;28:595–607.

53. Rameez S, Mostafa SS, Miller C, Shukla AA. High-throughput miniaturized bioreactors for cell culture process development: reproducibility, scalability, and control. Biotechnol Prog 2014;30:718–27.

54. Reyes SJ, Durocher Y, Pham PL, Henry O. Modern sensor tools and techniques for monitoring, controlling, and improving cell culture processes. Processes 2022;10:189.

55. Bauer I, Dreher T, Eibl D, Glöckler R, Husemann U, John GT, et al. Recommendations for process engineering characterisation of single-use bioreactors and mixing systems by using experimental methods, 2nd ed. Frankfurt am Main: DECHEMA Gesellschaft für Chemische Technik und Biotechnologie e.V; 2020.

56. Kuchemüller KB. Evaluation modellgestützter Design of Experiments-Methoden zur Auslegung biopharmazeutischer Prozesse. Hamburg: Hamburg University of Technology; 2022.
57. Möller J, Pörtner R. Computational efforts for the development and scale-up of antibody-producing cell culture processes. In: Cell culture engineering and technology. Cham: Springer; 2021:467–84 pp.
58. Moser A, Kuchemüller KB, Deppe S, Hernández Rodríguez T, Frahm B, Pörtner R, et al. Model-assisted DoE software: optimization of growth and biocatalysis in Saccharomyces cerevisiae bioprocesses. Bioproc Biosyst Eng 2021;44:683–700.

Axel Schmidt, Alina Hengelbrock and Jochen Strube*

6 Continuous biomanufacturing in upstream and downstream processing

Abstract: Continuous bioprocesses have become a significant technological change in regulated industries, with process analytical technology (PAT) and quality-by-design (QbD) being essential for enabling continuous biomanufacturing. PAT and QbD are associated with process automation and control, providing real-time key process information. Continuous manufacturing eliminates hold times and reduces processing times, providing benefits such as improved product quality, reduced waste, lower costs, and increased manufacturing flexibility and agility. Over the past decade, advancements in science and engineering, along with the adoption of QbD and the advancement of PAT, have progressed the scientific and regulatory readiness for continuous manufacturing. Regulatory authorities support the implementation of continuous manufacturing using science- and risk-based approaches, providing a great deal of potential to address issues of agility, flexibility, cost, and robustness in the development of pharmaceutical manufacturing processes.

Keywords: continuous biomanufacturing (CBM); process intensification; Biopharma 4.0; automation; digital twins; machine learning

6.1 Introduction

Continuous biomanufacturing (CBM) for biopharmaceuticals has significant advantages over traditional batch processing, including agility, flexibility, quality, cost, and societal benefits [1]. However, the pharmaceutical industry is still primarily reliant on batch processes, which has led regulatory agencies such as the FDA to encourage the adoption of continuous processes in pharmaceutical manufacturing [2]. The lack of agility, flexibility, and robustness in the pharmaceutical manufacturing sector poses a potential public health threat as failures within manufacturing facilities can lead to drug shortages [3].

One of the most significant advantages of continuous manufacturing is the ability to expand production volumes without the current problems related to batch scale-up, making it a more agile approach to production [4]. This is particularly important

---

***Corresponding author: Jochen Strube**, Clausthal University of Technology, Institute for Separation and Process Technology, Leibnizstr. 15, D-385678 Clausthal-Zellerfeld, Germany, E-mail: strube@itv.tu-clausthal.de
Axel Schmidt and Alina Hengelbrock, Clausthal University of Technology, Institute for Separation and Process Technology, Leibnizstr. 15, D-385678 Clausthal-Zellerfeld, Germany

As per De Gruyter's policy this article has previously been published in the journal Physical Sciences Reviews. Please cite as:
A. Schmidt, A. Hengelbrock and J. Strube "Continuous biomanufacturing in upstream and downstream processing" *Physical Sciences Reviews* [Online] 2023. DOI: 10.1515/psr-2022-0106 | https://doi.org/10.1515/9783110760330-006

in situations where there is a need to rapidly increase production in the event of shortages or emergencies [5]. Traditional batch processes have supply chains that are spread globally, making them vulnerable in many ways, whereas CBM enables regional and intranational manufacturing, which can reduce these vulnerabilities [6].

CBM also allows for the introduction of more sophisticated, advanced process control, which can enhance product quality and decrease initial investment costs [7]. Although this is of course also possible for batch processes, for an automized continuous process, a reliable control strategy is a prerequisite for operation. The environmental impact of CBM is generally lower, and it also triggers the demand for highly trained staff [8, 9]. Freed resources can be invested in new products, thus creating more job opportunities. Furthermore, due to the small volume of materials needed to run continuous manufacturing systems, it may be possible to design and optimize a continuous system on commercial scale equipment, thus eliminating the need for scale-up [10]. This would increase the agility to facilitate rapid clinical development of breakthrough drugs [11].

Shortening supply chains is another potential advantage of CBP [12]. Under the current batch manufacturing process, intermediates may not be immediately processed. Instead, they are stored in containers and shipped around the world to the next manufacturing facility [13]. Continuous manufacturing provides opportunities to shorten supply chains, facilitating regional or in-country manufacturing.

To promote the transition from batch to continuous processing, regulatory agencies such as the FDA and EMA have started initiatives and published guidance documents. The most prominent example of such guidance is the QbD associated ICH Q8 to Q11 [14, 15]. These guidances aim to explore key focus areas, such as pharmaceutical quality systems (PQS), real time release testing (RTRT), quality-by-design (QbD), and process analytical technology (PAT), which are needed to enable the transition from batch to continuous processing [16].

However, there is a lack of precise examples and strategies that address the linkage of QbD principles and PAT in more detail, especially regarding dynamic batch processes in contrast to semi-stationary continuous processes. It is crucial to demonstrate the applicability of these guidances to enable the transition from batch to continuous processing [17].

Recognizing that shortages commonly begin with a supply disruption related to product or facility quality, the FDA is focusing on encouraging and sustaining advancements in pharmaceutical manufacturing [18, 19].

Automation is key to successful continuous biomanufacturing, as water, personnel, and expensive media such as enzymes and nutrients are crucial resources [20, 21]. Recycling and process integration can help reduce resource usage and increase efficiency [22, 23]. Process control systems(PCS) from vendors such as ABB, Siemens, and Honeywell are commonly used in manufacturing [24]. However, there are currently no standardized protocols or connectivity between labscale single vendor solutions, thus highlighting the need for standards such as OPC-UA [25].

6.1.1 Biopharma 4.0 and sustainability

Innovative APIs and new entities present challenges for process development within Biopharma 4.0, as there is no platform, and therefore different methods need to be utilized [26–28]. Continuous biomanufacturing also offers a more sustainable approach to manufacturing in the biopharmaceutical industry [29]. Water is a crucial resource in biomanufacturing, and while it is a renewable resource, it is also a finite resource in many regions [30]. Additionally, personnel resources can also be rare, which is where automation can play a crucial role. Continuous manufacturing systems can help to conserve both water and personnel resources through recycling and process integration [31] (Figure 6.1). The potential savings in water has been extensively shown in literature by life-cycle-assessments and analysis of process data [32, 33]. The concrete optimization potential depends on the product and its respective market, however, given that a big-part of personnel is often attributed to quality control, the personnel saving potential increases, when process-analytical-technology in combination with a robust QbD-strategy is applied, that reduces the in-between off-line quality control demands.

The history of continuous biomanufacturing can be traced back to the production of factor 8 rFVII in BHK (baby hamster kidney) cells [34]. This initial success sparked a boom in new monoclonal antibody (mAb) studies in labscale and industrialization. However, the transfer of continuous biomanufacturing to other biologics remains unclear. This is due in part to the process development challenges that arise when working

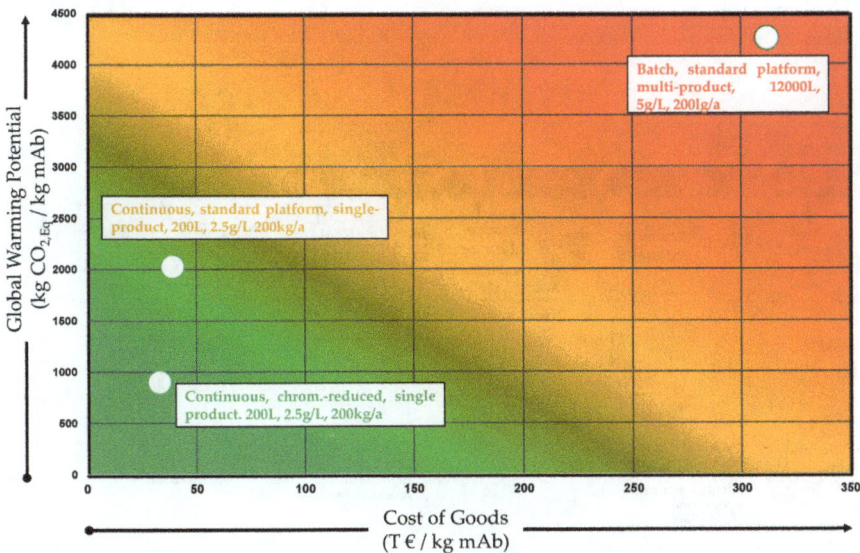

Figure 6.1: Biologicals' global warming potential (GWP) versus cost of goods (COGs) portfolio. Monoclonal antibody (mAb) [31].

with innovative active pharmaceutical ingredients (APIs) and new entities. Since there is no one-size-fits-all platform, different methods must be employed [35].

One feasible approach to continuous biomanufacturing is within the context of Biopharma 4.0 [36]. Biopharma 4.0 relates to the fourth industrial revolution based on digitalization [37]. Digitalization has reached biopharmaceutical industry along the whole development and value chain, starting with drug discovery [38], *in-silico* patient studies [39–42], toxicology studies [43], big data mining technologies for clinical patient data evaluation [44], product design [45, 46] up to recently biomanufacturing [47, 48] based on process engineering tools [49–52]. As all other steps and methods have been broadly reviewed still biomanufacturing is a traditional bottleneck in sufficient supply of drugs under economic constraints of ageing society and limited health care systems.

This concept is built upon the integration of advanced technologies such as process simulation, automation, and data analytics to improve biomanufacturing processes. By implementing process simulation software such as process modeling and using machine learning (ML) (Figure 6.3) algorithms, the learning curve for new processes can be steeper. This can lead to a quicker and more efficient manufacturing process.

One of the key components of Biopharma 4.0 is the use of digital twins (DT) and scalable DT [55]. DTs are virtual replicas of the physical processes and systems that allow for real-time monitoring and optimization of the manufacturing process [56, 57]. This technology enables autonomous operation, which can improve both product quality and manufacturing efficiency (Figure 6.2).

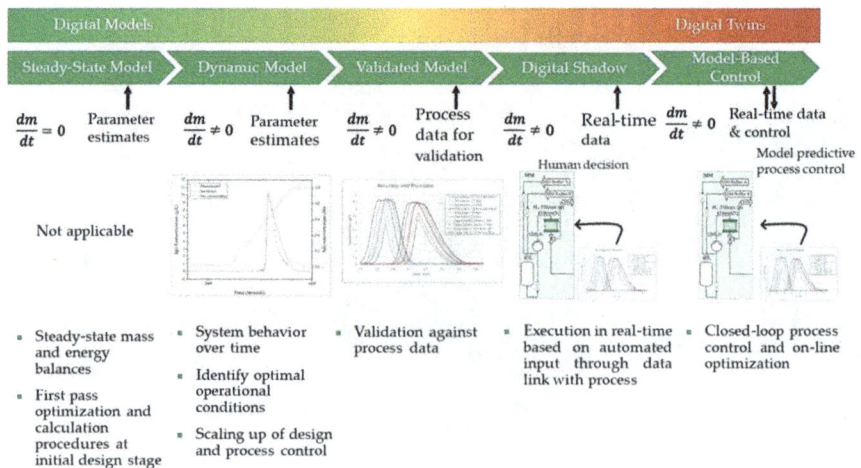

Figure 6.2: Levels of digital twins based on Udugama et al. [53, 54].

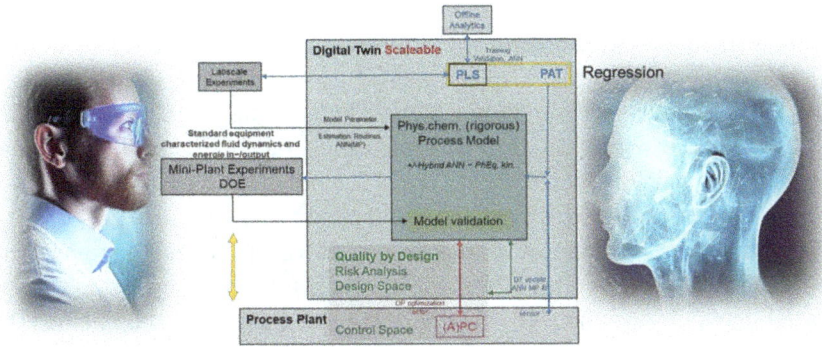

Figure 6.3: Concept of a scalable digital twin, based on the concepts of Sixt et al. [58], Zobel-Roos et al. [59], and Uhl et al. [60]. It is based on an experimentally validated mechanistic model with possibly hybrid parts while considering the quality by design (QbD) [61] approach.

6.1.2 Challenges and opportunities in continuous manufacturing

However, implementing continuous biomanufacturing systems comes with its own set of challenges, including the high cost of resources such as expensive media like enzymes and nutrients, as well as inline buffer preparation. Automation can help to mitigate some of these costs, but it is still important to find cost-effective solutions to these challenges with appropriately equipped biomanufacturing facilities (Figure 6.4) [62, 63]. Integration and scale-up can be challenging, as systems must be robust over long runs. Quality control can be difficult due to the need for real-time monitoring and maintaining product consistency. Regulatory frameworks are still evolving, requiring extensive documentation. Economically, the high initial costs and uncertain ROI are barriers. Product-

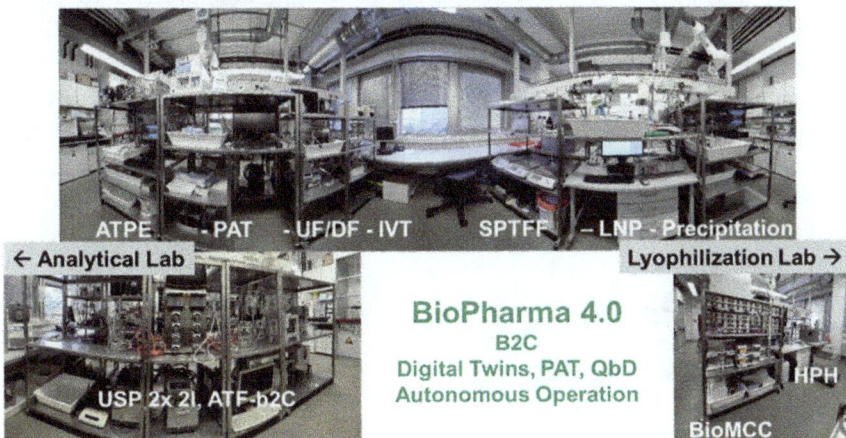

Figure 6.4: Biomanufacturing development facility [10].

specific issues like biological variability and sensitivity to process conditions complicate matters further. Lastly, waste management and energy use can also be sustainability concerns.

Despite these disadvantages, continuous biomanufacturing offers several advantages over traditional batch processing, including agility, flexibility, quality, cost, and societal benefits. Regulatory agencies such as the FDA and CDER have encouraged the adoption of continuous processes in pharmaceutical manufacturing due to the potential public health threat posed by failures in batch manufacturing facilities. To promote the transition from batch to continuous processing, these agencies have started initiatives and published guidance documents. While there are still challenges to implementing continuous biomanufacturing systems, technologies such as Biopharma 4.0, digital twins, and process simulation can help to address these challenges and improve the efficiency and sustainability of biopharmaceutical manufacturing.

This chapter provides an in-depth examination of continuous up- and downstream manufacturing across various drug and vaccine systems. The chapter is organized as follows: The first case study focuses on monoclonal antibodies, which serve as a benchmark process. This section delves into alternative unit operations and assesses their impact on both global warming potential and the cost of goods. This is followed by a case study on *Escherichia coli* fermentation for antibody fragments production, which exemplifies microbial fermentation of recombinant proteins. The focus is on showcasing the effectiveness of alternative unit operations like precipitation in a continuous downstream process. The subsequent section discusses peptides and emphasizes on continuous chromatography, contrasting different processing modes to provide a comprehensive understanding of peptide manufacturing. The following study on HI-VLPs (human immunodeficiency virus like particles) focuses on metabolic modeling and optimized scheduling strategies for continuous operation in the production of virus-like particles. The next example is pDNA/mRNA vaccines, as a state-of-the-art example for new and innovative vaccines, and explores how digital twins can facilitate control strategies for quick, continuous supply, especially in pandemic scenarios. The chapter concludes with a brief overview of cell and gene therapeutics these emerging medicines, setting the stage for future discussions and research. By covering these diverse case studies, the chapter aims to offer a multifaceted perspective on the current state and future directions of continuous manufacturing in pharmaceuticals.

6.2 Case studies

6.2.1 Monoclonal antibodies

Biopharmaceutical proteins are currently produced mainly through batch processes, but continuous bioprocessing has been shown to offer significant benefits, including improvements in agility, flexibility, quality, cost, and society. The biopharmaceutical market is largely dominated by oncology drugs, followed by antirheumatic and antidiabetic medications. By 2024, two of the five most successful oncological products are anticipated to be monoclonal antibodies (Pembrolizumab and Nivolumab) [64].

The current state of the art in monoclonal antibody manufacturing is based on a batch platform process [65, 66]. This process involves upstream and downstream processing, with the recombinant target protein produced through cell cultivation in bioreactors during upstream processing. The downstream processing then isolates the target protein from side components, such as host cell proteins (HCPs), host cell DNA (hDNA), media components, viruses, and endotoxins, through various unit operations, including centrifugation, filtration, and chromatography [66–71].

The established platform process includes fed-batch suspension cultivation of mammalian cells up to 20,000 L bioreactor volume, centrifugation, and depth filtration as cell harvest, protein A affinity chromatography as capture, cation exchange chromatography (CIEX) as intermediate purification, and hydrophobic interaction chromatography (HIC) as a polishing step [72–74]. To reduce immunogenicity, the process incorporates multiple virus inactivation approaches, such as a low pH hold and virus filtration, applied in an orthogonal manner. After completing the protein A affinity chromatography and subjecting the material to a low pH environment for virus deactivation, diafiltration is the next step that must be carried out before proceeding to the cation-exchange chromatography (CIEX) [65].

However, with the increasing concentration of the target protein achieved through upstream processing optimization, the chromatographic steps in the platform process will reach their capacity limit. This limitation is widely known as downstream bottleneck [69, 75–79]. As the product concentration increases, the specific costs (€/kg) of the upstream and downstream processing decline. However, with higher product concentrations, the platform downstream process will reach its efficiency optimum, and further increasing the product concentration will lead to a significant shift in the cost of goods from upstream to downstream processing [66, 80–82].

Continuous bioprocessing circumvents the downstream bottleneck and batch scale-up problems by increasing the productivity and flexibility of each unit operation with a simultaneous increase in product quality resulting from continuous product processing [8, 68, 75, 82–93]. This productivity increase reduces the overall specific COGs by about 88 %. An additional benefit could be gained of about another 15 % decrease by implementing alternative process concepts like ATPE (aqueous two-phase extraction) combined with iCCC (integrated counter-current chromatography) (Figure 6.5).

Furthermore, the use of continuous bioprocessing allows for reduced metabolomics modeling for upstream optimization, such as media and operating point [94]. Continuous formulation, such as lyophilization, can also be implemented, which offers several advantages over traditional batch formulation, including higher product stability, better

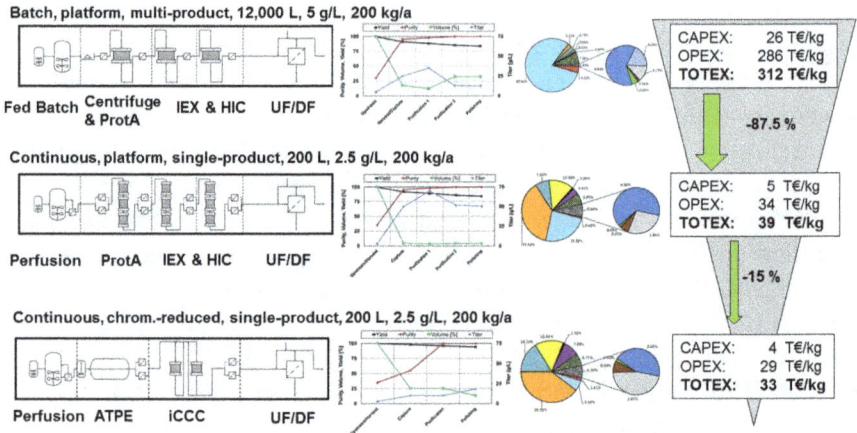

Figure 6.5: Schematic cost comparison between batch and continuous operational mode for the manufacturing of monoclonal antibodies [68].

process control, and reduced cycle times [95]. By adopting continuous bioprocessing and continuous formulation, biopharmaceutical manufacturers can reduce costs, improve product quality, and increase manufacturing flexibility and agility, ultimately benefiting patients and society as a whole. The process flowsheet for continuous antibody manufacturing is depicted in Figure 6.6.

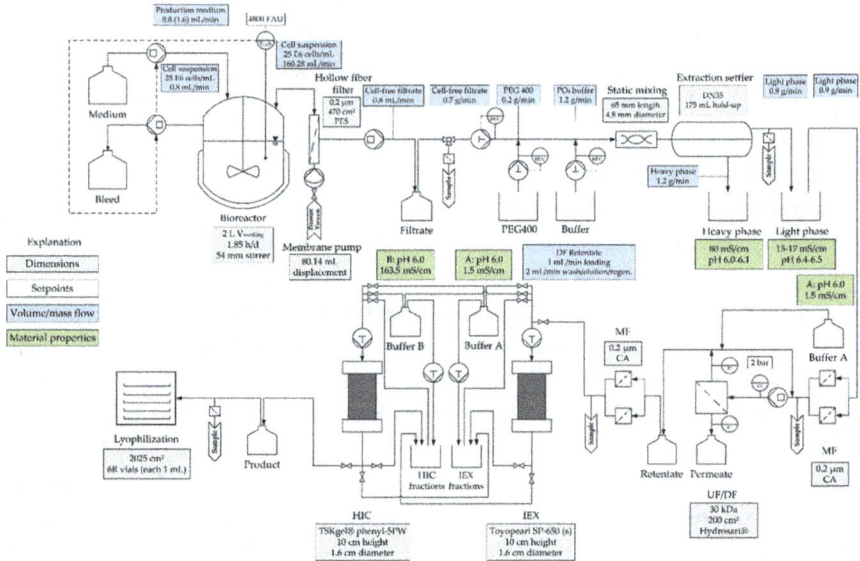

Figure 6.6: Process flowsheet of the continuous production of a monoclonal antibody [96].

6.2.1.1 ATPE and precipitation for continuous DSP

Even with fluctuations in the upstream process titer of up to 50 %, ATPE can consistently hold the concentration of monoclonal antibodies stable. This is made possible by adjusting the polymer and salt levels, all based on phase equilibrium insights. Serving as the principal sensor for both the entry and exit streams of the product phase, Raman spectroscopy is the go-to method for measuring mAb concentrations in USP and ATPE (Figure 6.7) [97].

A continuous precipitation process offers several advantages in terms of agility, flexibility, quality, and cost-effectiveness, making it the preferred process for industrial-scale production. In the field of precipitation, works on the continuous precipitation of monoclonal antibodies have already been carried out. Different precipitation agents, such as PEG 6000 [98] and PEG 6000 in combination with zinc chloride [99], have been used. Another work uses only zinc chloride for the selective precipitation of the antibody [100]. Two-stage processes have also been developed, in which secondary components are first precipitated, followed by the main component [98]. Li et al. integrate the dissolution of the precipitates [101].

The continuous process developed in the work of Lohmann et al. [102–104] is based on the processes of Burgstaller et al. and Li et al. and combines the advantages of both processes. The process of Burgstaller et al. consists of a two-stage process of precipitation and washing of the precipitates. The advantages of the process are the continuous build-up with a production capacity of 0.32 L/h. Furthermore, the feed-and-bleed process control is well-suited for concentrating the precipitates and has the advantage that the filter area is less heavily loaded due to the two-stage setup. Disadvantages include an additional retentate tank, the lack of dissolution step, and the long start-up time of > 60 min. Li et al. also developed a continuous process, compared it directly with the process of Burgstaller et al., and made specific improvement suggestions [101, 105]. Subsequently, the dissolution of the antibody is integrated. The setup is fully continuous,

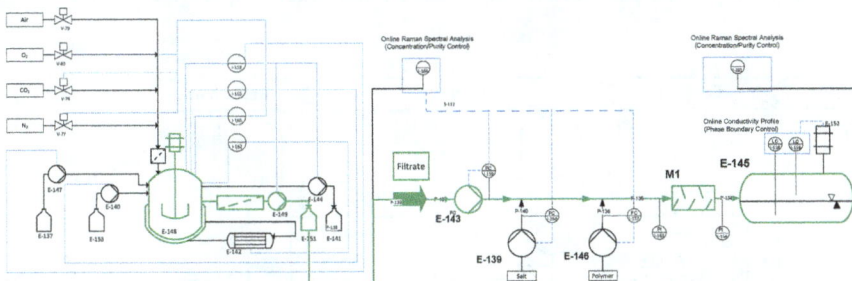

Figure 6.7: Proposed control strategy for USP and ATPE. A Raman probe is used as PAT at the filtrate outlet of USP and forwards mAb and side component concentrations to the ATPE, which adjusts the polymer and salt concentrations to produce light phase with a constant mAb concentration which is then forwarded to the precipitation unit [97].

has a short start-up phase of about 2 min, and can be operated at a comparable flow rate of 0.3 L/h. It dispenses with the additional retentate tanks. A disadvantage is the washing step, which consists of two hollow fiber modules and the resulting high product losses (>20 %). For this reason, a return of the washing solution has been added. One advantage of both processes is the feed-and-bleed process control of filtration. This low-pressure filtration is particularly gentle, resulting in low compression and accelerated dissolution of the precipitates [99, 105]. Furthermore, the filtration time in the feed-and-bleed is significantly shorter than in the dead-end operation. The process developed at the Institute of Thermal Process Engineering and Process Technology at the TU Clausthal by Lohmann et al. combines the advantages of both processes [104]. It includes the process steps of precipitation, washing, and dissolution, as shown schematically in Figure 6.8.

The feed-and-bleed process for filtration, using a PEG 4000 solution (40 wt%) as a precipitation agent, which is diluted to 12 wt% to set the defined precipitation conditions. The same PEG 4000 solution (12 wt%) is used as a washing solution to maintain the precipitation condition and prevent product loss during washing. The resolution buffer is the running buffer for the subsequent chromatography, and an additional buffer change after the precipitation unit is unnecessary. This process has the advantage of significantly reducing processing times compared to batch processing. In laboratory-scale, the average residence time until the dissolved product is about 8–10 min, with comparable flow rates (0.3 L/h) achieved by Burgstaller et al. and Li et al. Additionally, the process can easily be adapted to higher throughputs by simply resizing the membrane surface area of the hollow fiber modules. Moreover, the control and regulation effort is much lower compared to the semi-continuous process, as the timing of individual process phases is eliminated. However, the disadvantage is lower yield at >80 % and a reduced purity increase to >60 %. Figure 6.9 shows a comparison of feed and product, as well as feed and permeate 1 and permeate 2. The selectivity achieved in the batch process is also achieved in the continuous process. The lower purity increase is acceptable since the optimization of the entire manufacturing process is being sought. The subsequent ion exchange chromatography provides the necessary purity increase (Vetter 2021).

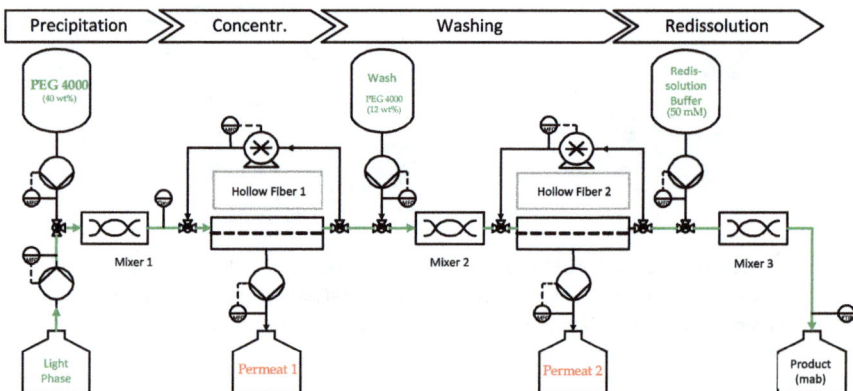

Figure 6.8: Setup of the continuous precipitation process [104].

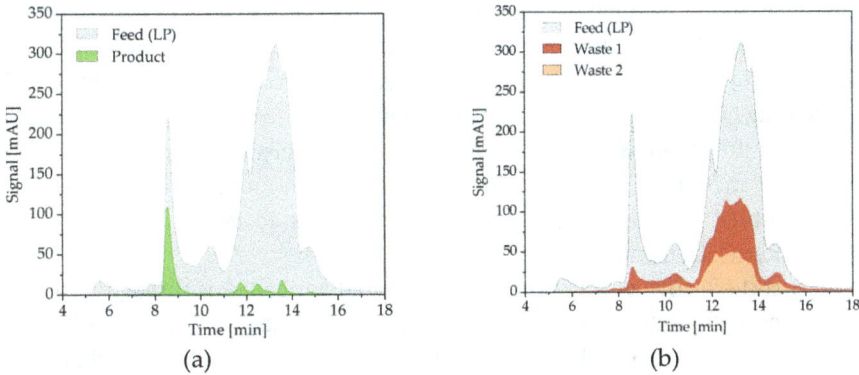

Figure 6.9: SEC chromatogram of (a) feed and product. (b) Shows the comparison of feed to permeate containers 1 and 2. It should be noted that the product is diluted by a factor of 1.5 [104].

The statistical evaluation of the experimental results, confirm the validity of the operating range during the experiments. Contour plots in Figure 6.10 are used to exemplify this. These plots illustrate the influence on (a) yield, (b) purity, and (c) PEG

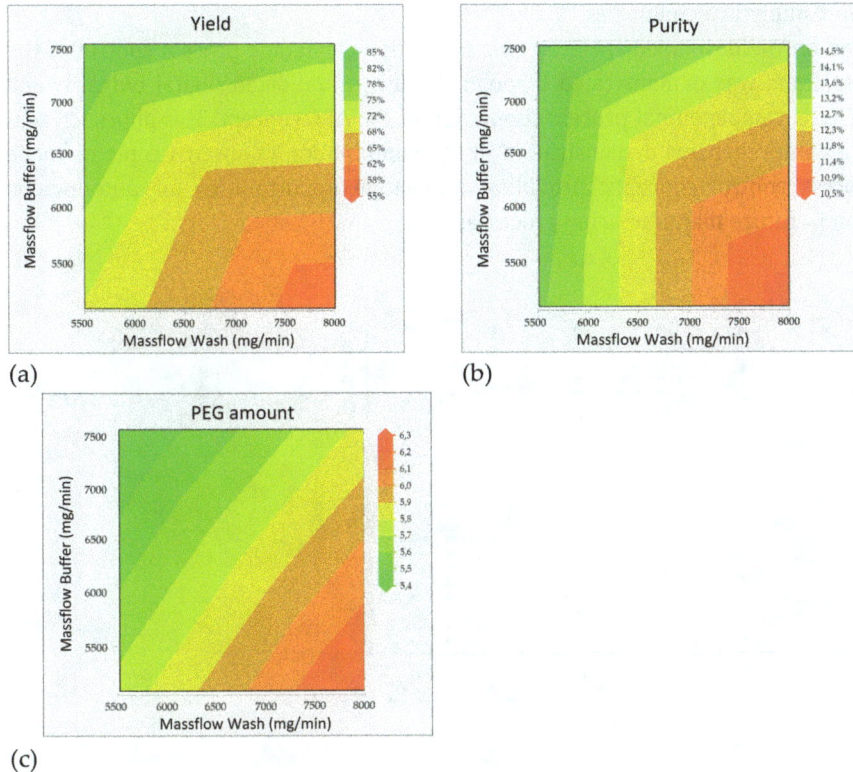

Figure 6.10: Contour plots of the operating spaces [104].

content in the product. The input parameters used are the mass flow rates of the washing solution and the dissolution buffer. These flow rates were chosen because they represent the variable input parameters that significantly affect the process volume at the end and the dimensioning of the second filter module. It is important to avoid both an increased process volume after the basic operation and an enlarged membrane due to an unfavorable choice of these flow rates. A product yield of 85 % is achieved at a low mass flow rate of the washing solution (5500 mg/min) and an increased mass flow rate of the dissolution buffer (7500 mg/min).

The most cost-effective way to produce monoclonal antibodies is through a continuous, chromatography-reduced process, as demonstrated by Kornecki et al. [96]. Robust process control can be achieved through advanced process control with digital twin technology. To enable process control, an inline concentration measurement must be implemented. Real-time process control is achievable in all unit operations using PLSR-based spectroscopic methods [97, 106]. The proposed process overview is given Figure 6.11, showing the course of purity, yield, titer, and DNA concentration. Chromatography mainly increases product titer, while UF/DF adjusts it. Purity steadily increases throughout the process, with most high molecular weight side components being eliminated in ATPE capture. Precipitation is used for purification, and after chromatographic polishing, over 99 % purity is achieved, with side components below the detection limit of the applied analytic technologies.

Based on the developed spectroscopic predictions, dynamic process control of the unit operations was demonstrated through the use of validated digital twins, which allowed for a complete total process simulation of the APC concept. This approach represents a comprehensive demonstration of the combination of spectroscopic methods and process control concepts, providing innovative opportunities for autonomous operation in future manufacturing processes.

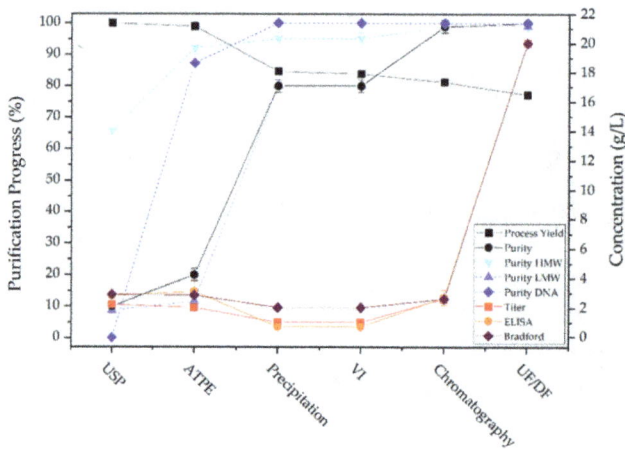

Figure 6.11: Course of purity, yield, and titer during the process [97].

6.2.1.2 Conclusion mabs

Continuous manufacturing for mabs has started the continuous bioprocessing CBP hype due to the need for increasing 100 kg to lower tons amounts and has been established among other by Samsung Biologics [107, 108] as CDMO for flexible production at lowest COGs intended. Platform processes are transferred from batch to continuous operation. Cost efficient alternatives have been proven by pilot studies.

6.2.2 Fragments

Escherichia coli is the most commonly used prokaryotic organism for producing commercially relevant and pharmaceutically active proteins due to its high growth rates, ability to grow to high cell densities on inexpensive, chemically defined media, and extensive study [109–111]. However, *E. coli* is unable to perform post-translational modifications such as glycosylation. Nonetheless, non-glycosylated therapeutic proteins such as insulin, growth factors, interferons, and antibody fragments can still be efficiently produced by *E. coli* [112–114].

Formation of disulfide bridges is an essential aspect of protein production. Reducing properties prevent the formation of disulfide bridges, which often lead to agglomeration of the target protein into insoluble inclusion bodies (IB), resulting in biologically inactive proteins [115, 116]. To overcome this issue, the target protein can be expressed as a fusion protein with a signal sequence that directs the protein into the periplasmic space, where oxidizing conditions promote efficient disulfide bridge formation, resulting in the target protein's soluble form. Furthermore, the periplasmic space has fewer host-specific impurities, allowing for higher purity of the target protein with a suitable lysis strategy.

The choice of expression vector also plays a crucial role in productivity. The promoter, selection marker, fusion proteins, and origin of replication all affect productivity [117, 118]. The origin of replication influences the number of plasmid copies, which does not always equate to high expression. In contrast, low plasmid copy numbers are also usually disadvantageous for productivity [119]. The use of inducible, strong promoters allows for high productivity and simultaneous control over the onset of expression [120, 121]. The overall process flowsheet of scFv manufacturing is depicted in Figure 6.12.

Continuous monitoring of the process within the proven acceptable ranges (PAR) is necessary to ensure robust process control. Spectroscopic techniques such as Raman and Fourier transform infrared (FTIR) spectroscopy allow for real-time process monitoring of critical process parameters, which can be used to predict process performance and control the process using digital twins. The study demonstrates the applicability of process modeling for the entire production of single-chain variable fragment (scFv) in *E. coli* from fermentation to formulation, showing accurate and precise predictions of biomass, glucose, and product concentration in fermentation. The model can also be used for pH and pO2 control, allowing for the adjustment of operating parameters accordingly.

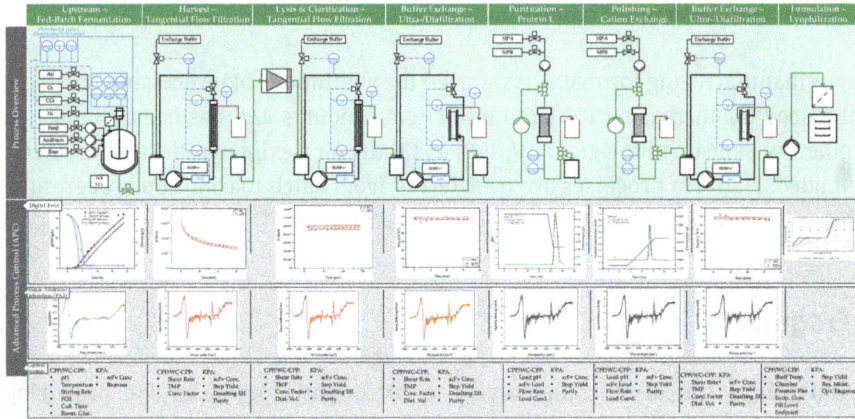

Figure 6.12: Process flowsheet including APC by digital twins and optimized PAT. Critical process parameter (CPP), well-controlled critical process parameter (WC-CPP), key process attribute (KPA), particle size distribution (PSD), viable cell density (VCD), transmembrane pressure (TMP) [122].

After fermentation, tangential flow filtration is employed to concentrate the fermentation broth and wash it with PBS (UF/DF), and the digital twin model can predict the flux across the membrane as a function of TMP and shear rate. Based on the model predictions, filtration time can be adjusted by adjusting the operating parameters, and the blockage can be minimized by adjusting the TMP and shear rate. Furthermore, real-time biomass concentration fluctuations from the fermentation can be taken into account, and operating parameters can be adjusted accordingly to achieve consistent concentration and defined washing.

The scFv is clarified by tangential flow filtration and then subjected to mechanical cell disruption using a high-pressure homogenizer. The digital twin model is used to predict, monitor, and control the washing, and the experimentally determined permeate fluxes can be predicted. Moreover, the model can adjust the exchange volumes used based on the current desalination level and purity, achieving optimal product purity or saving buffers.

The scFv is captured by affinity in protein L chromatography, and the process model can predict the experimentally determined process course based on the manufacturer data on the adsorbent used, as well as the experimentally determined parameters for the isotherms. Increasing the flow rate and reducing the wash step to 3 DV can result in a productivity of 5.3 g/L/d at a purity of 96 %. The model can also accurately predict elution time points, which can control fractionation of the product. Polishing using CEX chromatography follows protein L chromatography, and the model can be used to increase efficiency in terms of throughput, resulting in a productivity of 2167 g/L/d at a yield of 96 %.

Freeze drying is performed after polishing to ensure long-term stability, and the digital twin model can control the drying time by predicting the primary variables influencing drying such as shelf temperature, temperature gradient, chamber pressure, and residual moisture in the dried product. The model also allows for the prediction of the heterogeneity of the vials depending on the position in the freeze dryer (corner vs. center).

By employing digital twins, it is possible to predict the exact end points of the process steps, which can be optimally prepared and scheduled for subsequent operations, and PAT in combination with a predefined normal operating range eliminates the need for time-consuming quality controls. Ultimately, the use of digital twins can reduce the total time for downstream processing by a factor of two, enabling the downstream to be performed within one working day instead of two, resulting in maximum savings potential.

6.2.2.1 Continuous scFv DSP by precipitation

Precipitation is implemented in a similar manner as with antibodies (Figure 6.13). The precipitants are screened to determine whether major or minor component precipitation is needed, as well as whether process control adjustments are necessary.

Results demonstrate that the precipitation process developed can be applied to both primary and secondary component precipitation. The laboratory process is flexible due to its modular design and can be easily adapted to a different substance system. The process models components, such as tanks, mixers, hollow fibers, are structured in submodels, like the laboratory components, and can be combined as desired depending on the process control. The model is generally valid and accurately reflects the physical effects, as it can be adjusted to a new stock system by varying a few model parameters.

Figure 6.14(a) indicates that there is a solubility difference between the target and side components. At a weight fraction of about 4 wt% of PEG 8000, the scFv fragments are 96 % soluble, while the minor components only have a solubility of 55 %. This solubility difference makes minor component precipitation viable. PEG 12,000 is not suitable as a precipitant because it does not selectively, as shown in Figure 6.14(b). PEG 8000 with a

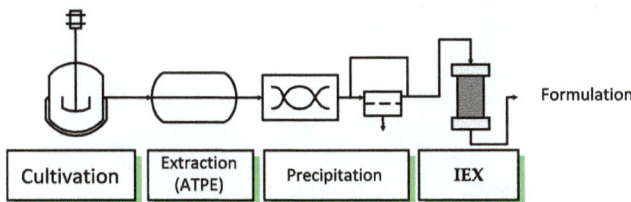

Figure 6.13: Process overview for scFv manufacturing [104].

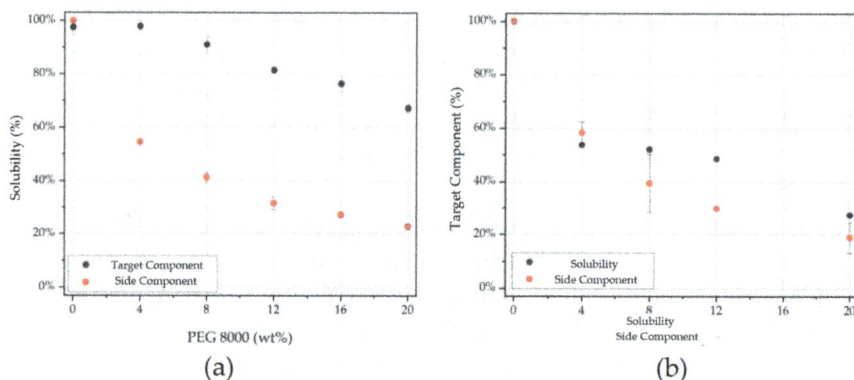

Figure 6.14: Solubility curves of antibody fragments (black) and minor components (red). The equilibrium curves for (a) PEG 8000 and in (b) for PEG 12,000 are shown.

weight fraction of 4 wt% was chosen as a suitable precipitating agent based on the screening results. Additionally, since the product is in the permeate of the first filtration stage, the continuous process can be simplified to a single-stage process, as depicted in Figure 6.15.

From a process engineering perspective, integrating precipitation after ATPE is not a sensible approach, as the scFv-fragments protein concentration in the feed is very low and is heavily contaminated by side components.

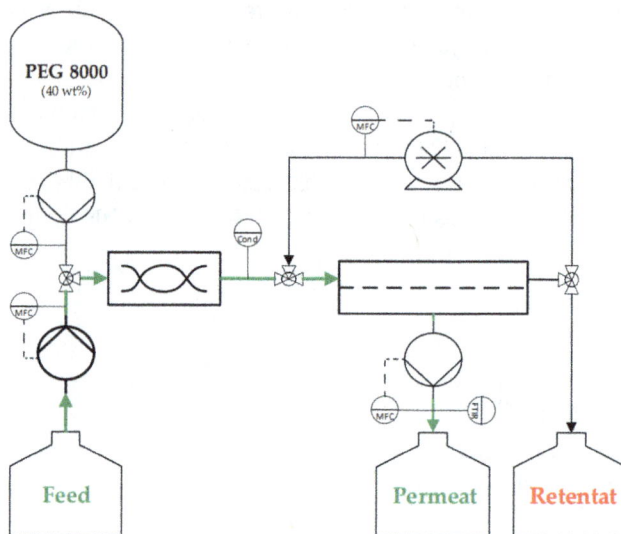

Figure 6.15: scFv precipitation flow sheet.

6.2.2.2 Conclusion fragments

This study showcases the use of digital twins for advanced process control in the entire production and purification process of scFv in *E. coli*, from fermentation to freeze-drying formulation. The digital twin not only predicts the concentration of glucose during fermentation but also aids in optimization. Using the digital twins knowledge of the ideal harvest time and resulting product concentration, optimal harvest preparation and concentration via tangential flow filtration are achieved, increasing concentration by a factor of three and washing with five diafiltration volumes. Tangential flow filtration is also utilized for clarification after high-pressure homogenization, where the digital twin accurately predicts flux decrease at different operating points within a DoE plan, identifying shear rate and biomass concentrations as critical process parameters. In the following steps of Protein L purification and cation exchange chromatography polishing, the digital twins help improve duration and productivity. Finally, in the lyophilization step, the digital twin determines the critical residual moisture and endpoint of the drying, depending on the vials position in the freeze dryer. This approach offers continuous monitoring and optimization of the entire process, leading to improved efficiency and productivity.

6.2.3 Peptides

Peptide therapeutics have become an essential part of medical practice since the discovery of insulin therapy in the 1920s [123]. Today, more than 70 peptide drugs are approved worldwide, and research into peptide-based therapeutics continues at a steady pace [124]. They are positioned between small molecules and proteins, with possible application for therapeutic intervention that mimics natural pathways [123]. Unlike small molecule therapeutics, peptide-based therapeutics have fewer off-target effects, and compared to larger biologics, they have higher tissue penetration, greater activity per unit mass, and lower manufacturing costs [125]. Continuous up- and down-stream processing in biomanufacturing is an area where innovation and development of sustainable and robust processes in the production of therapeutic peptides are needed [126, 127].

Two types of therapeutic peptides can be distinguished; fermentation derived and synthetically derived peptides. They have been applied to a wide range of diseases, such as diabetes mellitus, cancer, infectious diseases, and vaccine development [128, 129].

The most prominent fermentation derived peptide today is human insulin. It is fermented in bacteria (*E. coli* for Humulin® and Insuman®, saccharomyces cerevisiae for Novolin®) [130]. Recombinant human insulin production begins with the insertion of a DNA vector containing a gene encoding pre–pro-insulin precursor protein into a host organism. The upstream process demands tight control over culture and fermentation operating parameters in order to optimize the yield [131]. Insulin pre-cursor are typically expressed as intracellular inclusion bodies, therefor common continuous upstream methods such as perfusion are not applicable. In the subsequent downstream a multi-step purification, including reversed-chromatography, is applied.

Biotechnology first peptide produced is regarded as recombinant human insulin in response to the need for a steady and sufficient supply worldwide. This replaced the animal insulins and semisynthetic insulins modified from animal insulins.

Synthetic peptides are becoming increasingly important in the treatment of numerous diseases, and they present synthetic challenges and opportunities for the development and application of novel technologies [132]. The number of synthetically derived peptides has grown significantly, with three methodologies making significant contributions to the field: classical solution peptide synthesis (CSPS), solid-phase peptide synthesis (SPPS), and liquid-phase peptide synthesis (LPPS) [133]. SPPS involves the attachment of the first amino acid to a solid support, followed by the addition of successive amino acids to the growing peptide chain. LPPS, on the other hand, is performed in solution phase and involves the use of protecting groups to prevent unwanted side reactions [134].

Recent advances in flow chemistry and continuous manufacturing have made them attractive options for the synthesis of peptides [134]. In flow chemistry, reactions are performed in a continuous flow of reagents and solvents, allowing for better control of reaction conditions and faster reaction times. Continuous manufacturing, on the other hand, involves the use of automated processes to produce a product in a continuous stream, which can improve efficiency and reduce costs. The use of flow chemistry and continuous manufacturing in peptide synthesis offers several advantages over traditional batch methods. For example, the continuous flow of reagents and solvents in flow chemistry can reduce the amount of waste generated and improve yield. Additionally, the ability to perform reactions continuously in a controlled environment can lead to increased reproducibility and scalability.

Several studies have reported successful application of flow chemistry in peptide synthesis for SPPS and LPPS [135, 136]. Continuous manufacturing has also been applied to the synthesis of peptides, with companies offering commercial continuous peptide synthesizers [137].

Overall, the use of flow chemistry and continuous manufacturing in peptide synthesis shows great promise for improving the efficiency, reproducibility, and scalability of the process. However, further research is needed to optimize these methods for different types of peptides and to evaluate their applicability in large-scale production.

6.2.3.1 Continuous chromatography in peptide manufacturing

Nevertheless, other peptides may be manufactured by Marryfield solid phase synthesis up to few tons scales [138, 139]. This processes may be in general be transferred to continuous operation by classical synthesis flow chemistry [140–144], which is not content of this biologics chapter. In contrast, *E. coli* generated peptides as inclusion bodies which contradict continuous manufacturing as it prevents the ability fur perfusion fermentation application. Here, schedules batch fermentations need to be a hybrid start into continuous downstream processing [145]. As all peptide processes have isoforms as by-products a crucial key technology would be to transfer the standard RP (reversed phase) platform purifications into efficient continuous operation mode, which is discussed and exemplified in the following in general for peptides.

The current polishing step that needs improvement is a batch reversed-phase chromatography process, which exhibits pronounced Langmuir behavior when overloaded, as shown in Figure 6.16. The side components are highlighted by zooming in on the concentration axis, whereas the mean product concentration is significantly higher. The vertical green lines indicate the cut points to achieve >99 % product purity. Since this step is near the end of a long process consisting of over a dozen steps, any loss in yield would be particularly unfavorable. Therefore, the optimization task is to minimize yield loss and increase productivity. However, due to cost restrictions, the new process should not require too many additional units, which makes simulated moving bed chromatography unsuitable. Moreover, implementing gradient separation in SMB (simulated moving bed) is challenging. To minimize the time-consuming and expensive laboratory work, the optimization process is performed using a digital twin. The underlying process model, modeling approach, parameter determination, and validation concept were explained in a previous work [147].

The general rate model was employed to model the chromatography process, which represents five components, including a modifier, target protein, and three impurities. The impurities are mostly eluted before or at the beginning of the target component peak, and they are classified into two groups.

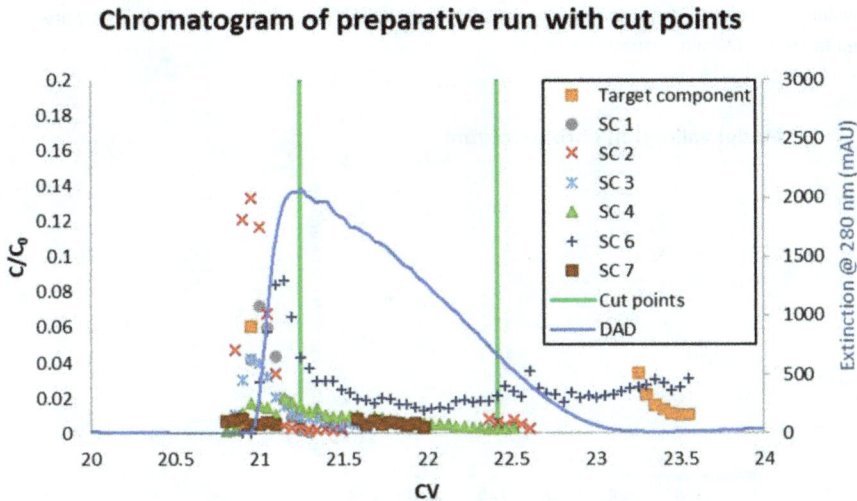

Figure 6.16: Batch chromatogram of the peptide polishing step. The solid blue line is the chromatogram at 280 nm measured with a diode array detector (DAD), assigned to the right axis. The left concentration axis is zoomed in to show the side components (SC), which are represented with dots [146].

To verify the fluid dynamics, tracer experiments were conducted to determine the fluid dynamic parameters, and the results were compared with simulations. Figures 6.17 and 6.18 illustrate that the simulations and experiments matched well, with a negligible deviation of 0.16 % in mean residence times between measurements and simulations. The coefficient of determination R^2 was found to be 0.99. To validate the whole batch process, three runs were carried out with different gradients, and one of the runs is presented in Figure 6.3. The simulations showed excellent agreement with the experimental results, with an overall deviation in mean residence time of 0.24 %, and an R^2 greater than 0.95 for all runs. Thus, the simulations accurately represent the actual process.

6.2.3.2 MCSGP for continuous peptide purification

The MCSGP process involves the simultaneous operation of two identical columns in order to achieve high yield through recycling. Figure 6.19 shows that the product peak is divided into three fractions, with the middle fraction being the high purity product

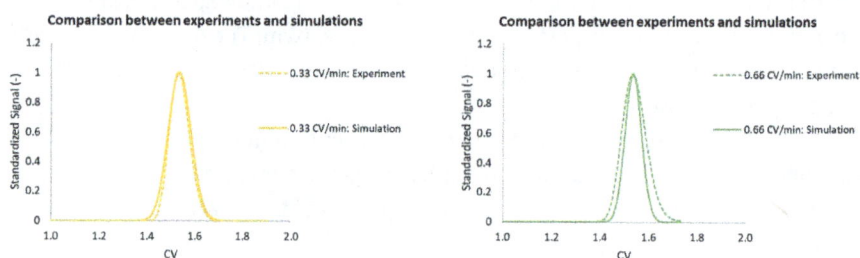

Figure 6.17: Comparison between measurements (dotted lines) and simulations (solid lines) for two different volumetric flows (0.33 CV/min, yellow; 0.66 CV/min, green). Time scale and volumetric flow are normalized to CV and CV/min, respectively [146].

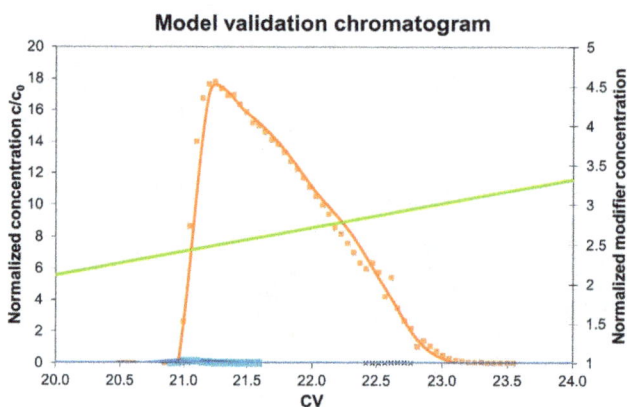

Figure 6.18: Comparison between measurements (dotted lines) and simulations (solid lines) for a batch chromatography run [146].

MCSGP Scheme, one Column

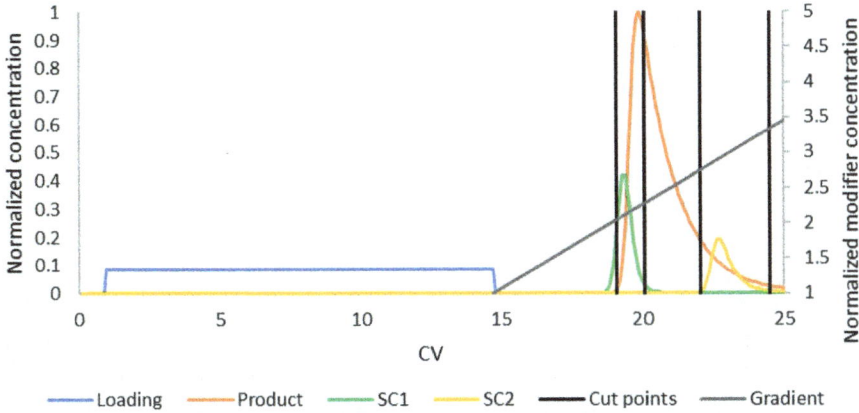

Figure 6.19: Chromatogram of a batch run with relevant cut points (black lines) for MCSGP operation [146].

fraction (F2) and the fractions before and after containing an overlap of product and side components (F1 and F3). These fractions are loaded into the other column before and after feed loading, respectively, and are diluted back to binding conditions due to the gradient. While the MCSGP process guarantees high yields without sacrificing purity, it may result in productivity losses.

To synchronize the two columns, one column must wait for the other if one process step takes longer than the other, or flow rates must be adjusted to slow down the faster process step. Figure 6.20 shows an MCSGP schedule where the gradient is shallow due to

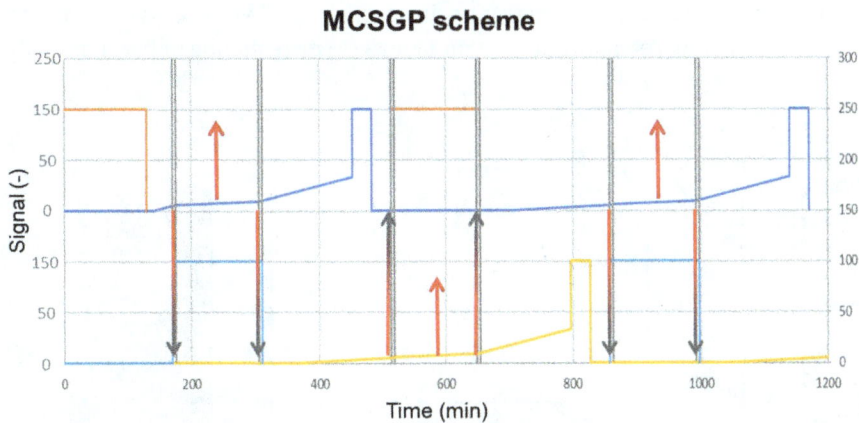

MCSGP scheme

Figure 6.20: MCSGP scheduling for column 1 (upper) and 2 (lower). Gray lines indicate fraction cut points. The color shifting arrows (red to gray) indicate the transfer of fraction 1 or 3 from one column to the other. Pure red arrows indicate pure product elution. Orange (above) and blue (below) lines indicate feed loading [146].

reduced flow rate to slow down product elution for fraction loading. Regeneration and equilibration also need to be completed before fraction 1 elutes, resulting in possible delays or flow rate adjustments. Despite achieving over 99 % yield during the first simulation studies, MCSGP was not pursued for further investigation due to the major delays it would cause, nearly doubling the process duration.

6.2.3.3 CTCC for continuous peptide purification

The MCSGP process offers high yield by recycling the product, but suffers from longer process times due to column synchronization. One solution is to decouple the two columns by storing fractions F1 and F3 in separate tanks, which would also simplify the necessary dilution steps. Initial process simulations revealed that storing F1 and F3 in either two individual tanks or in a single tank per column had no effect on purity. As a result, the proposed process scheme, depicted in Figure 6.21, would utilize additional tanks to allow for individual column speeds. The only requirement is that the total cycle time for both columns must remain the same. In this case, loading took 1 CV longer than the gradient, regeneration, and equilibration steps, which were compensated for with 1 CV longer equilibration. This resulted in a theoretical decrease in productivity of approximately 3 %.

The utilization of tanks enables individual speed control of both columns. However, it is important to maintain the same total cycle time for both columns. During the loading phase, it was observed that loading took longer than gradient, regeneration, and equilibration phases by 1 CV. To compensate for this, the equilibration phase was extended by 1 CV, which led to a theoretical decrease of approximately 3 % in productivity.

Nonetheless, it is possible to increase the overall yield significantly by considering the reloading options for fractions. The product peak has a sharp start and a tailing end, which results in low product concentration in fraction 3. Therefore, two reloading options were considered: reloading of fraction 1 and 3 (A) or reloading of fraction 1 only

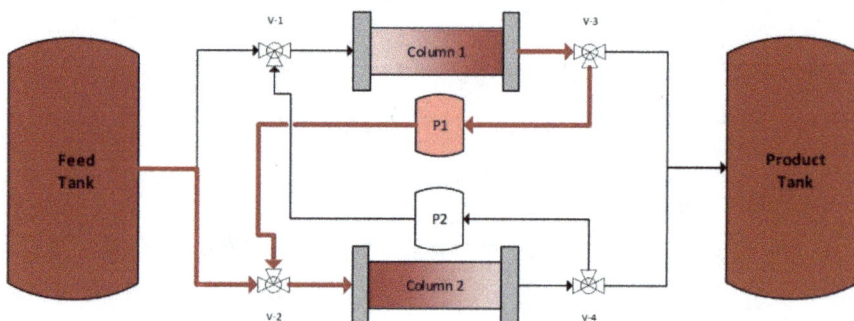

Figure 6.21: Process scheme for the CTCC process [146].

(B). Option A achieved higher yields but required more time due to the high elution strength of fraction 2, which needs more dilution before reloading, ultimately adding more time to the loading phase.

The yield was evaluated based on three factors: Loading yield, cycle yield, and overall yield. The Loading yield indicated the amount of product in the product fraction compared to the overall amount loaded in this cycle. The cycle yield indicated the amount of product in the product fraction compared to the amount of feed loaded in this cycle. Option A achieved a cycle yield of 100 % after the first cycle, whereas Option B lost 7 % in fraction 3. The overall yield indicated the overall amount of product gained compared to the overall amount of feed loaded. This started at the batch yield and approached the cycle yield with an infinite number of cycles. The gap between the cycle yield and the overall yield was caused by the amount of product stored in the fractions. However, the yield loss caused by this becomes unimportant compared to the overall amount of protein produced.

Figure 6.22 depicts the yield increase achieved by the CTCC process, which was 36.8 % when both fractions were reloaded (A) and 28.5 % when only fraction 1 was reprocessed (B). Option B had higher productivity increase of 27.6 % compared to 25.1 % for Option A, as it took less time. Moreover, eluent consumption decreased significantly by 20.2 % for Option B and 23.6 % for Option A, related to the amount of product. Furthermore, the purity of the corresponding batch process could be slightly increased due to slight changes in the cut points, specifically a narrower product fraction. Figure 6.23.

Figure 6.23 presents a comparison between the measured and simulated chromatograms, revealing that the digital twin effectively captures changes in buffer composition. The R^2 values range from 0.858 to 0.998 with an average of 0.952, indicating good accuracy in describing yield and productivity.

Utilizing the digital twin for buffer composition optimization is highly recommended, as evidenced by the contour plots in Figure 6.24. The top two rows depict the buffer and counter ion values (top) and pH over counter ion (middle), with a "more is better" trend identified for all components. However, the optimal spots for the combination of pH and buffer (bottom) are on the two end points: high pH with low buffer concentration or high buffer concentration with low pH. This opposing trend was predicted by the pareto charts. Further optimization studies reveal that the best buffer composition is 100 mM counter ion and 150 mM buffer at pH 3.35, resulting in a 29 % increase in yield and a 27 % increase in productivity compared to the center point.

In early process development, it is recommended to use state-of-the-art design-of-experiments plans to estimate the best buffer composition, which can be easily set up and executed. Coupling the DoE with a digital twin provides additional insights into the dependency of performance values on buffer composition. In this study, fitting isotherm parameters to the DoE runs and feeding these values back into the DoE as target values

(A)

Yield increase factor over numbers of cycles

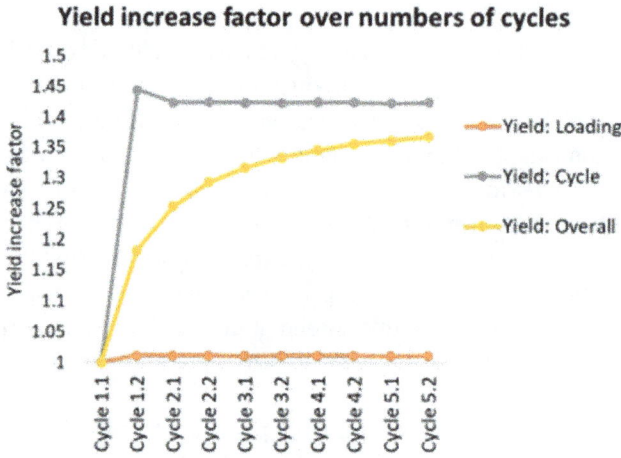

(B)

Yield increase factor over numbers of cycles

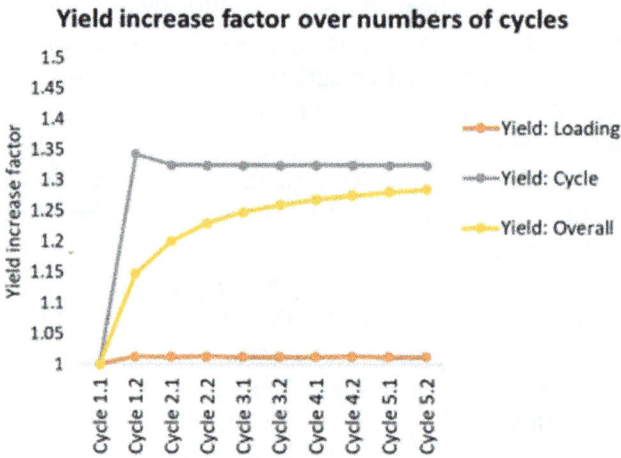

Figure 6.22: Purity (blue line) and yield for (a) reloading of fraction 1 and 3 and (b) reloading fraction 1 only [146].

generate good correlations, which can be integrated into the digital twin for buffer composition optimization.

Furthermore, the digital twin can be used for advanced process control, detecting fluctuations in buffer composition and optimizing the system, especially the cut points, to maintain maximum performance and purity.

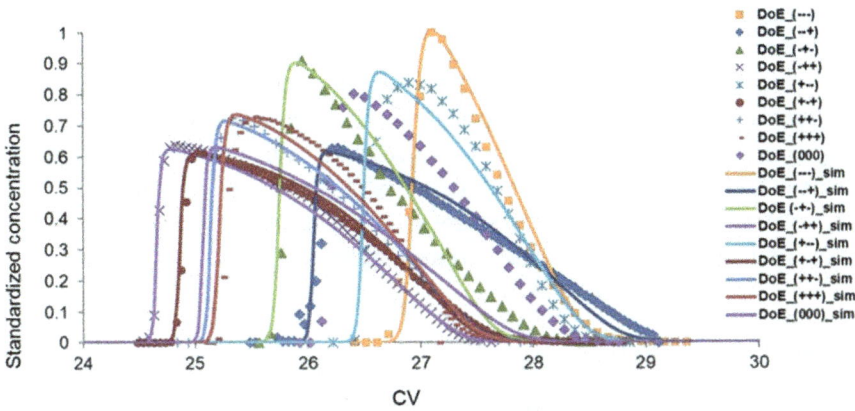

Figure 6.23: Comparison between DoE-experiments (dots) and digital twin simulations (solid lines) [148].

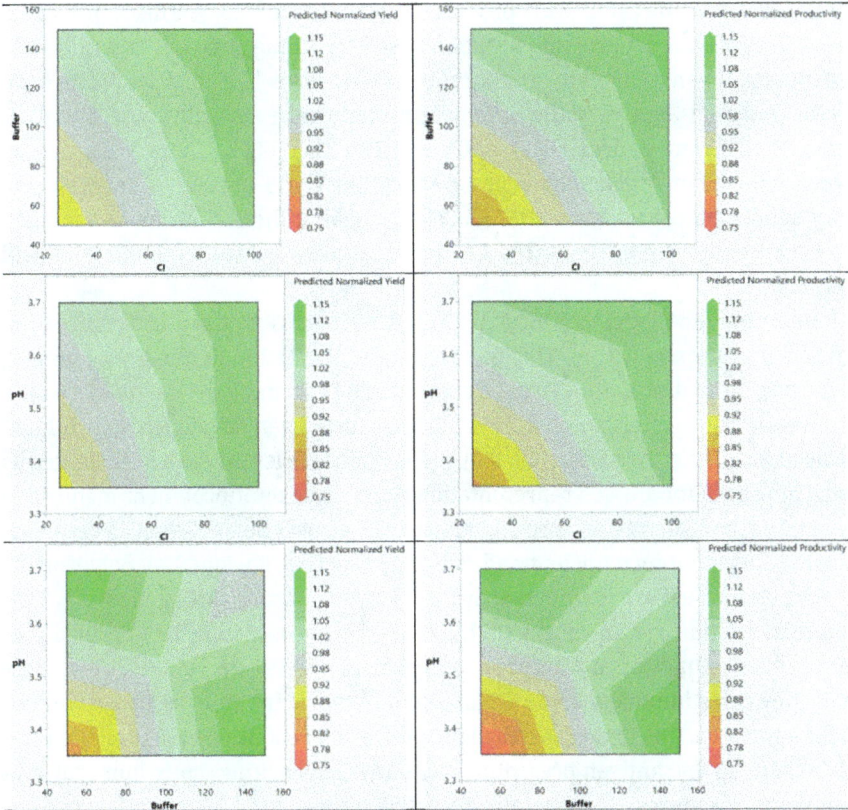

Figure 6.24: Contour plots showing the influence of two factors on target values, normalized yield on the left and normalized productivity on the right side. Top row: Buffer over counter ion, middle row: pH over counter ion, bottom row: pH over buffer. Due to the stepwise change in color instead of a steady color gradient, there are rounding errors leading to the display of more inflection points [148].

6.2.3.4 Conclusion peptides

Peptides are in general biotechnologically efficient manufactured by *E. coli* fermentation which generates inclusion bodies. This prevents perfusion fermentation and hybrid processing of scheduled batch fermentation to continuous downs stream processing would be need for any CBP approach. Filtration bay SPTFF and pH shifted refolding as well as chromatography steps like IEX or RP are in the meantime proven by simulation as well as pilot studies to b efficiently transferrable towards CBP.

6.2.4 Virus-like-particles

One of the main challenges in modern biotechnology is the development and adoption of digital twins for quality-by-design-based processes. These processes require flexible operating points within a proven acceptable range (PAR) and automation through advanced process control (APC) with process analytical technology. This approach is superior to the conventional process execution based on offline analytics and inflexible process set points. When it comes to drug substances (DS), virus-like particles (VLPs) have shown great potential as a versatile vaccination platform, especially those based on human immunodeficiency virus (HIV), such as HIV-1 Gag VLPs. These VLPs can be pseudotyped with heterologous envelope proteins, like the S protein of severe acute respiratory syndrome coronavirus 2 (SARS-CoV-2), making them even more versatile. However, because they are enveloped VLPs, optimal process control with minimal hold times is crucial.

The human immunodeficiency virus (HIV) is a retrovirus that causes acquired im-munodeficiency syndrome (AIDS). This disease leaves affected individuals at a greater risk of acquiring other diseases that can be prevented by vaccination. Despite more than 25 years of research, no effective vaccine candidates have been developed, which high-lights the need for further research [149, 150]. Virus-like particles (VLPs) have emerged as a promising approach for antigen representation [151]. These multiprotein or membrane structures mimic the structure and organization of real viruses but lack the viral genome, making them replication-incompetent and safe [152]. The Gag, Pol, and Env polyproteins form HIV particles, which are surrounded by a lipid layer and carry RNA genomes within the Gag-formed capsid core. In contrast, VLPs may also be immature HIV-derived par-ticles formed by the uncleaved Gag precursor proteins. VLPs are superior to soluble antigens as they can stimulate a better cellular and humoral immune response without the need for adjuvants. Due to their particulate nature and repetitive structure, VLPs are efficiently taken up by antigen-presenting cells (APCs), eliciting both humoral and cellular immune responses. The overall process flowsheet for VLP manufacturing is shown in Figure 6.25.

Virus-like particles (VLPs) can be produced using different expression systems such as bacterial, yeast, insect, mammalian, and plant cells. The production of HIV-1 Gag VLPs

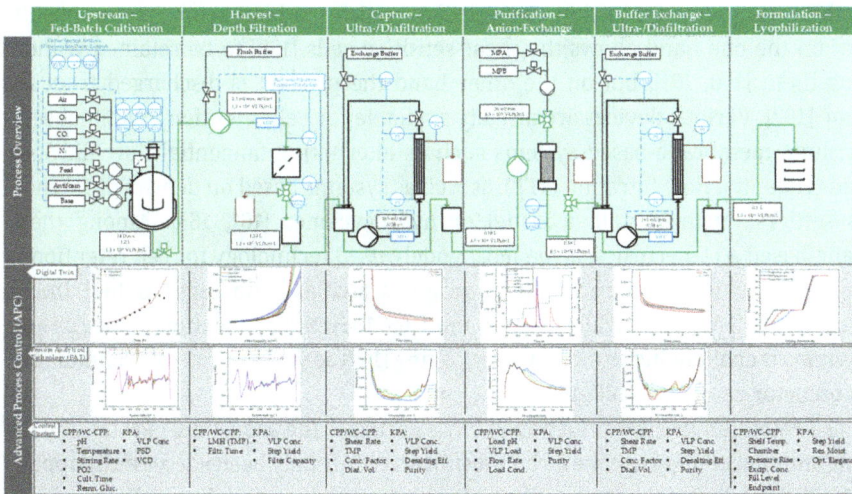

Figure 6.25: Process flowsheet including APC by digital twins and optimized PAT. Critical process parameter (CPP), well-controlled critical process parameter (WC-CPP), key process attribute (KPA), particle size distribution (PSD), viable cell density (VCD), transmembrane pressure (TMP), virus-like particle (VLP) [153].

on a large scale has primarily relied on the baculovirus expression system and insect cell lines [154–157]. Although mammalian cells have the capacity to produce more complex enveloped VLPs, such as HIV-1-Gag VLPs, their productivity is lower. Nonetheless, the human embryonic kidney 293 cells (HEK293) are suitable for producing VLPs in mammalian cells. These cells have good genetic manipulability, can grow in suspension, and can be cultivated at high cell densities. Additionally, 293F suspension cultures are already widely used in the industry for producing virus-based products such as viral vaccines and most viral vectors. Therefore, 293 cells are quickly accepted in the industry [158].

Producing HIV-based VLPs in mammalian cells, especially for suspension cultures, is challenging since enveloped nanoparticles are sensitive to shear stress, pH variation, and osmotic pressure. To address this challenge, quality-by-design methods can establish the relationship between process parameters and product quality characteristics. QbD-based process development is becoming the standard in the biopharmaceutical industry and is required by regulatory authorities. Developing a control strategy as part of QbD-based process development is necessary to achieve the quality target product profile (QTPP). To avoid out-of-specification batches, design spaces can be defined using validated process models. Advanced process control can also be achieved using a validated process model developed as a digital twin. By applying the holistic QbD approach, consistent product quality can be ensured from development to production. Process models can predict quality attributes in real-time, allowing for optimization even after submission.

For the continuous production of virus-like particles, a bioreactor system must be used that on the one hand allows the shear-sensitive cells [159] to be retained without damaging them [160, 161], but on the other hand the product is discharged from the bioreactor [162]. Various devices are already available for cell retention of animal cells. These include membrane-based systems such as alternating tangential flow filtration (ATF) and tangential flow filtration (TFF), as well as systems based on density differences such as settlers (inclined and acoustic) or hydrocyclones [161, 163]. Among these, membrane-based ATF perfusion is the most widely used technology for cell retention in the continuous production of recombinant proteins such as antibodies [96, 164]. Due to the large size of HI-VLPs, typically 100–150 nm [165, 166], the application of membrane-based systems is challenging as membrane clogging [167] as well as product accumulation in the bioreactor can occur [160, 168–171].

For cell retention in the production of virus particles, the acoustic separator and the novel combination of depth filter and hollow fiber have already been successfully applied [160, 162, 171–173]. To accelerate a settling of cells during density-driven cell separation, the g-force can be increased by using an acoustic resonant field [161, 174]. The acoustic separator enables separation of the product as well as dead cells, host cell proteins, and dsRNA, allowing higher cell numbers to be achieved [160, 161, 174]. However, for higher throughputs required in scale-up, higher input power is needed to maintain high separation efficiency [161, 174]. Furthermore, increased temperatures may occur in the settler, necessitating efficient temperature control. In combination with the decreasing dissolved oxygen concentration in the acoustic settler, a reduced viability of the cells can occur [160, 161, 174].

In size-dependent separation, a macroporous tubular membrane is preferred, which is constructed like a hollow fiber module, but the membrane used is constructed like a depth filter. This allows efficient cell retention and separation of toxic metabolites, dsDNA and host cell proteins, enabling cell concentrations of up to 80 million cells/mL [162, 171, 173]. In contrast to the acoustic separator, a complete recirculation of the cells into the bioreactor is also possible, since they do not pass the membrane and thus no cell loss occurs. In addition, the recirculation time is shorter, since no settling of the cells is required for cell separation [160, 171, 175, 176].

Due to the elimination of a cooling system and an obsolete complex pumping strategy, as well as the complete cell retention [171], the use of a macroporous tubular membrane is preferable to the acoustic separator when performing perfusion.

6.2.4.1 Study on HI-VLP metabolic modeling

The main objective of this study was to develop a metabolic model that could predict the production of HI-VLPs in HEK293 cells using the process optimized by Helgers et al. [177]. The model was designed to meet the qualitative and quantitative requirements of QbD-compliant model validation, making it suitable as a predictive process model and a basis for a digital twin. First, the model's accuracy was verified by comparing its predictions to experimental data. Then, sensitivity analysis was conducted to identify the main factors influencing the process on the basis of a statistical analysis, which identifies significant main effects and interactions within the investigated design space. Finally, the experimental data was compared with the simulation results to verify the accuracy and precision of the model.

The plausibility analysis aimed to compare the simulation results with the experimental data qualitatively. The simulation results should show qualitatively similar courses to the experimental data if the effect strengths and directions from the sensitivity analysis were correct. Figure 6.5 displays the results of fed-batch cultivation as well as the simulation results. The simulation results show spikes that correspond to concentration changes caused by the daily addition of feed, starting from day three. Experimentally

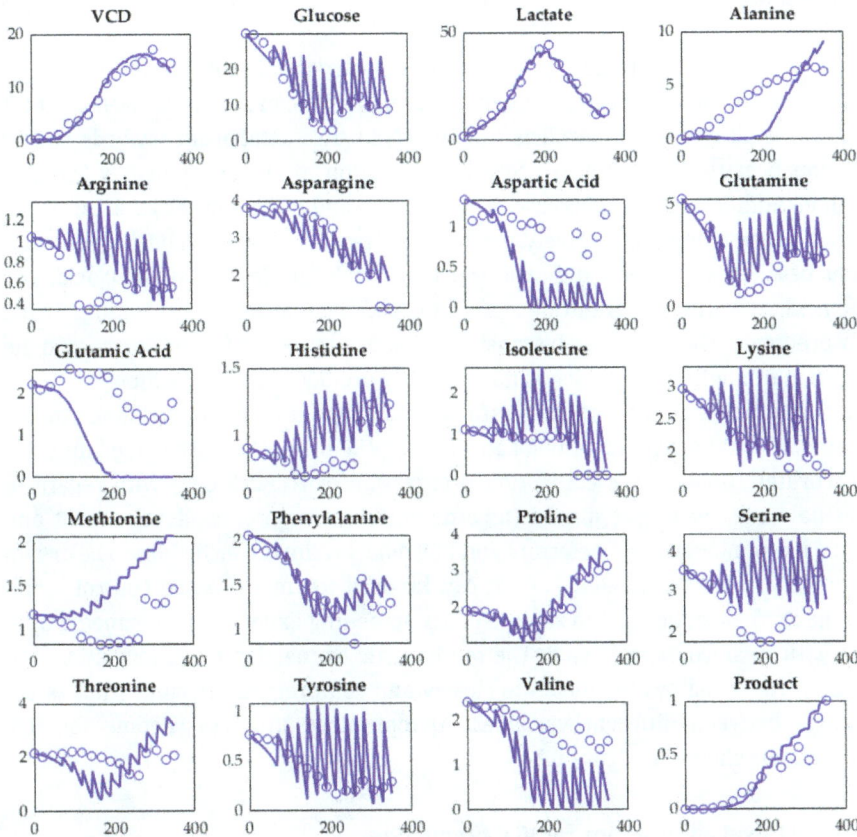

Figure 6.26: Experimentally measured concentration curves (dots) and simulated concentration curves (solid line) over the processing time in hours. Live cell count (VCD) in 1×10^6 cells/mL, the normalized product concentration is dimensionless, all other concentrations in mM [94].

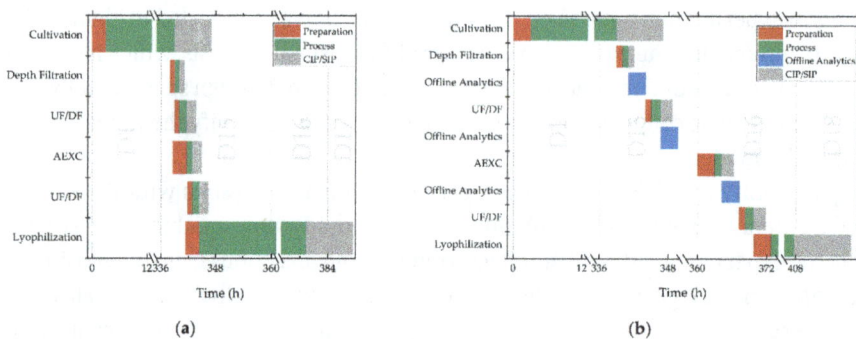

(a)   (b)

Figure 6.27: Optimized process schedule enabled by PAT for APC and DT (a) versus conventional process schedule based on offline analytics (b) [153].

determined concentrations should intersect the simulation results at their low points (Figure 6.26).

The model accurately predicts the concentration curves of live cell count (VCD), VLP (product), glucose, lactate, asparagine, and glutamine concentration, as well as most amino acids. It also correctly predicts the switch in the metabolism towards lactate consumption because of limiting glucose concentration after approximately 200 h of process time. Additionally, the decrease in growth rate and increased cell death after 250 and 300 h of process time, respectively, are correctly predicted by the model. The largest deviations between simulation and experiment are seen for alanine, aspartic acid, and glutamic acid, but with the trend of increasing and decreasing concentrations being correctly predicted. The model can be considered plausible, except for the decrease in the concentration of methionine, which is not correctly predicted in its direction.

This study demonstrates the successful validation of a dynamic metabolic model used to simulate fed-batch cultivation of an HIV Gag-VLP-producing HEK293 cell line. The model accurately predicts the cultivation progression of HEK293 cells with regard to experimental data, making it suitable as a digital twin for process development and design. The use of modern PAT detectors and PLS models allows conclusions to be drawn about critical process parameters, which can be used for model-based control. Additionally, the model can be used to replace costly screening experiments for media optimization with in silico experiments. The mechanistic connection of all formation and consumption rates enables the model to clearly and comprehensibly contextualize the relationships between different variables. An optimized process schedule for HIV manufacturing is shown in Figure 6.27.

6.2.4.2 Digital twin applications for VLP production

The digital twins showcased have the capability to replicate the physical process in real-time. With the addition of PAT, the process can be forecasted and adjusted based on the control strategy within the design space, as demonstrated by Hengelbrock et al. [153]. During cultivation, modern PAT detectors such as Raman, FTIR, and MALS/DLS can be utilized to determine concentrations of VLP, glucose, lactate, and other minor components in real-time. By using this information, the digital twin can regulate feeding and process parameters, including gassing, and provide crucial information for downstream processes such as optimal harvest time, total cell density (TCD) at harvest time, product concentration, purity, and size distribution.

Product concentration and size distribution play a crucial role in determining the maximum filter capacity and the associated blocking mechanism. Knowledge of these variables enables avoiding oversizing while ensuring that sufficient filter area is available. Digital twins in UF/DF enable the prediction of further development of filter resistance through the measured permeate flow and determine the expected endpoint. Additionally, the shear rate can be adjusted to load AEX with the desired product concentration, which can be done as gently as possible for the product.

In QbD-based process development, shear stress can be regarded as a critical process parameter (CPP) as part of the control strategy to ensure the process is operating within the proven acceptable range (PAR). Combining digital twins with PAT (e.g., FTIR) creates flexibility to set exchange volumes in UF/DF based on actual desalination and purity levels, something that is not feasible with conventional process development and offline analytics. Achieving optimal product purity or saving exchange buffers can be enabled only by digital twins and QbD-based process design.

In the anion-exchange stage, the integration of multi-angle light scattering/dynamic light scattering (MALS/DLS) with the digital twin allows for real-time modifications of the cut-off points for the product fraction. This enables a balance between yield and the purity or concentration of the product. Additionally, the digital twin enhances various aspects like the utilization of key raw materials, equipment technology, and process regulation. When fused with PAT approaches, it lays the groundwork for a seamless transition from batch-based to continuous operations.

Digital twins in combination with online CIP/SIP measurement techniques (e.g., real-time TOC, bioburden, and conductivity) provide the freedom to move away from commonly used single-use technology. Automated execution and validation make this possible and eliminate the advantage of single-use technology, which is simple and fast switching of campaigns. Sustainability benefits can also be realized.

Shifting from batch production to a continuous approach becomes feasible when integrating state-of-the-art PAT-driven control tactics with a digital twin. This integration facilitates key benefits of continuous production, such as RTRT for quicker market entry and reduced chances of batch failures. When pitted against traditional processes reliant on offline analysis, the advantages of automated production using PAT for APC and digital twins become evident.

The predictive capabilities of the digital twin enable follow-up processes to proceed with minimal delays. This is assured by real-time PAT, working together with a QbD oriented control strategy, to maintain operation within the normal operating range (NOR). This synergy allows for the completion of the entire downstream process within a single workday. In contrast, traditional methods necessitate offline analytics, extending the downstream process to two workdays and risking quality degradation for sensitive materials like enveloped VLPs.

6.2.4.3 Conclusion VLPs

This study demonstrates how digital twins can be used for batch to continuous manufacturing transfer in the context of quality by design (QbD)-based process development for the production and purification of HIV-1 Gag VLPs in HEK293 cells. The digital twin allows for optimized control, which in turn enables optimized feeding strategies and prediction of important key process attributes (KPAs) such as yield, VLP concentration, and particle size distribution by MALS/DLS. In depth filtration, the digital twin can predict real filter capacity at an early stage, which helps to optimize the process time. In ultrafiltration/diafiltration, the digital twin can predict the necessary process time based on product concentration.

One of the benefits of using digital twins with PAT and APC is the optimization of the process schedule. By avoiding hold times between process steps, the downstream process can be fully started within one working day instead of two, reducing the hold time until lyophilization by a factor of 2. Additionally, the digital twin can be used for scale-up, which could increase productivity by up to a factor of 2.

Overall, this study shows that digital twins can be used as process models with a maximum deviation of 4.5 % from experimental data. This, in combination with real-time PAT and a QbD-based control strategy, enables the automation of the process up to real-time release testing (RTRT).

6.2.5 pDNA to mRNA

Vaccine supply faces a bottleneck due to the requirement for operation personnel and chemicals in the manufacturing process. Continuous manufacturing processes offer a solution to this issue, and strict application of the regulatory demands for quality by design process based on digital twins, process analytical technology, and control automation strategies can improve process transfer for manufacturing capacity, reduce out-of-specification batch failures, optimize personnel training and numbers, utilization of buffers and chemicals, and speed-up product release. This work explains the necessary process control concepts for achieving autonomous, continuous manufacturing for mRNA manufacturing and demonstrates their readiness for industrialization, enabling the benefits mentioned above. By switching from batch-wise to continuous mRNA production, cost reduction by a factor of 5 is achievable. However, improvements in process control are possible, and future research may focus on model-based predictive control to reduce the potential for batch failure and out-of-specification numbers [10, 35, 178–180].

Messenger ribonucleic acid (mRNA) and other biologics are traditionally produced in batch mode, resulting in lower production throughput and relatively high capital costs. To address this issue, this study demonstrates an efficient solution for biologics manufacturing, namely the qualification and validation of a plant setup for clinical trial doses of approximately 1000 doses and a production scale-up of around 10 million doses. The production of Comirnaty BNT162b2 mRNA vaccine was optimized in batch mode, achieving a yield of 12 g/L mRNA, and then transferred successfully to continuous production in the segmented plug flow reactor. To realize automated process control and real-time product release, the use of appropriate process analytical technology is essential. Switching to continuous production reduces the production footprint and lowers the cost of goods. A direct industrialization of production-scale operation could be supported and managed by data-driven decisions, in collaboration with appropriate industry partners, their regulatory affairs, and quality assurance [10]. Process strategies for mRNA manufacturing are shown in Figure 6.28.

To synthesize mRNA, a starting point is template DNA that contains the genetic code for the respective spike protein. Although complete cell-free DNA manufacturing is feasible, the state-of-the-art approach is plasmid manufacturing by *E. coli* fermentation, followed by purification and linearization. In the next phase, the linearized DNA

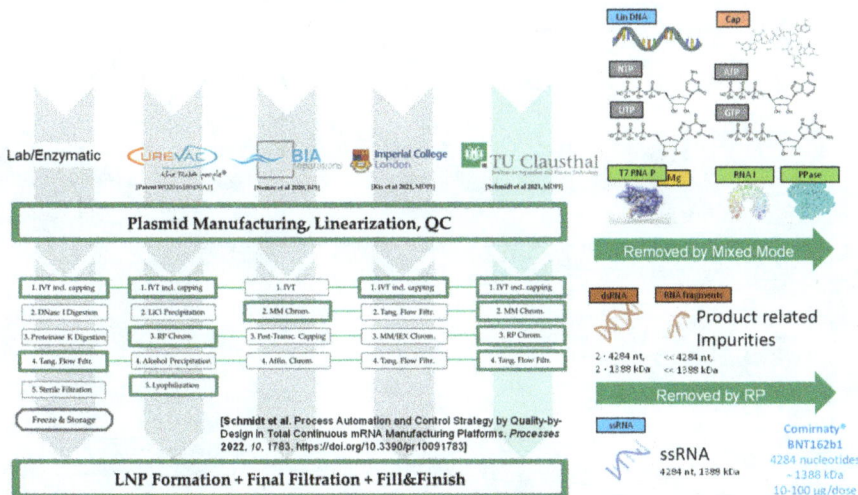

Figure 6.28: Overview of published and new mRNA manufacturing processes (RP, Reversed-Phase; MM, Mixed-Mode; IEX, Ion-Exchange). Numbers indicate the order of flow. The green arrow marks the process sequence as discussed by Schmidt et al. [178].

serves as a template during *in vitro* transcription. Modern technology enables co-transcriptional capping, which combines the traditionally separate steps of transcription and capping [35].

Purification processes of the obtained mRNA typically rely on platform technology, similar to monoclonal antibody production, where the same process steps can be used to manufacture different derivatives [181]. However, unlike antibody manufacturing, the purification strategy for mRNA vaccines is not standardized. While the key steps of transcription, purification, and final encapsulation into lipid nanoparticles are mostly the same for known mRNA vaccine candidates, the purification steps in-between vary widely, and there are many different separation technologies and overall purification strategies published in the literature [182]. A detailed flowsheet is depicted in Figure 6.29.

Studies on continuous biomanufacturing have highlighted the potential benefits of maintaining high-quality product supply in a timely and reliable manner [183]. Automation of the process is a key aspect of continuous production [33], which can be accomplished with advanced process control strategies [184]. Although continuous biomanufacturing is shorter in duration compared to bulk and petrochemical operations, which can take several months, it can lead to reduced variance in product quality and maintain constant operational state around an optimum level. Consequently, there are benefits to be gained from reducing operational costs, manufacturing time, and significant reductions in quality assurance (QA) costs through real-time release testing. To implement advanced process control (APC) strategies, models based on artificial neural networks (ANN) [54], rigorous or hybrid approaches are employed. The process models

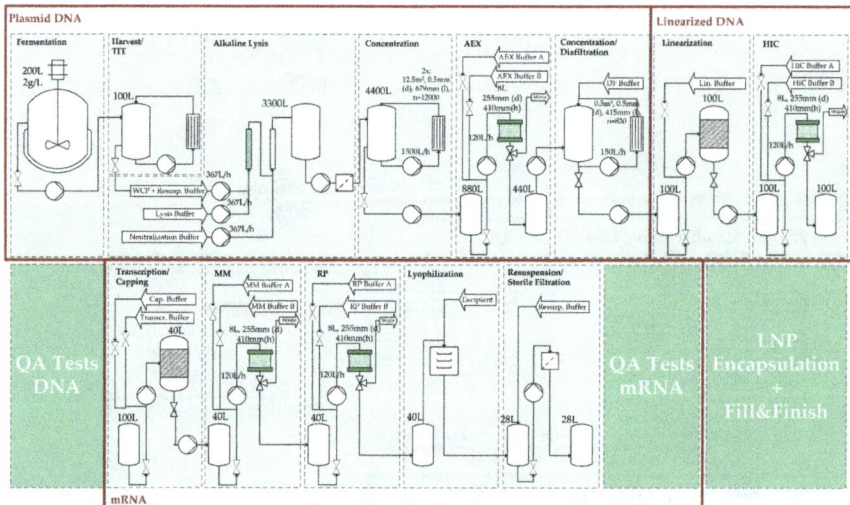

Figure 6.29: Detailed flowsheet for the production of mRNA. Starting with the generation and purification of the template DNA, followed by the *in vitro* transcription and purification of the mRNA, and ending with the encapsulation in lipid nanoparticles. Adapted from Schmidt et al. [178].

are combined with process analytical technology (PAT) via statistical-based data evaluation toward process control strategies. The resulting digital twins require specific technologies and concepts such as PAT and quality by design.

For example, in the case of mRNA vaccine production, lengthy quality controls are necessary before the purified mRNA drug can be encapsulated. This leads to holding times of up to several weeks [185, 186]. PAT is a key enabling technology for continuous biomanufacturing. Spectroscopic sensors based on chemometric calculations, such as partial least square regression and principal component analysis, are already widely used in the literature. Additionally, model-based sensors, based on mass and energy balances as well as (extended) Kalman filters, can provide an extended process understanding, as they can be based on physicochemical effects.

6.2.5.1 Digital twin enabled control studies for continuous mRNA manufacturing

In continuous bioproduction, digital twins operate on real-time data that is constantly refreshing the input to the process simulations. The trustworthiness of the data obtained from the digital twin is high when it encompasses basic process indicators like temperature, pressure, pH, conductivity, and levels of both the desired product and primary contaminants. For monitoring different biomanufacturing procedures, a range of spectroscopic methods including Raman, FTIR, UV-visible, fluorescence, and circular dichroism have been demonstrated to be effective. An overall continuous manufacturing process is shown in Figure 6.30.

In order to assess the potential risks to the space-time yield (STY) of capped mRNA during *in vitro* transcription, a qualitative risk analysis was conducted using an Ishikawa diagram. Several factors, including some process parameters, materials used, and equipment, have an impact on STY, while others such as mRNA sequence and staff-related risk factors are not easily quantifiable. Parameters that pose a significant risk to the critical process attribute (CQA) of STY include enzyme and template concentration,

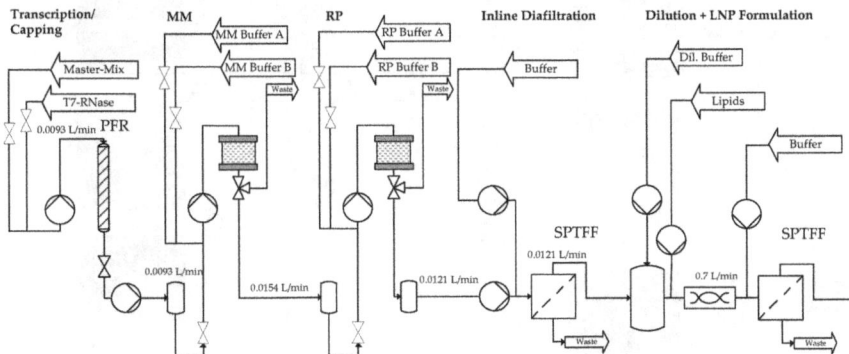

Figure 6.30: Process overview of continuous mRNA manufacturing [179].

the amount of nucleotide and cap analog used, the reactor length-to-diameter ratio, and process parameters such as temperature, pH, and volume flow rate. These risk factors were assessed using one factor at a time (OFAT) analysis.

In the case of continuous ultrafiltration, the process is mainly influenced by pressure and flow rate. Higher flow rates can increase the throughput, but also lead to higher feed pressures and pressure losses. The choice of equipment sets the boundaries for possible throughput, and membrane configurations used can impact feed pressure resulting from the feed flow. Additionally, fluctuating process parameters or other solubility reducers can impact buffer exchange efficiency by increasing membrane resistance, which can result in fouling.

The Ishikawa diagram was also used to qualitatively represent the risk factors in LNP formulation (see Figure 6.31). Factors such as personnel (training levels, stress load, motivation) and product characteristics (lipid composition, mRNA content) are difficult to quantify or are influenced by materials, equipment, and process parameters. Risks such as substrate amounts, mRNA concentration, and factors affecting hydrodynamics such as volume flow rate and volume flow ratio will be quantified using OFAT analysis. The number of diafiltration volumes and the transmembrane pressure (TMP) will also be considered in the analysis.

In the process of inline diafiltration, the critical quality attribute of utmost importance is achieving a required degree of desalination through buffer exchange before formulation. The volumetric concentration factor needs to be maintained within technically feasible and economically viable limits as a process attribute. Concentrations that are too high after diafiltration result in unnecessary complications during dilution before LNP formation, whereas it is equally important to avoid excessive reduction of concentration during diafiltration to ensure optimal concentration before formulation through dilution.

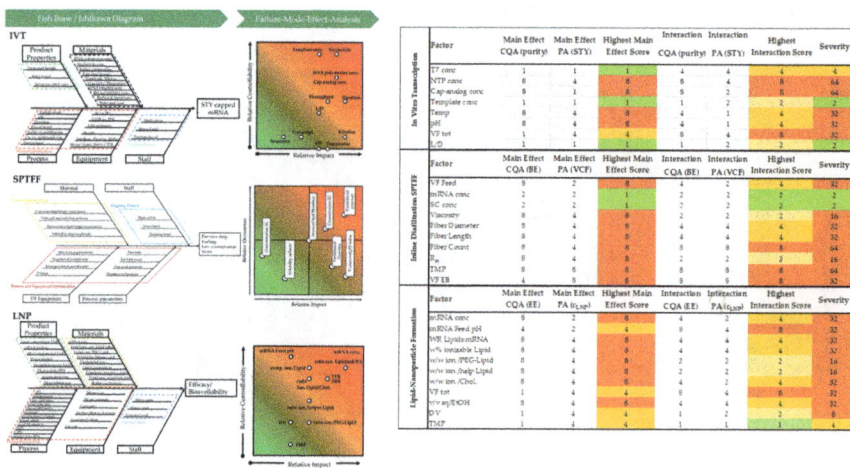

Figure 6.31: Ishikawa, graphical FMEA analysis and final impact assessment after OFAT studies and overview of main effect, interaction, and overall severity score [179].

Most process parameters, module parameters (membrane permeability, diameter, length, and number of fibers), and feed solution viscosity have a significant effect on the degree of desalination within the studied limits. The component concentrations are unlikely to have a significant influence on desalination efficiency due to high product purity after MMC and RPC, which is consistent with similar DF operations. The volumetric concentration factor shows varied effects strengths. The primary process variables, TMP and volumetric flow rate of the exchange buffer, are significant. The remaining parameters (feed and module properties) are less significant. All parameters are included in the multivariate investigation due to the severity limits defined in the planning of the investigation studies, resulting in 1025 simulations for a full factorial investigation and a center point. The OFAT study (40 simulations) identifies TMP and volume flow of the exchange buffer as the strongest effects.

The analysis of effects shows that TMP and volumetric flow rate of the exchange buffer have the most significant effects on buffer exchange (BE) and the volumetric concentration factor (VCF) for all three regression methods. The OLS and PLS statements are easily accessible through the log-p value (OLS) and the VIP score (PLS), respectively. The analysis of effect sizes in the neural network is more complex, increasing in complexity with the number of nodes and layers.

Comparing the resulting optimizations (Figure 6.32) of OLS, PLS, and neural network, the two critical process parameters show similar curves. However, the desirabilities differ between simple OLS and PLS or NN. The influence of the TMP is linearly positive for all three regressions, whereas PLS and NN can detect a nonlinear influence on the exchange buffer flow rate.

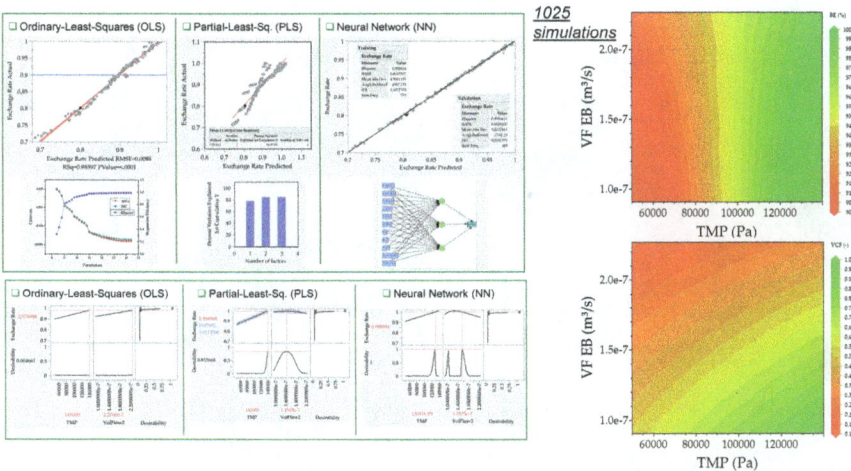

Figure 6.32: Comparison of regression efficiency and probability of OLS, PLS, and ANN in SPTFF as well as contour plot for buffer exchange and volumetric concentration factor in SPTFF based on OLS-regression mode [179].

The use of a QbD-based process control design, together with digital twins for the entire manufacturing process, supports the notion that a standard PID controller is generally sufficient for the industrialization of an autonomous continuous mRNA vaccine manufacturing platform. Nonetheless, additional benefits can be achieved by incorporating advanced process control methods based on existing digital twins, which can integrate available PAT strategies [94, 122, 177]. Simulation studies have demonstrated that such improvements can result in a 15–20 % increase in productivity and a reduction in personnel and chemical use by around 30 % [178, 187]. While it is important to acknowledge the benefits of autonomous operation, which is a key technology for future container-based manufacturing concepts [188, 189], it is worth noting that such approaches require less local manpower and can be supported remotely by the main manufacturer. Therefore, it is technically feasible to provide efficient and cost-effective worldwide supply of mRNA therapeutics for industrialization.

6.2.5.2 Process analytical technology

In order to achieve automatic process control, it is crucial to have real-time information about the product. This requires the use of process analysis technology, which was utilized in this study. The PAT station contained a variety of devices, including DAD, Raman, FTIR, MALS/DLS, and fluorescence, as described in literature [97, 106, 177] (Figures 6.33 and 6.34).

Although fluorescence and Raman spectroscopy did not show sufficient sensitivity to detect differences in mRNA concentration, DAD and FTIR were able to provide

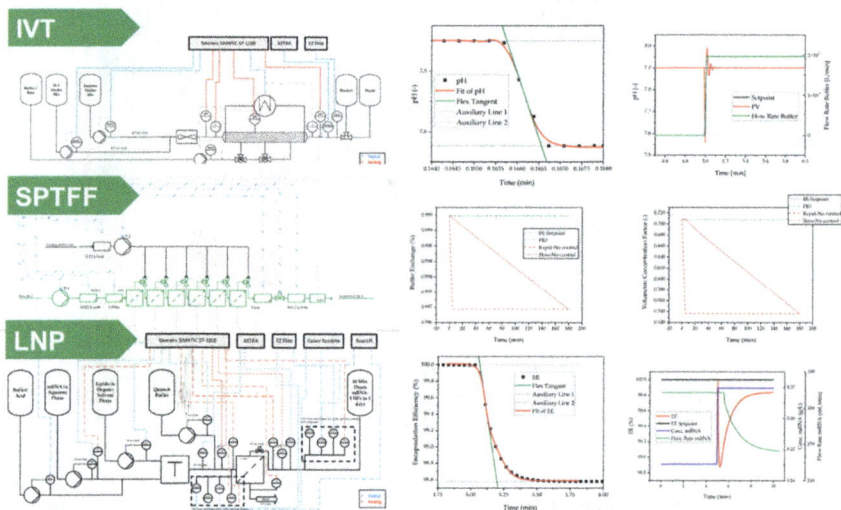

Figure 6.33: Process flowsheet and overview of control studies in continuous mRNA biomanufacturing [179].

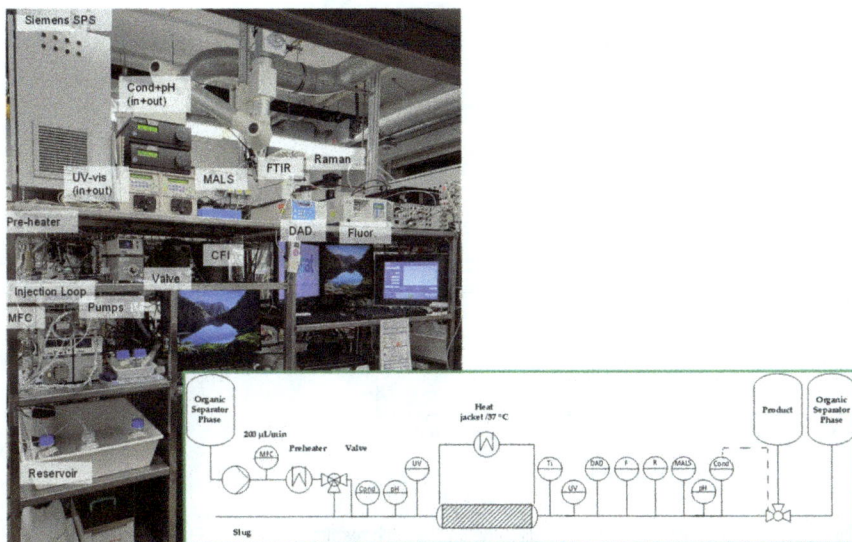

Figure 6.34: Setup of the plant including PAT [10].

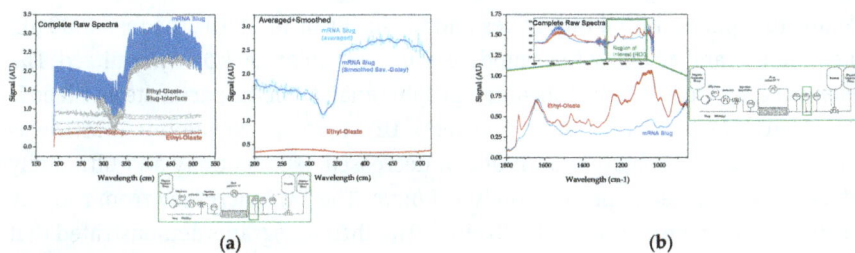

(a)

(b)

Figure 6.35: Spectra of ethyl oleate and slug in (a) DAD and (b) FTIR [10].

information for both high and low concentrations. Ethyl oleate and the slug were easily distinguishable through DAD and FTIR spectra, as shown in Figure 6.35.

To further analyze the hydrodynamic radius of the product, a MALS/DLS detector was utilized. The hydrodynamic radius was monitored over time, as shown in Figure 6.36, with the detector located behind the RP chromatograph. The radius ranged from approximately 10 to 25 nm, which was consistent with previous literature [190]. Additionally, the second peak was larger than the first peak, consistent with the functionality of RP, which retains larger molecules more strongly.

These PAT techniques are crucial for both batch and continuous manufacturing processes, allowing for real-time monitoring and control.

Following the IVT process in the PFR, a buffer exchange was conducted via ultra- and diafiltration for two purposes. Firstly, to prepare for subsequent chromatographic steps

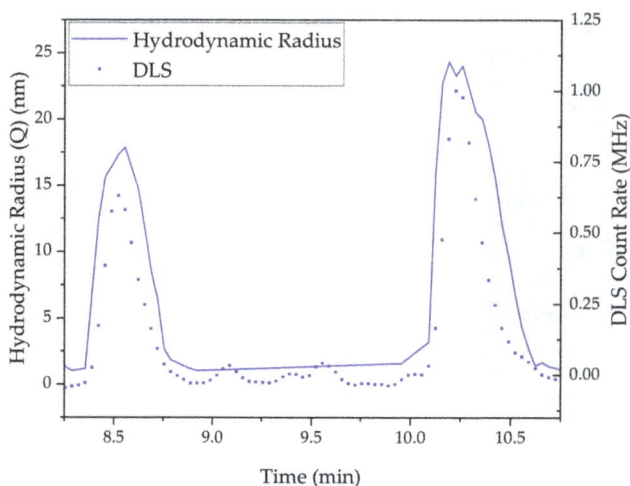

Figure 6.36: Course of the hydrodynamic radius (blue line), which results from the DLS signal (blue dots), of the slug over time [10].

and secondly, to separate initial impurities and media. Prior to UF/DF, the feed (800 μL) was diluted to 2 mL, and a hollow-fiber module with a cut-off of 300 kDa was utilized. The progression of buffer exchange and chromatographic analysis of the permeate and feed is depicted in Figure 6.23a, b, and c. At the outset of UF/DF, the permeate flux (LMH) was 50 L/m²/h. As the buffer exchange continued, it decreased exponentially and ultimately settled at a consistent rate of approximately 15 L/m²/h. The TMP increased from roughly 0.05 bar to 0.8 bar over the course of diafiltration. The chromatograms demonstrated that the mRNA was wholly retained and that both the NTPs and the cap analog could be separated. As a result, the mRNA in the retentate was purer, and recycling of the costly cap analog was feasible (Figure 6.37).

As mRNA product removal and separation of all smaller nutrients such as NTP, GTP, polymerase, cap-analog, and salts was highly efficient, a subsequent multicolumn countercurrent solvent gradient purification with mixed mode or ion exchange was not required. This was because the PCC based on the RP separation mechanisms worked effectively on its own.

6.2.5.3 Conclusion mRNA

This experimental study demonstrates the practicality of the newly invented slug-flow concept to scale up 1000 clinical doses to 10 million doses in one qualified and validated plant. The slug generation enables accurate signal-based fractionation of the mixture without using expensive reaction materials. Conductivity detectors and UV detectors can

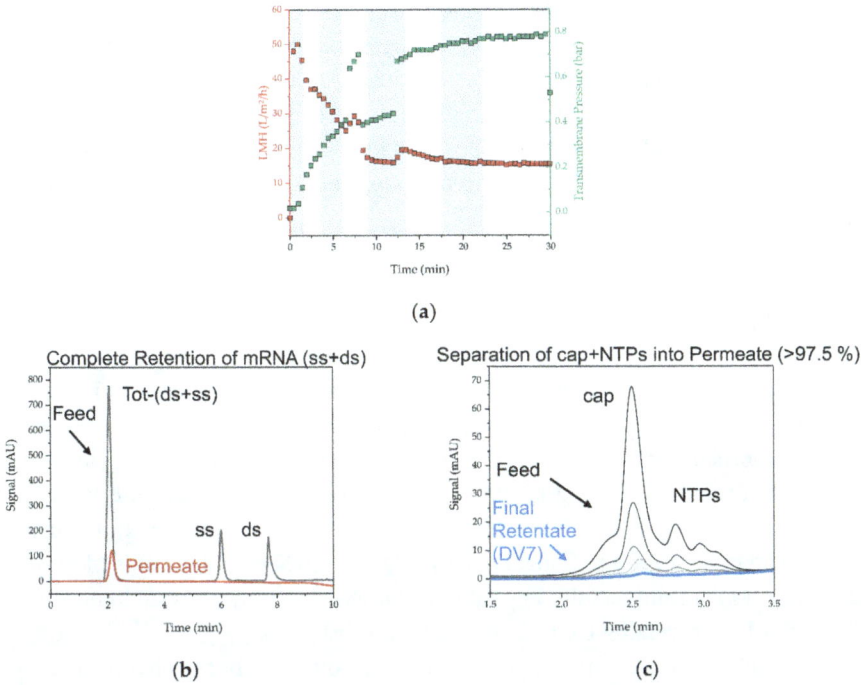

(a)

Complete Retention of mRNA (ss+ds)

Separation of cap+NTPs into Permeate (>97.5 %)

(b)

(c)

Figure 6.37: Overview of ultrafiltration performance. (a) Course of UFDF (LMH and TMP); analysis of permeate and feed by (b) RP; (c) PrimaS chromatography [10].

identify the slug and isolate the product fraction with a yield of up to 100 % and a purity of over 98 %. The feasibility of generating and fractionating such a slug flow with a variation of less than or equal to 1 % in a 10-million-dose plant is demonstrated. However, the process faces some challenges, such as potential yield losses due to impurities and an optimal operating point for the system's continuous production.

The slug-flow process can significantly reduce production time and cost compared to traditional batch processing methods. Moreover, the slug-flow process provides an opportunity for continuous production of mRNA in one facility, from small doses for clinical scale-up to 10 million doses at the production scale. To improve the purity of the product, the addition of urea is necessary, but this reduces the mRNA yield. However, optimization of the urea concentration used, as well as other process parameters, such as temperature, is desirable. The buffer exchange is performed using a hollow-fiber module with a cut-off of 300 kDa, which successfully separates unreacted nucleotides from the mRNA. Additionally, expensive cap analog can be recovered from the permeate and recycled back into the process.

The slug-flow process is suitable for autonomous process control, which can reduce personnel and chemical use, improve productivity, and reduce costs. The use of digital twins, DAD and FTIR spectrometers, and PID controllers enables the detection and fractionation of slugs and tracking of substrate consumption. The autonomous operation is an essential technology for decentralized, local, container-based manufacturing concepts of the future. The scalable mRNA machine based on slug-flow operation is technically feasible, and it is ready for industrialization. Of course any intended platform even reduced processing like enzymatic digestions followed by filtration steps are even much simpler to implement in this integrated continuous flow chemistry concept proposed.

6.2.6 Cell therapies

The use of mesenchymal stromal cell (MSC) therapies for various indications has been met with mixed clinical efficacy reports, causing some skepticism towards this approach [191]. By now reports of successful clinical outcomes as well as regulatory approvals for diseases such as Crohn's disease, graft-versus-host disease, and critical limb ischemia increased interest in MSC. Given the increasing demand for large-scale manufacturing protocols and the large number of ongoing clinical trials involving MSCs, it is of utmost importance to establish universally agreed-upon processing standards and consensus assays for both MSC processing control and product release [192].

The field of cell therapy continues to gain attention, and MSC therapy has been at the forefront for more than two decades. Although progress has been slower than expected, recent approvals and changing perceptions suggest that 10 MSC therapeutics will be approved by 2030. At projected low to high doses, an estimated 300 trillion MSCs per year will be needed to meet demand [193].

To meet this increasing demand for MSCs, it is essential to develop advanced manufacturing technologies that pave the way from autologous, patient-specific, to allogeneic cell therapy, in which MSCs are derived from a universal donor and delivered to a master cell bank [194, 195]. A major challenge for scale-up is adherent cell growth [195–197], which requires the development of scalable bioreactor technologies.

Perfusion-based reactors involve the attachment of MSCs to an immobilized substrate, such as beads or other surfaces, in a packed bed. This technology provides a continuous surface area for cells to grow but must be perfused at low-flow velocity to minimize shear stress. The limitation of this technology is that the harvesting efficiency from packed bed reactors is typically low, resulting in yields limited to about 100 million cells. Moreover, concentration gradients can arise across the packed beds, which could affect cell potency and performance [192].

Hollow-fiber bioreactors, on the other hand, offer a large surface area to volume ratio with an estimated average surface area of $2.1\,m^2$. These bioreactors use stacked semipermeable hollow fibers that allow separation between cells grown on the intra-capillary surface of the fiber with the flow of nutrients and oxygen on the other surface. MSCs are typically grown on fibronectin-coated hollow-fiber reactors and have shown comparable expansion and cell surface marker expression to cells expanded on planar systems. The bioreactor systems also allow for in-line monitoring of metabolite concentrations and control of gas and glucose in real-time. The limitation of this technology is that the economic factors, based on high consumables, equipment costs, and low harvesting density, make it one of the most expensive MSC manufacturing technologies available. The scalability of the hollow-fiber bioreactor is also uncertain, requiring multiple disposable bioreactor units to supply sufficient MSCs for late-phase clinical trials or commercial distribution [192].

Continuous upstream in MSC manufacturing allows for increased production of cell products using smaller reactor vessels, leading to faster production of final products and less downtime due to changeover processes. Additionally, utilizing smaller scale reactors as part of a multi-reactor system serves as an effective risk management strategy, reducing potential loss in case of contamination or other manufacturing errors [198]. In summary, each of these bioreactor technologies has its advantages and limitations, so careful consideration must be given to which is best suited for each application. The conclusion is that while MSCs show clinical promise, economies of scale have not yet been achieved despite the numerous bioreactor technologies available for the scalable production of MSCs. Given the growing demand for MSC-based therapies, continuous manufacturing in bioreactors offers a solution to meet the increasing demand while maintaining high-quality and consistent cell products.

Additionally, the exponential growth of the cell and gene therapy industry has resulted in an increasing demand for lentiviral-vectors (LVV). However, current LVV production methods, which involve transient transfection, are limited by volume and time constraints. To overcome these limitations, alternative production processes that involve stable clonal cell lines may be required to meet future demands. With stable producer cell lines, it will be possible to continuously produce lentiviral vectors at a scalable rate for multiple days. Continuous manufacturing of drug products for autologous therapies presents an ideal opportunity, which could extend beyond decentralized manufacturing to point-of-care or even bedside manufacturing [199].

Currently, there are limited hardware options that can efficiently meet the downstream processing needs of cell therapies in a closed and scalable manner. Some of the existing options include counter-flow centrifugation continuous centrifugation and TFF. To overcome these challenges and effectively cater to the diverse downstream processing requirements of cell therapy products, new and innovative separation solutions may need to be developed, possibly by adapting technologies from other fields [200].

Tangential flow filtration (TFF) is a critical downstream process in the manufacturing of mesenchymal stem cells (MSCs). The need for TFF in this process is well-established, and it is widely used to remove impurities and concentrate the target product. TFF has traditionally been operated in a batch mode, where a fixed volume of feed is processed in a single run. However, the continuous mode of operation for TFF has recently gained interest in the MSC manufacturing industry due to several advantages, including increased process efficiency, reduced variability, and improved product quality.

One of the key advantages of continuous TFF is the ability to increase process efficiency. Continuous TFF eliminates the need for multiple start-up and shut-down procedures required for batch operation, leading to reduced processing time and an increase in overall productivity. Additionally, continuous TFF enables the processing of large volumes of feed without the need for significant equipment scaling. This reduces capital and operational costs, making the process more economically feasible.

Continuous TFF is also ideal for maintaining product quality and consistency. The continuous mode of operation allows for real-time monitoring of the process, enabling faster adjustments to maintain product quality and consistency. This is particularly important in the manufacturing of MSCs, where the quality and consistency of the final product are critical for their clinical use.

The continuous mode of operation for TFF can be achieved using single-pass tangential flow filtration (SPTFF), which is an innovative technology that enables continuous processing. SPTFF is a continuous flow-through system that is designed to process feed streams with minimal hold-up volume, making it ideal for continuous TFF. SPTFF can be integrated into a fully automated continuous manufacturing process, enabling real-time monitoring and control of the process.

Process control concepts such as digital twins and process analytical technology (PAT) are critical for utilizing the full potential of continuous TFF in MSC manufacturing. In the case of continuous TFF, digital twins can be used to optimize the process design and predict the behavior of the system under different operating conditions. PAT, on the other hand, involves the use of real-time measurements and analytics to control and optimize the manufacturing process. PAT can be used to monitor key process parameters such as pressure, flow rate, and temperature, enabling faster adjustments to maintain product quality and consistency. Continuous TFF using SPTFF is an ideal unit operation for MSC manufacturing due to its potential to increase process efficiency, maintain product quality and consistency, and reduce capital and operational costs. Process control concepts such as digital twins and PAT are essential for utilizing the full potential of continuous TFF in MSC manufacturing. Several case studies have demonstrated the feasibility and benefits of continuous TFF, highlighting the potential for its widespread adoption in the future.

The pharma industry and its regulatory authorities are shifting towards closed, automated systems in many drug manufacturing areas, including cell culture production, due to the many advantages it offers such as improved safety, reduced risk of contamination, and significant reductions in space, energy consumption, staff numbers, and approval time. There are three approaches to cell culture production, which are open and manual, semi-closed and automated, and fully closed and automated. Each approach has its advantages and disadvantages in terms of safety, risk, costs, scalability, and suitability for specific types of cells. Although fully closed and automated cell culture production offers the greatest benefits in terms of safety and reduced costs, it may not be suitable for all product types, especially those requiring large-scale production. The best approach depends on the specific needs of the cell therapy production process [201].

Equipment development for tailored therapies is necessary to move away from manual handling and towards a fully closed and fully automated process, using isolator technologies for continuous processing. The lack of cooperation between pharma/biotech companies and equipment suppliers is a missing link in the development towards a more sustainable concept [202].

In addition to reducing the need for switching connectors, continuous manufacturing also offers the potential for increased automation. Automated processes can help ensure that equipment operates consistently and as intended, reducing the risk of operator error or variability in processing steps. This can lead to more consistent and reliable product quality. Furthermore, automation can enable real-time monitoring and control of the manufacturing process, allowing for rapid adjustments if any deviations from specifications are detected. This not only improves the efficiency of the manufacturing process, but also reduces the risk of producing suboptimal or non-conforming products. Overall, combining continuous manufacturing with automation can lead to a more efficient and reliable manufacturing process, with reduced need for manual intervention and potential errors [203].

6.2.6.1 Conclusion cell therapies

This overview points out the potential for even small product amounts dedicated to e.g. allogenic cell therapy manufacturing to be of benefit under continuous manufacturing mode. Mesenchymal stromal cell (MSC) therapies show clinical promise, achieving economies of scale has not yet been possible despite the numerous bioreactor technologies available for scalable MSC production. Given the growing demand for MSC-based therapies, continuous manufacturing in bioreactors offers a solution to meet the increasing demand while maintaining high-quality and consistent cell products.

With stable producer cell lines, it will be possible to continuously produce lentiviral vectors at a scalable rate for multiple days. Continuous manufacturing of drug products for autologous therapies presents an ideal opportunity, which could extend beyond decentralized manufacturing to point-of-care or even bedside manufacturing.

To efficiently meet the downstream processing needs of cell therapies in a closed and scalable manner, there are several opportunities in both up and downstream processing, as highlighted in this chapter, which can be explored to improve cell therapy manufacturing.

6.3 Conclusions

After exploring the various aspects of upstream and downstream processing in continuous biomanufacturing, it is clear that Biopharma 4.0 is essential to modernize the biopharmaceutical industry. With the increasing demand for faster and more affordable medicines and vaccines, the industry must adapt to continuous bioprocessing techniques that offer greater efficiency and productivity.

The chapter showcases different case studies, demonstrating how continuous bioprocessing has been implemented in different entities, including monoclonal antibody production and chromatography-based downstream processing. These examples showcase the benefits of using continuous bioprocessing, such as reduced production times, improved product quality, and increased yield.

Digital twins and process analytical technology (PAT) are key enablers in quality by design (QbD)-based process development. Digital twins provide a virtual representation of the manufacturing process, allowing for optimization and monitoring in real-time. Meanwhile, PAT enables real-time monitoring of critical quality attributes, ensuring that the process stays within acceptable limits.

Continuous bioprocessing is the only solution to meet the demands of the current pandemic times. The advantages of continuous bioprocessing include reduced production times, improved flexibility, and higher productivity, allowing for faster and more cost-effective production of vital medicines and vaccines. With the COVID-19 pandemic, the need for rapid and efficient vaccine production has become even more crucial, and continuous bioprocessing has played a significant role in meeting this challenge.

In conclusion, the implementation of Biopharma 4.0 and continuous bioprocessing techniques is essential for the biopharmaceutical industry's future success. The chapter showcases how continuous bioprocessing has been successfully implemented in various case studies, highlighting the benefits of this approach. The use of digital twins and PAT also enables QbD-based process development, further improving process optimization and product quality. Finally, the advantages of continuous bioprocessing, including faster and more affordable medicines and vaccines, make it the only solution to meet the demands of the current pandemic times.

Acknowledgments: The authors would like to thank the editor Dirk Holtmann for their guidance and review of this article before its publication.

References

1. Fisher AC, Lee SL, Harris DP, Buhse L, Kozlowski S, Yu L, et al. Advancing pharmaceutical quality: an overview of science and research in the U.S. FDA's Office of Pharmaceutical Quality. Int J Pharm 2016;515: 390–402.
2. Fisher AC, Liu W, Schick A, Ramanadham M, Chatterjee S, Brykman R, et al. An audit of pharmaceutical continuous manufacturing regulatory submissions and outcomes in the US. Int J Pharm 2022;622:121778.
3. Lee SL, O'Connor TF, Yang X, Cruz CN, Chatterjee S, Madurawe RD, et al. Modernizing pharmaceutical manufacturing: from batch to continuous production. J Pharm Innov 2015;10:191–9.
4. ISPE | International Society for Pharmaceutical Engineering. Continuous manufacturing as a tool for accelerated development. https://ispe.org/pharmaceutical-engineering/july-august-2021/continuous-manufacturing-tool-accelerated-development [Accessed 19 Apr 2023].

5. Jarrett S, Pagliusi S, Park R, Wilmansyah T, Jadhav S, Santana PC, et al. The importance of vaccine stockpiling to respond to epidemics and remediate global supply shortages affecting immunization: strategic challenges and risks identified by manufacturers. Vaccine X 2021;9:100119.
6. Zhu G, Chou MC, Tsai CW. Lessons learned from the COVID-19 pandemic exposing the shortcomings of current supply chain operations: a long-term prescriptive offering. Sustainability 2020;12:5858.
7. Kleinebudde P, Khinast J, Rantanen J. Continuous manufacturing of pharmaceuticals, 1st. Hoboken, New Jersey: John Wiley & Sons; 2017.
8. Konstantinov KB, Cooney CL. White paper on continuous bioprocessing May 20–21 2014 continuous manufacturing symposium. J Pharmaceut Sci 2015;104:813–20.
9. Woodcock J. Modernizing pharmaceutical manufacturing–continuous manufacturing as a key enabler. In: Proceedings of the MIT-CMAC International symposium on continuous manufacturing of pharmaceuticals. Cambridge, MA, USA; 2014, 20.
10. Hengelbrock A, Schmidt A, Helgers H, Vetter FL, Strube J. Scalable mRNA machine for regulatory approval of variable scale between 1000 clinical doses to 10 million manufacturing scale doses. Processes 2023;11:745.
11. Croughan MS, Konstantinov KB, Cooney C. The future of industrial bioprocessing: batch or continuous? Biotechnol Bioeng 2015;112:648–51.
12. Ding B. Pharma Industry 4.0: literature review and research opportunities in sustainable pharmaceutical supply chains. Process Saf Environ Protect 2018;119:115–30.
13. Yu YB, Briggs KT, Taraban MB, Brinson RG, Marino JP. Grand challenges in pharmaceutical research series: ridding the cold chain for biologics. Pharm Res 2021;38:3–7.
14. Ghani S. ICH quality guidelines. In: Chan CC, Chow K, Mckay B, Fung M, editors. Therapeutic delivery solutions. Hoboken, New Jersey: John Wiley & Sons; 2014:367–80 pp.
15. Teasdale A, Elder D, Nims RW. ICH quality guidelines: an implementation guide. In: Teasdale A, Elder D, Nims RW, editors, 1st ed. Hoboken, New Jersey: John Wiley & Sons; 2017.
16. Kornecki M, Schmidt A, Strube J. PAT as key-enabling technology for QbD in pharmaceutical manufacturing a conceptual review on upstream and downstream processing. Chimica Oggi Chem Today 2018;36:44–8.
17. Godawat R, Konstantinov K, Rohani M, Warikoo V. End-to-end integrated fully continuous production of recombinant monoclonal antibodies. J Biotechnol 2015;213:13–19.
18. FDA's comprehensive effort to advance new innovations: initiatives to modernize for innovation. FDA [Online], Thu, March 10, 2022 – 10:09. https://www.fda.gov/news-events/fda-voices/fdas-comprehensive-effort-advance-new-innovations-initiatives-modernize-innovation [Accessed 19 Apr 2023].
19. FDA's new efforts to advance biotechnology innovation. FDA [Online], Thu, March 10, 2022 – 12:16. https://www.fda.gov/news-events/fda-voices/fdas-new-efforts-advance-biotechnology-innovation [Accessed 19 Apr 2023].
20. Udugama IA, Lopez PC, Gargalo CL, Li X, Bayer C, Gernaey KV. Digital Twin in biomanufacturing: challenges and opportunities towards its implementation. Syst Microbiol Biomanuf 2021;1:257–74.
21. Subramanian G., editor. Process control, intensification, and digitalisation in continuous biomanufacturing. Weinheim: Wiley-VCH; 2022.
22. Khanal O, Lenhoff AM. Developments and opportunities in continuous biopharmaceutical manufacturing. mAbs 2021;13:1903664.
23. Boodhoo K, Flickinger MC, Woodley JM, Emanuelsson E. Bioprocess intensification: a route to efficient and sustainable biocatalytic transformations for the future. Chem Eng Process Process Intensif 2022;172:108793.
24. Mahmoud MS, Sabih M, Elshafei M. Using OPC technology to support the study of advanced process control. ISA Trans 2015;55:155–67.

25. Shao G, Latif H, Martin-Villalba C, Denno P. Standards-based integration of advanced process control and optimization. J Ind Inf Integr 2019;13:1–12.
26. European Pharmaceutical Review. Key challenges for bio/pharmaceutical manufacturing 2022. https://www.europeanpharmaceuticalreview.com/article/167733/key-challenges-for-bio-pharmaceutical-manufacturing-2022/ [Accessed 19 Apr 2023].
27. ISPE | International Society for Pharmaceutical Engineering. Data science for pharma 4.0™, drug development, & production—Part 1. https://ispe.org/pharmaceutical-engineering/march-april-2021/data-science-pharma-40tm-drug-development-production [Accessed 19 Apr 2023].
28. Otto Ralf, et al. Rapid growth in biopharma: challenges and opportunities. McKinsey & Company; 2014;1.
29. Erickson J, Baker J, Barrett S, Brady C, Brower M, Carbonell R, et al. End-to-end collaboration to transform biopharmaceutical development and manufacturing. Biotechnol Bioeng 2021;118:3302–12.
30. Mathur AK, Vyas A. Greening of management: social responsibility of Indian pharmaceutical industry. Int J Pharm Sci Res 2013;4:502–5.
31. Schmidt A, Uhlenbrock L, Strube J. Technical potential for energy and GWP reduction in chemical–pharmaceutical industry in Germany and EU—focused on biologics and botanicals manufacturing. Processes 2020;8:818.
32. Madabhushi SR, Pinto NDS, Lin H. Comparison of process mass intensity (PMI) of continuous and batch manufacturing processes for biologics. Nat Biotechnol 2022;72:122–7.
33. Kumar A, Udugama IA, Gargalo CL, Gernaey KV. Why is batch processing still dominating the biologics landscape? Towards an integrated continuous bioprocessing alternative. Processes 2020;8:1641.
34. Boedeker, Berthold GD. "Recombinant factor VIII (Kogenate®) for the treatment of hemophilia a: the first and only World-Wide licensed recombinant protein produced in high-throughput perfusion culture." In: *Modern biopharmaceuticals: recent success stories*. Weinheim Wiley-VCH; 2013: 429–43 pp.
35. Schmidt A, Helgers H, Vetter FL, Juckers A, Strube J. Digital twin of mRNA-based SARS-COVID-19 vaccine manufacturing towards autonomous operation for improvements in speed, scale, robustness, flexibility and real-time release testing. Processes 2021;9:748.
36. Reinhardt IC, Oliveira JC, Ring DT. Current perspectives on the development of industry 4.0 in the pharmaceutical sector. J Ind Inf Integr 2020;18:100131.
37. Chen Y, Yang O, Sampat C, Bhalode P, Ramachandran R, Ierapetritou M. Digital twins in pharmaceutical and biopharmaceutical manufacturing: a literature review. Processes 2020;8:1088.
38. Rao V, Srinivas K. Modern drug discovery process. An in silico approach J Bioinform Sequence Anal 2011;2:89–94.
39. Erol T, Mendi AF, Dogan D. The digital twin revolution in Healthcare. In: 2020 4th International Symposium on Multidisciplinary Studies and Innovative Technologies (ISMSIT), Istanbul, Turkey, 22–24 Oct. 2020. IEEE; 2020:1–7 pp.
40. Coorey G, Figtree GA, Fletcher DF, Redfern J. The health digital twin: advancing precision cardiovascular medicine. Nat Rev Cardiol 2021;18:803–4.
41. Fraunhofer-Gesellschaft. med2icin. https://websites.fraunhofer.de/med2icin/ [Accessed 5 Jan 2023].
42. Venkatesh KP, Raza MM, Kvedar JC. Health digital twins as tools for precision medicine: considerations for computation, implementation, and regulation. NPJ Digit Med 2022;5:150.
43. Valerio LG. Application of advanced in silico methods for predictive modeling and information integration. Expet Opin Drug Metabol Toxicol 2012;8:395–8.
44. Rønning-Andersson A, Pedersen DIK. Novo Nordisk and Novartis will participate in large Danish project on decentralized clinical trials. JP/Politikens Hus A/S [Online]. September 28, 2022. https://medwatch.com/News/Pharma___Biotech/article14445325.ece [Accessed 5 Jan 2023].
45. Lorenz S, Amsel A-K, Puhlmann N, Reich M, Olsson O, Kümmerer K. Toward application and implementation of in silico tools and workflows within benign by design approaches. ACS Sustainable Chem Eng 2021;9:12461–75.

46. Boetker J, Raijada D, Aho J, Khorasani M, Søgaard SV, Arnfast L, et al. In silico product design of pharmaceuticals. Asian J Pharm Sci 2016;11:492–9.
47. Roush D, Asthagiri D, Babi DK, Benner S, Bilodeau C, Carta G, et al. Toward in silico CMC: an industrial collaborative approach to model-based process development. Biotechnol Bioeng 2020;117:3986–4000.
48. Zobel-Roos S, Schmidt A, Uhlenbrock L, Ditz R, Köster D, Strube J. Digital twins in biomanufacturing. Adv Biochem Eng Biotechnol 2021;176:181–262.
49. Klatt K-U, Marquardt W. Perspectives for process systems engineering – a personal view from academia and industry. In: 17th European symposium on computer aided process engineering, Elsevier; 2007:19–32 pp.
50. Thon C, Finke B, Kwade A, Schilde C. Artificial intelligence in process engineering. Adv Intell Syst 2021;3: 2000261.
51. Mowbray M, Vallerio M, Perez-Galvan C, Zhang D, Del Rio Chanona A, Navarro-Brull FJ. Industrial data science – a review of machine learning applications for chemical and process industries. React Chem Eng 2022;7:1471–509.
52. SKF. Measuring unscheduled downtime. Paper Advance [Online]. September 2, 2021. https://www. paperadvance.com/mills-technologies/process-optimization/measuring-unscheduled-downtime.html [Accessed 5 Jan 2023].
53. Udugama IA, Bayer C, Baroutian S, Gernaey KV, Yu W, Young BR. Digitalisation in chemical engineering: industrial needs, academic best practice, and curriculum limitations. Educ Chem Eng 2022;39:94–107.
54. Mouellef M, Vetter FL, Strube J. Benefits and limitations of artificial neural networks in process chromatography design and operation. Processes 2023;11:1115.
55. Leng J, Wang D, Shen W, Li X, Liu Q, Chen X. Digital twins-based smart manufacturing system design in Industry 4.0: a review. J Manuf Syst 2021;60:119–37.
56. Cimino C, Negri E, Fumagalli L. Review of digital twin applications in manufacturing. Comput Ind 2019;113: 103130.
57. Gargalo CL, de Las Heras SC, Jones MN, Udugama I, Mansouri SS, Krühne U, et al. Towards the development of digital twins for the bio-manufacturing industry. Adv Biochem Eng Biotechnol 2021;176: 1–34.
58. Sixt M, Uhlenbrock L, Strube J. Toward a distinct and quantitative validation method for predictive process modelling—on the example of solid-liquid extraction processes of complex plant extracts. Processes 2018;6:66.
59. Zobel-Roos S, Schmidt A, Mestmäcker F, Mouellef M, Huter M, Uhlenbrock L, et al. Accelerating biologics manufacturing by modeling or: is approval under the QbD and PAT approaches demanded by authorities acceptable without a digital-twin? Processes 2019;7:94.
60. Uhl A, Schmidt A, Hlawitschka MW, Strube J. Autonomous liquid–liquid extraction operation in biologics manufacturing with aid of a digital twin including process analytical technology. Processes 2023;11:553.
61. ICH. Technical and regulatory considerations for pharmaceutical product lifecycle management Q12. In: International conference on harmonisation of technical requirements for registration of pharmaceuticals for human use; 2017.
62. Reardon KF. Practical monitoring technologies for cells and substrates in biomanufacturing. Curr Opin Biotechnol 2021;71:225–30.
63. Silva TC, Eppink M, Ottens M. Automation and miniaturization: enabling tools for fast, high-throughput process development in integrated continuous biomanufacturing. J Chem Tech Biotechnol 2022;97: 2365–75.
64. Evaluate Pharmaceutical. World preview 2018, outlook to 2024. Evaluate Ltd. 2018;1–47.
65. Sommerfeld S, Strube J. Challenges in biotechnology production – generic processes and process optimization for monoclonal antibodies. Chem Eng Process Process Intensif 2005;44:1123–37.
66. Gronemeyer P, Ditz R, Strube J. Trends in upstream and downstream process development for antibody manufacturing. Bioengineering 2014;1:188–212.

67. Kelley B. Very large scale monoclonal antibody purification: the case for conventional unit operations. Biotechnol Prog 2007;23:995–1008.
68. Strube J, Ditz R, Kornecki M, Huter M, Schmidt A, Thiess H, et al. Process intensification in biologics manufacturing. Chem Eng Process Process Intensif 2018;133:278–93.
69. Strube J, Grote F, Ditz R. Bioprocess design and production technology for the future. Biopharmaceutical production technology. Wiley-VCH Verlag GmbH & Co. KGaA; 2012:657–705 pp.
70. Singh N, Arunkumar A, Chollangi S, Tan ZG, Borys M, Li ZJ. Clarification technologies for monoclonal antibody manufacturing processes: current state and future perspectives. Biotechnol Bioeng 2016;113: 698–716.
71. Jain E, Kumar A. Upstream processes in antibody production: evaluation of critical parameters. Biotechnol Adv 2008;26:46–72.
72. Pollock J, Ho SV, Farid SS. Fed-batch and perfusion culture processes: economic, environmental, and operational feasibility under uncertainty. Biotechnol Bioeng 2013;110:206–19.
73. Birch JR, Racher AJ. Antibody production. Adv Drug Deliv Rev 2006;58:671–85.
74. Al-Rubeai M, editor. Animal cell culture. Cham: Springer International Publishing; 2015.
75. Subramanian G, editor. Continuous biomanufacturing - innovative technologies and methods. Weinheim, Germany: Wiley-VCH Verlag GmbH & Co. KGaA; 2017.
76. Subramanian G, editor. Biopharmaceutical production technology. Weinheim, Germany: Wiley-VCH Verlag GmbH & Co. KGaA; 2012.
77. Subramanian G, editor. Bioseparation and bioprocessing, 2nd ed. Weinheim, Germany: Wiley VCH; 2007.
78. Rita Costa A, Elisa Rodrigues M, Henriques M, Azeredo J, Oliveira R. Guidelines to cell engineering for monoclonal antibody production. Eur J Pharm Biopharm 2010;74:127–38.
79. Gstraunthaler G, Lindl T. Zell- und Gewebekultur: Allgemeine Grundlagen und spezielle Anwendungen, 7. Auflage. Berlin, Heidelberg, s.l.: Springer Berlin Heidelberg; 2013.
80. Gronemeyer P, Ditz R, Strube J. Implementation of aqueous two-phase extraction combined with precipitation in a monoclonal antibody manufacturing process. Chimica Oggi Chem Today 2016;34:66–70.
81. Gronemeyer P, Schmidt A, Zobel S, Strube J. Efficient manufacturing of biologics. Chem Ing Tech 2016;88: 1329.
82. Gronemeyer P, Thiess H, Zobel-Roos S, Ditz R, Strube J. Integration of upstream and downstream in continuous biomanufacturing. In: Subramanian G, editor. Continuous biomanufacturing – innovative technologies and methods. Weinheim, Germany: Wiley-VCH Verlag GmbH & Co. KGaA; 2017:481–510 pp.
83. Jungbauer A. Continuous downstream processing of biopharmaceuticals. Trends Biotechnol 2013;31: 479–92.
84. Karst DJ, Steinebach F, Soos M, Morbidelli M. Process performance and product quality in an integrated continuous antibody production process. Biotechnol Bioeng 2017;114:298–307.
85. Karst DJ, Serra E, Villiger TK, Soos M, Morbidelli M. Characterization and comparison of ATF and TFF in stirred bioreactors for continuous mammalian cell culture processes. Biochem Eng J 2016;110:17–26.
86. Warikoo V, Godawat R, Brower K, Jain S, Cummings D, Simons E, et al. Integrated continuous production of recombinant therapeutic proteins. Biotechnol Bioeng 2012;109:3018–29.
87. Zydney AL. Continuous downstream processing for high value biological products: a review. Biotechnol Bioeng 2016;113:465–75.
88. Hernandez R. Continuous manufacturing: a changing processing paradigm. Biopharm Int 2015;2015: 20–7.
89. Montgomery A, Scott C, Center A. Continuous Chromatography: experts weigh in on the possibilities and the reality. Boston, MA, USA: BioProcess International; 2019.
90. Thiess H, Zobel-Roos S, Gronemeyer P, Ditz R, Strube J. Engineering challenges of continuous biomanufacturing processes (CBP). In: Subramanian G, editor. Continuous biomanufacturing – innovative technologies and methods. Weinheim, Germany: Wiley-VCH Verlag GmbH & Co. KGaA; 2017: 69–106 pp.

91. Steinebach F, Ulmer N, Wolf M, Decker L, Schneider V, Walchli R, et al. Design and operation of a continuous integrated monoclonal antibody production process. Biotechnol Prog 2017;33:1303–13.
92. Kateja N, Kumar D, Sethi S, Rathore AS. Non-protein A purification platform for continuous processing of monoclonal antibody therapeutics. J Chromatogr A 2018;1579:60–72.
93. Whitford W. GE Healthcare. Continuous biomanufacturing: 10 reasons sponsors Hesitate. https://www.bioprocessonline.com/doc/continuous-biomanufacturing-reason-sponsors-hesitate-0001 [Accessed 16 May 2019].
94. Helgers H, Schmidt A, Strube J. Towards autonomous process control—digital twin for CHO cell-based antibody manufacturing using a dynamic metabolic model. Processes 2022;10:316.
95. Pisano R, Arsiccio A, Capozzi LC, Trout BL. Achieving continuous manufacturing in lyophilization: technologies and approaches. Eur J Pharm Biopharm 2019;142:265–79.
96. Kornecki M, Schmidt A, Lohmann L, Huter M, Mestmäcker F, Klepzig L, et al. Accelerating biomanufacturing by modeling of continuous bioprocessing—piloting case study of monoclonal antibody manufacturing. Processes 2019;7:495.
97. Helgers H, Schmidt A, Lohmann LJ, Vetter FL, Juckers A, Jensch C, et al. Towards autonomous operation by advanced process control—process analytical technology for continuous biologics antibody manufacturing. Processes 2021;9:172.
98. Hammerschmidt N, Hobiger S, Jungbauer A. Continuous polyethylene glycol precipitation of recombinant antibodies: sequential precipitation and resolubilization. Process Biochem 2016;51:325–32.
99. Burgstaller D, Jungbauer A, Satzer P. Continuous integrated antibody precipitation with two-stage tangential flow microfiltration enables constant mass flow. Biotechnol Bioeng 2019;116:1053–65.
100. Dutra G, Komuczki D, Jungbauer A, Satzer P. Continuous capture of recombinant antibodies by ZnCl$_2$ precipitation without polyethylene glycol. Eng Life Sci 2020;20:265–74.
101. Li Z, Gu Q, Coffman JL, Przybycien T, Zydney AL. Continuous precipitation for monoclonal antibody capture using countercurrent washing by microfiltration. Biotechnol Prog 2019;35:e2886.
102. Lohmann LJ, Strube J. Accelerating biologics manufacturing by modeling: process integration of precipitation in mAb downstream processing. Processes 2020;8:58.
103. Lohmann LJ, Strube J. Process analytical technology for precipitation process integration into biologics manufacturing towards autonomous operation—mAb case study. Processes 2021;9:488.
104. Lohmann LJ. Quality by Design basierte Prozessintegration der Präzipitati-on von Monoklonalen Antikörpern sowie die Entwicklung eines Digitalen Zwillings zur Unterstützung der PAT-gestützten autonomen Prozessführung: Shaker Verlag. Clausthal-Zellerfeld: University of Technology Clausthal; 2022.
105. Li Z. Development of an integrated precipitation-filtration process for initial purification of recombinant proteins - Dissertation. PennState University Libraries; 2020.
106. Schmidt A, Helgers H, Lohmann LJ, Vetter F, Juckers A, Mouellef M, et al. Process analytical technology as key-enabler for digital twins in continuous biomanufacturing. J Chem Tech Biotechnol 2022;97:2336–46.
107. Biologics S. Large-scale continuous biomanufacturing: utilizing N-1 technology for higher productivity and quality. https://samsungbiologics.com/media/science-technology/large-scale-continuous-biomanufacturing-utilizing-n-1-technology-for-higher-productivity-and-quality [Accessed 19 Apr 2023].
108. Shanley A. Technology redefines continuous processing efficiency. Biopharm Int 2019;32:12–16.
109. Casali N. Escherichia coli host strains. Methods Mol Biol 2003;235:27–48.
110. Sørensen HP, Mortensen KK. Advanced genetic strategies for recombinant protein expression in Escherichia coli. J Biotechnol 2005;115:113–28.
111. Terpe K. Overview of bacterial expression systems for heterologous protein production: from molecular and biochemical fundamentals to commercial systems. Appl Microbiol Biotechnol 2006;72:211–22.
112. Sanchez-Garcia L, Martín L, Mangues R, Ferrer-Miralles N, Vázquez E, Villaverde A. Recombinant pharmaceuticals from microbial cells: a 2015 update. Microb Cell Factories 2016;15:33.
113. Walsh G. Biopharmaceutical benchmarks 2014. Nat Biotechnol 2014;32:992–1000.

114. Baeshen NA, Baeshen MN, Sheikh A, Bora RS, Ahmed MMM, Ramadan HAI, et al. Cell factories for insulin production. Microb Cell Factories 2014;13:141.
115. Ferrer-Miralles N, Domingo-Espín J, Corchero JL, Vázquez E, Villaverde A. Microbial factories for recombinant pharmaceuticals. Microb Cell Factories 2009;8:17.
116. Swartz JR. Advances in Escherichia coli production of therapeutic proteins. Curr Opin Biotechnol 2001;12: 195–201.
117. Tripathi NK. Production and purification of recombinant proteins from Escherichia coli. ChemBioEng Rev 2016;3:116–33.
118. Rosano GL, Ceccarelli EA. Recombinant protein expression in Escherichia coli: advances and challenges. Front Microbiol 2014;5:172.
119. Baneyx F. Recombinant protein expression in Escherichia coli. Curr Opin Biotechnol 1999;10:411–21.
120. Studier F, Moffatt BA. Use of bacteriophage T7 RNA polymerase to direct selective high-level expression of cloned genes. J Mol Biol 1986;189:113–30.
121. Weickert MJ, Doherty DH, Best EA, Olins PO. Optimization of heterologous protein production in Escherichia coli. Curr Opin Biotechnol 1996;7:494–9.
122. Helgers H, Hengelbrock A, Schmidt A, Vetter FL, Juckers A, Strube J. Digital twins for scFv production in Escherichia coli. Processes 2022;10:809.
123. Lau JL, Dunn MK. Therapeutic peptides: Historical perspectives, current development trends, and future directions. Bioorg Med Chem 2018;26:2700–7.
124. Pennington MW, Zell B, Bai CJ. Commercial manufacturing of current good manufacturing practice peptides spanning the gamut from neoantigen to commercial large-scale products. Med Drug Discov 2021;9:100071.
125. Wang L, Wang N, Zhang W, Cheng X, Yan Z, Shao G, et al. Therapeutic peptides: current applications and future directions. Signal Transduct Targeted Ther 2022;7:48.
126. Ferrazzano L, Catani M, Cavazzini A, Martelli G, Corbisiero D, Cantelmi P, et al. Sustainability in peptide chemistry: current synthesis and purification technologies and future challenges. Green Chem 2022;24: 975–1020.
127. Jadhav S, Seufert W, Lechner C, Schönleber R. Bachem – insights into innovative and sustainable peptide chemistry and technology by the leading independent manufacturer of TIDES. Chimia (Aarau) 2021;75: 476–9.
128. Malonis RJ, Lai JR, Vergnolle O. Peptide-based vaccines: current progress and future challenges. Chem Rev 2020;120:3210–29.
129. Muttenthaler M, King GF, Adams DJ, Alewood PF. Trends in peptide drug discovery. Nat Rev Drug Discov 2021;20:309–25.
130. Landgraf W, Sandow J. Recombinant human insulins – clinical efficacy and safety in diabetes therapy. Eur Endocrinol 2016;12:12–17.
131. Sandow J, Landgraf W, Becker R, Seipke G. Equivalent recombinant human insulin preparations and their place in therapy. Eur Endocrinol 2015;11:10–16.
132. Ucar B, Acar T, Pelit Arayici P, Sen M, Derman S, Mustafaeva Z. Synthesis and applications of synthetic peptides. In: Varkey JT, editor. Peptide synthesis. London: IntechOpen; 2019.
133. Sharma A, Kumar A, de La Torre BG, Albericio F. Liquid-phase peptide synthesis (LPPS): a third wave for the preparation of peptides. Chem Rev 2022;122:13516–46.
134. Martin V, Egelund PHG, Johansson H, Le Thordal Quement S, Wojcik F, Sejer Pedersen D. Greening the synthesis of peptide therapeutics: an industrial perspective. RSC Adv 2020;10:42457–92.
135. Simon MD, Heider PL, Adamo A, Vinogradov AA, Mong SK, Li X, et al. Rapid flow-based peptide synthesis. Chembiochem 2014;15:713–20.
136. Isidro-Llobet A, Kenworthy MN, Mukherjee S, Kopach ME, Wegner K, Gallou F, et al. Sustainability challenges in peptide synthesis and purification: from R&D to production. J Org Chem 2019;84:4615–28.

137. Mijalisi AJ. Pushing the limits of solid phase peptide synthesis with continuous flow. California: UC Berkeley; 2018.

138. GMP production. https://www.bachem.com/products/services-and-capabilities/gmp-production/ [Accessed 19 Apr 2023].

139. Custom peptide synthesis. https://www.bachem.com/products/research-and-specialties/custom-peptide-synthesis/ [Accessed 19 Apr 2023].

140. Hessel V, Löwe H. Mikroverfahrenstechnik: Komponenten – Anlagenkonzeption – Anwenderakzeptanz – Teil 1. Chem Ing Tech 2002;74:17–30.

141. Ehrfeld Mikrotechnik. Miprowa | ehrfeld mikrotechnik. https://www.ehrfeld.com/en/miprowa [Accessed 19 Apr 2023].

142. Plutschack MB, Pieber B, Gilmore K, Seeberger PH. The Hitchhiker's guide to flow chemistry. Chem Rev 2017;117:11796–893.

143. Rogers L, Jensen KF. Continuous manufacturing – the green chemistry promise? Green Chem 2019;21: 3481–98.

144. Bornscheuer UT, Huisman GW, Kazlauskas RJ, Lutz S, Moore JC, Robins K. Engineering the third wave of biocatalysis. Nature 2012;485:185–94.

145. Johansson K, Frederiksen SS, Degerman M, Breil MP, Mollerup JM, Nilsson B. Combined effects of potassium chloride and ethanol as mobile phase modulators on hydrophobic interaction and reversed-phase chromatography of three insulin variants. J Chromatogr A 2015;1381:64–73.

146. Zobel-Roos S, Vetter F, Scheps D, Pfeiffer M, Gunne M, Boscheinen O, et al. Digital twin based design and experimental validation of a continuous peptide polishing step. Processes 2023;11:1401.

147. Zobel-Roos S, Mouellef M, Ditz R, Strube J. Distinct and quantitative validation method for predictive process modelling in preparative chromatography of synthetic and bio-based feed mixtures following a quality-by-design (QbD) approach. Processes 2019;7:580.

148. Zobel-Roos S, Vetter F, Scheps D, Pfeiffer M, Gunne M, Boscheinen O, et al. Multivariate parameter determination of multi-component isotherms for chromatography digital twins. Processes 2023;11:1480. https://doi.org/10.3390/pr11051480.

149. Flynn NM, Forthal DN, Harro CD, Judson FN, Mayer KH, Para MF. Placebo-controlled phase 3 trial of a recombinant glycoprotein 120 vaccine to prevent HIV-1 infection. J Infect Dis 2005;191:654–65.

150. Hammonds J, Chen X, Zhang X, Lee F, Spearman P. Advances in methods for the production, purification, and characterization of HIV-1 Gag-Env pseudovirion vaccines. Vaccine 2007;25:8036–48.

151. Deml L, Speth C, Dierich MP, Wolf H, Wagner R. Recombinant HIV-1 Pr55gag virus-like particles: potent stimulators of innate and acquired immune responses. Mol Immunol 2005;42:259–77.

152. Roldão A, Mellado MCM, Castilho LR, Carrondo MJT, Alves PM. Virus-like particles in vaccine development. Expert Rev Vaccines 2010;9:1149–76.

153. Hengelbrock A, Helgers H, Schmidt A, Vetter FL, Juckers A, Rosengarten JF, et al. Digital twin for HIV-gag VLP production in HEK293 cells. Processes 2022;10:866.

154. Cruz PE, Cunha A, Peixoto CC, Clemente J, Moreira JL, Carrondo MJ. Optimization of the production of virus-like particles in insect cells. Biotechnol Bioeng 1998;60:408–18.

155. Pillay S, Meyers A, Williamson A-L, Rybicki EP. Optimization of chimeric HIV-1 virus-like particle production in a baculovirus-insect cell expression system. Biotechnol Prog 2009;25:1153–60.

156. Puente-Massaguer E, Grau-Garcia P, Strobl F, Grabherr R, Striedner G, Lecina M, et al. Accelerating HIV-1 VLP production using stable High Five insect cell pools. Biotechnol J 2021;16:e2000391.

157. Visciano ML, Diomede L, Tagliamonte M, Tornesello ML, Asti V, Bomsel M, et al. Generation of HIV-1 Virus-Like Particles expressing different HIV-1 glycoproteins. Vaccine 2011;29:4903–12.

158. Cervera L, Gutiérrez-Granados S, Martínez M, Blanco J, Gòdia F, Segura MM. Generation of HIV-1 Gag VLPs by transient transfection of HEK 293 suspension cell cultures using an optimized animal-derived component free medium. J Biotechnol 2013;166:152–65.

159. Vickroy B, Lorenz K, Kelly W. Modeling shear damage to suspended CHO cells during cross-flow filtration. Biotechnol Prog 2007;23:194–9.
160. Gränicher G, Coronel J, Trampler F, Jordan I, Genzel Y, Reichl U. Performance of an acoustic settler versus a hollow fiber-based ATF technology for influenza virus production in perfusion. Appl Microbiol Biotechnol 2020;104:4877–88.
161. Voisard D, Meuwly F, Ruffieux P-A, Baer G, Kadouri A. Potential of cell retention techniques for large-scale high-density perfusion culture of suspended mammalian cells. Biotechnol Bioeng 2003;82:751–65.
162. Cattaneo MV, Spanjaard RA. Perfusion filtration systems. In: U.S. Patent No. 10,358,626. 23 Jul. 2019.
163. Patil R, Walther J. Continuous manufacturing of recombinant therapeutic proteins: upstream and downstream technologies. Adv Biochem Eng Biotechnol 2018;165:277–322.
164. Bielser J-M, Wolf M, Souquet J, Broly H, Morbidelli M. Perfusion mammalian cell culture for recombinant protein manufacturing – a critical review. Biotechnol Adv 2018;36:1328–40.
165. Pereira Aguilar P, Reiter K, Wetter V, Steppert P, Maresch D, Ling WL, et al. Capture and purification of Human Immunodeficiency Virus-1 virus-like particles: convective media vs porous beads. J Chromatogr A 2020;1627:461378.
166. Floderer C, Masson J-B, Boilley E, Georgeault S, Merida P, El Beheiry M, et al. Single molecule localisation microscopy reveals how HIV-1 Gag proteins sense membrane virus assembly sites in living host CD4 T cells. Sci Rep 2018;8:16283.
167. Cortin V, Thibault J, Jacob D, Garnier A. High-titer adenovirus vector production in 293S cell perfusion culture. Biotechnol Prog 2004;20:858–63.
168. Genzel Y, Vogel T, Buck J, Behrendt I, Ramirez DV, Schiedner G, et al. High cell density cultivations by alternating tangential flow (ATF) perfusion for influenza A virus production using suspension cells. Vaccine 2014;32:2770–81.
169. Vázquez-Ramírez D, Genzel Y, Jordan I, Sandig V, Reichl U. High-cell-density cultivations to increase MVA virus production. Vaccine 2018;36:3124–33.
170. Vázquez-Ramírez D, Jordan I, Sandig V, Genzel Y, Reichl U. High titer MVA and influenza A virus production using a hybrid fed-batch/perfusion strategy with an ATF system. Appl Microbiol Biotechnol 2019;103: 3025–35.
171. Hein MD, Chawla A, Cattaneo M, Kupke SY, Genzel Y, Reichl U. Cell culture-based production of defective interfering influenza A virus particles in perfusion mode using an alternating tangential flow filtration system. Appl Microbiol Biotechnol 2021;105:7251–64.
172. Fuenmayor J, Cervera L, Gòdia F, Kamen A. Extended gene expression for Gag VLP production achieved at bioreactor scale. J Chem Tech Biotechnol 2019;94:302–8.
173. Hock L, Poirier M, Palasek A, Cattaneo MV, Spanjaard RA, McNally D, et al. High-density perfusion bioreactor platform for suspension HEK293 cells. https://www.artemisbiosystems.com/wp-content/uploads/2021/02/Poster_MassBiologics_ASGCT2020.pdf [Accessed 19 Apr 2023].
174. Shirgaonkar IZ, Lanthier S, Kamen A. Acoustic cell filter: a proven cell retention technology for perfusion of animal cell cultures. Biotechnol Adv 2004;22:433–44.
175. Chotteau V. Perfusion processes. In: Al-Rubeai M, editor. Animal cell culture. Cham: Springer International Publishing; 2015:407–43 pp.
176. Coronel J, Gränicher G, Sandig V, Noll T, Genzel Y, Reichl U. Application of an inclined settler for cell culture-based influenza A virus production in perfusion mode. Front Bioeng Biotechnol 2020;8:672.
177. Helgers H, Hengelbrock A, Schmidt A, Rosengarten J, Stitz J, Strube J. Process design and optimization towards digital twins for HIV-gag VLP production in HEK293 cells, including purification. Processes 2022; 10:419.
178. Schmidt A, Helgers H, Vetter FL, Juckers A, Strube J. Fast and flexible mRNA vaccine manufacturing as a solution to pandemic situations by adopting chemical engineering good practice—continuous autonomous operation in stainless steel equipment concepts. Processes 2021;9:1874.

179. Schmidt A, Helgers H, Vetter FL, Zobel-Roos S, Hengelbrock A, Strube J. Process automation and control strategy by quality-by-design in total continuous mRNA manufacturing platforms. Processes 2022;10: 1783.
180. Vetter FL, Zobel-Roos S, Mota JPB, Nilsson B, Schmidt A, Strube J. Toward autonomous production of mRNA-therapeutics in the light of advanced process control and traditional control strategies for chromatography. Processes 2022;10:1868.
181. Geipel-Kern A. Covid-19-Impfstoff – warum bei den Rohstoff-Herstellern die Kasse klingelt; 2021. https://www.process.vogel.de/covid-19-impfstoff-warum-bei-den-rohstoff-herstellern-die-kasse-klingelt-a-1004361/ [Accessed 6 Mar 2021].
182. Kis Z, Kontoravdi C, Shattock R, Shah N. Resources, production scales and time required for producing RNA vaccines for the global pandemic demand. Vaccines 2020;9. https://doi.org/10.3390/vaccines9010003.
183. Rathore AS, Thakur G, Kateja N. Continuous integrated manufacturing for biopharmaceuticals: a new paradigm or an empty promise? Biotechnol Bioeng 2023;120:333–51.
184. Chopda V, Gyorgypal A, Yang O, Singh R, Ramachandran R, Zhang H, et al. Recent advances in integrated process analytical techniques, modeling, and control strategies to enable continuous biomanufacturing of monoclonal antibodies. J Chem Tech Biotechnol 2022;97:2317–35.
185. Christensen J. Quality issue at Baltimore vaccine plant delays some of Johnson & Johnson's vaccine. https://edition.cnn.com/2021/03/31/health/johnson–johnson-vaccine-manufacturing-problem/index.html [Accessed 2021].
186. Rees V. EDQM releases new guidelines for COVID-19 vaccine quality testing. https://www.europeanpharmaceuticalreview.com/news/133928/edqm-releases-new-guidelines-for-covid-19-vaccine-quality-testing/ [Accessed 2021].
187. Helgers H, Hengelbrock A, Schmidt A, Strube J. Digital twins for continuous mRNA production. Processes 2021;9:1967.
188. Uli Beisel. BioNTainer – A manufacturing solution for Africa or circumventing capacity? Available from: https://www.medizinethnologie.net/biontainer-a-manufacturing-solution-for-africa-or-circumventing-capacity/.
189. Saied AA, Metwally AA, Dhawan M, Choudhary OP, Aiash H. Strengthening vaccines and medicines manufacturing capabilities in Africa: challenges and perspectives. EMBO Mol Med 2022;14:e16287.
190. Wang P, Akula R, Chen M, Legaspi K. AN1616: SEC-MALS method for characterizing mRNA biophysical attributes. California: WYATT Technology; 2020.
191. Wang L-T, Ting C-H, Yen M-L, Liu K-J, Sytwu H-K, Wu KK, et al. Human mesenchymal stem cells (MSCs) for treatment towards immune- and inflammation-mediated diseases: review of current clinical trials. J Biomed Sci 2016;23:76.
192. Robb KP, Fitzgerald JC, Barry F, Viswanathan S. Mesenchymal stromal cell therapy: progress in manufacturing and assessments of potency. Cytotherapy 2019;21:289–306.
193. Olsen TR, Ng KS, Lock LT, Ahsan T, Rowley JA. Peak MSC – are we there yet? Front Med 2018;5:178.
194. Wall IB, Brindley DA. Commercial manufacture of cell therapies. In: Standardisation in cell and tissue engineering. Elsevier; 2013:212–39a pp.
195. George H, Freeman-Cook L. Addressing the challenge of scalability in viral vectors. Philadelphia, PA, USA: BioprocessOnline 2021. Available from: https://www.cellandgene.com/doc/addressing-the-challenge-of-scalability-in-viral-vectors-0002.
196. Colao IL, Corteling R, Bracewell D, Wall I. Manufacturing exosomes: a promising therapeutic platform. Trends Mol Med 2018;24:242–56.
197. Lenzini S, Brennan J, Zakhem E, Rowley J. Developing a microcarrier stirred tank process for large-scale hMSC-EV production; 2022. https://info.roosterbio.com/hubfs/Posters/ISEV-2022_POSTER_Developing-a-microcarrier-stirred-tank-process.pdf [Accessed 15 Feb 2023].

198. Werner S, Thompson S, Day R, Hawkins B, Petrosky J. Possibilities for continuous closed-system processing of cell therapies. Cell Gene Therapy Insights 2022;8:799–807.
199. Abu-Absi S, Chan L, Kotz K. Application of continuous processing in cell and gene therapy: current state and future opportunities, 2019.
200. Pigeau GM, Csaszar E, Dulgar-Tulloch A. Commercial scale manufacturing of allogeneic cell therapy. Front Med 2018;5:233.
201. NNE. Closed, automated cell therapy manufacturing – is your facility ready? https://www.nne.com/techtalk/closed-automated-cell-therapy-manufacturing/ [Accessed 20 Apr 2023].
202. Henriette S. Facilities for personalised medicine in the most personal form – today and tomorrow. BioPharma Asia; 2017. Available from: https://biopharma-asia.com.
203. Lin-Gibson S, Hanrahan B, Matosevic S, Jiwen Zhang AS, Zylberberg C. Points to consider for cell manufacturing equipment and components. Cell Gene Therapy Insights 2017;3:793–805.

Markus Stöckl*, André Gemünde and Dirk Holtmann

7 Microbial electrotechnology – Intensification of bioprocesses through the combination of electrochemistry and biotechnology

Abstract: Both biotechnological and electrochemical processes have economic and environmental significance. In particular, biotechnological processes are very specific and stable, while electrochemical processes are generally very atom-and energy-efficient. A combination of these processes is therefore a potentially important approach to intensify biotechnological processes. In this paper, the relevant options for process integration are presented, key performance indicators for quantitative evaluation are given, and an evaluation based on performance indicators is carried out using the example of the electrochemical reduction of CO_2 to formate and the subsequent biotechnological conversion to the biopolymer polyhydroxybutyrate.

Keywords: bioelectrochemical process intensification; key performance indicators; quantitative data

7.1 Introduction – how can electrochemistry be used to intensify bioprocesses?

The intensification of bioprocess is an important step toward more efficient and sustainable production of bio-based products. In particular, bioprocessing refers here to processes that use microorganisms to produce chemicals or generate electrical energy. Overall, the intensification of bioprocesses should help to increase both the productivity and efficiency of bio-based production systems. This is crucial for creating sustainable and environmentally friendly alternatives to conventional production methods. In electrochemical engineering the process intensification methods include the use of centrifugal force fields, ultrasonics, rotating electrodes, and rotating cells, which operate

*Corresponding author: Markus Stöckl, Sustainable Electrochemistry, DECHEMA Research Institute, Theodor-Heuss-Allee 25, 60486 Frankfurt am Main, Germany, E-mail: markus.stoeckl@dechema.de. https://orcid.org/0000-0002-1372-7642
André Gemünde, Institute of Bioprocess Engineering and Pharmaceutical Technology, University of Applied Sciences Mittelhessen, Wiesenstrasse 14, 35390 Giessen, Germany. https://orcid.org/0000-0003-3825-7365
Dirk Holtmann, Institute of Bioprocess Engineering and Pharmaceutical Technology, University of Applied Sciences Mittelhessen, Wiesenstrasse 14, 35390 Giessen, Germany; and Karlsruhe Institute of Technology (KIT), Institute of Process Engineering in Life Sciences, Fritz-Haber-Weg 4, 76131 Karlsruhe, Germany. https://orcid.org/0000-0001-5540-3550

As per De Gruyter's policy this article has previously been published in the journal Physical Sciences Reviews. Please cite as: M. Stöckl, A. Gemünde and D. Holtmann "Microbial electrotechnology – Intensification of bioprocesses through the combination of electrochemistry and biotechnology" *Physical Sciences Reviews* [Online] 2023. DOI: 10.1515/psr-2022-0108 | https://doi.org/10.1515/9783110760330-007

with and without the application of magnetic fields [1]. Electrochemistry and process intensification could work synergistically to improve bioprocess performance. For example, electrochemical methods can be used for process intensification by allowing specific control of physicochemical processes in bioreactors. The directed application of electric fields, currents, or electrochemical reactions to microorganisms in bioreactors can influence and optimize their selectivity, activity, and growth. Overall, electrochemistry and process intensification can help improve the performance of bioprocesses by providing new ways to control and optimize processes in bioreactors and to interact directly with the metabolism of the organisms. In general, bioelectrochemical systems (BES) are hybrid systems that use (electroactive) organisms or enzymes and electrochemical techniques. Solid electrodes serve as electron donors or acceptors for microorganisms or enzymes to produce electricity and/or valuable substances. By combining microbiological with electrochemical processes, it is possible to merge the advantages of both disciplines. Characteristics of microbiological catalysts such as high reaction specificity, high stability and robustness, and self-replication are combined with the reaction control by applied current or potential and high coulombic efficiencies typical for electrochemical systems. This combination in BES offers the opportunity to develop innovative, sustainable, and efficient processes for power generation and energy storage via a range of valuable products [2]. The combination of electrochemical and microbial as well as enzymatic reactions is well established in the field of biosensors. In the bioeconomy in general, this combination is considered to be highly effective for optimizing established processes or establishing new production routes [3]. In recent years, the first processes with high technology readiness levels (TRL) have been demonstrated (see overview in [4, 5]). In addition, BES is highly relevant to the achievement of the United Nations Sustainable Development Goals [6].

The aim of this manuscript is to briefly present the various options of BES. Furthermore, since PI always focuses on quantitative statements, the relevant performance indicators will be summarized. A case study on the electrochemical formate synthesis coupled with the synthesis of PHB on formate is then used to demonstrate the value of the performance indicators.

7.2 Fundamental mechanisms of the microbial electrotechnology (MET)

The key function behind the working principle of a microbial BES is the ability of a group of microorganisms (archaea, bacteria, and also eukaryotes) to exchange electrons with an electrode. In general, electroactive microorganisms (EAMs) can be defined as having the ability to transport electrons across biological membranes to or from an electrode. In the last decade, some of the underlying mechanisms for the interaction of microorganisms with solid electrodes have been discovered, providing a new insight into microbial activity. The ability to exchange electrons with extracellular compounds is one of the key principles of microbial life, and a variety of mechanisms have been discovered for microorganisms under aerobic and anaerobic conditions. Well-known mechanisms under anaerobic conditions are denitrification, sulfate and sulfur respiration, acetogenesis, or methanogenesis. In addition, bacterial interactions with metal ions play an important ecological role. The dissimilatory reduction of metal ions such as manganese and iron [7], uranium [8] or cobalt [9] has been reported in aquatic sediments. Bacteria such as *Geobacter metallireducens* [7] or *Shewanella oneidensis* [9] transfer electrons to metal ions in the absence of oxygen to complete their respiratory chain. Besides reduction, microbial oxidation of metals is another ecological niche occupied by bacteria. Leaching bacteria use the autotrophic oxidation of metal ions to generate energy through electron uptake. Prominent examples of electron uptake are *Acidithiobacillus ferrooxidans* or *Leptospirillum ferrooxidans* [10, 11]. Using the thiosulfate or polysulfide mechanism, these organisms take up electrons from the oxidation of iron to ferric iron [12]. Based on this observation, *A. ferrooxidans*, for example, has been used in a BES by drawing current from a graphite cathode [13] and has been identified as a promising organism for electrification of microorganisms for chemical production [14]. However, in BES so far leaching bacteria play only a minor role. More prominent and frequently applied bacteria in BES are *S. oneidensis* [15, 16], *Geobacter sulfurreducens* [17, 18] or *Sporomusa ovata* [19, 20] for example. Even more sophisticated approaches address the introduction electroactivity to established production strains such as *E. coli* [21]. The mechanism of extracellular electron transfer (EET) varies depending on the type of organism and can generally be divided into direct electron transfer (DET) and mediated electron transfer (MET) or indirect electron transfer (IET) ([22, 23], Figure 7.1). Table 7.1 provides a brief summary of the extracellular electron transfer pathways and a reference for further reading.

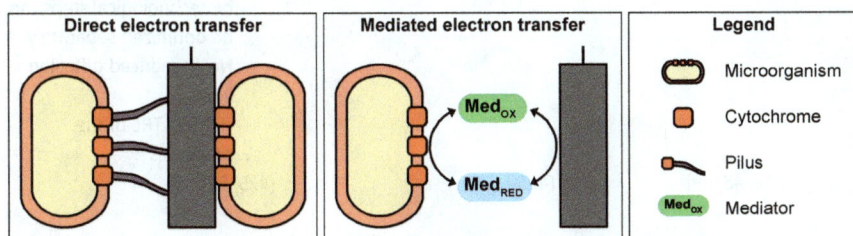

Figure 7.1: Mechanisms of extracellular electron transfer (EET): direct electron transfer (DET), mediated electron transfer (MET). Indirect electron transfer can be compared to MET, without circulation of the mediator.

Table 7.1: Basic electron transfer mechanism in microbial electrotechnology.

	Direct electron transfer (DET)	Mediated electron transfer (MET)	Indirect electron transfer (IET)
Short description	– Typically requires physical contact between cells and electrode – The majority of organism that use the DET form biofilms to adhere to the electrode – Conductive pili, nanowires, or cytochromes are responsible for electron transfer	– Cells are mainly present in the liquid phase and electrons are exchanged via redox-active mediators that shuttle electrons between the cells and the electrode – Mediators can be naturally excreted by microorganisms or artificial mediators can be added to the culture	– An electrochemical step produces intermediates such as hydrogen or formate, which are used in a second biological step – Microbes and reactions usually take place in the culture solution
Typical organisms using this pathway	– *Geobacter sulfurreducens* [24–28] – *Sporomusa ovata* [18, 20] – *Clostridium acetobutylicum* [29]	– *Shewanella oneidensis* [16, 30–33] – *Pseudomonas aeruginosa* [34–36] – *Pseudomonas putida* [37–39] – *Corynebacterium glutamicum* [31, 40] – *Vibrio natriegens* [41]	– *Cupriavidus necator* [42–45] – *Methylorubrum extorquens* [46] – *Clostridium ljungdahlii* [47]
Most important advantages	– Retention of the cells/biofilms on the electrode enables continuous processes – Underlying mechanisms largely unexplored, but with high future potential	– No physical contact between organisms and electrode is required – Advantageous in terms of high space time yields – No or reduced diffusion limitations	– The processes can be integrated in one reactor or separated in two devices (see chapter secondary microbial electrochemical technologies) – In the second case, the electrochemical and biotechnological steps can be optimized separately – No or reduced diffusion limits – Highest TRL of the processes
Literature for further reading	[4, 18, 48, 49]	[4, 50, 51]	[4, 5, 52]

7.3 Applications of microorganism in BES

In general, BES and all electrochemical system are at least comprised of two electrodes, an anode and a cathode, which are often separated by a membrane. Electrical contact between the two electrodes can thereby be realized via a resistor, a potentiostat, or another power supply. A schematic of an electrochemical reaction system is presented in Figure 7.2. As all electrochemical systems, BES can be divided into current consuming system (electrolysis, $\Delta G > 0$, endergonic, non-spontaneous reaction) or current generating systems (galvanic element, $\Delta G < 0$, exergonic, spontaneous reaction). Electrolysis can be described as a technique that uses electric current to drive an otherwise non-spontaneous chemical reaction. For example, electrolysis of water uses electricity to split water into oxygen and hydrogen. In a fuel cell, for example, hydrogen and oxygen can be used to generate electrical energy. A fuel cell uses chemical reactions to produce electricity. In electrochemistry, the electrode at which the reaction of interest takes place is called the working electrode. Depending on whether the desired reaction at the working electrode is a reduction or an oxidation, the working electrode can be either the anode or the cathode. The counter or auxiliary electrode acts as a cathode when the working electrode is acting as an anode, and *vice versa*.

7.3.1 Microbial fuel cells

Microbial fuel cells (MFCs) are a type of BES that use electroactive microorganisms at the anode to catalyze the oxidation of organic substances while generating electricity (Figure 7.3A). They operate on the principle of biologically catalyzed conversion of chemical energy into electrical work. Reduced organic molecules of high chemical energy (fuel/waste) are oxidized by electroactive microorganisms to molecules of low energy (waste/CO_2). The electroactive microorganisms transfer (respire) the electrons to the anode in the absence of alternative electron acceptors such as oxygen. The electrons are then transferred to the cathode of the MFC and the terminal electron acceptor is reduced. The theoretical redox potential of an MFC is defined by the difference in the standard redox potentials (ΔE^0) between the oxidized compound (anode) and the reduced

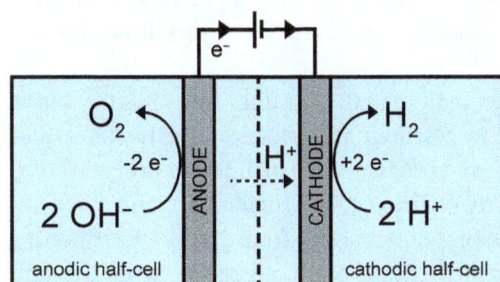

Figure 7.2: Schematic of an electrochemical system composed of anodic and cathodic half-cell reaction. Exemplarily, anodic and cathodic half-cell reactions are water splitting reactions. Abbreviations: e⁻: electron, H⁺: proton, membrane: dashed line.

compound (cathode). For example, the theoretical maximum potential of an MFC oxidizing acetate (ΔE^0 = −0.28 V) coupled with the oxygen reduction reaction (ORR) (ΔE^0 = 0.82 V) is 1.1 V. There is a wide variety of electroactive microbial catalysts used in MFCs. By far the most prominent model organisms belong to the Geobacter and Shewanella families and have been used in a large number of lab studies. In real application with complex substrates such as municipal waste water, usually mixed cultures are used.

The first MFC was reported in 1911 by Potter, who used yeast to oxidize glucose in an electrochemical cell, creating a potential difference between electrodes [53]. Since then, there have been several "waves" of MFC research (e.g., initiated by NASA as part of the space program, or during the oil crisis of the 1970s). It has been shown that the overall efficiency of an MFC process based on energy-rich materials produced for this purpose is low. For example, considering the production, transport and processing of carbohydrates, energy is wasted in the overall process. Therefore, these applications are justified only in niche applications. Today, in the majority of MFCs use wastewater streams for power generation. Integrations of MFCs into wastewater treatment plants seem to be the most likely application of this technology. Generating electrical energy could improve the economics of the wastewater treatment process, since the energy-rich substrate must be degraded anyway. In conclusion, MFCs are an economical technology for combining wastewater treatment with power generation. Integrated into the purification process of a wastewater treatment plant (WWTP), large-scale MFC would significantly reduce the overall energy demand of the WWTP [54–56]. PI is specifically referring to the wastewater treatment industry, where significant amounts of aeration energy can be saved by using the MFC. Overall, a more energy efficient process is possible. At the same time, it is ensured that the required effluent values are met.

7.3.2 Cathodic microbial electrosynthesis

The most recent technology emerging from the field of BES is microbial electrosynthesis (MES, Figure 7.3B). It offers the possibility of converting electrical energy into chemicals and biofuels. Electric current is applied to the cathode of a bioelectrochemical cell and is converted into high-energy (product) and high-value substances by reducing carbon-based molecules (substrate/CO_2). Comparable to the theoretical cell potential of an MFC, standard redox potentials of the desired products determine the theoretically applied cathode potential.

So far, the spectrum of products generated with MES is rather in its infancy but is quickly emerging. In the late-1980 different research groups were able to show, that applied potentials could improve the yields and productivities in different processes (e.g. butanol production with *Clostridium acetobutylicum* [57], propionate formation with *Propionibacterium freudenreichii* [58]). The first MES starting from CO_2 as substrate was reported by Nevin in 2010 [20]. The organism *S. ovata* produced acetate from CO_2 and applied current. Nevin showed the generation of acetate, 2-oxobutyrate and formate with *Clostridium ljungdahlii* and *Clostridium aceticum* [19]. A further approach of MES is the generation of methane catalyzed by methanogenic bacteria, representing the "power to gas" approach. Cheng reported the generation of methane from CO_2 at a cathode polarized to −700 mV versus SHE (standard hydrogen electrode) or even more negative values with a mixed culture biofilm dominated by *Methanobacterium palustre* [59]. Comparable to these findings, Villano presented autotrophic methane production by hydrogenophilic methanogenic bacteria at a potential more negative than −650 mV versus SHE [60]. However, due to the relatively high negative potentials applied in these studies, hydrogen production is assumed to serve as precursor for microbial methanogenesis. In contrast to this, Beese–Vasbender presented in their publication that already at a cathode potential of −160 mV versus SHE methane production with the *Methanobacterium*-like archaeon strain IM1 was observed. Since hydrogen production appears only at potentials below −400 mV versus SHE, the organism was shown to utilize electrons directly from the cathode without hydrogen as electron carrier [61]. It should be noted that the distinction between direct and indirect electron transfer is not clear-cut, since the underlying mechanisms are often unknown. In the field of future energy management MES can be regarded as a promising technology to produce biochemicals and biofuels to store excessive energy (positive residual loads) provided from renewable energy. In the context of PI, the unique ability to defossilize chemical industry should also be mentioned here.

7.3.3 Anodic microbial electrosynthesis

In nature as well as in industrial microbiology, all microorganisms must achieve redox equilibrium. Their redox state and energy conservation strongly depend on the availability of a terminal electron acceptor, e.g. oxygen in aerobic production processes. Under anaerobic conditions, when no electron acceptor is available, redox equilibrium is achieved by the production of reduced carbon compounds (fermentation). An alternative strategy to artificially stabilize the microbial redox and energy state is the use of anodic microbial electrosynthesis or anodic electrofermentations (see also chapter *Electrofermentation*) [51]. From a thermodynamic point of view the processes are similar to those of fuel cells. These processes have been demonstrated in a variety of organisms and products (e.g. *S. oneidensis* [16, 30], *Pseudomonas aeruginosa* [34–36], *Pseudomonas putida* [37–39], *Corynebacterium glutamicum* [31, 40] or *Vibrio natriegens* [41], see also Table 7.1 in [51]). The benefits of the anodic microbial electrosynthesis, and thus the reference to PI, are the savings in aeration energy, the avoidance of foaming, the increase in carbon yield, and the generation of electrical energy.

7.3.4 Electrofermentation

The availability of electron donors and acceptors plays a critical role in microbial production processes, especially for soluble components, as it is reflected in the oxidation–reduction potential of the medium. Electrofermentation systems (Figure 7.3C) enhance the "classical" fermentation process by introducing an additional electron source or sink through an electrode, without the electric current being the primary substrate or final product. Consequently, electrofermentation enables the process to operate under unbalanced redox conditions and with a higher carbon yield [4, 62–64]. The exact way of how the cell metabolism is influenced electrochemically is not yet known. Most probably, by changing the extracellular oxidation–reduction potential, the NADH/NAD$^+$ balance inside the cell is also altered, which can impact the product spectrum of fermentations [63]. These processes can also be named as electrochemical assisted bio production. In electrofermentation the working electrode can be the anode as well as the cathode (see e.g. Table 7.1 in [64]).

7.3.5 Paired electrolysis cells or microbial electrolysis cells

As mentioned above, in electrochemical reactions, the working electrode is where the oxidation or reduction of interest occurs and is typically the focus of the process. The reaction at the counter electrode is usually less important and sometimes completely ignored. In paired electrolysis, both electrochemical reactions are used to produce valuable products. In this process, 100 % of the electrons supplied by the cathode and 100 % of the electrons absorbed at the anode contribute to the production of the final product or products, making it theoretically possible to achieve a combined current efficiency of 200 %. With such systems, the energy is used in an optimal way.

Microbial electrolysis cells are based on microbial fuel cells. The electroactive microorganisms are present in the anode chamber and convert the substrate/wastewater components to produce electricity and protons. The electrons are then transferred to the cathode where they reduce the protons to produce H$_2$. The reduction current generated at the anode is too low to allow this reaction. Therefore, a small voltage is applied between the two electrodes to drive the reaction. However, the applied voltage is only about 0.2–0.8 V. This technology can produce hydrogen by consuming less electrical energy than water electrolysis does [65, 66]. By combining anode and cathode reactions, other higher value products can be produced. Teetz et al. showed a combination of electrochemical synthesis (Kolbe electrolysis) at the anode with microbial conversion of the cathodically produced hydrogen [45]. Kolbe electrolysis of valeric acid yields the liquid drop-in fuel additive n-octane and CO$_2$. At the same time, isopropanol is produced by *Cupriavidus necator* using gaseous electrolysis products CO$_2$ and H$_2$.

7.3.6 Secondary microbial electrochemical technologies

Secondary microbial electrochemical technologies (Figure 7.3D and E) are based on the combination of microbial conversions and electrosynthesis, i.e. only abiotic electrochemical reactions take place. Secondary microbial electrosynthesis can bed distinguished in two strategies. First, electrochemical synthesis to produce chemical feedstocks for subsequent microbial conversions. Second, the upgrading of products obtained from microbial processes by subsequent electrosynthesis. Examples of the first strategy are:

- The coupling of an electrolyser to produce hydrogen and the subsequent conversion of CO_2 to methane is established on industrial scale by the company Electrochaea (see www.electrochaea.com).

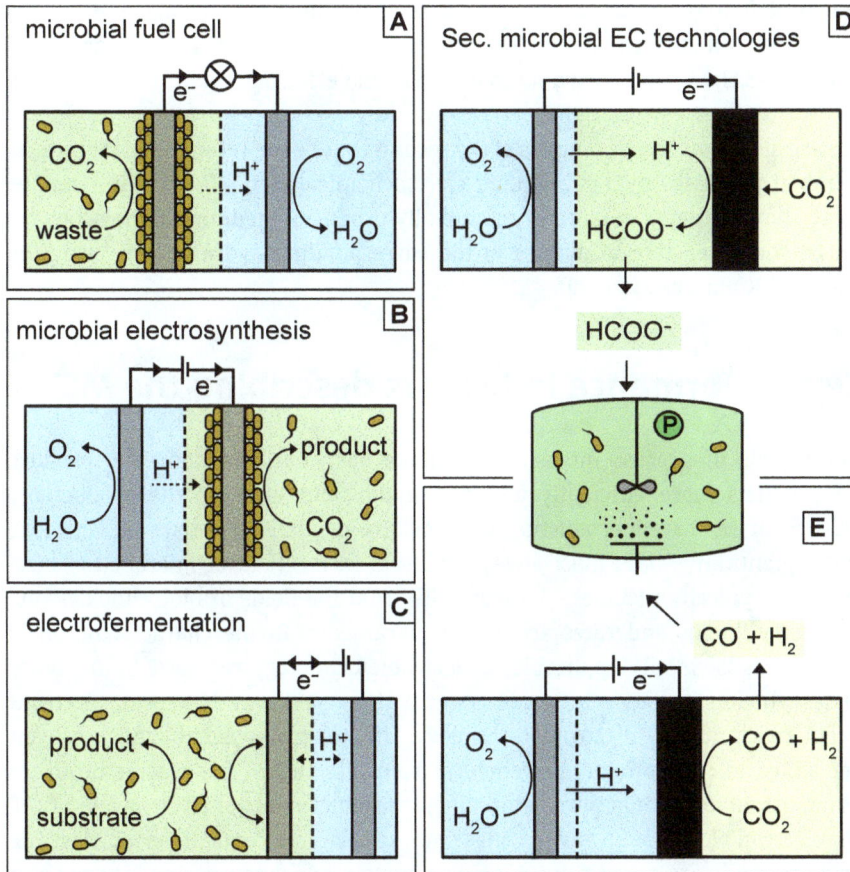

Figure 7.3: Composition of different bioelectrochemcial systems: A. microbial fuel cell; B: microbial electrosynthesis cell; C: elecrofermentation; D and E: secondary microbial electrochemical systems: D: Formic acid-based fermentation; E: Syngas based fermentation.

– Hass et al. coupled a CO_2 electrolyser with fermentation (Figure 7.3E). In the gas fermentation the electrochemical product syngas (H_2/CO mixture) was converted to butanol and hexanol [67]. In the meantime the process was scaled up into 2.000 L scale at Evonik's plant in Marl (Press release from Siemens: https://press.siemens.com/global/en/pressrelease/research-project-rheticus).

– The electrochemical conversion of CO_2 to formic acid or formate and the subsequent microbial conversion (Figure 7.3D) have been demonstrated by several authors using different organisms [42, 44, 46]. In particular, gas diffusion electrodes have proven to be very efficient for this reaction [52].

The combination of electrochemical synthesis of reduced C1 molecules (such as syngas or formate) with subsequent biological product formation in secondary microbial systems offers significant advantages by combining the high selectivity and energy efficiency of electrochemical fixation with the wide range of biotechnological products [68]. This process transforms a surface-related reaction, such as direct electron transfer, into a volumetric process by introducing a soluble carbon-and electron-donating intermediate [69].

There are also promising examples of the second route – the refinement of products from microbial processes by a subsequent electrochemical step [70–72]. For example, Urban et al. showed that a mixture of microbially produced medium-chain carboxylic acids can be converted into a mixture of the corresponding hydrocarbons using the electrochemical Kolbe reaction [70].

7.4 Key performance indicators describing the MET

The general goals of process intensification have been outlined above: to produce bio-based products more efficiently and sustainably and increase the productivity and efficiency of bio-based production systems. To evaluate processes and achieve these goals, quantitative data must always be used. In the field of microbial electro-technology, this typically includes characteristics from the fields of biotechnology and electrochemistry. Yields and rates are critical parameters in the characterization of bioprocesses. This is equally applicable to electrobiotechnology reactors. Unlike other bioprocesses which primarily rely on substrates such as carbon and nitrogen, electrons are crucial reactants in electrobiotechnology. In electrobiotechnology, electrons are either added to (via cathode) or removed from (via anode) the reactor broth. As such, it is important to consider parameters such as product yield per electron and/or cell yield per electron. Electron transfer rates and cellular concentrations of electron-transferring proteins are other helpful key parameters when comparing electroactive organisms in bioelectrochemical systems. Overall, it has to be considered that all reactions within METs are limited by thermodynamic conditions. Hence, the evaluation of the electromotive force based on the Gibbs free energy can be used to assess the potential difference between anode and cathode for a specific reaction, setting the theoretical boundaries for the cell voltage. These values can help to evaluate the feasibility of the electrobiotechnological process beforehand.

While several microbial mechanisms can facilitate electron transfer, all require an electrode to interface with the reactor liquid. Consequently, the hybrid nature of bioelectrochemical systems arises from the need to create an electrode-liquid interface and separate the anodic and cathodic chambers to prevent undesired side or cross reactions. These systems must satisfy the requirements of conventional bioreactors while incorporating the characteristics of electrochemical reactors. At the same time, capital costs must be competitive with traditional/established technologies. For example, in domestic wastewater treatment, the capital costs of a MFC exceed the conventional treatment by 30-fold [73]. Economic performance indicators are therefore as important as process performance indicators. In Table 7.2 the most important

Table 7.2: Performance indicators for the evaluation of microbial electrotechnology (adapted and expanded from [74, 75]).

Parameters	Equation	Explanation and Units	Description
Coulombic efficiency (CE)	$\frac{F \times z \times n_{eff}}{Q}$	n_{eff} – total amount of product produced, mol Q – charge spent, A s z – electron stoichiometry, mol F – Faraday constant, A s mol^{-1}	Relates the number of electrons released/consumed by the substrate/educt in an ideal scenario to that achieved in practice
Power (P), W	$I \times U$	I – current, A U – (cell/clamping) voltage, V	Amount of energy, which is required to run or is obtained from an electrochemical process
Energy efficiency (η)	$\frac{P_{out}}{P_{in}}$	P_{out} – power output P_{in} – power input: electrical, gassing, heating	Evaluates sustainability of energy production via MET
Specific electrode surface area (SSA)	$\frac{A_e}{V_e}$	A_e – surface area of the electrode, m^2; V_e – volume of the electrode, m^3	Measure of the "accessible"/ "useful" surface area of the electrode (under debate on how it is determined)
Surface-to-volume ratio (SVA)	$\frac{A_e}{V_R}$	A_e – surface area of the electrode, m^2 V_R – volume of the reactor, m^3	Used for process scaling and engineering
Current density (j)	$\frac{I}{A_e}$	I – current, A j – C] urrent per (specific)surface area, A m^{-2} or reactor volume, A m^{-3}	Used for comparing materials and designs as well as scaling and engineering
Cell yield per electron (YX/e–)	$\frac{\Delta X \times F}{Q}$	ΔX– amount of grown biomass, g Q– charge spent, A s F – Faraday constant, A s mol^{-1}	Used for process description and benchmarking

Table 7.2: (continued)

Parameters	Equation	Explanation and Units	Description
Product yield per electron (YP/e−)	$\frac{P \times F}{Q}$	P– amount of product, g Q– charge spent, A s F – Faraday constant, A s mol^{-1}	Used for process description and benchmarking
COD removal ($\eta\ COD$)	$\left(1 - \frac{COD_{out}}{COD_{in}}\right) \times 100$	COD – chemical oxygen demand, g L^{-1}	Evaluates the performance of wastewater treatment [76]
Hydrogen/ methane production rate ($r_{H2/CH4}$)	$\frac{P}{V_R} dt$	P – amount of product, m^3 V_R – volume of the reactor, m^3 t – time, h	Describes the productivity of a microbial electrolysis cell [77, 78]
Relative anodic electron uptake	$\frac{q_a}{X \times t \times z_{O2} \times q_{O2}}$	q_a – electrons transferred to the anode, mol X – amount of biomass, g$_{CDW}$ t – time, h z_{O2} – electron equivalent of the oxygen molecule q_{O2} – biomass specific oxygen uptake rate, mol$_{O2}$ g$_{CDW}^{-1}$ h^{-1}	Describes the percentage of oxygen replaced by an anode [41]
Electromotive force (E_{emf})	$-\frac{\Delta G_r}{Q}$	Q– charge spent, A s; ΔG_r– Gibbs free energy, J	Evaluates the thermodynamical driving force of the redox reaction (anodic/cathodic) with a potential difference [76]
Electron transfer rate coefficient (k_f)	$k_0 \times e^{-\frac{\alpha z F}{RT}\Delta E}$	k_0 – standard rate constant z – electrons transferred α – electron transfer coefficient ΔE – potential difference between reaction partners	Based on Arrhenius equation. The resulting current density can be estimated based on the Butler-Volmer equation [79]
Economic feasibility	$\frac{T}{I}$	T – total cost of MFC operation, € I – current generated, A	Used to determine and compare the economic feasibility of MFCs [80]

performance indicators in microbial electrotechnology are summarized. Still, there are KPIs that are important for describing MET but cannot be expressed as a standardized equation. Among those are two important factors to mention: the long-term stability of the system over extended operation periods and the reactor scale-up potential. These factors have to be considered for the specific application since they differ greatly for various factors such as electrode materials, reactor design, and microbial community dynamics.

7.5 Case study – secondary microbial electrosynthesis based in formic acid

Within the following chapter we provide a theoretical case study on a secondary microbial electrosynthesis. In a 2-step process, CO_2 is initially reduced to formic acid (HCOOH), which is then further used as sole feedstock for the synthesis of the model biopolymer polyhydroxybutyrate (PHB) (Figures 7.3D and 7.4). For the sake of simplification and demonstration, the case study only implements the energy costs, which are needed for the substrate synthesis. Capital expenditures (CAPEX) are neglected as well as initial biomass production, energy to run the fermentation as and down-stream processing of the final biopolymer.

For the initial electrosynthesis of formic acid, the following process characteristics shall be given: CO_2 is reduced to formic acid in a stacked gas diffusion electrode-based electrolyzer (total geometrical electrode surface A_e = 1 m^2) with a current density j of 200 mA cm^{-1} and a current efficiency CE of 90 %. The resulting cell/clamping voltage U, which is required to run the electrolysis at the given current density shall be 4 V. A total of 10 kg of formic acid (n = 217 mol) is to be produced as a substrate for the subsequent biosynthesis of PHB. Initially, the electrolysis time t to obtain the 10 kg formic acid is calculated via formula 7.1 based on Faraday's law:

$$CE = \frac{F \times z \times n}{I \times t} \tag{7.1}$$

with: CE = 0.9; F = Faraday-constant (96,485 A s mol^{-1}); z = number of transferred electrons (2); n = amount of produced formate [mol]; I = current [A] = 2000 A (via: $A_e \times j$); t = time

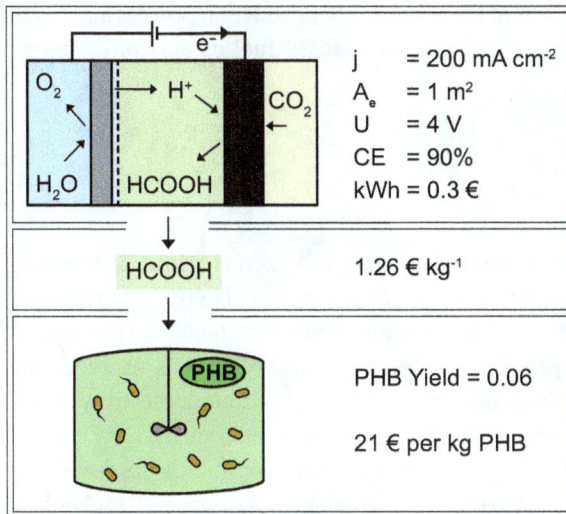

j = 200 mA cm^{-2}
A_e = 1 m^2
U = 4 V
CE = 90%
kWh = 0.3 €

1.26 € kg^{-1}

PHB Yield = 0.06

21 € per kg PHB

Figure 7.4: Schematic of a theoretical case study on a secondary microbial electrosynthesis. In a 2-step process, CO_2 is initially electrochemically reduced to formic acid (HCOOH), which is then further used as sole feedstock for the microbial synthesis of the biopolymer polyhydroxybutyrate (PHB). The respective parameters of both the electrochemical and the biological process step are given as examples.

of electrolysis [s]. Inserting the given the values (formula 7.2), the electrolysis would need to run for 5.14 h to obtain 10 kg of formic acid as substrate.

$$\frac{96485 \, \text{As mol}^{-1} \times 2 \times 217 \, \text{mol}}{2000 \, \text{A} \times 0.9} = 18865 \, \text{s} \tag{7.2}$$

The amount of energy, which is needed to produce the formic acid is then derived by implementing the cell voltage of the electrolyzer via formula 7.3:

$$E = I \times U \times t \tag{7.3}$$

with I = 2000 A, U = 4 V and t = 18,865 s, the amount of electrical energy to produce 10 kg formic acid would be 41 kWh. Taking an assumed price of 0.3 € per kWh for an industrial process, the electricity costs would be 12.58 € (1.26 € per kg formic acid).

In the second step of the case study, the microbial PHB synthesis is conducted by *C. necator* WT-strain with formic acid as sole substrate (carbon and energy source). It shall be assumed, that resting *C. necator* cells utilize the formic acid with the theoretical PHB yield (g PHB per g formic acid) of 0.06 [81]. Following this assumption, 10 kg of formic acid are consumed to synthesize 0.6 kg PHB. The final price of the PHB, which is only derived from the energy costs of the electrochemical substrate synthesis would be around 21 € kg^{-1}.

To intensify the PHB production and reduce costs in our case study, two main parameters can be addressed. One option is the increase of the theoretical PHB yield. Instead of the energy intensive Calvin cycle, which leads to the rather poor theoretical PHB yield, less energy intensive pathways such as the reductive glycine pathway can be employed [81, 82]. The second option is the reduction of electricity costs for the initial formic acid synthesis, which can only be realized by either a further increase of the CE, or a decrease of the cell voltage U. In an ideal scenario, the improvement of both individual steps leads to a significant improvement of the overall PHB production process and result in product costs, which can provide an alternative to fossil fuel or agriculture-based substrates.

7.6 Conclusions

The combination of biotechnology and electrochemistry is a powerful tool for intensification of established production processes as well as development of new sustainable and bio-based synthesis routes within a circular economy. Bioelectrochemical systems can improve wastewater purification processes, directly provide electrons as reduction equivalents in microbial electrosynthesis or combine established abiotic electrolysis and fermentation processes. However, there are also limitations for BES such as low current densities and high CAPEX. Therefore, we highlight key performance indicators describing BES and present a case study on energy costs of a secondary microbial electrochemical process. Since the establishment of a new (bioelectrochemical) process in industry finally depends on their rentability, the comprehensive collection of quantitative data, which allows the determination the mentioned key performance indicators is crucial. It must be emphasized again that process intensification is always based on quantitative and comparable data – therefore the relevant data must always be provided in the publications. Otherwise, the processes will probably never find their way into the application.

Acknowledgment: The authors would like to thank the editor for their guidance and review of this article before its publication.

References

1. Scott K. Process intensification: an electrochemical perspective. Renew Sustain Energy Rev 2018;81: 1406–26.
2. Krieg T, Sydow A, Schröder U, Schrader J, Holtmann D. Reactor concepts for bioelectrochemical syntheses and energy conversion. Trends Biotechnol 2014;32:645–55.
3. Harnisch F, Urban C. Electrobiorefineries: unlocking the synergy of electrochemical and microbial conversions. Angew Chem Int Ed 2018;57:10016–23.
4. Fruehauf HM, Enzmann F, Harnisch F, Ulber R, Holtmann D. Microbial electrosynthesis—an inventory on technology readiness level and performance of different process variants. Biotechnol J 2020;15:2000066.
5. Stöckl M, Claassens N, Lindner S, Klemm E, Holtmann D. Coupling electrochemical CO2 reduction to microbial product generation – identification of the gaps and opportunities. Curr Opin Biotechnol 2022;74: 154–63.
6. Gizewski J, Sande Lv.d., Holtmann D. Contribution of electrobiotechnology to sustainable development goals. Trends Biotechnol 2023. https://doi.org/10.1016/j.tibtech.2023.02.009.
7. Lovley DR, Phillips EJ. Novel mode of microbial energy metabolism: organic carbon oxidation coupled to dissimilatory reduction of iron or manganese. Appl Environ Microbiol 1988;54:1472–80.
8. Lovley DR, Giovannoni SJ, White DC, Champine JE, Phillips EJP, Gorby YA, et al. Geobacter metallireducens gen. nov. sp. nov., a microorganism capable of coupling the complete oxidation of organic compounds to the reduction of iron and other metals. Arch Microbiol 1993;159:336–44.
9. Hau HH, Gilbert A, Coursolle D, Gralnick JA. Mechanism and Consequences of anaerobic respiration of cobalt by Shewanella oneidensis strain MR-1. Appl Environ Microbiol 2008;74:6880–6.
10. Sand W, Rohde K, Sobotke B, Zenneck C. Evaluation of Leptospirillum ferrooxidans for leaching. Appl Environ Microbiol 1992;58:85–92.
11. Schippers A, Jozsa P, Sand W. Sulfur chemistry in bacterial leaching of pyrite. Appl Environ Microbiol 1996; 62:3424–31.
12. Rohwerder T, Gehrke T, Kinzler K, Sand W. Bioleaching review part A: progress in bioleaching: fundamentals and mechanisms of bacterial metal sulfide oxidation. Appl Microbiol Biotechnol 2003;63: 239–48.
13. Carbajosa S, Malki M, Caillard R, Lopez MF, Palomares FJ, Martín-Gago JA, et al. Electrochemical growth of Acidithiobacillus ferrooxidans on a graphite electrode for obtaining a biocathode for direct electrocatalytic reduction of oxygen. Biosens Bioelectron 2010;26:877–80.
14. Tremblay PL, Zhang T. Electrifying microbes for the production of chemicals. Front Microbiol 2015;6:201.
15. Logan BE, Murano C, Scott K, Gray ND, Head IM. Electricity generation from cysteine in a microbial fuel cell. Water Res 2005;39:942–52.
16. Marsili E, Baron DB, Shikhare ID, Coursolle D, Gralnick JA, Bond DR. Shewanella secretes flavins that mediate extracellular electron transfer. Proc Natl Acad Sci U S A 2008;105:3968–73.

17. Bond DR, Holmes DE, Tender LM, Lovley DR. Electrode-reducing microorganisms that harvest energy from marine sediments. Science 2002;295:483–5.
18. Lovley DR. Powering microbes with electricity: direct electron transfer from electrodes to microbes. Environ Microbiol Rep 2011;3:27–35.
19. Nevin KP, Hensley SA, Franks AE, Summers ZM, Ou J, Woodard TL, et al. Electrosynthesis of organic compounds from carbon dioxide is catalyzed by a diversity of acetogenic microorganisms. Appl Environ Microbiol 2011;77:2882–6.
20. Nevin K, Woodard T, Franks A, Summers Z, Lovley D. Microbial electrosynthesis: feeding microbes electricity to convert carbon dioxide and water to multicarbon extracellular organic compounds. mBio 2010;1:1–4.
21. Bird LJ, Kundu BB, Tschirhart T, Corts AD, Su L, Gralnick JA, et al. Engineering wired life: synthetic biology for electroactive bacteria. ACS Synth Biol 2021;10:2808–23.
22. Sydow A, Krieg T, Mayer F, Schrader J, Holtmann D. Electroactive bacteria—molecular mechanisms and genetic tools. Appl Microbiol Biotechnol 2014;98:8481–95.
23. Karthikeyan R, Singh R, Bose A. Microbial electron uptake in microbial electrosynthesis: a mini-review. J Ind Microbiol Biotechnol 2019;46:1419–26.
24. Strycharz SM, Glaven RH, Coppi MV, Gannon SM, Perpetua LA, Liu A, et al. Gene expression and deletion analysis of mechanisms for electron transfer from electrodes to Geobacter sulfurreducens. Bioelectrochemistry 2011;80:142–50.
25. Holmes DE, Chaudhuri SK, Nevin KP, Mehta T, Methe BA, Liu A, et al. Microarray and genetic analysis of electron transfer to electrodes in Geobacter sulfurreducens. Environ Microbiol 2006;8:1805–15.
26. Bond DR, Lovley DR. Electricity production by Geobacter sulfurreducens attached to electrodes. Appl Environ Microbiol 2003;69:1548–55.
27. Stöckl M, Teubner NC, Holtmann D, Mangold KM, Sand W. Extracellular polymeric substances from geobacter sulfurreducens biofilms in microbial fuel cells. ACS Appl Mater Interfaces 2019;11:8961–8.
28. Frühauf HM, Holtmann D, Stöckl M. Influence of electrode surface charge on current production by Geobacter sulfurreducens microbial anodes. Bioelectrochemistry 2022;147:108213.
29. Engel M, Gemünde A, Holtmann D, Müller-Renno C, Ziegler C, Tippkötter N, et al. Clostridium acetobutylicum's connecting world: cell appendage formation in bioelectrochemical systems. Chemelectrochem 2020;7:414–20.
30. Carmona-Martinez AA, Harnisch F, Fitzgerald LA, Biffinger JC, Ringeisen BR, Schröder U. Cyclic voltammetric analysis of the electron transfer of Shewanella oneidensis MR-1 and nanofilament and cytochrome knock-out mutants. Bioelectrochemistry 2011;81:74–80.
31. Krieg T, Phan LMP, Wood JA, Sydow A, Vassilev I, Krömer JO, et al. Characterization of a membrane-separated and a membrane-less electrobioreactor for bioelectrochemical syntheses. Biotechnol Bioeng 2018;115:1705–16.
32. Stöckl M, Schlegel C, Sydow A, Holtmann D, Ulber R, Mangold KM. Membrane separated flow cell for parallelized electrochemical impedance spectroscopy and confocal laser scanning microscopy to characterize electro-active microorganisms. Electrochim Acta 2016;220:444–52.
33. Arinda T, Philipp LA, Rehnlund D, Edel M, Chodorski J, Stöckl M, et al. Addition of riboflavin-coupled magnetic beads increases current production in bioelectrochemical systems via the increased formation of anode-biofilms. Front Microbiol 2019;10:1–8.
34. Rabaey K, Boon N, Siciliano SD, Verhaege M, Verstraete W. Biofuel cells select for microbial consortia that self-mediate electron transfer. Appl Environ Microbiol 2004;70:5373–82.
35. Saunders SH, Tse EC, Yates MD, Otero FJ, Trammell SA, Stemp ED, et al. Extracellular DNA promotes efficient extracellular electron transfer by pyocyanin in Pseudomonas aeruginosa biofilms. Cell 2020;182: 919–32.e19.

36. Bosire EM, Rosenbaum MA. Electrochemical potential influences phenazine production, electron transfer and consequently electric current generation by Pseudomonas aeruginosa. Front Microbiol 2017;8:1–11. https://doi.org/10.3389/fmicb.2017.00892.

37. Schmitz S, Nies S, Wierckx N, Blank LM, Rosenbaum MA. Engineering mediator-based electroactivity in the obligate aerobic bacterium Pseudomonas putida KT2440. Front Microbiol 2015;6:1–13.

38. Hintermayer S, Yu S, Krömer JO, Weuster-Botz D. Anodic respiration of Pseudomonas putida KT2440 in a stirred-tank bioreactor. Biochem Eng J 2016;115:1–13.

39. Lai B, Yu S, Bernhardt PV, Rabaey K, Virdis B, Krömer JO. Anoxic metabolism and biochemical production in Pseudomonas putida F1 driven by a bioelectrochemical system. Biotechnol Biofuels 2016;9:39.

40. Vassilev I, Gießelmann G, Schwechheimer SK, Wittmann C, Virdis B, Krömer JO. Anodic electro-fermentation: anaerobic production of L-Lysine by recombinant Corynebacterium glutamicum. Biotechnol Bioeng 2018;115:1499–508.

41. Gemünde A, Gail J, Holtmann D. Anodic respiration of Vibrio natriegens in a bioelectrochemical system. ChemSusChem 2023:e202300181. https://doi.org/10.1002/cssc.202300181.

42. Stöckl M, Harms S, Dinges I, Dimitrova S, Holtmann D. From CO2 to bioplastic – coupling the electrochemical CO2 reduction with a microbial product generation by drop-in electrolysis. ChemSusChem 2020;13:4086–93.

43. Krieg T, Sydow A, Faust S, Huth I, Holtmann D. CO2 to terpenes: autotrophic and electroautotrophic α-humulene production with Cupriavidus necator. Angew Chem Int Ed 2018;57:1879–82.

44. Li H, Opgenorth PH, Wernick DG, Rogers S, Wu TY, Higashide W, et al. Integrated electromicrobial conversion of CO2 to higher alcohols. Science 2012;335:1596.

45. Teetz N, Holtmann D, Harnisch F, Stöckl M. Upgrading Kolbe electrolysis—highly efficient production of green fuels and solvents by coupling biosynthesis and electrosynthesis. Angew Chem Int Ed 2022;61: e202210596.

46. Hegner R, Neubert K, Kroner C, Holtmann D, Harnisch F. Coupled electrochemical and microbial catalysis for the production of polymer bricks. ChemSusChem 2020;13:5295–300.

47. Im C, Valgepea K, Modin O, Nygård Y. Clostridium ljungdahlii as a biocatalyst in microbial electrosynthesis – effect of culture conditions on product formation. Bioresour Technol Rep 2022;19:101156.

48. Ueki T. Cytochromes in extracellular electron transfer in geobacter. Appl Environ Microbiol 2021;87:1–16. https://doi.org/10.1128/aem.03109-20.

49. Madjarov J, Soares R, Paquete CM, Louro RO. Sporomusa ovata as catalyst for bioelectrochemical carbon dioxide reduction: a review across disciplines from microbiology to process engineering. Front Microbiol 2022;13:913311.

50. Gemünde A, Lai B, Pause L, Krömer J, Holtmann D. Redox mediators in microbial electrochemical systems. Chemelectrochem 2022;9:e202200216.

51. Vassilev I, Averesch NJ, Ledezma P, Kokko M. Anodic electro-fermentation: empowering anaerobic production processes via anodic respiration. Biotechnol Adv 2021;48:107728.

52. Stöckl M, Lange T, Izadi P, Bolat S, Teetz N, Harnisch F, et al. Application of gas diffusion electrodes in bioeconomy: an update. Biotechnol Bioeng 2023;120:1465–77.

53. Potter MC. Electrical effects accompanying the decomposition of organic compounds. Proc R Soc Lond – Ser B Contain Pap a Biol Character 1911;84:260–76.

54. Hiegemann H, Herzer D, Nettmann E, Lübken M, Schulte P, Schmelz KG, et al. An integrated 45L pilot microbial fuel cell system at a full-scale wastewater treatment plant. Bioresour Technol 2016;218:115–22.

55. Hiegemann H, Littfinski T, Krimmler S, Lübken M, Klein D, Schmelz KG, et al. Performance and inorganic fouling of a submergible 255 L prototype microbial fuel cell module during continuous long-term operation with real municipal wastewater under practical conditions. Bioresour Technol 2019;294:122227.

56. Krieg T, Mayer F, Sell D, Holtmann D. Insights into the applicability of microbial fuel cells in wastewater treatment plants for a sustainable generation of electricity. Environ Technol 2019;40:1101–9.

57. Kim TS, Kim BH. Electron flow shift in Clostridium acetobutylicum fermentation by electrochemically introduced reducing equivalent. Biotechnol Lett 1988;10:123–8.
58. Emde R, Schink B. Enhanced propionate formation by Propionibacterium freudenreichii subsp. freudenreichii in a three-electrode amperometric culture system. Appl Environ Microbiol 1990;56:2771–6.
59. Cheng S, Xing D, Call DF, Logan BE. Direct biological conversion of electric current into methane by electromethanogenesis. Environ Sci Technol 2009;43:3953–8.
60. Villano M, Aulenta F, Ciucci C, Ferri T, Giuliano A, Majone M. Bioelectrochemical reduction of CO2 to CH4 via direct and indirect extracellular electron transfer by a hydrogenophilic methanogenic culture. Bioresour Technol 2010;101:3085–90.
61. Beese-Vasbender PF, Grote JP, Garrelfs J, Stratmann M, Mayrhofer KJ. Selective microbial electrosynthesis of methane by a pure culture of a marine lithoautotrophic archaeon. Bioelectrochemistry 2015;102:50–5.
62. Schievano A, Pepé Sciarria T, Vanbroekhoven K, De Wever H, Puig S, Andersen SJ, et al. Electro-fermentation – merging electrochemistry with fermentation in industrial applications. Trends Biotechnol 2016;34:866–78.
63. Moscoviz R, Toledo-Alarcón J, Trably E, Bernet N. Electro-fermentation: how to drive fermentation using electrochemical systems. Trends Biotechnol 2016;34:856–65.
64. Yamada S, Takamatsu Y, Ikeda S, Kouzuma A, Watanabe K. Towards application of electro-fermentation for the production of value-added chemicals from biomass feedstocks. Front Chem 2021;9:805597.
65. Rousseau R, Etcheverry L, Roubaud E, Basséguy R, Délia ML, Bergel A. Microbial electrolysis cell (MEC): strengths, weaknesses and research needs from electrochemical engineering standpoint. Appl Energy 2020;257:113938.
66. Saravanan A, Karishma S, Senthil Kumar P, Yaashikaa P, Jeevantantham S, Gayathri B. Microbial electrolysis cells and microbial fuel cells for biohydrogen production: current advances and emerging challenges. Biomass Conv Bioref 2020;1–21. https://doi.org/10.1007/s13399-020-00973-x.
67. Haas T, Krause R, Weber R, Demler M, Schmid G. Technical photosynthesis involving CO2 electrolysis and fermentation. Nat Catal 2018;1:32–9.
68. Enzmann F, Stöckl M, Pfitzer M, Holtmann D. Empower C1: combination of electrochemistry and biology to convert C1 compounds. In: Zeng AP, Claassens NJ, editors. One-carbon feedstocks for sustainable bioproduction. Cham: Springer International Publishing; 2022:213–41 pp.
69. Enzmann F, Stöckl M, Zeng AP, Holtmann D. Same but different-Scale up and numbering up in electrobiotechnology and photobiotechnology. Eng Life Sci 2019;19:121–32.
70. Urban C, Xu J, Sträuber H, dos Santos Dantas TR, Mühlenberg J, Härtig C, et al. Production of drop-in fuels from biomass at high selectivity by combined microbial and electrochemical conversion. Energy Environ Sci 2017;10:2231–44.
71. Suastegui M, Matthiesen JE, Carraher JM, Hernandez N, Rodriguez Quiroz N, Okerlund A, et al. Combining metabolic engineering and electrocatalysis: application to the production of polyamides from sugar. Angew Chem Int Ed 2016;55:2368–73.
72. Holzhäuser FJ, Artz J, Palkovits S, Kreyenschulte D, Büchs J, Palkovits R. Electrocatalytic upgrading of itaconic acid to methylsuccinic acid using fermentation broth as a substrate solution. Green Chem 2017;19:2390–7.
73. He L, Du P, Chen Y, Lu H, Cheng X, Chang B, et al. Advances in microbial fuel cells for wastewater treatment. Renew Sustain Energy Rev 2017;71:388–403.
74. Krieg T, Madjarov J, Rosa L, Enzmann F, Harnisch F, Holtmann D, et al. Reactors for microbial electrobiotechnology. In: Harnisch F, Holtmann D, editors. Bioelectrosynthesis. Cham: Springer International Publishing; 2019:231–71 pp.
75. Enzmann F, Gronemeier D, Holtmann D. Evaluation of bioelectromethanogenesis part I: energy calculations. Chem Ing Tech 2020;92:137–43.
76. Logan BE, Hamelers B, Rozendal R, Schröder U, Keller J, Freguia S, et al. Microbial fuel cells: methodology and technology. Environ Sci Technol 2006;40:5181–92.

77. Jeremiasse AW, Hamelers HV, Croese E, Buisman CJ. Acetate enhances startup of a H2-producing microbial biocathode. Biotechnol Bioeng 2012;109:657–64.
78. Seelajaroen H, Spiess S, Haberbauer M, Hassel MM, Aljabour A, Thallner S, et al. Enhanced methane producing microbial electrolysis cells for wastewater treatment using poly(neutral red) and chitosan modified electrodes. Sustain Energy Fuels 2020;4:4238–48.
79. Okamoto A, Hashimoto K, Nealson KH, Nakamura R. Rate enhancement of bacterial extracellular electron transport involves bound flavin semiquinones. Proc Natl Acad Sci U S A 2013;110:7856–61.
80. Modin O, Gustavsson DJ. Opportunities for microbial electrochemistry in municipal wastewater treatment- an overview. Water Sci Technol 2014;69:1359–72.
81. Vlaeminck E, Quataert K, Uitterhaegen E, De Winter K, Soetaert WK. Advanced PHB fermentation strategies with CO2-derived organic acids. J Biotechnol 2022;343:102–9.
82. Claassens NJ, Bordanaba-Florit G, Cotton CA, De Maria A, Finger-Bou M, Friedeheim L, et al. Replacing the Calvin cycle with the reductive glycine pathway in Cupriavidus necator. Metab Eng 2020;62:30–41.

Michael E. Runda and Sandy Schmidt*

8 Light-driven bioprocesses

Abstract: Enzyme catalysis and photocatalysis are two research areas that have become of major interest in organic synthesis. This is mainly because both represent attractive strategies for making chemical synthesis more efficient and sustainable. Because enzyme catalysis offers several inherent advantages, such as high substrate specificity, regio-, and stereoselectivity, and activity under environmentally benign reaction conditions, biocatalysts are increasingly being adopted by the pharmaceutical and chemical industries. In addition, photocatalysis has proven to be a powerful approach for accessing unique reactivities upon light irradiation and performing reactions with an extended substrate range under milder conditions compared to light-independent alternatives. It is therefore not surprising that bio- and photocatalytic approaches are now often combined to exploit the exquisite selectivity of enzymes and the unique chemical transformations accessible to photocatalysis. In this chapter, we provide an overview of the wide variety of light-driven bioprocesses, ranging from photochemical delivery of reducing equivalents to redox enzymes, photochemical cofactor regeneration, to direct photoactivation of enzymes. We also highlight the possibility of catalyzing non-natural reactions via photo-induced enzyme promiscuity and the combination of photo- and biocatalytic reactions used to create new synthetic methodologies.

Keywords: cyanobacteria; nicotinamide cofactors; oxidoreductases; photobiocatalysis; photocatalysis; redox reactions.

8.1 Introduction

Light represents an ideal 'reagent' in chemical synthesis as it is safe to use and readily available [1]. Inspired by the way plants use sunlight to build complex molecules, the chemist Ciamician already speculated in 1912 that chemicals could be synthesized by utilizing sunlight as an abundant and renewable energy source [2]. While only a vision in the beginning, significant progress has been achieved in the past decade to implement light-driven catalytic processes in sustainable organic synthesis.

*Corresponding author: Sandy Schmidt, Department of Chemical and Pharmaceutical Biology, Groningen Research Institute of Pharmacy, University of Groningen, Antonius Deusinglaan 1, 9713AV Groningen, The Netherlands, E-mail: s.schmidt@rug.nl. https://orcid.org/0000-0002-8443-8805
Michael E. Runda, Department of Chemical and Pharmaceutical Biology, Groningen Research Institute of Pharmacy, University of Groningen, Antonius Deusinglaan 1, 9713AV Groningen, The Netherlands

As per De Gruyter's policy this article has previously been published in the journal Physical Sciences Reviews. Please cite as: M. E. Runda and S. Schmidt "Light-driven bioprocesses" *Physical Sciences Reviews* [Online] 2023. DOI: 10.1515/psr-2022-0109 | https://doi.org/10.1515/9783110760330-008

Enzymes are natural catalysts in the living world, catalyzing reactions with high chemo-, regio-, and stereoselectivity under mild reaction conditions. Thus, over the last two decades, much effort has been made to further extend the range of biocatalytic reactions available to synthetic chemists, in particular by increasing the available toolbox of biocatalysts, leading to novel catalytic functionalities [3]. Accelerated by the rapid development of genetic engineering tools in combination with protein engineering and advanced screening methods, biocatalytic processes are increasingly implemented as an alternative synthetic strategy in both academia and industry [4–6]. Here, the intrinsic advantages of enzymes for accessing enantiopure products are especially exploited in the pharmaceutical industries, further emphasizing the powerfulness of biocatalysts for organic synthesis [4]. Among the range of enzyme candidates that are highly desired for industrial implementation, redox enzymes or oxidoreductases have gained significant attention [7]. This is because of their outstanding ability to catalyze complex redox reactions under mild conditions, which cannot be performed by classical organic synthesis approaches or only by using harmful and toxic chemicals.

Photocatalysis, on the other hand, has proven itself as a truly powerful approach to perform reactions in the presence of light that would be very difficult or even impossible to conduct under dark conditions [8]. While earlier ultraviolet (UV) light was predominantly used for the direct activation of organic molecules, modern photochemical activation techniques are predominantly based on selective excitation of photocatalysts with visible light. As such, visible light-mediated activation of organic molecules represents a mild, sustainable and clean method for chemical activation and has been applied to various challenging reactions such as C–H bond activation, C–C bond cross-coupling, C–N bond formation, cycloadditions, and halogenations [9–11].

While nature invented the intriguing concept of combining enzyme catalysis with photocatalysis, chemists nowadays strive to mimic this approach by coupling the catalytic reactivity and outstanding selectivity of enzymes with photocatalytic reactions. In such a

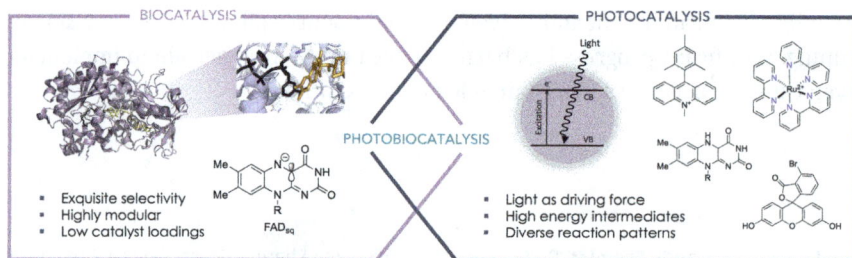

Figure 8.1: Photobiocatalysis, possessing advantages of the reactivity of photocatalysts and the selectivity of enzymes, is increasingly realized for the green synthesis of value-added chemicals. Adapted with permission from Ref. [12]. Copyright 2022 American Chemical Society.

dual system, referred to as 'photobiocatalysis' (Figure 8.1), the exquisite selectivity of enzymes is combined with the unique chemical transformations accessible to photocatalysis. In such a dual system, light is absorbed by a photoactive compound to activate an organic substrate, and the subsequent enzymatic reaction catalyzes a selective transformation, synergistically leading to the synthesis of complex molecules that are difficult to obtain by conventional methods. While early efforts in the field predominantly focused on developing light-driven systems to provide reduced cofactors (NADPH, FAD) or oxidants (e.g. H_2O_2) to redox enzymes, researchers intensively explore now the possibility of catalyzing non-natural reactions via photoinduced enzyme promiscuity, and the combination of photocatalysis and biocatalysis that can be applied to create novel synthetic methodologies.

This book chapter provides an overview of the recent progress and application of light-driven bioprocesses (photobiocatalysis) in organic synthesis, ranging from photobiocatalytic cofactor regeneration, natural and artificial photoenzymes, non-natural reactions via photoinduced enzyme promiscuity, and cascades involving photobiocatalytic transformations.

8.2 Basic principles and terms in light-driven bioprocesses

Photobiocatalysis adopts natural or artificial light for exploiting the advantages of photocatalyst reactivity and enzyme selectivity (Figure 8.1) [13–17]. In the search for efficient strategies that exploit the potential of light in biocatalysis, promising basic concepts have emerged that are fundamental to the meaningful implementation of both whole-cell and *in vitro* photobiocatalytic applications [15, 18, 19]. These concepts can be divided according to the following classification: (1) photocatalytic regeneration cascades, (2) photoenzymatic reactions utilizing natural or artificial photoenzymes or photoinduced enzyme promiscuity and (3) photoenzymatic systems (Figure 8.2) [20]. In photocatalytic regeneration cascades, light can be used to generate photoexcited electrons to fuel enzyme-catalyzed redox reactions, i.e. photocatalytic regeneration of redox enzymes (*in vitro* and *in vivo*, Figure 8.2A and B) [21]. In photoenzymatic reactions, natural or artificial cofactor-dependent photoenzymes are applied that directly need light to perform their catalytic reaction (Figure 8.2C) [12]. On the other hand, photoenzymatic systems refer to the combination of photo(organo)catalytic reactions that use light to directly drive small molecule conversions in combination further enzymatic functionalization steps in one-pot (tandem) photobiocatalytic reactions (Figure 8.2D) [16].

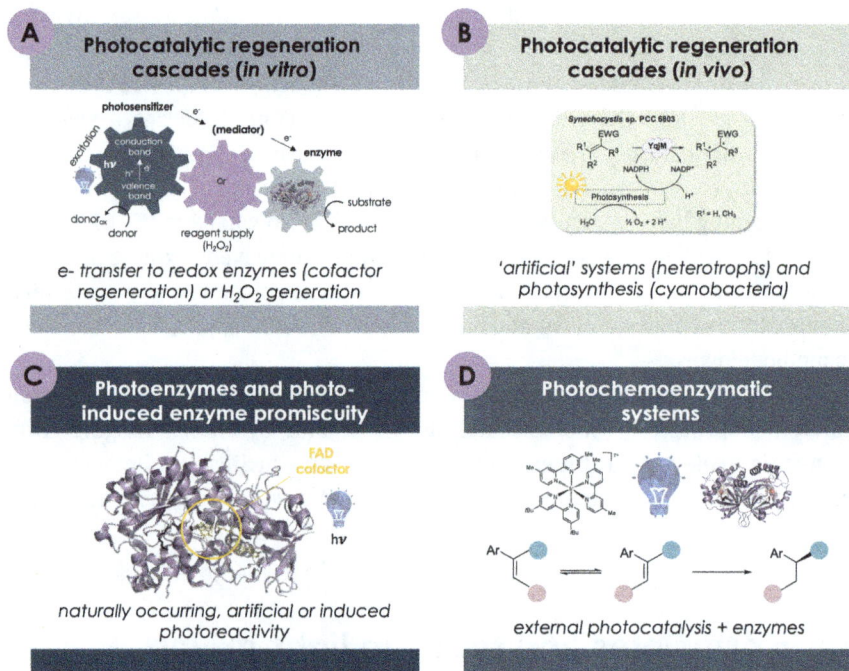

Figure 8.2: The main design principles in photobiocatalysis. A) Photocatalytic regeneration cascades via direct or indirect activation of redox enzymes (*in vitro*); B) light-driven cofactor supply using photoautotrophs (*in vivo*); C); natural and artificial photoenzymes for (non-natural) photo-biocatalytic reactions; D) photoenzymatic cascades combining photo- and biocatalytic transformations. Adapted with permission from Refs. [12, 14, 16]. Copyright 2022 American Chemical Society.

Photobiocatalytic approaches aiming at the direct or indirect photoactivation of redox enzymes gained increasing attention in the past decade, in particular to tackle common bottlenecks related to the application of oxidoreductases in organic synthesis. Independent on the reaction system i.e. *in vitro*, or whole cells, light can be used to generate photoexcited electrons to fuel enzyme-catalyzed redox reactions. Dependent on the reaction setup, these electrons can be subsequently utilized for regenerating or even circumventing the need for nicotinamide cofactors. The basic concept behind this light-induced electron transfer relies on a photochemical effect defined as charge separation (Box 8.1).

In such light-induced electron transfer processes, the difference between the redox potentials between an electron donor and an oxidized photoactive compound is crucial. In the context of photobiocatalysis, these photoactive compounds are commonly defined as photosensitizers and determine the feasibility of a light-driven reaction system. Photosensitizers can absorb light at specific wavelengths in their excited states [22]. The non-absorbed visible light portion is reflected, which we discern as colors. The reason plants or algae appear green is because they contain chlorophyll. For instance, chlorophyll *b* within the P680 reaction center of photosystem II (PSII) shows absorption maxima at around 460 and 650 nm, which overlaps with the visible spectrum of blue and red light, respectively [23]. Light-irradiation of P680 and subsequent charge separation yield in P680$^+$ and the reduction of pheophytin functioning as an electron acceptor. The formed P680$^{\bullet+}$ proton radical acts as a strong oxidant with a redox potential of about 1.2–1.4 V [24]. Emphasized as the strongest oxidant generated in a natural reaction [25], P680$^{\bullet+}$ can initiate the oxidation of H_2O via the oxygen-evolving complex [26]. Via multiple transition states, four electrons and protons are subsequently subtracted from two water molecules releasing O_2 and four protons, a reaction defined as water splitting. Analog to the described light-driven water-splitting reaction in PSII, current photobiocatalytic approaches often rely on the exploitation of the photo-excitability of a photosensitizer to realize light-induced electron transfers. Especially, organic molecules such as dyes or pigments with delocalized conjugated π-systems have revealed suitable charge separation properties in various photobiocatalytic applications [27, 28]. Their redox potentials during photoexcitation and charge separation determine their compatibility within the electron donor/acceptor framework.

The overall aim of implementing photosensitizers in enzyme-catalyzed reaction systems is either the injection or withdrawal of photoexcited electrons. In that sense, photosensitizers act as 'converters' of light into redox potential. This change in redox potential is the fundamental driving force for the transfer of electrons from an electron donor to an electron acceptor. As indicated, this mechanism follows a linear reaction mode rather than a closed reaction cycle. Thus, the supply of electrons must be maintained over time to avoid implementing an undesired bottleneck or rate-limiting step within the reaction system. As the initial source of electrons is continuously consumed, it is conventional to indicate its consumption using the term 'sacrificial' electron donor. Compounds exhibiting low redox potentials at operational pH, including organic acids (formic acid or ascorbic acid) or tertiary amines (triethanolamine (TEOA), 2-ethanesulfonic acid (MES), 3-(N-morpholino)propane-sulfonic acid (MOPS) or ethylenediaminetetraacetic acid (EDTA)), are typically considered as suitable sacrificial electron donors [15, 29]. However, choosing a suitable sacrificial electron donor/photosensitizer combination for a desired light-driven reaction setup is often based on the preceded empirical evaluation and optimization studies [27].

BOX 8.1: Charge separation.

Photoinduced charge separation describes the light-driven transfer of electrons between donor and acceptor molecules resulting in the formation of respective charge transfer states. The driving force behind this mechanism is the excitability of electrons in a ground state to higher energy levels upon light irradiation. In homogeneous photocatalysis, the electrons within the Highest Occupied Molecular Orbital (HOMO) of the photocatalyst undergo excitation to the Lowest Unoccupied Molecular Orbital (LUMO). In the presence of an acceptor molecule, the excited electron can be subsequently mediated which results in photoinduced charge separation:

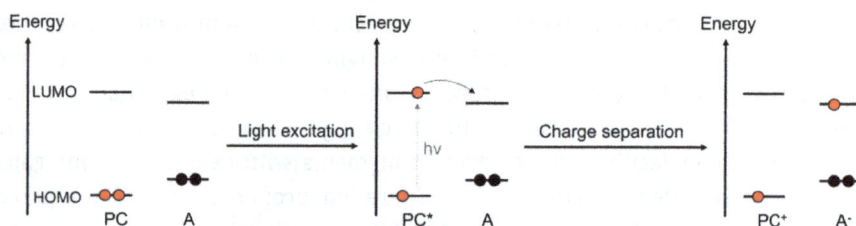

Light irradiation of the photocatalyst (PC) results in its excited state (PC*). Thereby an electron from the Highest Occupied Molecular Orbital (HOMO) is excited to a higher energy level, the Lowest Unoccupied Molecular Orbital (LUMO). The excited electron can be subsequently mediated to the LUMO of an electron acceptor (A) which results in photoinduced charge separation and the formation of PC^+ and A^-.

In heterogeneous catalysis, electrons in the valence band can be exited into the conduction band upon light irradiation. Unlike in homogeneous catalysis, photoinduced charge separation occurs within the semiconductor material without the need for an acceptor or donor molecule. The charge separation results in the formation of negative charges electrons and positively charged holes within the semiconductor. Mediated to the surface of the photocatalyst, positively charged holes or electrons can participate in oxidation and reduction reactions, respectively.

8.3 Photocatalytic regeneration cascades via direct or indirect activation of redox enzymes

The reactivity of most oxidoreductases (EC1, over 80 %) involves the transfer of electrons between different atoms or molecules, while during the catalyzed redox reaction one compound is being oxidized, another one is being reduced. Due to that, oxidoreductases require additional electron donor or acceptor molecules that have to be supplied in stoichiometric amounts. Typical electron donors used by oxidoreductases are the nicotinamide cofactors NADH and NADPH. Since NAD(P)H plays a central role in biocatalytic processes utilizing oxidoreductases, researchers have developed various *in situ* regeneration strategies, allowing the use of the costly cofactor in catalytic amounts (Box 8.2). Overall, this contributes to a significant cost reduction of a biocatalytic process [30]. While enzymatic regeneration systems nowadays prevail in synthetic applications [31], several photochemical regeneration systems have been developed in the past decade. Reductive regeneration of redox enzymes can be achieved either directly, i.e. by direct reduction of the active sites of the enzymes, or indirectly, i.e. involving the nicotinamide cofactors (Scheme 8.1).

BOX 8.2: Conventional cofactor regeneration in biotechnology.

Conventional *in vitro* applications aiming at the regeneration of cofactors commonly rely on the coupled substrate approach [134], or the *in situ* regeneration via enzyme coupling [30]. In the substrate-coupled approach, the enzyme of interest accepts the desired substrate as well as a cosubstrate. As applied in the reduction of ketones catalyzed by alcohol dehydrogenases, isopropanol can be used as a cosubstrate for the oxidative back reaction generating reduced nicotinamide cofactor and acetone as a by-product [134]. By following the enzyme-coupled approach, the reaction system is extended by an additional enzymatic conversion, which catalyzes the *in situ* cofactor regeneration. For instance, in the enzyme-catalyzed reduction of ketones, NAD(P)H could be successfully regenerated by using glucose dehydrogenase or formate dehydrogenase in combination with glucose or formic acid as cosubstrates [135–137].

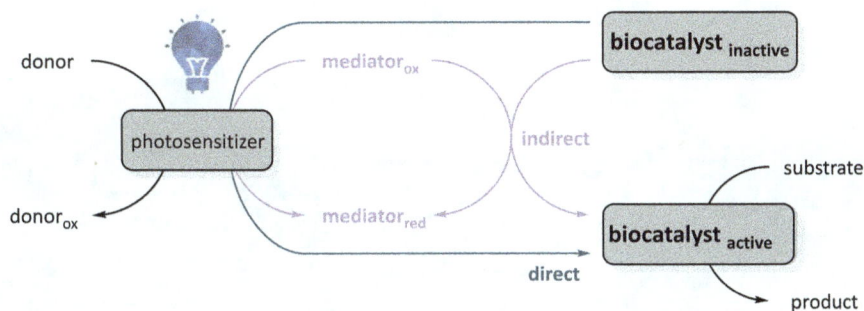

Scheme 8.1: Artificial light-induced electron transfer for the direct or indirect photoactivation of oxidoreductases. Adapted with permission from Ref. [16].

8.3.1 Indirect photoactivation of redox enzymes

In photobiocatalysis, several concepts have been reported that demonstrate the feasibility of reducing NAD(P)$^+$ by exploiting the reduction potential of photosensitizers in their excited state [32]. Analogous to oxygenic photosynthesis, light is converted into chemical energy stored in NAD(P)H. Unlike the oxidation of NAD(P)H to NAD(P)⁺, the back-reaction to regenerate reduced nicotinamide has to be catalyzed regioselectively, as the direct reduction of NAD(P)⁺ can yield inactive isomers or undesired dimerization [33, 34]. Other studies suggest the use of additional organometallic compounds as electron mediators, which facilitate the formation of the active 1,4-NAD(P)H, subsequently consumed in enzymatic reactions [35, 36]. Therein, Rh^{III} and Ir^{III} complexes turned out to be highly potent in performing the mediation of electrons between an electrode and the cofactor, while a hydride ion (H⁻) serves as a reducing equivalent in the selective reduction of NAD(P)⁺ [36]. Based on these observations, indirect photochemical regeneration strategies of reduced NAD(P) have been developed (Scheme 8.2A), wherein the photochemical nicotinamide cofactor recycling using electron mediators is for instance coupled to the activity of cytochrome P450s [37] or dehydrogenases [28].

A. General concept for the indirect photoactivation of redox enzymes

B. Photoactivation via a redox mediator

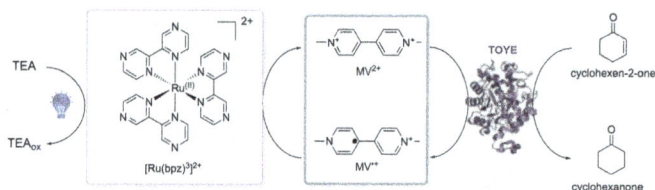

C. Photoactivation via light-driven cofactor regeneration

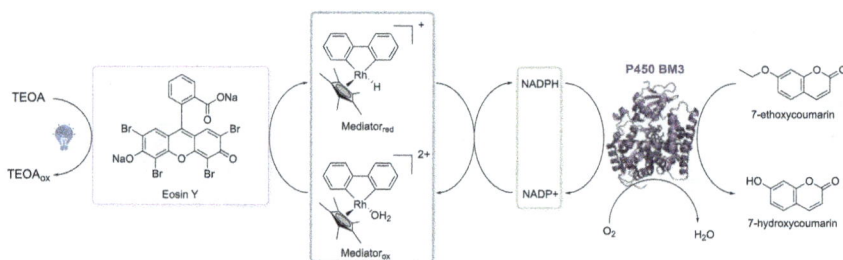

Scheme 8.2: General photobiocatalytic concept for indirect activation of redox enzymes. A) General scheme of the indirect photoactivation redox enzymes. B) A thermophilic ene-reductase (TOYE) catalyzes the enzymatic reduction of an α,β-unsaturated substrate. The reaction is driven by photoexcited electrons deriving from the sacrificial electron donor triethanolamine (TEA). [Ru/bpz)₃]²⁺ is used as photosensitizer and methyl viologen (MV²⁺) as mediator [39]. C) Light-driven activation of cytochrome P450 BM3 catalyzing the O-dealkylation of 7-ethoxycoumarin [37]. Eosin Y is used as photosensitizer transferring electrons from the sacrificial electron donor TEOA to an organometallic mediator for regenerating NADPH. Adapted with permission from Ref. [16].

Early studies on photochemical NAD(P)H regeneration systems rely on the use of Ru [(bpy)$_3$]$^{2+}$ or ZnTMPyP4 as photosensitizers in combination with methyl viologen (MV) as a primary electron acceptor [38]. In the respective photoinduced charge separation at the photosensitizer, electrons are initially transferred to MV^{2+}, using either EDTA, 2-mercaptoethanol, triethylamine (TEA) or TEOA as sacrificial electron donors. The generated MV$^{+\bullet}$ radical acts as the electron donor for the subsequent enzymatic reduction of NAD(P)$^+$ catalyzed by an NAD(P)$^+$-dependent dehydrogenase (Scheme 8.2B) [39]. By coupling the light-driven cofactor regeneration system to a second enzymatic conversion, CO$_2$ conversion into keto acids [40], reduction of ketones [41], and reduction and amination of keto acids was demonstrated [38, 42]. However, the need for an additional biocatalyst to catalyze the reduction of NAD(P)$^+$ does not offer a significant advantage over conventional enzymatic cofactor regeneration systems [31].

In searching for more safe and straightforward solutions for the photochemical regeneration of NAD(P)H, reaction systems implementing organic dyes [43], flavins [44], quantum dots [45], graphene-based photocatalysts [46] and porphyrins [47] as photosensitizers in combination with [Cp*Rh(bpy)H$_2$O]$^{2+}$ as electron mediator were developed. [Cp*Rh(bpy)H$_2$O]$^{2+}$ represents a versatile organometallic mediator exhibiting high activities and stability for the highly selective for generating the active 1,4-cofactor variants [15, 19, 44]. For example, Lee et al. used a light-driven approach to catalyze the O-dealkylation of 7-ethoxycoumarin with cytochrome P450 BM3 (Scheme 8.2C) [37]. The xanthene dye eosin Y acts as a photosensitizer to transfer photoexcited electrons from TEOA to an organometallic mediator molecule for NADPH regeneration. Lee et al. also proposed the light-driven regeneration of NADH by using the sacrificial electron donor TEOA and eosin Y as photosensitizer [28]. Eosin Y, an organic dye previously used for cell staining in histological applications, revealed its potential as a photosensitizer in various photochemical applications [48]. In a photobiocatalytic setup for the conversion of α-ketoglutarate to L-glutamate catalyzed by L-glutamate dehydrogenase, [Cp*Rh(bpy) H$_2$O]$^{2+}$ was used as an electron mediator between photoexcited eosin Y and NAD$^+$ yielding NADH [28]. Thereby, a turnover number of about 25 NAD(P)$^+$ h^{-1} was estimated for the photosensitizer-mediator driven cofactor regeneration system, yielding about 90 % substrate conversion. Driven by these findings, the scope of potential photosensitizers was further explored by screening different xanthene dyes [49]. As a result, rose bengal, fluorescein, erythrosine B and phloxine B show similar or even better performance in the light-driven cofactor supply for the GDH compared to the previously reported eosin Y.

Other examples report the use of graphene-based photosensitizers harboring covalently bound porphyrin derivates [46], or difluoro-boraindacene (BODIPY) [50], as chromophores that can donate electrons to [Cp*Rh(bpy)H$_2$O]$^{2+}$ upon visible light exposure. Analogous to previously described examples, [Cp*Rh(bpy)H$_2$O]$^{2+}$ catalyzes the reduction of NAD$^+$ to NADH, the final step in the photocatalytic cofactor recycling mechanism. Also quantum dots have been used as photosensitizers in photochemical NADH cofactor regeneration systems with [Cp*Rh(bpy)H$_2$O]$^{2+}$ [51].

The previous examples emphasize the feasibility of light-driven NAD(P)H regeneration combined with the enzymatic reduction of organic compounds. Similarly, light-induced charge separation and electron transfer can be utilized for the *in situ* generation of oxidized NAD(P)$^+$. In this case, NAD(P)H generated during enzyme-catalyzed substrate oxidation serves as an initial electron donor quenching the oxidized photosensitizer. Consequently, an electron acceptor, often O$_2$, represents the final component in the light-driven electron transfer chain. Gargiulo et al. successfully coupled the photochemical generation of NAD(P)$^+$ to the oxidation of alcohols into ketones and lactones catalyzed by alcohol dehydrogenase [52]. In this case, flavins were applied as photosensitizers for the light-driven reoxidation of NAD(P)H to NAD(P)$^+$. Since oxygen serves as the terminal electron acceptor for the regeneration of the reduced flavin, the strong oxidant H$_2$O$_2$ is formed as a byproduct. While Gargiulo et al. used catalase to avoid possible side effects caused by H$_2$O$_2$, the use of dithionite as a photocatalytic component in the regeneration of NADP$^+$ yields in the final formation of H$_2$O as proposed by Rickus et al. [53]. The generation of NADP$^+$ was determined by the enzymatic conversion of isocitrate to α-ketoglutarate. The reaction catalyzed by an NADP$^+$-specific isocitrate dehydrogenase revealed turnover frequencies of about 35 h^{-1} by providing the photosensitizer in solution.

8.3.2 Direct photoactivation of redox enzymes

Reaction systems with the purpose of direct photoactivation of oxidoreductases have the following aims: activating cofactor-independent enzymes, circumventing the need for cofactors and decreasing the complexity of multicomponent reaction systems (Scheme 8.3A). Therefore, the focus is on the redox-active prosthetic groups, including flavins, heme moieties, and non-heme bond metal ions within the polypeptide framework of oxidoreductases. In nature, these enzymes receive electrons directly from reduced cofactors or indirectly via redox partners. These mechanisms provide a site for injecting photoexcited electrons resulting in light-driven enzyme activation.

In this context, members of the ene-reductase (ERED) enzyme family have been studied for regeneration by chromophores that are being utilized as photosensitizers, which, upon light excitation, can transfer electrons from a sacrificial electron donor such as TEOA to the oxidized FMN within the active site of the EREDs (Scheme 8.3B) [54, 55]. Instead of indirectly regenerating the ERED via regeneration of reduced nicotinamide cofactors, reducing equivalents from simple sacrificial electron donors such as TEOA, EDTA, formate, or phosphite via were provided via photocatalytic oxidation. The photoenzymatic conversion of, for instance, 2-methylcyclohexenone by TsOYE in combination with rose bengal (RB) resulted in enantiopure (*R*)-2-methylcyclohexanone (*ee* >99 %) with a yield of up to 53 % and total turnover numbers of 235 for TsOYE (Scheme 8.3B) [56]. The substitution of halogen atoms in xanthene dyes significantly affected the turnover frequency.

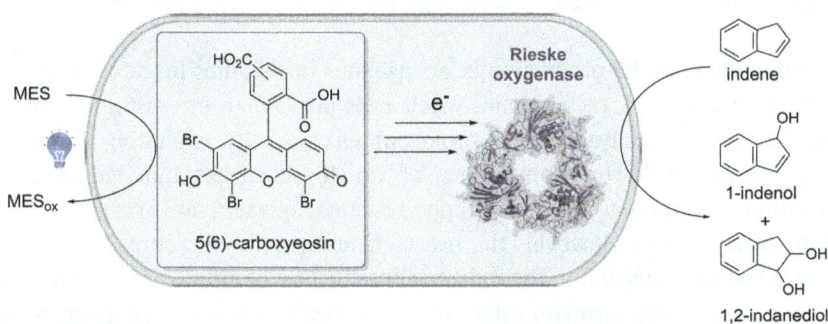

Scheme 8.3: Photobiocatalytic concepts for the direct activation of redox enzymes. A) General scheme for the direct photoactivation of redox enzymes via electron transfer from a sacrificial electron donor. B) *In vitro* direct photoactivation of an OYE variant from *Thermus scotoductus* (TsOYE) catalyzing the asymmetric reduction of C=C bonds [54]. Rose bengal as photosensitizer transfers photoexcited electrons to the prosthetic FMN group of TsOYE. C) The combination of a photosensitizer (e.g. 5(6)-carboxyeosin) with sacrificial electron donors (e.g. MES buffer) fuels Rieske oxygenase (RO)-catalyzed hydroxylations in whole cells [27]. In this example, photoexcited electrons are transferred to the active site (non-heme iron center) of the RO [27]. Adapted with permission from Ref. [16].

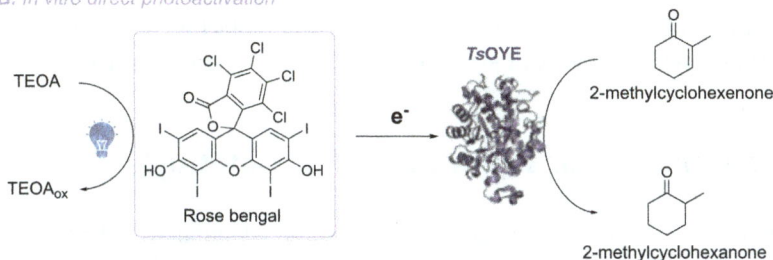

Also heme-containing enzymes such as cytochrome P540s (CYPs) have been utilized for direct photoactivations. For instance, Park et al. demonstrated the light-induced electron transfer from TEOA to the heme-prosthetic group of various CYPs [57]. Emphasized as a cofactor-independent reaction system, the hydroxylation of various drugs and steroid molecules by using resting *Escherichia coli* cells containing overexpressed CYP variants and eosin Y as photosensitizer has been shown. In reactions performed for over 20 h under visible light irradiation, total turnover numbers of up to

180 of various CYPs were achieved. Analogously, a study conducted by Özgen et al. indicates the versatility of the described whole-cell approach using dye derivates as organic photosensitizers in combination with EDTA, MES or MOPS buffer as sacrificial electron donors (Scheme 8.3C) [27]. The asymmetric hydroxylation of olefins catalyzed by four Rieske non-heme iron-dependent oxygenase (RO) variants has been performed to confirm the applicability of the proposed reaction system to further multi-component oxygenases. By that, the costly supply for reduced NAD(P)H cofactors and limitations arising from the assumed instability of ROs in cell-free systems were circumvented.

8.3.3 Light-driven cofactor supply using photoautotrophs

The application of whole-cells as biocatalysts offers unique advantages and has been widely used in the biosynthesis of high-value fine and bulk chemicals, as well as active pharmaceutical ingredients [58, 59]. The continued identification of new microorganisms as well as recent advances in synthetic biology and metabolic engineering, together with the rapid development of molecular genetic tools, have led to a renaissance of whole-cell biocatalysis.

Besides upregulating the production of endogenous metabolites in industrial-scale fermentation processes [60], recombinant whole cells provide an interesting platform for biotransformations catalyzed by heterologous enzymes. By exploiting the cell's metabolism, one-pot multi-step conversions of simple substrates into value-added chemicals without additional cofactor supply systems represent an overall goal in industrial biocatalysis [61]. However, the use of heterotrophic microorganisms such as *Corynebacterium glutamicum* in the industrial production of amino acids, or recombinant *E. coli* as a model organism for establishing e.g. enzyme cascades, is dependent on the sufficient supply of sugars (e.g. glucose) as carbon fuel for their metabolism. From an environmental perspective, glucose represents a renewable carbon source obtained from biomass and is thus considered a green alternative to petrochemical raw materials for producing fuels, bulk or fine chemicals. From a socio-economical perspective, however, glucose production is mainly based on the enzymatic hydrolysis of edible biomass, which is questioned as it overlaps with the global feedstock and food supply [62]. To counteract this debate, the development of alternative routes to increase the accessibility of fermentable carbon sources from non-edible biomass such as lignocellulose-rich materials has gained increased attention [63].

Unlike heterotrophs as whole-cell platforms for bioconversions, photoautotrophs can utilize CO_2 as a sole carbon source to fuel their metabolism. The energy needed to assimilate CO_2 is derived from natural light during photosynthesis [36]. The light-induced oxidation of water (referred to as water-splitting) yields hydrogen protons (H^+) and electrons, and the light-generated electrons are directly used to reduce $NADP^+$ to NADPH. This makes photosynthetic organisms a particularly interesting chassis for coupling heterologously expressed oxidoreductases to the rich NADPH pool, thereby driving whole-cell light-driven biotransformations [64]. Recently, cyanobacteria have emerged as highly interesting whole-cell biocatalysts in such light-driven redox reactions [64], and have been applied for instance in the asymmetric reduction of ketones [65] or aldehydes [66]. The similarity of cyanobacteria to other Gram-negative bacteria such as *E. coli* enables the generation of recombinant variants by applying available genetic engineering techniques [67]. By that, biotransformations can be conducted that go beyond the endogeneous reaction scope of cyanobacteria [68]. Considering the enhanced atom economy enabling water as a sacrificial electron donor to fuel enzymatic redox reactions, the use of cyanobacteria appears to have a significant advantage over recombinant heterotrophs.

Following these considerations, Köninger et al. coupled the photosynthetic apparatus of recombinant cyanobacteria to the activity of oxidoreductases (Figure 8.3A) [69]. Thereby, the asymmetric reduction of activated C=C bonds catalyzed by the NADPH-dependent ERED YqjM from *Bacillus subtilis* expressed in recombinant *Synechocystis* sp. PCC 6803 cells was successfully shown. On a semi-preparative scale using 100 mg substrate, an isolated product yield of up to 80 % of enantiopure (*R*)-2-methylsuccinimide was achieved, which is comparable to the efficiency of typical whole-cell biotransformations in *E. coli* [39]. Büchsenschütz et al. reported that recombinant *Synechocystis* cells can also be applied for the asymmetric reduction of cyclic amines [70]. Thereby, reduced NADPH cofactors, directly or indirectly provided by photosynthesis, can fuel the catalytic activity of imine reductases (IREDs) by supplying reducing equivalents. By optimizing reaction conditions, full conversions of prochiral substrates could be achieved, yielding optically pure secondary amines.

Hoschek et al. coupled the formation of O_2, driven by the initial water-splitting reaction of oxygenic photosynthesis, to enzyme-catalyzed oxyfunctionalization reactions (Figure 8.3B) [71]. It is assumed that the light-driven *in situ* generations of O_2 has the potential to counteract mass-transfer limitations between the gas and liquid phase, a crucial factor in the upscaling of O_2-dependent processes. In a corresponding study,

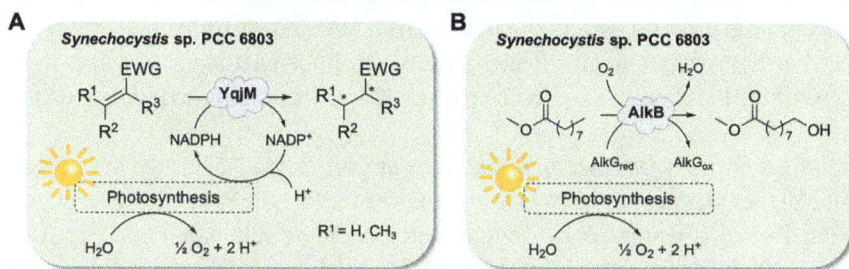

Figure 8.3: The cyanobacterium *Synechocystis* sp. PCC 6803 is used in whole-cell biotransformations for A) the C=C double bond reduction catalyzed by the ene reductase YqjM from *Bacillus subtilis* [69, 74]. B) The hydroxylation of nonanoic acid methyl ester catalyzed by the monooxygenase AlkB from *Pseudomonas putida* Gpo1 [71, 72]. Adapted with permission from Refs. [69, 71].

engineered *Synechocystis* sp. PCC 6803 was used as a host organism for the heterologous expression of an NADH-dependent alkane monooxygenase (AlkB) from *Pseudomonas putida* GPo1 [72]. As revealed in whole-cell reactions performed under anaerobic conditions, the hydroxylation of nonanoic acid methyl ester into the corresponding ω-hydroxylated product strictly depends on the light-driven water-oxidation yielding a sufficient amount of intracellular O_2. Driven by these findings, the same research group evaluated the scalability of hydroxylation reactions without requiring external aeration [73]. The *in vivo* conversion of cyclohexane into the corresponding mono-hydroxylated alcohol catalyzed by a CYP from *Acidovorax* sp. in a two-liquid phase setup resulted in 2.6 g of cyclohexanol after 52 h.

8.3.4 Photocatalytic generation of hydrogen peroxide *in situ*

While the obtained turnover numbers for photobiocatalytic cofactor regeneration (e.g. NADPH, FAD) are in many cases too low for synthetic applications, the developed examples of combining a photosensitizer with a sacrificial electron donor can also be used for the *in situ* generation of H_2O_2. This is particularly interesting as many oxidoreductases such as CYPs and ROs rely on a complex electron-transfer chain and are cofactor-dependent, limiting their application in organic synthesis. In contrast, peroxygenases use H_2O_2 as oxidant, however they usually show poor robustness against high H_2O_2 concentrations. As such, the concentration of H_2O_2 to be carefully controlled to avoid rapid deactivation of the enzymes while maintaining high reactivity. In this context, several photocatalytic strategies for *in situ* H_2O_2 generation have been developed. For instance, Zhang et al. used gold–titanium dioxide (Au–TiO_2) to generate H_2O_2 through the methanol-driven reductive activation of ambient O_2 to drive an unspecific peroxygenase from *Agrocybe aegerita* (*rAae*UPO) [75]. Using this approach, the stereoselective hydroxylation of ethylbenzene to (*R*)-1-phenylethanol was achieved with high enantioselectivity (>98 % *ee*) and high turnover numbers for the biocatalyst (>71,000). This reaction system was further explored by using water instead of methanol as electron donor to produce H_2O_2 *in situ* to drive *rAae*UPO [76]. By using this system, the stereoselective hydroxylation of ethylbenzene to (*R*)-1-phenylethanol has also been achieved, resulting in 110 mg product with an *ee* of >97 %. This system was later on further optimized.

In addition to using TiO_2-based photocatalysts or even flavin [77] for the *in situ* H_2O_2 generation, further photocatalysts have been explored for the same purpose and enzyme class. In this regard, a water-soluble sodium anthraquinone sulfonate (SAS) was used to generate H_2O_2 to drive a vanadium-dependent chloroperoxidase from *Curvularia inaequalis* (*Ci*V-CPO)-catalyzed halogenation reaction [78]. With this system, a yield of 91 % and turnover number of 318,000 was achieved. Interestingly, this approach eliminates the diffusion limitations of the heterogeneous photocatalyst and protects the enzyme from long-lived radicals. Other examples using nitrogen-doped carbon nanodots

[79, 80] also highlight that these photocatalysts are promising to drive peroxygenase-catalyzed hydroxylation reactions. In particular, it was shown that the spatial separation of the photocatalyst from the enzyme avoids the inactivation of the enzyme, resulting in promising turnover numbers of the biocatalyst of more than 60,000 [79]. In addition, these nitrogen-doped carbon nanodots can also be applied in neat reaction media. For instance, it was shown that the immobilization of *rAae*UPO increased enzyme stability and thus facilitated the reaction in neat substrate such as cyclohexane [80].

8.4 Natural and artificial photoenzymes for (non-natural) photo-biocatalytic reactions

While the last decade has seen a strong focus on the use of light-regenerated cofactors to achieve native enzymatic activity, recent developments indicate that the combination of biocatalysis and photocatalysis is even more powerful and can unlock non-natural enzyme reactivities [12]. In particular, the discovery and application of natural and artificial photoenzymes capable of directly converting light into chemical energy are well placed to further expand the applications of photobiocatalysis.

8.4.1 Natural photoenzymes

Yet, the scope of photoenzymes discovered in nature is very limited and the investigation of their biocatalytic applicability is often restricted to their natural substrate and reaction scope. One of the first natural photoenzymes was discovered in the 1950s by the physicist Claud S. Rupert [81]. This enzyme, a so-called photolyase, is a flavoprotein containing two light-harvesting cofactors, that is responsible for repairing UV light induced DNA damage (Scheme 8.4A) [82]. Although photolyases can catalyze DNA photoreactivation with considerable high efficiency, their implementation in biocatalytic applications has not yet been described in the literature [83].

Another class of natural photoenzymes are the light-dependent protochlorophyllide reductases (LPORs, Scheme 8.4B). These enzymes are involved in the biosynthesis of chlorophyll in both oxygenic and anoxygenic phototrophs, in which they catalyze the reduction of protochlorophyllide (pchlide) to chlorophyllide (chlide) [84]. Yet, the biocatalytic application of LPORs is limited, although it could be shown that *in vitro* they catalyze the conversion of various pchlide derivatives to the corresponding reduced products [85]. In addition, several LPOR homologs were successfully identified from different origins and biochemically characterized, and their activity was studied for the conversion of pchlide to chlide under different conditions and furthermore analyzed for their cofactor flexibility [86].

In 2017, Beisson et al. identified another natural photoenzyme from the microalgae *Chlorella variabilis* [87], and ever since its discovery, it is intensively studied for its biocatalytic potential. This enzyme is a fatty acid decarboxylase (FAP) catalyzing a light-mediated decarboxylation of fatty acids via a flavin cofactor (Scheme 8.4C). Upon light irradiation, the flavin cofactor mediates the first single electron transfer from the enzyme-bound carboxylic acid, initiating the sequence of CO_2 extrusion and back-transfer of the initially abstracted electron to the newly formed C-centered radical [87, 88]. Under light illumination at around 450 nm (blue light), *Cv*FAP efficiently converts long-chain fatty acids into the corresponding alkanes, while achieving high turnover numbers and almost full conversion [89]. An ever growing number of examples highlight the potential of this photoenzyme (and its engineered variants) for various biocatalytic applications, particularly for the synthesis of chiral compounds via kinetic resolution of racemic hydroxyl carboxylic acids [90], or implementation in cascade approaches for the synthesis of secondary alcohols [91] or polymer building blocks [92] from unsaturated fatty acids.

8.4.2 Artificial photoenzymes

While the chemistry that can be unlocked from natural photoenzymes is still limited, the combination of biocatalysis and photocatalysis opens up a golden window of opportunities to unlock abiological transformations. Although the concept of using light to harness abiological reaction chemistries from an existing and often engineered enzyme scaffold is a rather new development in photobiocatalysis, a handful of examples already highlight the powerfulness of this approach. The construction of these so-called artificial photoenzymes typically exploits two strategies: (1) the covalent linking of a photocatalysts via an (unnatural) amino acid and using the photocatalyst for direct single-electron transfer; and (2) the conjugation of a photosensitizer nearby an immobilized organometallic complex within a protein scaffold [10, 12, 22]. While artificial enzymes also have been constructed to supply electrons to naturally occurring enzyme cofactors to increase native activity [13, 15, 93], we herein focus on artificial enzymes that have been constructed for non-natural reactivities. Already in 2015, Lewis et al. used a click chemistry approach to conjugate a modified acridinium 9-mesityl-10-methyl (Acr$^+$-Mes) cofactor to an unnatural amino acid within a prolyl oligopeptidase to construct an artificial photoenzyme for the sulfoxidation of thioanisoles [94, 95].

Wang et al. constructed an artificial photoenzyme by conjugating a catalytically active organometallic complex in close vicinity to a genetically encoded photoactive chromophore in the superfolder yellow fluorescent protein scaffold [96]. They genetically encoded benzophenone-alanine into the protein scaffold and conjugated a nickel-terpyridine complex for the photocatalytic reduction of CO_2. With the help of protein engineering, the positions of the chromophore and catalyst were tuned, the photosensitizers' photochemical properties were modulated, and the microchemical environment

A. *Photolyase*

UV-damaged DNA

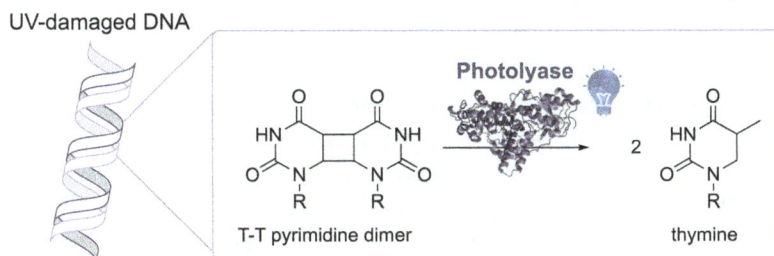

T-T pyrimidine dimer   thymine

B. *Protochlorophyllide reductase*

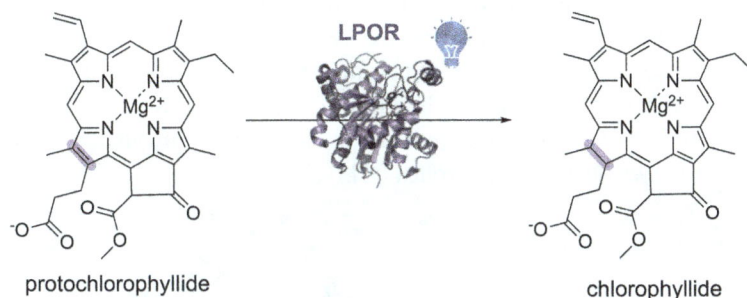

protochlorophyllide   chlorophyllide

C. *Fatty acid photodecarboxylase*

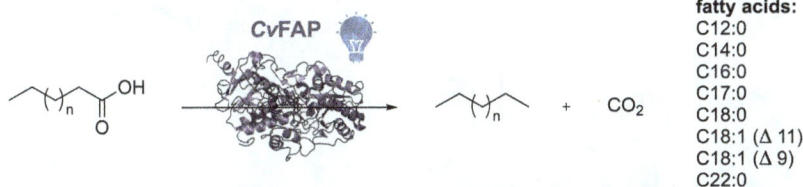

fatty acids:
C12:0
C14:0
C16:0
C17:0
C18:0
C18:1 (Δ 11)
C18:1 (Δ 9)
C22:0

Scheme 8.4: Overview on natural photoenzymes. A) Photoreactivation of pyrimidine dimers in UV-damage DNA catalyzed by a photolyase [82]. B) Light-dependent protochlorophyllide reductase (LPOR)-catalyzed reduction of protochlorophyllide to chlorophyllide [84]. C) The fatty acid photodecarboxylase from *Chlorella variabilis* (*Cv*FAP) catalyzes a decarboxylation reaction upon light illumination, resulting in the formation of hydrocarbons [87]. Adapted with permission from Refs. [12, 16]. Copyright 2022 American Chemical Society.

of the protein was adjusted to enable efficient CO_2 reduction [97]. This photoactive protein scaffold was further modified to enable accomplish cross-couplings of aryl halides via genetically encoded benzophenone chromophore and an adjacent artificial Ni^{II}(bpy) cofactor [98].

The incorporation of a photosensitizer into a protein scaffold via genetic code expansion for enantioselective [2+2]-cycloadditions via triplet energy transfer has been explored by the groups of Green as well as Chen, Zhong, and Wu [99, 100]. Both groups developed the artificial photoenzyme for [2+2]-cycloadditions based on a different protein scaffold, however, both artificial photobiocatalysts were constructed by genetic code expansion to incorporate a photosensitizer. Upon light irradiation of the photosensitizer, a [2+2]-cycloaddition is promoted. The chosen protein scaffold thereby delivers the needed selectivity, which was further optimized via protein engineering. For instance, the obtained first generation photoenzyme developed by Trimble et al. catalyzed the intramolecular [2+2]-cycloaddition of 4-(but-3-en-1-yloxy)quinolin-2(1H)-one to two regioisomeric products (straight and crossed, Figure 8.4) [99]. Via directed evolution, an efficient and enantioselective enzyme (up to 99 % *ee*) promoting selective cycloadditions with >300 turnovers under aerobic conditions was obtained.

These examples highlight the power of combining photocatalysis with biocatalysis by introducing photocatalysts via bioconjugation with unnatural amino acids or genetically encoding a photosensitizer into a protein scaffold. As such, novel chemical reactivities can be imparted to enzymes, while the design parameters of such biohybrid systems can be modulated to achieve more efficient chemical conversions. However, difficulties arising in the design and construction of artificial photoenzymes impede the chemical diversity that is yet accessible by this approach [12].

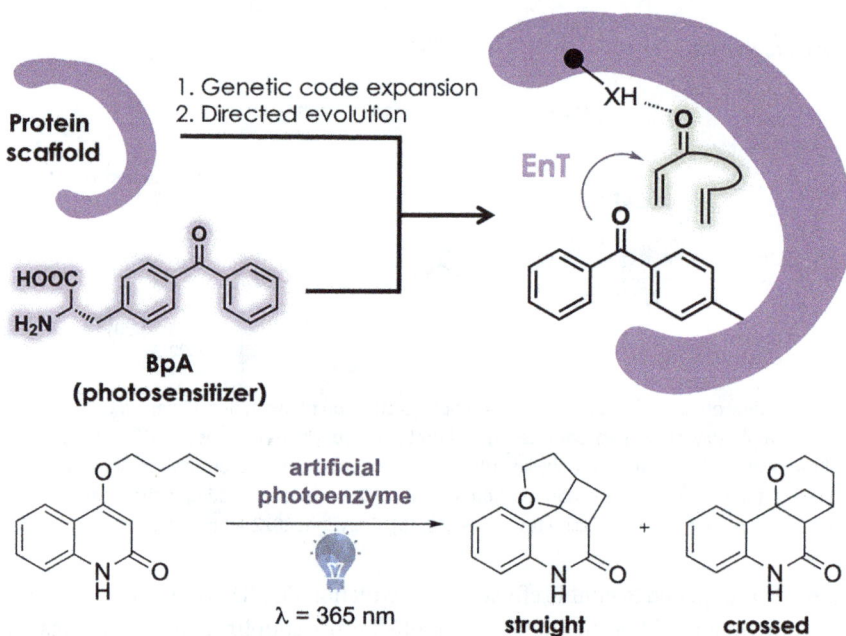

Figure 8.4: Design of an artificial photenzyme promoting enantioselective [2+2]-cycloadditions via triplet energy transfer (EnT) [99]. The photosensitizer 4-benzoylphenylalanine (BpA) is genetically encoded into the protein scaffold of a computationally designed Diels Alderase, followed by directed evolution, affording an enantioselective photoenzyme for the conversion of 4-(but-3-en-1-yloxy)quinolin-2(1H)-one to the straight or crossed product. Adapted from Ref. [99].

8.4.3 Photoinduced enzyme promiscuity

The protein machineries of living systems are often 'promiscuous' – that is, capable of catalyzing reactions other than its biological function. These promiscuous functions can be used to generate catalytic novelty, and promiscuous activities can even be induced or significantly improved by protein engineering [101–107]. Besides tailoring the substrate specificity or enzyme stability, changing the reaction mechanism of an enzyme is an integral approach that provides access to biocatalysts with a non-biological reaction scope [108, 109].

Recent studies have reported that light can induce catalytic promiscuity in oxido-reductases, which is driven by the excitation of flavin or nicotinamide cofactors as well as organic photosensitizers, leading to the generation of radical intermediates within the enzyme active sites, resulting in various biological transformations. This approach holds great promise for extending the synthetic capabilities of biocatalysts and, when combined with protein engineering, for addressing long-standing selectivity and reactivity challenges in chemical synthesis. This is particularly important as the stereochemical outcome of reactions involving radical intermediates is often difficult to control using existing small molecule catalysts [110].

Recently, Emmanuel et al. have shown that a ketoreductase (KRED) can catalyze the enantioselective dehalogenation of halolactones, a very different reaction from that they were evolved for (Scheme 8.5A) [111]. In this example, the authors use light to excite the cofactor NADPH bound to the active site of the KRED. Upon light irradiation, the photoexcited cofactor and halolactone substrate generates a charge-transfer complex, leading to the formation of the corresponding cofactor and substrate radicals. The KRED-catalyzed removal of a halogen atom from the halolactone results then in the formation of the dehalogenated chiral lactone product [111].

In follow-up studies, this concept was transferred to flavin-dependent ene-reductases (Scheme 8.5B) [112]. While intrinsically, these NAD(P)H-dependent enzymes catalyze the two-electron reduction of activated alkenes, the light-induced radical species formation enabled a hydroalkylation used as driving force for non-natural intramolecular cyclization [112], or an intermolecular radical hydroalkylation of terminal alkenes [114]. Unlike to the examples with KREDs, in which NAD(P)H served as the photoactive component, the light-induced single-electron transfer responsible for the promiscuity of ene-reductases is presumably driven by the photoexcitation of the electron-donor complex composed of reduced FMN- and substrate.

These examples rely on the photoinduced electron transfer from NADPH or ground state electron transfer from a flavin hydroquinone. Another approach is to use an exogenous reductant to facilitate electron transfer, while a protein scaffold serves as the chiral catalyst (Scheme 8.5C) [110, 113]. However, this strategy requires the development of gating strategies to ensure that the formation of the radical species occurs exclusively within the active site of the enzyme. This hypothesis was investigated on a reductive radical deacetoxylation reaction with a NADPH-dependent double bond reductase (DBR) as an enzymatic scaffold. Indeed, the deacetoxylation was observed with good levels of enantioselectivity upon addition of commonly used transition-metal and organic photocatalysts. Moreover, it was proposed that green-light

A.

KRED

rac-lactone   NAD(P)H   NAD(P)⁺   chiral lactone

B.

Ene-Reductase

α-chloroamide   NAD(P)H   NAD(P)⁺   γ-lactam

C.

Double-bond
Reductase

racemic
a-acetoxyketone   NADPH   NADP⁺   ketone

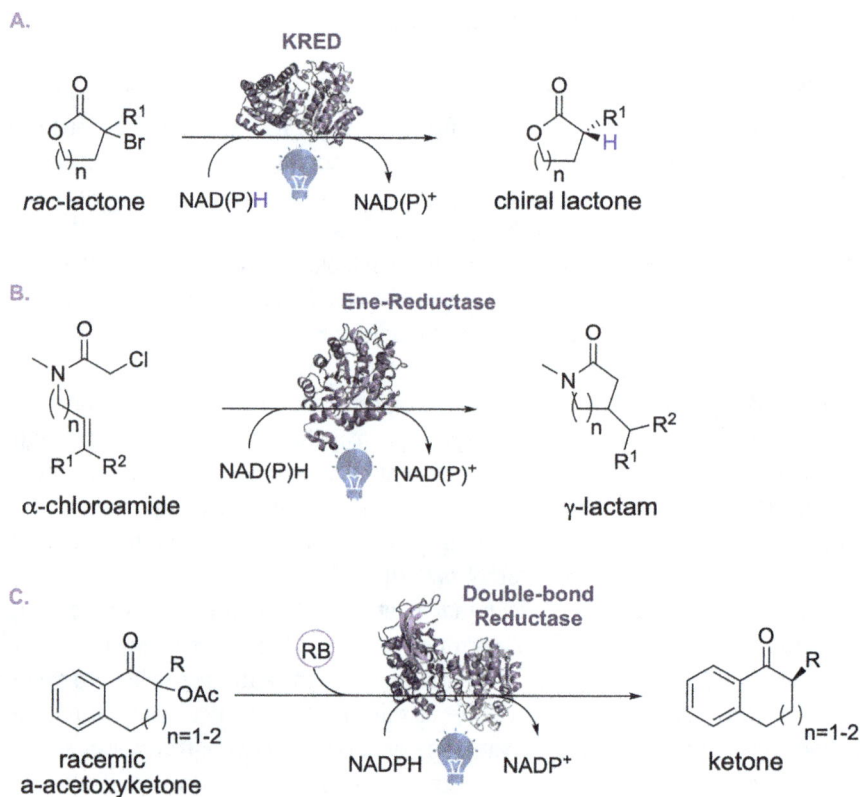

Scheme 8.5: Overview of different light-driven biocatalytic approaches that use light to induce promiscuity in NAD(P)H-dependent enzymes and flavoenzymes. A) Upon light irradiation, a ketoreductase (KRED) catalyzes an enantioselective dehalogenation resulting in the formation of chiral lactones [111]. B) A photoexcited ene-reductase (ERED) catalyzes the radical cyclization of α-chloroamides into γ-lactams [112]. C) Double-bond reductases catalyzing the deacetoxylation of α-acetoxyketones using photoexcited Rose Bengal (RB) [113]. Adapted with permission from Ref. [16].

irradiation of RB in the presence of NAD(P)H forms radical RB•⁻, an intermediate capable of reducing the enzyme-bond substrate yielding in the formation of a deacetoxylated α-acyl radical. The hydrogen transfer from reduced NADPH yields then the respective product in the final step [113].

8.5 Photoenzymatic cascades combining photo- and biocatalytic transformations

Next to photobiocatalytic applications that are designed to provide redox enzymes with reducing equivalents for catalysis, the combination of photo- with biocatalytic reactions steps is an emerging field of interest. These so-called photochemoenzymatic (PCE) cascades use light to directly drive small molecule interconversions and combine them with further enzymatic functionalization steps. Cascade reactions (Box 8.3) in general are of growing importance in biotechnological applications, thus it is not surprising that the development of cascades implementing photocatalytic reaction steps is also getting momentum. The increasing number of examples that have been reported recently further highlight these developments, particularly because the use of light can significantly extend the available reaction chemistries to synthesize various pharmaceuticals and fine chemicals [16].

While one-pot linear cascades (referred to as concurrent or tandem reactions) are much more preferable as no intermediate work-up is required [115], the possibility of performing a PCE cascade in a simultaneous reaction mode typically requires a careful evaluation of photo- and biocatalyst compatibility. Thereby, the type of photocatalyst, possible side reactions and the stability of the enzyme in the presence of the photocatalyst are crucial factors determining the efficiency of the simultaneous reaction mode [16]. Here, the combination of a photocatalyst with an enzyme can allow the design of powerful reactions that exploit the unique reactivity of photocatalysts and the effectiveness of enzymes [116]. However, critical challenges need to be overcome in order to operate such systems with high efficiency. Below are some examples of PCE cascades and the approaches used to overcome incompatibility issues.

BOX 8.3: Cascade biocatalysis – classifications.

As biocatalytic cascade, we define a reaction system that combines two or more chemical steps in one reaction vessel without isolation of intermediates. Thereby, enzymatic, chemical (metal or organo) or spontaneous catalytic reaction steps can occur. Moreover, these multi-step reactions can be either performed simultaneously or sequentially. Simultaneous cascades are also known as concurrent cascades or tandem reactions. In this case, all reaction steps take place at the same time, meaning that all enzymes and reagents are present from the beginning in the reaction vessel. In case of sequential cascades, the catalysis usually proceeds step-wise, while additional enzymes/reagents are added once another reaction step has been completed. According to Schrittwieser et al., cascades are generally classified according to their topology: (1) in a linear cascade the product of one chemical step serves as the substrate of the subsequent chemical step; (2) in an orthogonal cascade the conversion of a substrate to the product is coupled with a second reaction to remove one or more by-products; (3) in a cyclic cascade one enantiomer of a racemic substrate mixture is converted to an intermediate product which is then converted back to the racemic starting material, leaving the unreacted substrate enantiomer as the final product; and (4) a parallel linked cascade in which two separate biocatalytic reactions are linked by complementary cofactor requirements of the two enzymes [115].

8.5.1 Simultaneous and sequential photochemoenzymatic cascades

The realization of PCE cascades in simultaneous mode is often difficult, mainly because of the often opposite reaction conditions required by the photocatalyst and the enzyme. In addition, stability problems can arise due to the presence of the enzyme, substrate and product in the reaction mixture. Hartwig and colleagues reported a concurrent PCE cascade in which a photocatalyst is used for the isomerization of alkenes, which are further converted by an ERED to generate valuable enantio-enriched products (Scheme 8.6) [116]. Initially, a number of organometallic and organic photocatalysts were screened for their ability to isomerize the (Z)-configured substrates, and different photocatalysts such as FMN and Ir(III) were identified as most suitable. In the following, the enzymatic reduction of the (Z)-configured substrates was investigated. The cooperative reduction catalyzed by different EREDs and by using different photocatalysts such as FMN and Ir(III) under blue light illumination resulted in the formation of several products with high yields (up to 87 %) and ee's of >99 %. In this example, two features of the photocatalyst were crucial for performing the cascade in a concurrent reaction mode: (1) the photocatalytic reaction takes place at room temperature to match the enzyme requirements, and (2) the mechanism underlying the photocatalytic reaction involves intermediates that are stable towards water and the functional group in proteins.

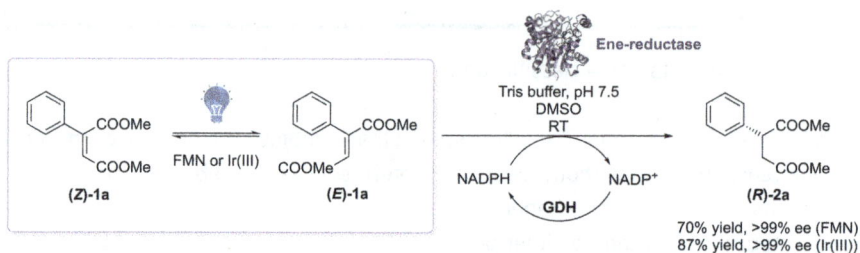

Scheme 8.6: Concurrent photochemoenzymatic cascade comprising a photocatalytic isomerization with the enzymatic reduction of alkenes to the corresponding enantio-enriched products under blue light illumination [116]. Adapted with permission from Ref. [16].

A second example emphasized the power of combining photocatalysis with biocatalysis for the selective functionalization of (non)-activated C–H bonds, which is an ongoing challenge in classical organic synthesis (Scheme 8.7) [117]. The applied photocatalyst, sodium anthraquinone sulfonate (SAS), was used for the oxidation of alkanes to the corresponding aldehydes or ketones, followed by the enzymatic transformation of the intermediary aldehyde/ketone to various high-value-added (chiral) products, such as formic esters, lactones, chiral cyanohydrins, chiral acyloins, carboxylic acids, chiral cyclohexanones and amines [117]. With this system, a range of chiral products featuring different functional groups were obtained. Additionally, (R)-benzoin and (R)-mandelonitrile were synthesized at gram-scale with high isolated yields and

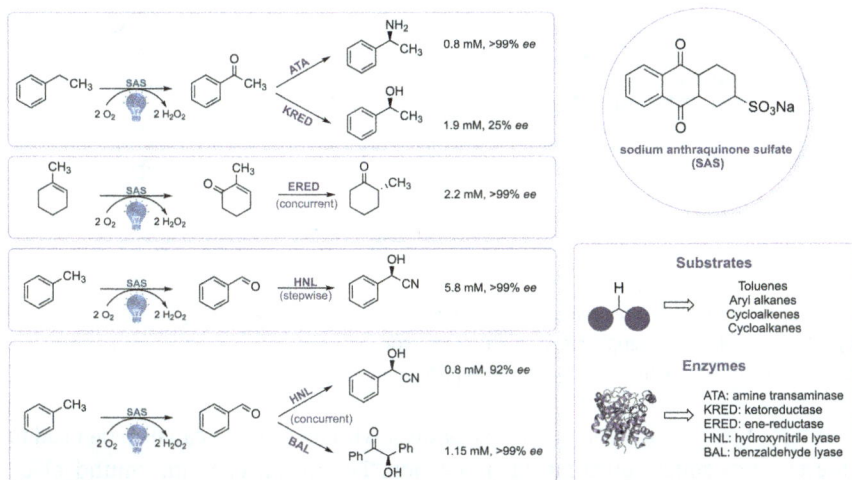

Scheme 8.7: Concurrent and step-wise photochemoenzymatic cascades comprising a photocatalytic oxyfunctionalization catalyzed by sodium anthraquinone sulfonate (SAS), followed by further functionalization of the aldehyde or ketone intermediate by different enzymes, such as amine transaminase (ATA), keto reductase (KRED), ene-reductase (ERED), hydroxynitrile lyase (HNL) and benzaldehyde lyase (BAL) [117]. Adapted with permisison from Ref. [16].

excellent *ee* (>99 %), highlighting the applicability of the developed PCE cascade. While the photochemoenzymatic synthesis of these two products was successfully demonstrated at gram-scale, the remaining cascades require further optimization in order to overcome limitations related to catalyst inhibition/deactivation due to the formation of reactive oxygen or radical species and cross-reactivity in the presence of light.

In addition to the highly desired functionalization of molecule scaffolds, building more complex molecules from cheap and readily available starting materials is a desirable feature in every synthetic endeavor [118]. The combination of enzymes with photocatalysts thereby offers an attractive approach for implementing multi-step synthetic processes to build molecular complexity. Ravelli and Schmidt recently showed that the merging of photocatalytic C–C bond formation with enzymatic asymmetric reduction enables the direct conversion of simple aldehydes and acrylates or unsaturated carboxylic acids into chiral γ-lactones (Scheme 8.8) [119]. In this example, the photocatalyst tetrabutylammonium decatungstate (TBADT) is used to catalyze the hydroacylation of the starting olefins, yielding the corresponding keto esters/acids. In the subsequent step, an alcohol dehydrogenase (ADH) converts the keto ester/acid to the corresponding chiral alcohol, which undergoes lactonization to the desired γ-lactone. It was demonstrated that the synthesis of several aliphatic and aromatic γ-lactones was thereby achieved with up to >99 % *ee* and >99 % yield. This example shows that building molecular complexity from simple, cheap and largely available starting materials is possible by merging photocatalysis with biocatalysis to access high-value added chiral compounds.

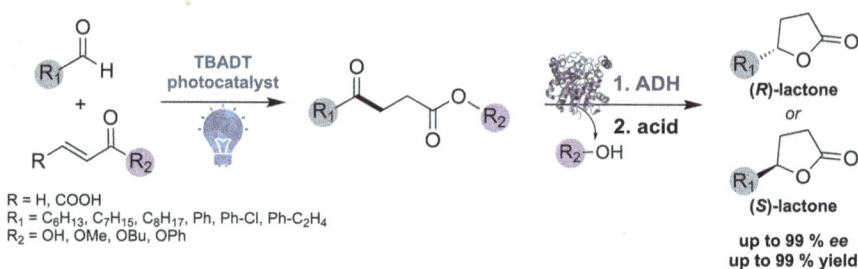

R = H, COOH
R_1 = C_6H_{13}, C_7H_{15}, C_8H_{17}, Ph, Ph-Cl, Ph-C_2H_4
R_2 = OH, OMe, OBu, OPh

Scheme 8.8: Step-wise photochemoenzymatic synthesis of chiral γ-lactones comprising decatungstate photocatalyzed hydroacylation of simple olefins with asymmetric reduction of the intermediate 1,4-keto esters/acids catalyzed by ADH. Adapted with permisison from Ref. [119].

Intriguingly, the combination of a photocatalyst with an enzyme can also be used to control the stereochemical outcome of a reaction. For instance, Schmermund et al. showed that the redox potential of a carbon nitride photocatalyst (CN-OA-m) can be tuned by changing the irradiation wavelength to generate electron holes with different oxidation potentials [120]. In combination with an unspecific peroxygenase from *A. aegerita*, the illumination with of CN-OA-m with green light led to the enantioselective hydroxylation of ethylbenzene to (*R*)-phenylethanol, whereas blue light irradiation led to the formation of acetophenone, which was further converted with an ADH to the desired (*S*)-phenylethanol.

8.5.2 Photoenzymes in biocatalytic cascades

While the majority of PCE cascades rely on the activation of a photocatalytic reaction step by light, also natural and artificial photoenzymes that only perform catalysis upon light irradiation have been implemented in multi-step processes. For instance, a hybrid P450 BM3 biocatalyst containing a covalently attached Ru(II)-diimine photosensitizer has been used in a PCE cascade reaction for the synthesis of trifluoro methylated/hydroxylated substituted arenes [121]. The photochemical properties of the Ru(II) imine photosensitizer allow for the initiation of single electron transfer events, which facilitate the addition of a CF_3 radical to arenes. In the hybrid enzyme, the Ru(II)-diimine photosensitizer provides the necessary electrons to carry out hydroxylation reactions on trifluoromethylated substrates upon visible light activation. While a range of different substrates could be converted, lower yields were obtained in most cases. Beneficially, the regio- and stereoselectivity of the hybrid P450 catalyst enables the differentiation between the trifluoromethylated isomers.

Also 'true' photoenzymes have been implemented in cascade approaches (Scheme 8.9) [89, 91, 122]. For example, Huijbers et al. demonstrated the use of *Cv*FAP in a cascade reaction for the conversion of saturated and unsaturated fatty acids to the corresponding long-chain alkanes or alkenes [89]. This approach for synthesizing long-chain alkenes directly from triglycerides via a two-step cascade reaction represents an interesting alternative strategy to the typical transesterification used in biofuel production. The two-step cascade thereby combines enzymatic hydrolysis of triglycerides with subsequent light-driven decarboxylation catalyzed by CvFAP. In addition to the production of alkanes from triglycerides, another bienzymatic cascade was developed to convert castor oil to (R,Z)-octadec-9-en-7-ol [123]. Therein, non-edible castor oil is hydrolyzed with the help of a lipase, yielding of free ricinoleic acid. Subsequently, CvFAP is catalyzing the decarboxylation under blue light illumination, yielding product concentrations of up to 60 mM [123]. The authors suggested that further optimization of the cascade is needed to compensate for the drop in pH caused by the accumulation of fatty acids in the reaction, and thus achieve even higher product yields [89, 123].

A. Light-driven enzymatic cascade converting lipids into long-chain alkanes

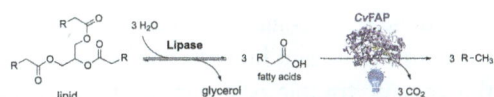

B. Light-driven enzymatic cascade converting non-edible oils into long-chain alcohols

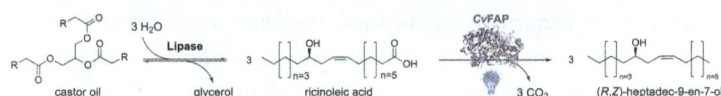

C. Light-driven enzymatic cascade converting unsaturated fatty acids into secondary alcohols

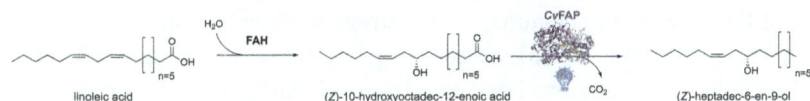

D. Light-driven enzymatic cascade converting unsaturated fatty acids into diols

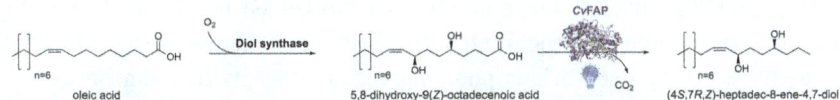

Scheme 8.9: Different enzymatic cascades comprising the photodecarboxylase CvFAP [89, 91, 122]. Adapted with permisison from Ref. [16].

8.6 Current challenges of applying light in bioprocesses

The use of light to drive photobiocatalytic reactions is an exciting development in applied biocatalysis, and holds great potential for overcoming current challenges in implementing enzyme-catalyzed processes in organic synthesis. However, as promising and revolutionary it might sound, the desire for the rapid development of novel photobiocatalytic reaction systems drives the distraction from the fundamental concept of reproducibility and implementation in an industrial setting.

This is a particular challenge when it comes to the equipment setup used for photocatalytic reactions. As recently emphasized by Edwards et al., ensuring reproducibility of light-driven bioprocesses and therewith standardization of photochemistry platforms is crucial to ensure a broad applicability of photobiocatalytic reactions [124]. Therefore, a deeper characterization of the used and often home-made light reactor setups is needed to enable a better understanding of the underlying photochemical principles and thus to enable the scale up photobiocatalytic approaches. The comprehensive characterization of the applied light system, such as light source used, photon stoichiometry, internal reaction temperature, light intensity, distance between the light source and reaction mixture, and path length, is crucial to avoid batch-to-batch variability that is often caused by differences of the light source. Encouragingly, a trend towards standardization of photochemistry platforms is there, which is expected to ease a broader implementation of photochemistry in both, academia and industry [124].

Next to the reproducibility challenge, the upscaling and implementation of photobiocatalytic platforms in an industrial setting remains troublesome. For instance, when a photosensitizer, absorbing light at specific wavelengths, is used, charge separation in the presence of an electron acceptor occurs. By implication, the light portion absorbed by a single chromophore is not accessible to another and restricts photoexcitation. This photo limitation not only has crucial impact on photobiocatalytic *in vitro* reactions, but also on establishing high-cell density cultures of cyanobacteria for respective *in vivo* applications. As the growth of photoautotrophs is strongly dependent on the availability and intensity of light, circumventing or restricting so-called self-shading effects by optimizing the cultivation conditions has a significant advantage for their implementation as whole-cell biocatalysts [125].

Recent developments in (scalable) photo(bio)reactors (PBRs) are already promising. They could overcome some of the limitations mentioned above. Recent reviews, for example, have highlighted various photobioreactor concepts that can be used to efficiently grow microalgae and perform biocatalytic processes, sometimes even on a large scale [126, 127]. The design and optimization of photobioreactors is an important aspect of photobiochemical processes, with the supply of light to the reactors being one of the major challenges, as it decreases exponentially with the distance from the light source in batch. In this context, continuous flow processes have an advantage as the light penetration is uniform regardless of the reactor scale [126]. Although there are examples of novel reactor designs for photocatalytic reactions in flow, photobiocatalytic applications using flow reactors are still limited. Next to continuous flow processes that are considered in photobiocatalysis, the use of an internal instead of external light source via wireless light emitters (WLEs) provides a promising approach to optimize light absorption in a PBR for cultures of phototrophic microorganisms [128]. A recent study by Hobisch et al. used WLEs to improve the light distribution and product formation in a bubble column reactor using cyanobacteria as whole-cell biocatalysts [129]. In addition, the internal illumination strategy also provides an interesting strategy for continuous flow applications, e.g., in a packed bed reactor.

When it comes to the robustness and catalytic performance of a photobiocatalytic reaction system, the choice of photosensitizer and electron donor is essential. Here, the stability of the photosensitizer often represents a bottleneck as photodegradation or photobleaching of organic dye photosensitizers is frequently observed [130, 131]. Due to that, the application of quantum dots in combination with metal oxides such as TiO_2 is emphasized for establishing more robust light-driven reaction setups [132]. On the other hand, the choice of sacrificial electron donor should be made under the consideration of atom economy, availability and compatibility with the reaction conditions. In this context, the use of H_2O as sacrificial electron donor is of utmost interest. However, only a few proof of concept studies reported the feasibility of using H_2O as direct electron donor in combination with inorganic photocatalysts such as titanium dioxide TiO_2 to perform light-driven reactions [133]. The problem with using H_2O as electron donor is the high stability and low oxidation potential, hampering general applicability in photobiocatalytic approaches [133].

8.7 Conclusions

Photobiocatalysis has emerged into an exciting research area with an ever growing impact in applied biocatalysis. Next to the large number of examples that report photobiocatalytic cofactor regeneration via direct or indirect activation of redox enzymes, it is expected that cyanobacteria as chassis for *in vivo* photobiocatalysis will be further explored and optimized in the near future. Furthermore, the development and application of photochemoenzymatic cascades to combine the outstanding reactivities accessible to photocatalysts with the high selectivity of enzyme-catalyzed reactions greatly highlights the diversity of photobiocatalysis. It is also anticipated that the discovery of additional natural photoenzymes or the design of new artificial photoenzymes, together with the further exploitation of light-induced enzyme promiscuity, will open up unexpected opportunities to extend the catalytic scope currently accessible to biocatalysis. With the recent developments of novel photobioreactor concepts for up-scaling, we expect that in the near future the potential of these newly developed photobiocatalytic strategies for organic synthesis can be fully assessed.

References

1. Yoon TP, Ischay MA, Du JN. Visible light photocatalysis as a greener approach to photochemical synthesis. Nat Chem 2010;2:527–32.
2. Ciamician G. The photochemistry of the future. Science 1912;36:385–94.

3. Pyser JB, Chakrabarty S, Romero EO, Narayan ARH. State-of-the-art biocatalysis. ACS Cent Sci 2021;7: 1105–16.
4. Wu S, Snajdrova R, Moore JC, Baldenius K, Bornscheuer UT. Biocatalysis: enzymatic synthesis for industrial applications. Angew Chem Int Ed 2021;60:88–119.
5. Sun H, Zhang H, Ang EL, Zhao H. Biocatalysis for the synthesis of pharmaceuticals and pharmaceutical intermediates. Bioorg Med Chem 2018;26:1275–84.
6. Bell EL, Finnigan W, France SP, Green AP, Hayes MA, Hepworth LJ, et al. Biocatalysis. Nat Rev Methods Prim 2021;1:46.
7. Dong JJ, Fernández-Fueyo E, Hollmann F, Paul CE, Pesic M, Schmidt S, et al. Biocatalytic oxidation reactions: a chemist's perspective. Angew Chem Int Ed 2018;57:9238–61.
8. Yang X, Wang D. Photocatalysis: from fundamental principles to materials and applications. ACS Appl Energy Mater 2018;1:6657–93.
9. Stephenson CRJ, Yoon TP, MacMillan DWC. Visible light photocatalysis in organic chemistry. Weinheim: Wiley-VCH; 2017.
10. Shaw MH, Twilton J, MacMillan DWC. Photoredox catalysis in organic chemistry. J Org Chem 2016;81: 6898–926.
11. Marzo L, Pagire SK, Reiser O, König B. Visible-light photocatalysis: does it make a difference in organic synthesis? Angew Chem Int Ed 2018;57:10034–72.
12. Harrison W, Huang X, Zhao H. Photobiocatalysis for abiological transformations. Acc Chem Res 2022;55: 1087–96.
13. Schmermund L, Jurkaš V, Özgen FF, Barone GD, Büchsenschütz HC, Winkler CK, et al. Photo-biocatalysis: biotransformations in the presence of light. ACS Catal 2019;9:4115–44.
14. Seel CJ, Gulder T. Biocatalysis fueled by light: on the versatile combination of photocatalysis and enzymes. ChemBioChem 2019;20:1871–97.
15. Lee SH, Choi DS, Kuk SK, Park CB. Photobiocatalysis: activating redox enzymes by direct or indirect transfer of photoinduced electrons. Angew Chem Int Ed 2018;57:7958–85.
16. Özgen FF, Runda ME, Schmidt S. Photo-biocatalytic cascades: combining chemical and enzymatic transformations fueled by light. ChemBioChem 2021;22:790–806.
17. Peng Y, Chen Z, Xu J, Wu Q. Recent advances in photobiocatalysis for selective organic synthesis. Org Process Res Dev 2022;26:1900–13.
18. Seel CJ, Gulder T. Biocatalysis fueled by light: on the versatile combination of photocatalysis and enzymes. ChemBioChem 2019;20:1871–97.
19. Maciá-Agulló JA, Corma A, Garcia H. Photobiocatalysis: the power of combining photocatalysis and enzymes. Chem – Eur J 2015;21:10940–59.
20. Höfler G, Hollmann F, Paul CE, Rauch M, van Schie M, Willot S. Chapter 9 Photocatalysis to promote cell-free biocatalytic reactions. In: Kourist R, Schmidt S, editors. The autotrophic biorefinery raw materials from biotechnology. Berlin, Boston: De Gruyter; 2021:247–76 pp.
21. Zhang W, Hollmann F. Nonconventional regeneration of redox enzymes – a practical approach for organic synthesis? Chem Commun 2018;54:7281–9.
22. Romero NA, Nicewicz DA. Organic photoredox catalysis. Chem Rev 2016;116:10075–166.
23. Chappelle EW, Kim MS, McMurtrey JE. Ratio analysis of reflectance spectra (RARS): an algorithm for the remote estimation of the concentrations of chlorophyll A, chlorophyll B, and carotenoids in soybean leaves. Remote Sens Environ 1992;39:239–47.
24. Cardona T, Sedoud A, Cox N, Rutherford AW. Charge separation in photosystem II: a comparative and evolutionary overview. Biochim Biophys Acta 2012;1817:26–43.
25. Shevela D, Kern JF, Govindjee G, Whitmarsh J, Messinger J. Photosystem II. ELS 2021;2:1–16.
26. Mandal M, Kawashima K, Saito K, Ishikita H. Redox potential of the oxygen-evolving complex in the electron transfer cascade of photosystem II. J Phys Chem Lett 2019;11:249–55.

27. Feyza Özgen F, Runda ME, Burek BO, Wied P, Bloh JZ, Kourist R, et al. Artificial light-harvesting complexes enable Rieske oxygenase catalyzed hydroxylations in non-photosynthetic cells. Angew Chem Int Ed 2020; 59:3982–7.

28. Lee SH, Nam DH, Kim JH, Baeg JO, Park CB. Eosin Y-sensitized artificial photosynthesis by highly efficient visible-light-driven regeneration of nicotinamide cofactor. ChemBioChem 2009;10:1621–4.

29. Pellegrin Y, Odobel F. Sacrificial electron donor reagents for solar fuel production. Compt Rendus Chim 2017;20:283–95.

30. Chenault HK, Simon ES, Whitesides GM. Cofactor regeneration for enzyme-catalysed synthesis. Biotechnol Genet Eng Rev 1988;6:221–70.

31. Mordhorst S, Andexer JN. Round, round we go – strategies for enzymatic cofactor regeneration. Nat Prod Rep 2020;37:1316–33.

32. Ni Y, Hollmann F. Artificial photosynthesis: hybrid systems. In: Jeuken LJC, editor. Biophotoelectrochemistry from bioelectrochemistry to biophotovoltaics. Cham: Springer International Publishing; 2016:137–58 pp.

33. Ke B. Electrolytic reduction of diphosphopyridine nucleotide at some solid metal electrodes. J Am Chem Soc 1956;78:3649–51.

34. Cunningham AJ, Underwood AL. Electrochemical reduction of triphosphopyridine nucleotide. Arch Biochem Biophys 1966;117:88–92.

35. Steckhan E, Herrmann S, Ruppert R, Dietz E, Frede M, Spika E. Analytical study of a series of substituted (2,2′-bipyridyl)(pentamethylcyclopentadienyl)rhodium and -iridium complexes with regard to their effectiveness as redox catalysts for the indirect electrochemical and chemical reduction of NAD(P)+. Organometallics 1991;10:1568–77.

36. Hollmann F, Schmid A, Steckhan E. The first synthetic application of a monooxygenase employing indirect electrochemical NADH regeneration. Angew Chem Int Ed 2001;40:169–71.

37. Lee SH, Kwon YC, Kim DM, Park CB. Cytochrome P450-catalyzed O-dealkylation coupled with photochemical NADPH regeneration. Biotechnol Bioeng 2013;110:383–90.

38. Mandler D, Willner I. Photosensitized NAD (P) H regeneration systems; application in the reduction of butan-2-one, pyruvic, and acetoacetic acids and in the reductive amination of pyruvic and oxoglutaric acid to amino acid. J Chem Soc, Perkin Trans 1986;2:805–11.

39. Peers MK, Toogood HS, Heyes DJ, Mansell D, Coe BJ, Scrutton NS. Light-driven biocatalytic reduction of α,β-unsaturated compounds by ene reductases employing transition metal complexes as photosensitizers. Catal Sci Technol 2016;6:169–77.

40. Willner I, Mandler D, Riklin A. Photoinduced carbon dioxide fixation forming malic and isocitric acid. J Chem Soc Chem Commun 1986:1022–4. https://doi.org/10.1039/C39860001022.

41. Mandler D, Willner I. Solar light induced formation of chiral 2-butanol in an enzyme-catalyzed chemical system. J Am Chem Soc 1984;106:5352–3.

42. Mandler D, Willner I. Photoinduced enzyme-catalysed synthesis of amino acids by visible light. J Chem Soc Chem Commun 1986:851–3. https://doi.org/10.1039/C39860000851.

43. Lee SH, Kwon YC, Kim DM, Park CB. Cytochrome P450-catalyzed O-dealkylation coupled with photochemical NADPH regeneration. Biotechnol Bioeng 2013;110:383–90.

44. Nam DH, Park CB. Visible light-driven NADH regeneration sensitized by proflavine for biocatalysis. ChemBioChem 2012;13:1278–82.

45. Nam DH, Lee SH, Park CB. CDTE, CDSE, and CDS nanocrystals for highly efficient regeneration of nicotinamide cofactor under visible light. Small 2010;6:922–6.

46. Yadav RK, Baeg JO, Oh GH, Park NJ, Kong KJ, Kim J, et al. A photocatalyst-enzyme coupled artificial photosynthesis system for solar energy in production of formic acid from CO_2. J Am Chem Soc 2012;134:11455–61.

47. van Esch JH, Hoffmann MAM, Nolte RJM. Reduction of nicotinamides, flavins, and manganese porphyrins by formate, catalyzed by membrane-bound rhodium complexes. J Org Chem 1995;60:1599–610.

48. Hari DP, König B. Synthetic applications of eosin Y in photoredox catalysis. Chem Commun 2014;50: 6688–99.
49. Lee SH, Nam DH, Park CB. Screening xanthene dyes for visible light-driven nicotinamide adenine dinucleotide regeneration and photoenzymatic synthesis. Adv Synth Catal 2009;351:2589–94.
50. Yadav RK, Baeg J-O, Kumar A, Kong K, Oh GH, Park N-J. Graphene–BODIPY as a photocatalyst in the photocatalytic–biocatalytic coupled system for solar fuel production from CO_2. J Mater Chem A 2014;2: 5068–76.
51. Ryu J, Lee SH, Nam DH, Park CB. Rational design and engineering of quantum-dot-sensitized TiO_2 nanotube arrays for artificial photosynthesis. Adv Mater 2011;23:1883–8.
52. Gargiulo S, Arends IWCE, Hollmann F. A photoenzymatic system for alcohol oxidation. ChemCatChem 2011;3:338–42.
53. Rickus JL, Chang PL, Tobin AJ, Zink JI, Dunn B. Photochemical coenzyme regeneration in an enzymatically active optical material. J Phys Chem B 2004;108:9325–32.
54. Lee SH, Choi DS, Pesic M, Lee YW, Paul CE, Hollmann F, et al. Cofactor-free, direct photoactivation of enoate reductases for the asymmetric reduction of C=C bonds. Angew Chem 2017;129:8807–11.
55. Grau MM, Van Der Toorn JC, Otten LG, Macheroux P, Taglieber A, Zilly FE, et al. Photoenzymatic reduction of C=C double bonds. Adv Synth Catal 2009;351:3279–86.
56. Lee SH, Choi DS, Pesic M, Lee YW, Paul CE, Hollmann F, et al. Cofactor-free, direct photoactivation of enoate reductases for the asymmetric reduction of C=C bonds. Angew Chem Int Ed 2017;56:8681–5.
57. Park JH, Lee SH, Cha GS, Choi DS, Nam DH, Lee JH, et al. Cofactor-free light-driven whole-cell cytochrome P450 catalysis. Angew Chem Int Ed 2015;54:969–73.
58. Lin B, Tao Y. Whole-cell biocatalysts by design. Microb Cell Factories 2017;16:1–12.
59. Wachtmeister J, Rother D. Recent advances in whole cell biocatalysis techniques bridging from investigative to industrial scale. Curr Opin Biotechnol 2016;42:169–77.
60. De Carvalho CCCR. Whole cell biocatalysts: essential workers from nature to the industry. Microb Biotechnol 2017;10:250–63.
61. Wu S, Li Z. Whole-cell cascade biotransformations for one-pot multistep organic synthesis. ChemCatChem 2018;10:2164–78.
62. Kumar B, Bhardwaj N, Agrawal K, Chaturvedi V, Verma P. Current perspective on pretreatment technologies using lignocellulosic biomass: an emerging biorefinery concept. Fuel Process Technol 2020; 199:106244.
63. Zhou Z, Liu D, Zhao X. Conversion of lignocellulose to biofuels and chemicals via sugar platform: an updated review on chemistry and mechanisms of acid hydrolysis of lignocellulose. Renew Sustain Energy Rev 2021;146:111169.
64. Jodlbauer J, Rohr T, Spadiut O, Mihovilovic MD, Rudroff F. Biocatalysis in green and blue: cyanobacteria. Trends Biotechnol 2021;39:875–89.
65. Nakamura K, Yamanaka R, Tohi K, Hamada H. Cyanobacterium-catalyzed asymmetric reduction of ketones. Tetrahedron Lett 2000;41:6799–802.
66. Yamanaka R, Nakamura K, Murakami M, Murakami A. Selective synthesis of cinnamyl alcohol by cyanobacterial photobiocatalysts. Tetrahedron Lett 2015;56:1089–91.
67. Pacheco CC, Ferreira EA, Oliveira P, Tamagnini P. Chapter 6 synthetic biology of cyanobacteria. In: Kourist R, Schmidt S, editors. The autotrophic biorefinery: raw materials from biotechnology. Berlin, Boston: De Gruyter; 2021:131–72 pp.
68. Grimm HC, Erdem E, Kourist R. Chapter 8 Biocatalytic applications of autotrophic organisms. In: Kourist R, Schmidt S, editors. The autotrophic biorefinery: raw materials from biotechnology. Berlin, Boston: De Gruyter; 2021:207–46 pp.
69. Köninger K, Gómez Baraibar Á, Mügge C, Paul CE, Hollmann F, Nowaczyk MM, et al. Recombinant cyanobacteria for the asymmetric reduction of C=C bonds fueled by the biocatalytic oxidation of water. Angew Chem Int Ed 2016;55:5582–5.

70. Büchsenschütz HC, Vidimce-Risteski V, Eggbauer B, Schmidt S, Winkler CK, Schrittwieser JH, et al. Stereoselective biotransformations of cyclic imines in recombinant cells of Synechocystis sp. PCC 6803. ChemCatChem 2020;12:726–30.

71. Hoschek A, Bühler B, Schmid A. Overcoming the gas–liquid mass transfer of oxygen by coupling photosynthetic water oxidation with biocatalytic oxyfunctionalization. Angew Chem Int Ed 2017;56: 15146–9.

72. Hoschek A, Bühler B, Schmid A. Stabilization and scale-up of photosynthesis-driven ω-hydroxylation of nonanoic acid methyl ester by two-liquid phase whole-cell biocatalysis. Biotechnol Bioeng 2019;116: 1887–900.

73. Hoschek A, Toepel J, Hochkeppel A, Karande R, Bühler B, Schmid A. Light-dependent and aeration-independent gram-scale hydroxylation of cyclohexane to cyclohexanol by CYP450 harboring Synechocystis sp. PCC 6803. Biotechnol J 2019;14:1–10.

74. Assil-Companioni L, Büchsenschütz HC, Solymosi D, Dyczmons-Nowaczyk NG, Bauer KKF, Wallner S, et al. Engineering of NADPH supply boosts photosynthesis-driven biotransformations. ACS Catal 2020;10: 11864–77.

75. Zhang W, Burek BO, Fernández-Fueyo E, Alcalde M, Bloh JZ, Hollmann F. Selective activation of C–H bonds in a cascade process combining photochemistry and biocatalysis. Angew Chem Int Ed 2017;56:15451–5.

76. Zhang W, Fernández-Fueyo E, Ni Y, Van Schie M, Gacs J, Renirie R, et al. Selective aerobic oxidation reactions using a combination of photocatalytic water oxidation and enzymatic oxyfunctionalizations. Nat Catal 2018;1:55–62.

77. Churakova E, Kluge M, Ullrich R, Arends I, Hofrichter M, Hollmann F. Specific photobiocatalytic oxyfunctionalization reactions. Angew Chem Int Ed 2011;50:10716–9.

78. Yuan B, Mahor D, Fei Q, Wever R, Alcalde M, Zhang W, et al. Water-soluble anthraquinone photocatalysts enable methanol-driven enzymatic halogenation and hydroxylation reactions. ACS Catal 2020;10: 8277–84.

79. Van Schie MMCH, Zhang W, Tieves F, Choi DS, Park CB, Burek BO, et al. Cascading g-C$_3$N$_4$ and peroxygenases for selective oxyfunctionalization reactions. ACS Catal 2019;9:7409–17.

80. Hobisch M, van Schie MMCH, Kim J, Røjkjær Andersen K, Alcalde M, Kourist R, et al. Solvent-free photobiocatalytic hydroxylation of cyclohexane. ChemCatChem 2020;12:1–6.

81. Rupert CS, Goodgal SH, Herriott RM. Photoreactivation *in vitro* of ultraviolet-inactivated Hemophilus influenzae transforming factor. J Gen Physiol 1958;41:451–71.

82. Dikbas UM, Tardu M, Canturk A, Gul S, Ozcelik G, Baris I, et al. Identification and characterization of a new class of (6-4) photolyase from Vibrio cholerae. Biochemistry 2019;58:4352–60.

83. Zhong D. Ultrafast catalytic processes in enzymes. Curr Opin Chem Biol 2007;11:174–81.

84. Kaschner M, Loeschcke A, Krause J, Minh BQ, Heck A, Endres S, et al. Discovery of the first light-dependent protochlorophyllide oxidoreductase in anoxygenic phototrophic bacteria. Mol Microbiol 2014;93:1066–78.

85. Klement H, Helfrich M, Oster U, Schoch S, Rüdiger W. Pigment-free NADPH: protochlorophyllide oxidoreductase from Avena sativa L. Purification and substrate specificity. Eur J Biochem 1999;265: 862–74.

86. Schmermund L, Bierbaumer S, Schein VK, Winkler CK, Kara S, Kroutil W. Extending the library of light-dependent protochlorophyllide oxidoreductases and their solvent tolerance, stability in light and cofactor flexibility. ChemCatChem 2020;12:4044–51.

87. Sorigué D, Légeret B, Cuiné S, Blangy S, Moulin S, Billon E, et al. An algal photoenzyme converts fatty acids to hydrocarbons. Science 2017;357:903–7.

88. Heyes DJ, Lakavath B, Hardman SJO, Sakuma M, Hedison TM, Scrutton NS. Photochemical mechanism of light-driven fatty acid photodecarboxylase. ACS Catal 2020;10:6691–6.

89. Huijbers MME, Zhang W, Tonin F, Hollmann F. Light-driven enzymatic decarboxylation of fatty acids. Angew Chem Int Ed 2018;57:13648–51.

90. Xu J, Hu Y, Fan J, Arkin M, Li D, Peng Y, et al. Light-driven kinetic resolution of α-functionalized carboxylic acids enabled by an engineered fatty acid photodecarboxylase. Angew Chem 2019;310027: ange.201903165.

91. Zhang W, Lee JH, Younes SHH, Tonin F, Hagedoorn PL, Pichler H, et al. Photobiocatalytic synthesis of chiral secondary fatty alcohols from renewable unsaturated fatty acids. Nat Commun 2020;11:1–8.

92. Cha H-J, Hwang S-Y, Lee D-S, Akula RK, Kwon Y-U, Voß M, et al. Whole-cell photoenzymatic cascades to synthesize long chain aliphatic amines and esters from renewable fatty acids. Angew Chem Int Ed 2020; 59:7024–28.

93. Zhang Y, Zhao Y, Li R, Liu J. Bioinspired NADH regeneration based on conjugated photocatalytic systems. Sol RRL 2021;5:2000339.

94. Gu Y, Ellis-Guardiola K, Srivastava P, Lewis JC. Preparation, characterization, and oxygenase activity of a photocatalytic artificial enzyme. ChemBioChem 2015;16:1880–3.

95. Zubi YS, Liu B, Gu Y, Sahoo D, Lewis JC. Controlling the optical and catalytic properties of artificial metalloenzyme photocatalysts using chemogenetic engineering. Chem Sci 2022;13:1459–68.

96. Liu X, Kang F, Hu C, Wang L, Xu Z, Zheng D, et al. A genetically encoded photosensitizer protein facilitates the rational design of a miniature photocatalytic CO_2-reducing enzyme. Nat Chem 2018;10:1201–6.

97. Kang F, Yu L, Xia Y, Yu M, Xia L, Wang Y, et al. Rational design of a miniature photocatalytic CO_2-reducing enzyme. ACS Catal 2021;11:5628–35.

98. Fu Y, Huang J, Wu Y, Liu X, Zhong F, Wang J. Biocatalytic cross-coupling of aryl halides with a genetically engineered photosensitizer artificial dehalogenase. J Am Chem Soc 2021;143:617–22.

99. Trimble JS, Crawshaw R, Hardy FJ, Levy CW, Brown MJB, Fuerst DE, et al. A designed photoenzyme promotes enantioselective [2+2]-cycloadditions via triplet energy transfer. Nature 2022;611:709–14.

100. Sun N, Huang J, Qian J, Zhou T-P, Guo J, Tang L, et al. Enantioselective [2+2]-cycloadditions with triplet photoenzymes. Nature 2022;611:715–20.

101. Van Der Meer JY, Poddar H, Baas BJ, Miao Y, Rahimi M, Kunzendorf A, et al. Using mutability landscapes of a promiscuous tautomerase to guide the engineering of enantioselective Michaelases. Nat Commun 2016;7:10911.

102. Renata H, Wang ZJ, Arnold FH. Expanding the enzyme universe: accessing non-natural reactions by mechanism-guided directed evolution. Angew Chem Int Ed 2015;54:3351–67.

103. Arnold FH. Directed evolution: bringing new chemistry to life. Angew Chem Int Ed 2018;57:4143–8.

104. Brandenberg OF, Fasan R, Arnold FH. Exploiting and engineering hemoproteins for abiological carbene and nitrene transfer reactions. Curr Opin Biotechnol 2017;47:102–11.

105. Coelho PS, Brustad EM, Kannan A, Arnold FH. Olefin cyclopropanation via carbene transfer catalyzed by engineered cytochrome P450 enzymes. Science 2013;339:307–10.

106. Zhang RK, Chen K, Huang X, Wohlschlager L, Renata H, Arnold FH. Enzymatic assembly of carbon–carbon bonds via iron-catalysed sp^3 C–H functionalization. Nature 2019;565:67–72.

107. Yang Y, Arnold FH. Navigating the unnatural reaction space: directed evolution of heme proteins for selective carbene and nitrene transfer. Acc Chem Res 2021;54:1209–25.

108. Dunham NP, Arnold FH. Nature's machinery, repurposed: expanding the repertoire of iron-dependent oxygenases. ACS Catal 2020;10:12239–55.

109. Ye Y, Fu H, Hyster TK. Activation modes in biocatalytic radical cyclization reactions. J Ind Microbiol Biotechnol 2021;48:1–18.

110. Hyster TK. Radical biocatalysis: using non-natural single electron transfer mechanisms to access new enzymatic functions. Synlett 2020;31:248–54.

111. Emmanuel MA, Greenberg NR, Oblinsky DG, Hyster TK. Accessing non-natural reactivity by irradiating nicotinamide-dependent enzymes with light. Nature 2016;540:414–7.

112. Biegasiewicz KF, Cooper SJ, Gao X, Oblinsky DG, Kim JH, Garfinkle SE, et al. Photoexcitation of flavoenzymes enables a stereoselective radical cyclization. Science 2019;364:1166–9.

113. Biegasiewicz KF, Cooper SJ, Emmanuel MA, Miller DC, Hyster TK. Catalytic promiscuity enabled by photoredox catalysis in nicotinamide-dependent oxidoreductases. Nat Chem 2018;10:770–5.
114. Huang X, Wang B, Wang Y, Jiang G, Feng J, Zhao H. Photoenzymatic enantioselective intermolecular radical hydroalkylation. Nature 2020;584:69–74.
115. Schrittwieser JH, Velikogne S, Hall M, Kroutil W. Artificial biocatalytic linear cascades for preparation of organic molecules. Chem Rev 2018;118:270–348.
116. Litman ZC, Wang Y, Zhao H, Hartwig JF. Cooperative asymmetric reactions combining photocatalysis and enzymatic catalysis. Nature 2018;560:355–9.
117. Zhang W, Fueyo EF, Hollmann F, Martin LL, Pesic M, Wardenga R, et al. Combining photo-organo redox- and enzyme catalysis facilitates asymmetric C–H bond functionalization. Eur J Org Chem 2019;2019:80–4.
118. Nicolaou KC, Hale CRH, Nilewski C, Ioannidou HA. Constructing molecular complexity and diversity: total synthesis of natural products of biological and medicinal importance. Chem Soc Rev 2012;41:5185–238.
119. Özgen FF, Jorea A, Capaldo L, Kourist R, Ravelli D, Schmidt S. The synthesis of chiral γ-lactones by merging decatungstate photocatalysis with biocatalysis. ChemCatChem 2022;14:e202200855.
120. Schmermund L, Reischauer S, Bierbaumer S, Winkler CK, Diaz-Rodriguez A, Edwards LJ, et al. Chromoselective photocatalysis enables stereocomplementary biocatalytic pathways**. Angew Chem 2021;133:7041–5.
121. Sosa V, Melkie M, Sulca C, Li JJ, Tang L, Li JJ, et al. Selective light-driven chemoenzymatic trifluoromethylation/hydroxylation of substituted arenes. ACS Catal 2018;8:2225–9.
122. Ma Y, Zhang X, Zhang W, Li P, Li Y, Hollmann F, et al. Photoenzymatic production of next generation biofuels from natural triglycerides combining a hydrolase and a photodecarboxylase. ChemPhotoChem 2020;4:39–44.
123. Ma Y, Zhang X, Li Y, Li P, Hollmann F, Wang Y. Production of fatty alcohols from non-edible oils by enzymatic cascade reactions. Sustain Energy Fuels 2020;4:1–6.
124. Bonfield HE, Knauber T, Lévesque F, Moschetta EG, Susanne F, Edwards LJ. Photons as a 21st century reagent. Nat Commun 2020;11:2–5.
125. Alagesan S, Gaudana SB, Krishnakumar S, Wangikar PP. Model based optimization of high cell density cultivation of nitrogen-fixing cyanobacteria. Bioresour Technol 2013;148:228–33.
126. Chanquia SN, Valotta A, Gruber-Woelfler H, Kara S. Photobiocatalysis in continuous flow. Front Catal 2022; 1:1–15.
127. Chanquia SN, Vernet G, Kara S. Photobioreactors for cultivation and synthesis: specifications, challenges, and perspectives. Eng Life Sci 2022;22:712–24.
128. Heining M, Sutor A, Stute SC, Lindenberger CP, Buchholz R. Internal illumination of photobioreactors via wireless light emitters: a proof of concept. J Appl Phycol 2015;27:59–66.
129. Hobisch M, Spasic J, Malihan-Yap L, Barone GD, Castiglione K, Tamagnini P, et al. Internal illumination to overcome the cell density limitation in the scale-up of whole-cell photobiocatalysis. ChemSusChem 2021; 14:3219–25.
130. Herculano LS, Malacarne LC, Zanuto VS, Lukasievicz GVB, Capeloto OA, Astrath NGC. Investigation of the photobleaching process of eosin Y in aqueous solution by thermal lens spectroscopy. J Phys Chem B 2013; 117:1932–7.
131. Demchenko AP. Photobleaching of organic fluorophores: quantitative characterization, mechanisms, protection. Methods Appl Fluoresc 2020;8:22001.
132. Kalanur SS, Hwang YJ, Joo O-S. Construction of efficient CdS–TiO$_2$ heterojunction for enhanced photocurrent, photostability, and photoelectron lifetimes. J Colloid Interface Sci 2013;402:94–9.
133. Mifsud M, Gargiulo S, Iborra S, Arends IWCE, Hollmann F, Corma A. Photobiocatalytic chemistry of oxidoreductases using water as the electron donor. Nat Commun 2014;5:1–6.
134. Ju X, Tang Y, Liang X, Hou M, Wan Z, Tao J. Development of a biocatalytic process to prepare (S)-N-Boc-3-hydroxypiperidine. Org Process Res Dev 2014;18:827–30.

135. Seelbach K, Riebel B, Hummel W, Kula M-R, Tishkov VI, Egorov AM, et al. A novel, efficient regenerating method of NADPH using a new formate dehydrogenase. Tetrahedron Lett 1996;37:1377–80.
136. Xu Z, Jing K, Liu Y, Cen P. High-level expression of recombinant glucose dehydrogenase and its application in NADPH regeneration. J Ind Microbiol Biotechnol 2007;34:83–90.
137. Shaked Z, Whitesides GM. Enzyme-catalyzed organic synthesis: NADH regeneration by using formate dehydrogenase. J Am Chem Soc 1980;102:7104–5.

Sera Bolat, Raphael Greifenstein, Matthias Franzreb and
Dirk Holtmann*

9 Process intensification using immobilized enzymes

Abstract: The application potential of enzymes is undoubtedly very high. However, despite the very large number of different enzymes and enzyme activities, the number of industrial enzyme processes is comparatively small. The particular challenge often lies in transferring promising laboratory processes to an industrial scale. Here, the required performance parameters, such as enzyme stability or productivity, must be achieved. On the one hand, this can be achieved by improving the enzymes. On the other hand, the key performance indicators can often only be achieved by using technical systems in the sense of process intensification. In enzymatic processes, immobilization of enzymes is often the means of choice to enable technical processes. The aim of this article is to outline the most important enzyme immobilization methods and to summarize the most important performance indicators of immobilized enzymes. Finally, the different immobilization methods and performance indicators are compared in a case study with unspecific peroxygenase.

Keywords: Immobilized enzymes; Performance indicators; Unspecific peroxygenase.

9.1 Introduction

In enzyme catalysis, great successes have been achieved in recent years in the identification of new enzymes, the optimization of existing enzymes with the help of rational and evolutionary methods and the establishment of new reactions, e.g. novel reactivities or enzymatic cascades [1]. If the enzyme reaction is transferred into a technical process, it often turns out that the performance of the resulting processes is not sufficient. For example, productivity, end product concentrations or stability of the

Sera Bolat and Raphael Greifenstein equally contributed to the publication.

*Corresponding author: Dirk Holtmann, Institute of Bioprocess Engineering and Pharmaceutical Technology, University of Applied Sciences Mittelhessen, Wiesenstrasse 14, 35390 Giessen, Germany; and Institute of Process Engineering in Life Sciences, Karlsruhe Institute of Technology, Karlsruhe, Fritz-Haber-Weg 4, 76131 Karlsruhe, Germany, E-mail: dirk.holtmann@kit.edu
Sera Bolat, Institute of Bioprocess Engineering and Pharmaceutical Technology, University of Applied Sciences Mittelhessen, Wiesenstrasse 14, 35390 Giessen, Germany
Raphael Greifenstein and Matthias Franzreb, Institute of Functional Interfaces, Karlsruhe Institute of Technology, Bld. 330, Hermann-von-Helmholtz-Platz 1, 76344 Eggenstein-Leopoldshafen, Germany

As per De Gruyter's policy this article has previously been published in the journal Physical Sciences Reviews. Please cite as: S. Bolat, R. Greifenstein, M. Franzreb and D. Holtmann "Process intensification using immobilized enzymes" *Physical Sciences Reviews* [Online] 2023. DOI: 10.1515/psr-2022-0110 | https://doi.org/10.1515/9783110760330-009

enzymes are not satisfactory. It is therefore necessary to optimize and intensify the process.

Optimization can be achieved by gradually improving the performance indicators, e.g. by changing the buffer composition, temperature or feeding method. Very often, these optimizations are not sufficient to achieve the required improvements [1]. Completely new reactions systems are often necessary to achieve the required economic and ecological process indicators. Here, process intensification has proven itself as a method that resulted very often in processes with at least doubled process performance [1]. An intensified process allows for fewer biotransformation's to be performed to produce a given amount of product or for easier product isolation or higher product quality to be achieved in fewer processing steps. Besides to economic advantages, this also results in ecological benefits. In addition to enzyme engineering, technical procedure can be used to realize the needed process indicators. Enzyme engineering can be used to improve, for instance, limited specific activity, insufficient selectivity's, unfavorable kinetics or limited enzyme stability at the reaction conditions. Engineering approaches can be used to address aspects such as unfavorable reaction thermodynamics, selectivity in multi-step reactions, or low water solubility of the substrates. New types of reactors are being developed to address challenges, resulting in improved mixing as well as an enhanced heat and mass transfer. Further methods of process intensification include integration of reaction and separation, multifunctional reactors and tailor-made solvents.

In addition, enzyme immobilization can be used for further dramatic improvements. Enzyme immobilization can be seen as the combination of the advantages of the enzyme (e.g. selectivity and kinetics) with the physical and chemical properties of the immobilization support. Enzyme immobilization methods could significantly improve the long-term stability of the enzymes, resulting in lower costs and an easier product separation [2]. This translates to reduced cost of the final product by reducing the cost contribution of the biocatalyst [3]. Immobilized enzymes could have a number of advantages over native/free enzymes. On the other hand, there are also several challenges related to the implementation of immobilized enzymes (Table 9.1).

The general objective of process intensification is to achieve significant benefits in terms of capital and operating costs as well as product quality, waste and process safety by applying innovative principles, or in other words "to do more with less" [1]. However, this implies that enzyme immobilization, like all other process intensification methods, is about improving the key performance indicators. To do this, the corresponding performance indicators must always be calculated and compared.

The aim of this article is on the one hand to briefly introduce the most important enzyme immobilization methods and on the other hand to summarise the most important performance indicators. Finally, the different immobilization methods are quantitatively compared in a case study.

Table 9.1: Advantages and disadvantages of immobilized enzymes (adapted from [4]).

Advantages	Disadvantages/challenges
– Better stability, especially towards organic solvents and higher temperatures	– Reduced enzyme activity compared to native enzyme
– Easy separation of biocatalyst after the process, resulting in reduced costs of downstream processing	– Additional costs for carriers and immobilizations procedure
– Multiple use of biocatalyst (recycling)	– Lower reaction rates compared to native enzymes
– Enable continuous processes without the need of a membrane	
– Co-immobilization of cofactors or enzymes for cofactor regeneration within the same compartment is possible	
– Immobilization enzymes are often more stable against the detrimental effects of shear	

9.2 Short introduction in enzyme immobilization procedures

Enzyme immobilization methods can generally be divided into (I) adsorption, (II) covalent binding, (III) entrapment and (IV) cross-linking (see Figure 9.1). These general immobilization methods all have their advantages and disadvantages, depending on the enzyme, the reaction and the process in which the immobilized enzymes are used.

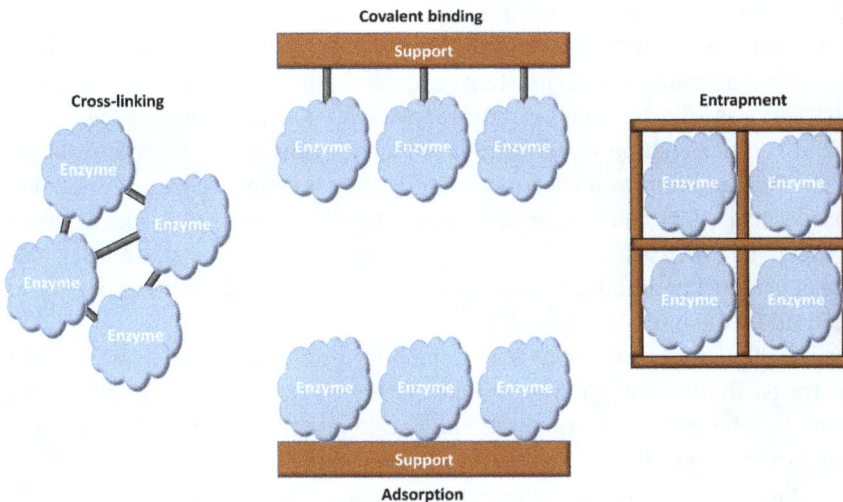

Figure 9.1: General enzyme immobilization methods.

(I) Adsorption

The mechanisms of the immobilization method known as adsorption are based on weak bonds such as Van der Waals forces, dipole–dipole interactions, hydrophobic interactions, hydrogen bonding, ionic interactions, and highly specific interactions. Immobilization by adsorption offers several advantages including ease of preparation and easy implementation. In addition, there is no need for functionalization or additional reagents, so there is little loss of enzyme activity. The active sites of the enzymes are also unaffected by the immobilization process, allowing the enzyme to retain its activity [5]. The adsorption process involves incubating a solid support with an enzyme solution of a specific concentration for a defined period of time. After incubation, the support is separated from the enzyme solution and washed with a buffer to remove any un-adsorbed enzymes. If necessary, the enzymes can be gently removed from the support and the support regenerated with fresh enzymes for the next immobilization. This reusability feature is of particular interest for expensive support materials. To utilize ionic interactions for enzyme immobilization, it is important to consider the pH of the reaction solution and the isoelectric point of the enzyme. Depending on the difference between the isoelectric point of the enzyme and the pH of the solution, the surface of the enzyme molecules can carry a positive or negative charge. Therefore, the support must be oppositely charged to achieve an attractive interaction. The ionic interaction is highly dependent on the salt concentrations in the surrounding solution. Consequently, enzymes can be easily unloaded by buffers of higher ionic strength [6]. The hydrophobic interaction is driven by an entropy gain that occurs when both the enzyme and the support have large hydrophobic surfaces. An enzyme molecule is immobilized by displacing a large number of water molecules from both the support and its own surface. For hydrophobic interactions, the most important experimental parameters are salt concentration, pH, and temperature [7]. Immobilization by highly specific interactions makes use of chelating compounds on the support to entrap metal ions, which serve as affinity ligands for the enzymes. A popular example is the His tag, consisting of a concatenation of six histidines bound to the N- or C-terminus of the enzyme. The His tag binds to a chelating agent such as nitrilotriacetic acid (NTA) or iminodiacetic acid (IDA) coupled to the surface of the support and loaded with a divalent metal ion such as Ni^{2+}, Cu^{2+}, or Zn^{2+}. This adsorption method offers the advantage of oriented immobilization, which can prevent or decrease a loss of enzyme activity caused by the immobilization [8].

However, the use of adsorption as an immobilization method has its drawbacks. Enzymes are easily desorbed due to the comparatively weak nature of adsorptive interactions by changes in temperature, pH, or ionic strength. In addition, most adsorption methods result in immobilized enzymes having no orientation, which can lead to blocking of the active site and steric hindrance of substrate binding, thereby reducing enzyme activity [9].

(II) Covalent binding

The most widely used immobilization method is the covalent binding of enzymes to supports [10]. This method creates stable chemical bonds between functional groups on the support surface and on enzyme molecules, which minimizes enzyme leaching and allows for reusable immobilized enzymes [11]. The strong covalent bonds formed in aqueous solutions increase the half-life and thermal stability of enzymes [12]. Furthermore, multiple covalent bonds can be formed between the enzyme and support, which reduces conformational flexibility and thermal vibrations, thereby preventing protein unfolding and denaturation [13]. The covalent binding process usually consists of two steps (Figure 9.2). First, the surface of the support is chemically modified with functional groups and second, the enzyme is incubated with the modified support. The binding mechanism can occur through two ways. One way is where amino groups on the enzyme act as nucleophiles and react with functional groups on the support, such as an aldehyde group. The other way is through the use of coupling agents, such as glutaraldehyde or carbodiimides, which convert carboxylic acids on the surface of the support to form amide or ester bonds with the enzyme. Other common covalent bonds include ether, thioether, amide, or carbamate bonds [11]. Common functional groups on enzymes used for covalent binding include amino groups, carboxyl groups, phenolic groups, sulfhydryl groups, thiol groups, imidazole groups, indole groups, and hydroxyl groups [14].

The strong covalent bonds also have some disadvantages. The reduced conformational flexibility can lead to a decrease in enzyme activity. In addition, covalent binding can cause chemical modification of the enzyme, which can also lead to loss of activity. In most cases, the covalent bonds are irreversible, so the enzyme and support cannot be recycled once the bound enzyme has lost its activity [15]. Typically, the

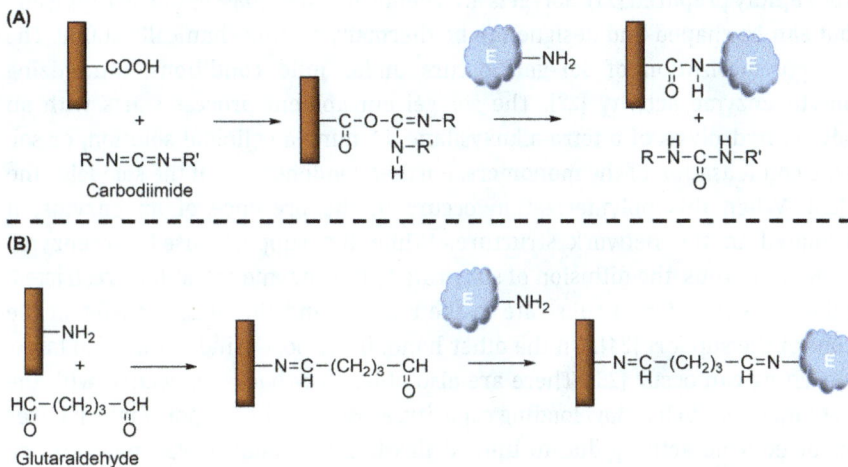

Figure 9.2: Covalent binding by (A) carbodiimide and (B) glutaraldehyde coupling.

orientation of the immobilized enzyme cannot be controlled and the functional groups in the active site may be involved in binding to the support. This can lead to reduced accessibility and modification of the active site, resulting in a significant loss of enzyme activity. To overcome this, the use of spacer arms on the support surface can help to avoid steric hindrance and increase enzyme activity [12]. Also, using a cleavable linker allows both the enzyme and the support to be reused multiple times while still retaining the advantages of covalent binding [16].

(III) Entrapment

Enzyme entrapment or encapsulation is the process of irreversibly immobilizing enzymes in a confined space, such as within a gel or fiber matrix [17, 18]. This physical retention allows substrates and products to move through the lattice structure while minimizing enzyme leaching. The two main methods of entrapment are matrix entrapment and membrane entrapment [19]. This method of immobilization typically results in minimal reduction in enzyme activity because there are no chemical interactions with the support and the enzyme retains its flexibility. In addition, entrapment can improve enzyme stability by limiting folding and reducing denaturation [10]. The small pore windows that retain the enzyme can also protect it from proteases. Furthermore, entrapment allows for the modification of the support material to create a microenvironment with optimal conditions such as polarity, pH, or amphiphilicity [20]. The immobilization process of entrapment involves mixing enzyme, monomer, and crosslinker, followed by polymerization through a chemical reaction or changes in experimental conditions. Common supports used for entrapment include Ca-alginate, agar, carrageen, polyacrylamide, collagen, activated carbon, porous ceramic, and diatomaceous earth [19]. The most widely used method of entrapment is the sol–gel method, which utilizes highly porous, silica-based materials that can be rapidly prepared [21]. Sol–gels are chemically inert glasses that are typically brittle, but can be shaped and designed to be thermally and mechanically stable. The hydrolytic polymerization of sol–gels occurs under mild conditions, minimizing reductions in enzyme activity [22]. The sol–gel entrapment process starts with an acid-catalysed hydrolysis of a tetra-alkoxysilane, forming a colloidal solution, or sol, through the condensation of the monomers. Further condensation of the sol yields the sol–gel [23]. When this polymerization occurs in the presence of an enzyme, it is encapsulated in the network structure. While the supports used for enzyme entrapment are porous, the diffusion of substrate to the enzyme can still be restricted, limiting the mass transfer of substrate to the enzyme and the mass transfer of the product out of the support [24]. On the other hand, if the pore windows are too large, enzyme leaching can occur [25]. There are also other drawbacks associated with the entrapment method, such as low loading capacity, erosion of the support material over time, loss of enzyme activity due to limited flexibility in small pores, and enzyme denaturation during the immobilization process. All of these factors can negatively affect the efficiency and effectiveness of the enzyme entrapment process [10, 20].

(IV) Cross-linking

Cross-linking is an immobilization method that does not involve the use of a support to bind the enzyme. Instead, enzymes are intermolecularly cross-linked by covalent bonds. Multifunctional reagents such as glutaraldehyde, bis(diazo)benzidine, and hexamethylene diisocyanate are used as linkers to create three-dimensional cross-linked enzyme networks [26]. Because it does not use a support, the advantages and disadvantages associated with supports, such as high costs and dilution of mass-specific activity and space-time yields, are eliminated [7, 27]. Cross-linking is a simple immobilization method with minimal enzyme leaching, due to the strong chemical bonds formed. Stabilizing agents can also be used to adjust the microenvironment for the enzyme and increase its stability [28]. There are two main types of cross-linked enzyme immobilizates: cross-linked enzyme crystals (CLEC) and cross-linked enzyme aggregates (CLEA). In the synthesis of both types, the amino groups of the lysine residues on the outer surface of the enzyme react with the cross-linking agents. In the CLEC process, the enzymes are crystallized and cross-linked with glutaraldehyde, which can result in improved resistance to heat, organic solvents, and proteolysis [29]. The particle size of cross-linked enzyme crystals (CLECs) is controllable, and can vary from 1 to 100 μm. However, the crystallization of enzymes is an elaborate and costly process because it requires enzymes of high purity [30]. The CLEA method involves the aggregation of enzymes by the addition of salts, organic solvents or non-ionic polymers, followed by cross-linking to maintain the pre-organized superstructure. Unlike the CLEC method, it does not require high purity enzymes and combines purification and immobilization in one step. Cross-linked enzyme aggregates (CLEAs) don't require crystallization and typically have similar stability and activity as CLECs [31]. One of the major drawbacks of the cross-linking method is that it can lead to a high activity loss due to the reaction of the cross-linking reagent with residues of the active site of the enzyme. Additionally, the harshness of the reagents used for cross-linking can structurally modify or denature the enzyme, which limits the selection of enzymes that can be used for this method [32]. Preparation of CLECs is a laborious process, and for CLEAs, a considerable effort must be made to optimize preparation conditions because an effective aggregation and cross-linking protocol must be established for each enzyme individually [31]. The advantages and disadvantages of the general enzyme immobilization methods introduced above are listed in Table 9.2.

9.3 Supports for enzyme immobilization

In addition to the immobilization method, the choice of the support is also important, with the exception of cross-linking. The support should ideally be inert or at least the reaction with the medium, reactants, or the enzyme should be as low as possible. Furthermore, a good mechanical and chemical stability is helpful for long-term use. It's

Table 9.2: Advantages and disadvantages of the general enzyme immobilization methods.

Immobilization method	Advantages	Disadvantages
Adsorption	– Simple preparation – No functionalization or reagents – Low denaturation – Stable enzyme activity – Expensive supports can be recycled	– Weak bonds can lead to enzyme desorption – Non-specific adsorption – Activity loss by unoriented binding
Covalent binding	– Minimal enzyme leaching due to stable bonds – Increased long term and thermal stability – Reduced denaturation	– Decreased enzyme activity due to reduced flexibility and chemical modification of the active site – Difficult recycling of support and enzyme
Entrapment	– Minimized enzyme leaching – No enzyme modification – Stable enzyme activity – Improved stability – Protection against harsh conditions	– Restricted substrate diffusion – Support erosion leads to enzyme loss – Low loading capacity
Cross-linking	– No expensive support – Simple – Minimal enzyme leaching – Enhanced stability due to agents	– Loss of activity, if the active center gets modified – Limited selection of enzymes – Laborious preparation

advantageous if the support is cheap and easy to obtain, while having a large surface area for a higher loading capacity. Supports are usually in the form of membranes, beads, fibers, discs, hollow spheres, or thin films. Table 9.3 demonstrates an overview of various support materials [32].

Novel substrates include DNA, MOFs, and magnetic nanoparticles. Deoxyribonucleic acid (DNA) allows the formation of spatially highly precise nanostructures, enabling phenomena such as substrate channeling [33]. Substrate channeling occurs in an enzyme cascade in which multiple enzymes are immobilized in close proximity. The product of one enzyme reaction is delivered directly to the active site of the next enzyme in the cascade. Metal-organic frameworks (MOFs) are reticular, highly porous crystals formed by the coordination of organic linker molecules at metal nodes. They provide a high surface area, and due to the wide variety of geometries, linkers, and nodes available, it's possible to tailor the pores and pore window size of MOFs [34]. The methods to immobilize enzymes in MOFs can be classified as (i) encapsulation, (ii) post synthetic infiltration, and (iii) surface adsorption [34]. The MOF structure acts like an armor that

Table 9.3: Support materials for enzyme immobilization.

Organic	Inorganic
Natural polymers	Minerals
– Agar and agarose	– Attapulgite clays
– Albumin	– Bentonite
– Alginate	– Diatomaceous earth (Kieselgur)
– Carrageenan	– Hornblend
– Cellulose	– Pumise stone
– Chitin and chitosan	– Sand
– Collagen	
– Dextran	
– Ferritin	
– Gelatin	
– Pectin	
– Polysaccharides	
– Proteins	
– Starch	
Synthetic polymers	Fabricated materials
– Aldehyde-based polymer	– Alumina
– Diethylamino ethyl cellulose	– Controlled pore glass
– Hydroxyalkyl methacrylate	– Controlled pore metal oxides
– Maleic anhydride polymer	– Iron oxide
– Polyacrylamides	– Non-porous glass
– Poly-acrylate and poly-methacrylate	– Porous silica
– Poly-ethyleneglycol	– Silochrome
– Polystyrene	– Stainless steel
– Vinyl polymer	

protects the enzyme from harsh conditions and denaturation, while allowing the control of the transport of reactants and products [35]. Magnetic nanoparticles (MNP) usually consists of iron oxide (Fe_3O_4) with an average size ranges between 1 and 100 nm [36]. They provide high surface area, large surface-to-volume ratio, and high mass transfer, next to an easy handling via an external magnetic field. With these properties, MNP can load a high capacity of enzymes, be easily separated and reused [37].

9.3.1 Whole cell biocatalysts

Instead of using a single enzyme molecule as a biocatalyst, a whole cell containing the enzyme can be immobilized. Therefore, the enzyme has already its optimal environment in the cell and can have higher stability. Moreover, there is no need for expensive extraction and purification of the enzyme. A whole cell can be filled with multiple enzymes, whereby multi-step reactions and cofactor regeneration can occur. This eliminates the need to immobilize all the enzymes and the use of an expensive cofactor [38]. In addition, complex reaction pathways can be exploited, allowing the use of inexpensive starting materials [39]. These pathways are usually optimized by metabolic engineering, as well as the cell itself [40]. Compared to the immobilization of purified enzyme molecules, the enzyme density is less for whole cell biocatalysts. Also, in a whole cell, the synthesis of undesired enzymes and products can occur, as well as the modification of the products [41]. The most common immobilization method for whole cell biocatalysts is entrapment. As support, porous gels based on polysaccharides containing charged functional groups with oppositely charged ions are applied [42]. Furthermore, the same general immobilization methods are used for whole cell biocatalysts as for enzymes [43].

9.3.2 Performance indicators

The enzyme immobilization process can be validated using performance indicators, independent of the immobilization method. The main performance indicators are presented in Table 9.4. First, the quality of the immobilization – or enzyme loading – step can be expressed by the loading yield. It indicates what percentage of the applied enzyme has been immobilized. The affinity of an enzyme for a substrate is indicated by the Michaelis constant. Another term of the Michaelis–Menten equation that can be used as a performance indicator is the turnover number. It's the maximum number of substrate

Table 9.4: Important performance indicators for immobilized enzymes.

Loading yield	$L_y = \frac{m_i}{m_a} \times 100$	L_y=loading yield (%) m_i=immobilized enzymes mass (g) m_a = applied enzymes mass (g)
Michaelis constant	$K_m = \frac{k_{-1}+k_2}{k_1}$ or $K_m = c_s$ at $\frac{v_{max}}{2}$	K_m = Michaelis constant ($\frac{mol}{L}$) k_1 = forward rate constant ($\frac{L}{mol \times s}$) k_{-1} = reverse rate constant (1s) k_2 = catalytic rate constant (1s) c_s = substrate concentration ($\frac{mol}{L}$) v_{max} = maximal velocity ($\frac{mol}{L \times s}$)
Turnover number	$k_{cat} = \frac{v_{max}}{c_0}$	k_{cat} = turnover numver (1s) c_0 = initial enzyme concentration ($\frac{mol}{L}$)
Effective specific activity	$A_{es} = \frac{A_{si}}{A_{sf}} \times 100$	A_{es} = effective specific activity (%) A_{si} = mass specific activity of immobilized enzymes ($\frac{U}{mg}$) A_{sf} = mass specific activity of free enzymes ($\frac{U}{mg}$)
Activity yield	$A_y = \frac{A_{si}}{A_{sf}} \times L_y$	A_y = activity yield (%)
Total turnover number	$TTN = \frac{k_{2, obs}}{k_{d, obs}}$	TTN = Total Turnover Number () $k_{2, obs}$ = observed catalytic rate constant (1s) $k_{d, obs}$ = observed deactivation rate constant (1s)

molecules that an enzyme molecule can convert to product per second [44]. The Michaelis constant and turnover number of immobilized enzymes can be compared with those of free enzymes to get a better idea of the influence of the immobilization process on the enzyme. Another performance indicator is the effective specific activity. It shows the percentage of the mass specific activity of the immobilized enzymes to that of the free enzymes. Since the mass specific activity depends on the substrate concentration applied, it is best to use the maximum rate – or turnover number – for the calculation. The effective specific activity can be combined with the loading yield to obtain the activity yield. This indicator describes the ratio between the total activity of the immobilized enzyme and the total activity of the free enzyme initially offered for immobilization. The dimensionless total turnover number (TTN) is an important indicator of productivity over the lifetime of enzymes. The TTN is the ratio of the total product generated by an enzyme molecule used in a reaction. Therefore, it expresses the number of substrate turnovers performed by an active site of an enzyme molecule during its lifetime [45].

9.3.3 Case study – Immobilization of unspecific peroxygenases

This case study aims to compare and discuss immobilization methods and metrics using a specific example. The regio-and stereo-selective oxyfunctionalization of non-activated organic molecules is one of the most difficult challenges in synthetic and organic chemistry. Cytochrome P450 monooxygenases (P450s for short) can catalyze a wide range of selective oxidation and oxyfunctionalization reactions. Therefore, P450s have been the focus of experiments for several decades [46, 47]. However, the industrial application of these enzymes is still limited. This may be due to the need for both expensive nicotinamide cofactors and electron transfer proteins to catalyze the selective reactions [48] or the limited operational stability. One further type of heme-thiolate enzyme, unspecific peroxygenase (UPO, E.C. 1.11.2.1), was discovered in 2004 in the basidiomycete *Agrocybe aegerita* [49]. These enzymes have a similar structure to the P450s and can also catalyze an enormous range of reactions [50]. Compared to P450s, UPOs do not require expensive redox cofactors or auxiliary flavoproteins to catalyze the reaction, but use hydrogen peroxide as a co-substrate [51].

UPOs are extracellular fungal enzymes and have been phylogenetically divided into two major groups: Group I–the short UPO sequences–and Group II–the long UPO sequences–[52]. The short UPOs, like a well-studied *Mro*UPO from *Marasmius rotula* [53], have molecular a weight of ~29 kDa and are found in all fungal phyla. The long UPOs with a molecular weight of ~44 kDa are found in Basidiomycetes and Ascomycetes. As an example for the long UPOs, the model enzyme of *A. aegerita* (*Aae*UPO) has been biochemically well characterized [52]. The UPO consists of ten α-helices, five short β-sheets, a cysteinate-ligated heme (iron protoporphyrin IX) as an active site prosthetic group, and a characteristic disulfide bridge [54, 55]. The disulfide bridge is located between Cys278 and Cys319 and stabilizes the C-terminal region. The short UPOs do not have an internal disulfide bridge like the long UPOs, but consist of two catalytically active UPO monomers connected by a disulfide bridge. Another difference between the two UPO groups lies in their active sites. In the short UPOs, the negative charge is stabilized by a histidine, whereas in the long UPOs an arginine acts as a charge stabilizer [52]. Molina-Espeja et al. evolved the *Aae*UPO mutant (called PaDa-I variant) to enhance heterologous expression [56]. PaDa-I could be recombinantly expressed in *Saccharomyces cerevisiae* and later further increase in the expression levels of PaDa-I was achieved in *Pichia pastoris* [57]. Examples of UPO-catalyzed reactions include hydroxylation of aromatic and aliphatic C–H bonds, epoxidation, sulfoxidation, etc. The spectrum of UPO-catalyzed reactions has been reported in detail in reviews [52, 58–60]. Although UPOs have a broad reaction spectrum and a simplified catalase system compared to P450s, there are limiting aspects, which need to be overcome for their use in the industrial processes. These are the low level of heterologous expression, the inactivation by H_2O_2, limited enzyme activity and stability under process conditions, the unwanted peroxidase activity, and the overoxidation of products [46, 59, 61].

Immobilization is a suitable method to overcome some limiting factors in the industrial use of UPOs. This technique has many advantages. These include increased robustness to reaction conditions, reusability, simplification of downstream processes, and flexibility in solvent selection. In order to improve the use of UPOs on a technical scale, several studies on the immobilization of UPOs have been carried out in recent years. Various supports with different immobilization methods have been used to immobilize UPOs. A comparison of immobilized UPOs in terms of immobilization yield and activity yield is shown in Table 9.5.

The recombinantly expressed *Aae*UPO was covalently immobilized on HA403/M resin and the hydroxylation of ethylbenzene to (R)-1-phenylethanol was studied under neat conditions, which means using substrate as sole reaction medium [64]. Under non-aqueous conditions, the activity and stability of the enzyme were low. Nevertheless, they have demonstrated a semi-preparative reaction set up on a 250-mL scale. 40.9 mM of (R)-1-phenylethanol was produced within 3 h of reaction time, with a TTN (r*Aae*UPO) of 90,000. The Kara-group demonstrated the hydroxylation of cyclohexan with alginate-entrapped *Aae*UPO PaDa-I in non-aqueous media [66]. H_2O_2 production occurred in-situ photocatalytically by nitrogen-doped carbon nanodots and UV LED illumination. During the immobilization process, the enzyme loss was 19.2 %. They also observed an increase in enzyme stability in neat cyclohexane due to immobilization. The activity of the immobilized enzyme lasted up to seven days under reaction conditions, and 2.5 mM cyclohexanol could be produced.

Furthermore, different immobilization methods of recombinantly expressed *Aae*UPO were investigated and compared [67]. For covalent immobilization, glyoxyl-, vinyl sulfone-or glutaraldehyde-activated agarose was tested at different pH and temperatures. For immobilization with vinyl sulfone-activated support, an immobilization yield of less than 20 % was observed at pH 5.0 and 7.0. In addition, the expressed activity decreased by more than 50 % after immobilization. The enzyme immobilized with vinyl

Table 9.5: Overview for immobilization of unspecific peroxygenases.

Enzym	Carrier	Material-functional group	Loading yield (%)	Activity yield (%)	Ref.
Adsorption immobilization					
*Aae*UPO-PaDa-I_his	Amber (EnginZyme EziG™)	CPG[a]-semi-hydrophilic	n.a.[b]	53	[62]
*Aae*UPO-PaDa-I_his	Coral (EnginZyme EziG™)	CPG-hydrophobic	45	5	[63]
*Aae*UPO-PaDa-I	Opal (EnginZyme EziG™)	CPG-hydrophilic	92	55	[63]
*Aae*UPO-PaDa-I	Amber (EnginZyme EziG™)	CPG-semi-hydrophilic	46	11	[63]
Covalent immobilization					
r*Aae*UPO	HA 403/M (Relizyme™)	PMMA[c]h-examethylenamino	100	54	[64]
*Aae*UPO-PaDa-I	ECR8285 (Purolite Life Science Ltd.)	PMMA-epoxy	54	0.44	[63]
*Aae*UPO-PaDa-I	ECR8215F (Purolite Life Science Ltd.)	PMMA-epoxy/butyl	54	4	[63]
*Aae*UPO-PaDa-I	ECR8304F (Purolite Life Science Ltd.)	PMMA-amino	57	0.2	[63]
*Aae*UPO-PaDa-I	ECR8404F (Purolite Life Science Ltd.)	PMMA-amino	32	n.d.[d]	[63]
*Aae*UPO-PaDa-I	ECR8409F (Purolite Life Science Ltd.)	PMMA-amino	35	n.d.	[63]
*Aae*UPO-PaDa-I	ECR8315F (Purolite Life Science Ltd.)	PMMA-amino	55	3	[63]
*Aae*UPO-PaDa-I	ECR8415F (Purolite Life Science Ltd.)	PMMA-amino	28	1	[63]
*Aae*UPO-PaDa-I	ECR8315F (Purolite Life Science Ltd.)	PMMA-amino	52	n.a.	[65]
Ionic immobilization					
*Aae*UPO-PaDa-I	ECR1508 (Purolite Life Science Ltd.)	Polystyrene-NR$_2$	38	1	[63]
*Aae*UPO-PaDa-I	ECR1604 (Purolite Life Science Ltd.)	Polystyrene-NR$_3^+$Cl$^-$	61	1	[63]
Entrapment					
*Aae*UPO	Calcium alginate (Sigma Aldrich)	Sodium alginate	81	n.a.	[66]

[a]Controlled porosity glass; [b]Not available; [c]Polymethacrylate; [d]Not detected.

sulfone-activated agarose showed a higher immobilization yield at pH 9.0 (more than 90 %). However, the enzyme was almost completely inactivated after immobilization. They also performed ion exchange immobilization on poly-ethlyenimine (PEI) or mon-oaminoethyl-N-aminoethyl (MANAE) activated agarose. In both cases, the immobilization yield was 100 % at pH 7.0. The activity on MANAE-activated agarose was well maintained, whereas the activity on PEI showed a significant decrease. Immobilization on MANAE-agarose at pH 5.0 and 9.0 slightly changed these parameters. However, the enzyme immobilized at pH 9.0 was more unstable than the free enzyme.

Bormann et al. studied the immobilization of the his-tagged AaeUPO-PaDa-I by metal affinity binding on porous glass carriers [62]. They produced the enzyme in *P. pastoris*, purified, and immobilized it. The hydroxylation of 4-ethylbenzoic acid (EBA) to 4-(1-hydroxyethyl)benzoic acid (HEBA) was used as a model reaction. With immobilized enzyme, 8.3 mM HEBA was produced in 500 min, while with free enzyme, 7.4 mM HEBA was obtained in 250 min. Compared with the free enzyme, the productivity of the immobilized enzyme was lower. However, they observed that the free enzyme was inactive after 250 min, while the immobilized enzyme was still active. Moreover, they showed the reusability of the his-tagged immobilized enzyme for six cycles by repeated fed-batch experiments. Furthermore, in another approach, they immobilized the biocatalyst directly in crude culture broth to omit the purification steps. Product formation occurred at a similar rate as when a free enzyme was used [62].

In another study three different immobilization techniques (covalent, ionic, and metal affinity) with a total of 12 carriers were compered [63]. AaeUPO-PaDa-I was used as a model enzyme, and the enzyme activity was measured by ABTS assay. To compare the immobilization methods, performance indicators like protein loading, immobilized enzyme activity, immobilization yield, and activity yield were determined. The most suitable method was metal affinity binding using the Opal EziG™ carrier. The immobilized enzyme exhibited a protein loading of 0.76 ± 0.02 mg$_{protein}$/g$_{carrier}$, an immobilization yield of 92 %, a specific activity of 67 ± 9 U/mg and an activity yield of 55 %. In addition, 75 percent of the enzyme activity of retained for up to 30 days. The covalently immobilized enzyme with the amino carrier ECR8315F had a protein loading of 0.616 ± 0.006 mg$_{protein}$/g$_{carrier}$, an immobilization yield of 55 %, a specific activity of 27.5 ± 0.5 U/mg and an activity yield of 3 %. In ionic immobilization, the enzyme showed a protein loading of 0.833 ± 0.003 mg$_{protein}$/g$_{carrier}$, immobilization yield of 61 %, specific activity of 7 ± 3 U/mg, and activity yield of 1 %. The covalently immobilized enzyme was used in the ethylbenzene hydroxy-functionalization reaction as a proof-of-concept, and repeated batch experiments were performed. The immobilized enzyme could be recycled up to seven times. The loss in product formation was 89 % after six batches. The same research group described in the article before, have investigated continuous processes with both free and immobilized AaeUPO-PaDa-I for the synthesis of (*R*)-1-phenylethanol [68]. For this purpose, they immobilized the enzyme covalently on ECR8315F carrier material and used it in a packed-bed reactor

(PBR). For the free enzyme, an enzyme membrane reactor (EMR) was used. They observed 1.5-times higher experimental residence times than theoretical residence times for the EMR and 1.2-times higher for the PBR. With the EMR, the requirements of a robust continuous synthesis could not be achieved. They performed the oxy-functionalization of ethylbenzene to (R)-1-phenylethanol in continuous mode in the PBR. They obtained a space-time-yield of 0.97 g/(L h) and a productivity of 0.254 mg/(L h) in a total run time of 6.2 days. They detected (almost) no by-product acetophenone or (S)-enantiomer. The covalently immobilized AaeUPO-PaDa-I was used in a rotating bed reactor for the hydroxylation of ethylbenzene [69]. Here, a two-liquid-phase system was used as a medium and ECR8315F carrier material was used for immobilization. They obtained an immobilization yield of 52 %, a protein loading of 1.6 $mg_{protein}/g_{carrier}$, and a recovered activity 22 U/g. To optimize the UPO-catalyzed reaction in the multi-phasic system, product formation in the organic phase and H_2O_2 concentration in the aqueous phase were monitored. In a classic fed batch, they produced a total of 414 mM (R)-1-phenylethanol in the organic phase over 58 h. This resulted in a productivity of 436 mg/(L h) and a TTN of 658 000. The selectivity for the target product over the overoxidation product (acetophenone) was 62 %. This selectivity was increased to 79 % in a repetitive batch operation. In addition, the reusability of the immobilized enzyme was observed over four batches.

Recently, an article reported the immobilization of UPO-PaDa-I in 3D printable synthetic hydrogel and its use in a biocatalytic continuous flow process. The enzyme showed an activity yield of 6.1 % after the immobilization. Relatively poor reusability of the enzyme-loaded hydrogel pellets was observed. The continuous flow system was in the form of random film and pellet packing. The random hydrogel film packing pro-duced 2.2 ± 1.6 mg/L of ABTS⁺ and the random pellet packing produced 1.6 ± 0.7 mg/L of ABTS⁺ [70]. Furthermore, the coupling of photocatalysis and enzyme catalysis has been investigated [71]. First, they constructed Pd-loaded three-dimensional ordered microporous titania (3DOM TiO_2) supports to generate H_2O_2 *in situ* under photocatalytic conditions. Then, the enzyme AaeUPO was immobilized as cross-linked enzyme aggregates (CLEAs) on 3DOM TiO_2–Pd. The kinetic parameters of the free enzyme, CLEAs-AaeUPO and AaeUPO-3DOM TiO_2–Pd were compared. While the free enzyme showed the highest V_{max} value (2.896 mM/min), the V_{max} values of CLEAs-AaeUPO (1.105 mM/min) and of AaeUPO-3DOM TiO_2–Pd (1.741 mM/min) were lower. However, CLEAs-AaeUPO showed a K_m value of 16.033 mM and AaeUPO-3DOM TiO_2–Pd exhibited a K_m value of 13.439 mM, while the free enzyme showed a K_m value of 7.061 mM. In the stability studies, they observed higher acid-base tolerance, thermostability, mechanical stability, and light stability of AaeUPO-3DOM TiO_2–Pd compared to CLEAs-AaeUPO and free enzyme. In addition, the hydroxylation of ethylbenzene to (R)-1-phenylethanol was carried out. 99 % ethylbenzene was converted with the AaeUPO-3DOM TiO2–Pd, while 38 % was converted with free enzyme and 49 % with CLEAs-AaeUPO ethylbenzene.

9.4 Conclusions

Obviously, the immobilization of enzymes often makes it possible to achieve the desired performance parameters in technical processes. Due to the wide variety of immobilization methods, it can be assumed that it is almost always possible to immobilize an enzyme in one way or another. Currently, the identification of the "best" method is extremely laborious; it is expected that automation, parallelization, and digitalization will significantly reduce process development times in the future.

Unfortunately, many studies on enzyme immobilization do not fully specify the relevant metrics, which often makes it difficult to compare methods and to identify the "best" method. However, in the view of the authors of this article, it is precisely the specification of the key parameters that is of enormous importance in order to bring more enzyme processes into technical application. Pure "proof-of-principle" studies on enzyme immobilization are only of limited use for process intensification.

Acknowledgement: The authors would like to thank the editors and reviewer for their guidance and review of this article before its publication.

References

1. Burek BO, Dawood AWH, Hollmann F, Liese A, Holtmann D. Process intensification as game changer in enzyme catalysis. Front Catal 2022;2. https://doi.org/10.3389/fctls.2022.858706.
2. Žnidaršič-Plazl P. Biocatalytic process intensification via efficient biocatalyst immobilization, miniaturization, and process integration. Curr Opin Green Sustain Chem 2021;32:100546.
3. Federsel HJ, Moody TS, Taylor SJC. Recent trends in enzyme immobilization—concepts for expanding the biocatalysis toolbox. Molecules 2021;26. https://doi.org/10.3390/molecules26092822.
4. Basso A, Serban S. Industrial applications of immobilized enzymes—a review. Mol Catal 2019;479:110607.
5. Ligler FS, Taitt CR. Optical biosensors – today and tomorrow. Amsterdam: Elsevier Science; 2011.
6. Cao L. Carrier-bound immobilized enzymes: principles, application and design. Hoboken, New Jersey: John Wiley & Sons; 2006.
7. Hanefeld U, Gardossi L, Magner E. Understanding enzyme immobilisation. Chem Soc Rev 2009;38:453–68.
8. Gaberc-Porekar V, Menart V. Perspectives of immobilized-metal affinity chromatography. J Biochem Biophys Methods 2001;49:335–60.
9. Rahman NZRA, Basri M. New lipases and proteases. Hauppauge, New York: Nova Publishers; 2006.
10. Datta S, Christena LR, Rajaram YR. Enzyme immobilization: an overview on techniques and support materials. 3 Biotech 2013;3:1–9.
11. Guisan JM. Immobilization of enzymes and cells. Heidelberg, Germany: Springer; 2006.
12. Isgrove FH, Williams R, Niven G, Andrews A. Enzyme immobilization on nylon–optimization and the steps used to prevent enzyme leakage from the support. Enzym Microb Technol 2001;28:225–32.
13. Ispas C, Sokolov I, Andreescu S. Enzyme-functionalized mesoporous silica for bioanalytical applications. Anal Bioanal Chem 2009;393:543–54.
14. Novick SJ, Rozzell JD. Immobilization of enzymes by covalent attachment. In: Barredo JL, editor. Microbial enzymes and biotransformations. Totowa, NJ: Humana Press; 2005:247–71 pp.

15. Marrazza G. Piezoelectric biosensors for organophosphate and carbamate pesticides: a review. Biosensors 2014;4:301–17.
16. Fraas R, Franzreb M. Reversible covalent enzyme immobilization methods for reuse of carriers. Biocatal Biotransform 2017;35:337–48.
17. Won K, Kim S, Kim KJ, Park HW, Moon SJ. Optimization of lipase entrapment in Ca-alginate gel beads. Process Biochem 2005;40:2149–54.
18. Aehle W. Enzymes in industry: production and applications. Hoboken, New Jersey: John Wiley & Sons; 2007.
19. Liu S. Chapter 7 – enzymes. In: Liu S, editor. Bioprocess engineering, 2nd ed. Amsterdam, the Netherlands: Elsevier; 2017:297–373 pp.
20. Mohamad NR, Marzuki NHC, Buang NA, Huyop F, Wahab RA. An overview of technologies for immobilization of enzymes and surface analysis techniques for immobilized enzymes. Biotechnol Biotechnol Equip 2015;29:205–20.
21. Mousavizadegan M, Roshani A, Hosseini M. Chapter 17 – novel paper-based diagnostic devices for early detection of cancer. In: Khan R, Parihar A, Sanghi SK, editors. Biosensor based advanced cancer diagnostics. Cambridge, Massachusetts: Academic Press; 2022:285–301 pp.
22. David AE, Yang AJ, Wang NS. Enzyme stabilization and immobilization by sol-gel entrapment. Methods Mol Biol 2011;679:49–66.
23. Braun S, Rappoport S, Zusman R, Avnir D, Ottolenghi M. Biochemically active sol-gel glasses: the trapping of enzymes. Mater Lett 1990;10:1–5.
24. Górecka E, Jastrzębska M. Immobilization techniques and biopolymer carriers. Biotechnol Food Sci 2011;75:65–86.
25. Sheldon RA. Cross-linked enzyme aggregates (CLEAs): stable and recyclable biocatalysts. Biochem Soc Trans 2007;35:1583–7.
26. Lee DH, Park CH, Yeo JM, Kim SW. Lipase immobilization on silica gel using a cross-linking method. J Ind Eng Chem 2006;12:777–82.
27. Cao L, Langen L, Sheldon RA. Immobilised enzymes: carrier-bound or carrier-free? Curr Opin Biotechnol 2003;14:387–94.
28. Chang BS, Mahoney RR. Enzyme thermostabilization by bovine serum albumin and other proteins: evidence for hydrophobic interactions. Biotechnol Appl Biochem 1995;22:203–14.
29. Quiocho FA, Richards FM. Intermolecular cross linking of a protein in the crystalline state: carboxypeptidase-a. Proc Natl Acad Sci U S A 1964;52:833–9.
30. Sheldon RA. Enzyme immobilization: the quest for optimum performance. Adv Synth Catal 2007;349:1289–307.
31. Velasco-Lozano S, López-Gallego F, Mateos-Díaz JC, Favela-Torres E. Cross-linked enzyme aggregates (CLEA) in enzyme improvement – a review. Biocatalysis 2016;1:166–77.
32. Elnashar M. Biotechnology of biopolymers. London, UK: IntechOpen; 2011.
33. Klein WP, Thomsen RP, Turner KB, Walper SA, Vranish J, Kjems J, et al. Enhanced catalysis from multienzyme cascades assembled on a DNA origami triangle. ACS Nano 2019;13:13677–89.
34. Ye N, Kou X, Shen J, Huang S, Chen G, Ouyang G. Metal-organic frameworks: a new platform for enzyme immobilization. Chembiochem 2020;21:2585–90.
35. Greifenstein R, Ballweg T, Hashem T, Gottwald E, Achauer D, Kirschhöfer F, et al. MOF-hosted enzymes for continuous flow catalysis in aqueous and organic solvents. Angew Chem Int Ed 2022;61:e202117144.
36. Ali A, Shah T, Ullah R, Zhou P, Guo M, Ovais M, et al. Review on recent progress in magnetic nanoparticles: synthesis, characterization, and diverse applications. Front Chem 2021;9. https://doi.org/10.3389/fchem.2021.629054.
37. Bilal M, Zhao Y, Rasheed T, Iqbal HM. Magnetic nanoparticles as versatile carriers for enzymes immobilization: a review. Int J Biol Macromol 2018;120:2530–44.
38. Lin B, Tao Y. Whole-cell biocatalysts by design. Microb Cell Factories 2017;16:106.

39. Gehring C, Wessel M, Schaffer S, Thum O. The power of biocatalysis: a one-pot total synthesis of rhamnolipids from butane as the sole carbon and energy source. ChemistryOpen 2016;5:513–6.
40. Wachtmeister J, Rother D. Recent advances in whole cell biocatalysis techniques bridging from investigative to industrial scale. Curr Opin Biotechnol 2016;42:169–77.
41. Garzón-Posse F, Becerra-Figueroa L, Hernández-Arias J, Gamba-Sánchez D. Whole cells as biocatalysts in organic transformations. Molecules 2018;23:1265.
42. Buchholz K, Kasche V, Bornscheuer UT. Biocatalysts and enzyme technology. Hoboken, New Jersey: John Wiley & Sons; 2012.
43. Jack TR, Zajic JE. The immobilization of whole cells. In: Advances in biochemical engineering, volume 5. Berlin, Heidelberg: Springer Berlin Heidelberg; 1977.
44. Roskoski R. Michaelis-menten kinetics☆. In: Reference module in biomedical sciences. Amsterdam, the Netherlands: Elsevier; 2015.
45. Rogers TA, Bommarius AS. Utilizing simple biochemical measurements to predict lifetime output of biocatalysts in continuous isothermal processes. Chem Eng Sci 2010;65:2118–24.
46. Beltran-Nogal A, Sánchez-Moreno I, Méndez-Sánchez D, Gómez de Santos P, Hollmann F, Alcalde M. Surfing the wave of oxyfunctionalization chemistry by engineering fungal unspecific peroxygenases. Curr Opin Struct Biol 2022;73:102342.
47. Fasan R. Tuning P450 enzymes as oxidation catalysts. ACS Catal 2012;2:647–66.
48. Grogan G. Hemoprotein catalyzed oxygenations: P450s, UPOs, and progress toward scalable reactions. JACS Au 2021;1:1312–29.
49. Ullrich R, Nüske J, Scheibner K, Spantzel J, Hofrichter M. Novel haloperoxidase from the agaric basidiomycete Agrocybe aegerita oxidizes aryl alcohols and aldehydes. Appl Environ Microbiol 2004;70: 4575–81.
50. Hofrichter M, Ullrich R. Heme-thiolate haloperoxidases: versatile biocatalysts with biotechnological and environmental significance. Appl Microbiol Biotechnol 2006;71:276–88.
51. Ullrich R, Hofrichter M. The haloperoxidase of the agaric fungus Agrocybe aegerita hydroxylates toluene and naphthalene. FEBS Lett 2005;579:6247–50.
52. Hofrichter M, Kellner H, Pecyna MJ, Ullrich R. Fungal unspecific peroxygenases: heme-thiolate proteins that combine peroxidase and cytochrome P450 properties. In: Hrycay EG, Bandiera SM, editors. Monooxygenase, peroxidase and peroxygenase properties and mechanisms of cytochrome P450. Cham: Springer International Publishing; 2015:341–68 pp.
53. Grobe G, Ullrich R, Pecyna MJ, Kapturska D, Friedrich S, Hofrichter M, et al. High-yield production of aromatic peroxygenase by the agaric fungus Marasmius rotula. Amb Express 2011;1:31.
54. Pecyna MJ, Ullrich R, Bittner B, Clemens A, Scheibner K, Schubert R, et al. Molecular characterization of aromatic peroxygenase from Agrocybe aegerita. Appl Microbiol Biotechnol 2009;84:885–97.
55. Piontek K, Strittmatter E, Ullrich R, Gröbe G, Pecyna MJ, Kluge M, et al. Structural basis of substrate conversion in a new aromatic peroxygenase: cytochrome P450 functionality with benefits. J Biol Chem 2013;288:34767–76.
56. Molina-Espeja P, Garcia-Ruiz E, Gonzalez-Perez D, Ullrich R, Hofrichter M, Alcalde M. Directed evolution of unspecific peroxygenase from Agrocybe aegerita. Appl Environ Microbiol 2014;80:3496–507.
57. Molina-Espeja P, Ma S, Mate DM, Ludwig R, Alcalde M. Tandem-yeast expression system for engineering and producing unspecific peroxygenase. Enzym Microb Technol 2015;73–74:29–33.
58. Burek BO, Bormann S, Hollmann F, Bloh JZ, Holtmann D. Hydrogen peroxide driven biocatalysis. Green Chem 2019;21:3232–49.
59. Markus H, Holtmann D, Gomez de Santos P, Alcalde M, Hollmann F, Kara S. Recent developments in the use of peroxygenases – exploring their high potential in selective oxyfunctionalisations. Biotechnol Adv 2021; 51:107615.
60. Aranda C, Carro J, González-Benjumea A, Babot ED, Olmedo A, Linde D, et al. Advances in enzymatic oxyfunctionalization of aliphatic compounds. Biotechnol Adv 2021;51:107703.

61. Bormann S, Baraibar AG, Ni Y, Holtmann D, Hollmann F. Specific oxyfunctionalisations catalysed by peroxygenases: opportunities, challenges and solutions. Catal Sci Technol 2015;5:2038–52.
62. Sebastian B, Burek BO, Ulber R, Holtmann D. Immobilization of unspecific peroxygenase expressed in Pichia pastoris by metal affinity binding. Mol Catal 2020;492:110999.
63. De Santis P, Petrovai N, Meyer LE, Hobisch M, Kara S. A holistic carrier-bound immobilization approach for unspecific peroxygenase. Front Chem 2022;10:985997.
64. Fernandez-Fueyo E, Ni Y, Baraibar AG, Alcalde M, van Langen LM, Hollmann F. Towards preparative peroxygenase-catalyzed oxyfunctionalization reactions in organic media. J Mol Catal B Enzym 2016;134: 347–52.
65. Hobisch M, De Santis P, Serban S, Basso A, Byström E, Kara S. Peroxygenase-driven ethylbenzene hydroxylation in a rotating bed reactor. Org Process Res Dev 2022;26:2761–5.
66. Hobisch M, Schie MMCH, Kim J, Andersen KR, Alcalde M, Kourist R, et al. Solvent-free photobiocatalytic hydroxylation of cyclohexane. ChemCatChem 2020;12:4009–13.
67. Carballares D, Morellon-Sterling R, Xu X, Hollmann F, Fernandez-Lafuente R. Immobilization of the peroxygenase from Agrocybe aegerita. The effect of the immobilization pH on the features of an ionically exchanged dimeric peroxygenase. Catalysts 2021;11:560.
68. Meyer LE, Hauge BF, Kvorning TM, De Santis P, Kara S. Continuous oxyfunctionalizations catalyzed by unspecific peroxygenase. Catal Sci Technol 2022;12:6473–85.
69. Hobisch M, De Santis P, Serban S, Basso A, Byström E, Kara S. Peroxygenase-driven ethylbenzene hydroxylation in a rotating bed reactor. Org Process Res Dev 2022;269:2761–5.
70. Meyer LE, Horváth D, Vaupel S, Meyer J, Alcalde M, Kara S. A 3D printable synthetic hydrogel as an immobilization matrix for continuous synthesis with fungal peroxygenases. React Chem Eng 2023;8:984–8.
71. Xuewu D, Zheng X, Jia F, Cao C, Song H, Jiang Y, et al. Unspecific peroxygenases immobilized on Pd-loaded three-dimensional ordered macroporous (3DOM) titania photocatalyst for photo-enzyme integrated catalysis. Appl Catal B: Environmental 2023;330:122622.

Ulises A. Salas-Villalobos and Oscar Aguilar*

10 *In situ* product removal

Abstract: During current years, the industrial biotechnology area has grown at giant steps, supported by the necessity of a sustainable supply chain and the inevitable depletion of petrochemical feedstocks. From this accelerated growth, the need for the development of more efficient bioprocesses in term of productivity and cost has emerged. A substantial number of bioprocesses have their potential hindered by product inhibition, a phenomenon that appears due to microbial metabolites produced in concentrations that become toxic even for the producing microorganism. *In situ* product recovery (ISPR) appears as a strategy to overcome such problems by primary recovery stage to the upstream, thus continuously extracting a desired or undesired target molecule from the fermentation broth as soon as it is produced. In this chapter, we will review the inherent advantages of implementing this technology in the production process, not only in terms of productivity, but also in equipment. A revision across the main the ISPR technologies can be found, explaining their main mechanisms and configurations, the appropriate scenarios to use each one and the main factors that must be considered that affect process efficiency. The chapter will be divided into three parts according to the types of ISPR that are reviewed, liquid–liquid, solid–liquid and gas–liquid techniques. Some recent trends and further perspectives for each method are also mentioned leaving space for further analysis of these technologies.

Keywords: *in situ* product recovery; process integration; downstream processing; product inhibition

During the last decades, the incidence of industrial biotechnology over the economy has increased, driven mainly by the need of a more sustainable supply chain and the inevitable depletion of petrochemical feedstocks [1]. This behavior has a direct effect over the worldwide demand of bioproducts, urging the development of more efficient processes and leading to a boost in productivity, most of them based on upstream enhancements, via media optimization and synthetic biology strategies [2]. Despite

*Corresponding author: Oscar Aguilar**, Escuela de Ingeniería y Ciencias, Tecnologico de Monterrey, Ave. Eugenio Garza Sada 2501, Monterrey, N.L., 64849, México, E-mail: alex.aguilar@tec.mx. https://orcid.org/0000-0002-5352-9579
Ulises A. Salas-Villalobos, Escuela de Ingeniería y Ciencias, Tecnologico de Monterrey, Ave. Eugenio Garza Sada 2501, Monterrey, N.L., 64849, México

As per De Gruyter's policy this article has previously been published in the journal Physical Sciences Reviews. Please cite as:
U. A. Salas-Villalobos and O. Aguilar "*In situ* product removal" *Physical Sciences Reviews* [Online] 2023. DOI: 10.1515/psr-2022-0111 |
https://doi.org/10.1515/9783110760330-010

this, the recovery of a specific molecule with a high level of purity from a fermentation, is still one of the major challenges for bioprocesses development, usually requiring a series of unit operations that account up to 80 % of the entire cost of the process, not to mention the requirements in terms of space and processing time [3].

With the new genetic engineering tools available, or even exploiting the natural adaptability of microorganisms to stress or elicitors, cell titers and productivities have been pushed to higher limits. However, many bioprocesses have their potential hindered by product inhibition, a phenomenon that appears due to microbial metabolites produced in concentrations that become toxic even for the producing microorganism [4]. *In situ* product recovery (ISPR) or "extractive fermentation" emerges as a strategy to overcome such problems by integrating a primary recovery stage to the upstream, thus continuously extracting a desired or undesired target molecule from the fermentation broth as soon as it is produced. Besides the evident advantage that this technology has in terms of product inhibition, the integration of unit operations also represents advantages in terms of space, process time and even costs. These advantages emerge from the reduction in the required downstream processing equipment, skipping the steps of harvest, separation and concentration [5].

The potential benefits of implementing ISPR to produce microbial metabolites have been known for a long time, and during the last 20 years, a steady increment in the number of scientific reports by the scientific community has been observed (Figure 10.1). This has allowed the growth of the current knowledge of ISPR techniques and the development of new and more complex technologies. The incremental trend highlights the continuous search for technologies that allow process engineers to increase productivity. Figure 10.1 was obtained by quantifying scientific reports from the Scopus database by entering terms: "ISPR", "extractive fermentation" and "*in situ* product removal" in the indicated time range. The major portion of research found was published

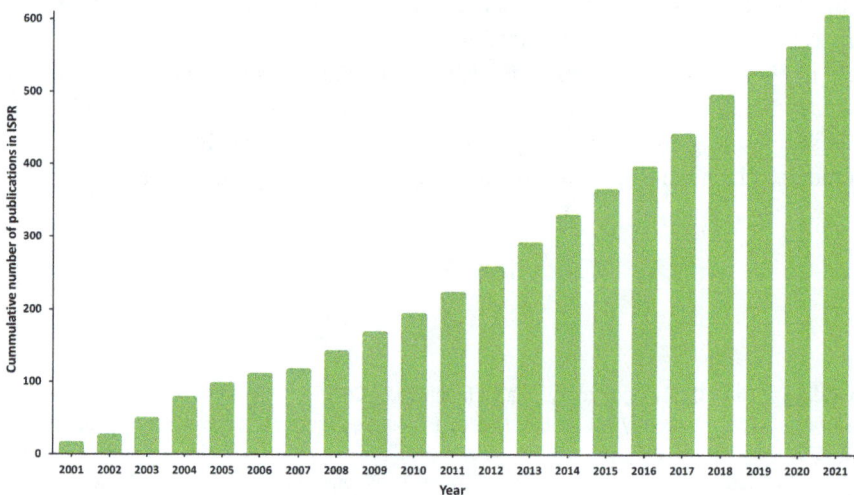

Figure 10.1: Scientific papers published during 2001–2021.

during the last decade, highlighting the recent trend into the implementation of ISPR aiming to boost specific bioprocesses, like the pattern described by Santos *et al.,* [3].

ISPR processes have many advantages compared to non-ISPR: (i) higher productivities for cells and metabolites, (ii) higher yields, (iii) less investment capital for capture processes, (iv) higher product enrichment, (v) lower feedback inhibition and (vi) lower process flows [5].

According to numerous authors [6–8], the selection of a suitable separation technique for the recovery of a specific target from its surrounding matrix is one of the most relevant factors to successfully implement an ISPR. The separation process performed by *in situ* recovery technologies exploit the differential properties between the target product and the fermentation broth, divergences in solubility, volatility, charge, size, hydrophobicity, and other physical/chemical parameters are the key to select the appropriate technique to obtain a successful separation. As example, for a process with a volatile product like acetone–butanol–ethanol (ABE) fermentation, gas stripping is an easy, cheap, and scalable technique to recover such metabolites [9, 10]. But for non-volatile products, extraction or adsorbents may be used instead, taking advantage of solubility and affinity/charge, respectively. A scheme for the technique selection based on differential properties is shown in Figure 10.2. This topic has been extensively boarded by some reviews that can be consulted in case of more specific interests [7].

Product recovery strategies have been studied in the production of different compounds such as solvents [8], organic acids [11], antibiotics [12], pigments [13], enzymes

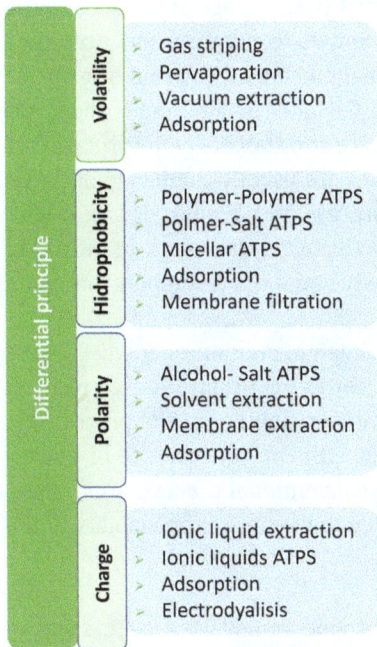

Differential principle

Volatility
- Gas striping
- Pervaporation
- Vacuum extraction
- Adsorption

Hidrophobicity
- Polymer-Polymer ATPS
- Polmer-Salt ATPS
- Micellar ATPS
- Adsorption
- Membrane filtration

Polarity
- Alcohol- Salt ATPS
- Solvent extraction
- Membrane extraction
- Adsorption

Charge
- Ionic liquid extraction
- Ionic liquids ATPS
- Adsorption
- Electrodyalisis

Figure 10.2: ISPR techniques classified by the differential principle that exploits.

[14] and food additives [15], demonstrating in general an improvement in terms of cell growth and process productivity. According to Santos *et al.*, in recent years the trends at products favored by extractive fermentation are solvents (fuels or ABE fermentation), fine chemicals, organic acid, and with minor frequency enzymes and fragrances. ISPR can be also used to remove by-products that negatively interact with some fermentation compounds (i.e., substrate, cell, product...), allowing to enhance the bioprocess [3]. However, an important disadvantage that must be present, even if it seems evident, is that ISPR is limited to extracellular products because it is improbable to release the products maintaining the cell viability.

Another problem that needs to be addressed by future research is the lack of high-scale studies in this field. The vast majority of the published ISPR research has been done at laboratory scale and few studies addressed its implementation at pilot level. Is important to mention that due to the increase in the complexity of diverse research at extractive fermentation, some projects end in failure in terms of being scaled to industry. Focused on this problem, Van Hecke established a number of requirements that a ISPR technology should fulfill to increase the chance for industrial implementation [5]:

– **Scalability**: An ISPR technology must be as simple as possible, facilitating its scale-up. Also, performance indicators of ISPR should be gathered and made available, making it easier to compare against normal processes and determine its viability.
– **Long term robustness and stability**: For a technique to be considered for industrial implementation, it must show good long-term robustness in bioconversion and stability of the kinetic parameters of the fermentation, the type of device used must not interfered negatively in the production process (i.e., leakage of substrate), also the integrity of the fermentation must not be committed (contamination). According to Stark and von Stockar, configuration of the method chosen (internal or external separation device) and the operation mode (batch, fed batch or continuous) may have a relevant influence over the long-term performance and stability of the process [16].
– **Decrease energy consumption**: ISPR technologies are usually claimed to decrease the energy consumption into a downstream process and produced wastewater, this is mainly accomplished through the product concentration and reduction of the volume in stream. Unfortunately, the respective analysis to support this claim is rarely addressed in the published studies.
– **Maximize product recovery**: Commonly, ISPR implementation only will allow a partial recovery of the product, resulting in a certain concentration of residual product at the effluent, that may affect the economic viability of the process due to product loss and, in case of recovering from effluent, an increase of energy consumption per kg of product compared to the conventional process. From these, the product recovery should be as high as possible to maximize the viability of the process.

From this information, it is possible to see that *in situ* product removal is a useful strategy to address the problems associated with product or by-product inhibition. By separating these metabolites from the fermentation broth, a concentrated extractive phase can be obtained, reducing the outflow and the costs associated with downstream processing. However, this strategy still presents challenges that need to be addressed to increase their attractiveness for industrial adoption. *In situ* product removal encompasses a series of techniques applicable in various scenarios, primarily depending on the target molecule. In the following sections, we will discuss the most relevant techniques and their classification, focusing on their technical aspects and examples of the processes that have been optimized from these strategies.

10.1 ISPR classification

ISPR techniques can be divided using different classifications systems, for this chapter they will be categorized based on the physical state of the extraction phases, dividing the different ISPR methods in liquid–liquid extraction, solid–liquid extraction, and gas–liquid extraction. Is necessary to highlight that even when selecting the appropriate separation technique is fundamental, the ISPR process does not only depend on it. Factors like the configuration of the system, which refers to whether the separation systems are inside the fermenter or outside it ("in-stream") plays an important role in the separation efficiency and must be evaluated to determine which configuration will adapt better to a certain process [17].

On one hand, at internal systems, the separation unit (i.e., membrane or adsorbent) is submerged in the fermentation medium, increasing the interaction between the medium and the ISPR system. On the other hand, in-stream systems consist of recovery units that are placed outside the reactor and where the fermentation media is recirculated. This system can increase the control over the separation process but limits the interaction

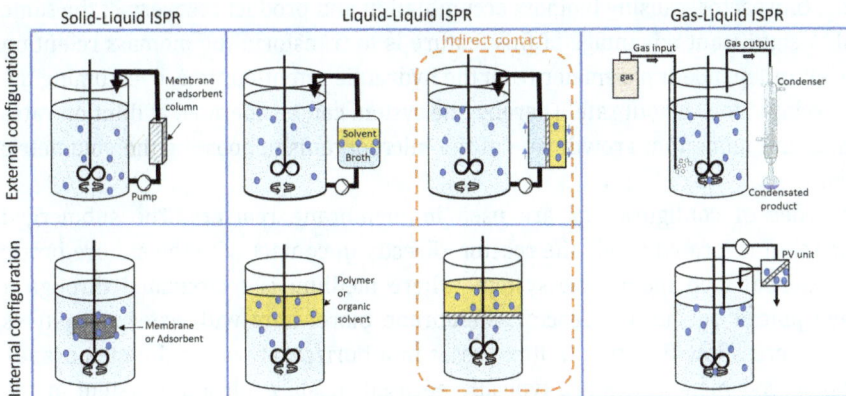

Figure 10.3: Schematic representation of the main ISPR techniques and its possible configuration considering the location of the separation device.

between the separation phase and the product in the fermented broth and may affect the fermentation productivity by deficiencies in oxygen transfer to the cells during media recirculation [18]. A representation of the main ISPR techniques and its classification according to the system configuration and type of contact can be seen in Figure 10.3. Apart from this characteristics, the operation mode of the fermenter (batch, fed-batch or continuous) and the contact mode (direct or indirect contact) between the culture broth and the extractive phases, are factors to be considered for ISPR election. Currently, there is a growing interest in the versatility of techniques that can be used at continuous fermentation (i.e., membranes extraction), taking advantage of the inherent benefits of this mode of operation, such as reduced of death time, major volumetric productivities, and a possible reduction in operating volumes.

10.2 Solid–liquid ISPR

10.2.1 Membrane bioreactors

Membranes have been widely used as a product separation system. This method uses semipermeable barriers to remove the product present in the culture broth. It is usually operated at continuous mode, by simultaneously feeding substrate to the bioreactor and separating the product through the membrane. In this way, two fractions resulted from the separation process, the permeate, that consists in the fraction of the broth that crosses the membrane (usually rich in product), and the retentate, which refers to the remaining fraction that is retained by the membrane.

Barriers are made of low reactive or inert materials such as ceramics, fibers, or steel, avoiding interactions with the medium components and productivity problems. The membranes tend to use a pore size up to 0.2 μm, which gives the opportunity to separate the biomass from the broth that permeates through the membrane, and recirculate the cells to the bioreactor, causing biomass accumulation and product recovery at the same time [18]. A significant advantage of this feature is to transform the biomass retention time into an independent parameter from the hydraulic retention time, overcoming the problems related to washout rate. Thereby, the system can be operated at dilution rates higher than the maximum growth rate of the microorganism, boosting the volumetric productivity.

Two types of configurations are used in membrane reactors. The submerged membrane that is located inside the reactor, directly in contact with the culture broth, and the external loop membrane system, where medium is recirculated through a membrane placed outside the reactor. Membrane bioreactors with external systems mostly use a crossflow filtration (a flux passes in a horizontal way to the membrane), while submerged membranes are typically encased inside a filtration system at the interior of the bioreactor. Due to these filtration modes, membranes bioreactors typically present fewer problems with cake layer formation compared against downstream traditional methods that use in-line filtration modes (flux entering vertically); either by the flux flow in external membranes or by the presence of air bubbles in the submerged systems [19].

According to recent publications, research using external loop membranes is more extensive compared to the submerged set-up, even when the last configuration presents significant advantages. External units present challenges like (i) problematic sterilization, (ii) high energy consumption, (iii) space requirement, and (iv) oxygen limitations. However, they do have advantages in terms of fouling control, cake layer formation and commercial availability of filtration units [6].

During separation, two driving forces are mainly involved in the removal of the molecules at membrane systems (Figure 10.2). (i) The pressure gradient, that is the one that governs the process; and (ii) the concentration gradient (diffusion), involved to a lesser degree [18]. Membrane filtration can be combined with other recovery strategies, resulting in hybrid methods such as pervaporation, a technology that merge membrane separation and evaporation techniques, allowing to reduce the energy requirements of conventional evaporation processes [20]. Another hybrid technique is perstraction (membrane extraction), in which a solvent extraction is performed with a membrane that physically separates the culture broth and the solvent, minimizing the undesired interaction with toxic and avoiding emulsion formation [19] (Figure 10.4).

Perstraction is generally used at membrane based ISPR and the compounds are transferred by affinity to the solvent. The concentration gradient, in contrast to normal perfusion processes, plays a more significant role. Also, the system's efficiency can be improved by changing the solvent for one with a higher partition coefficient and the product recovery from the solvent can be easily achieved by evaporation. Outram *et al.*, performed a simulation-based comparison between evaporation, liquid–liquid extraction, and membrane techniques (pervaporation and perstraction) for the production

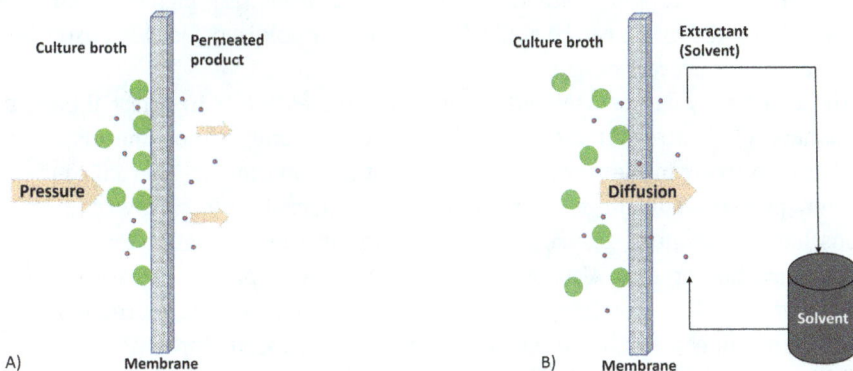

Figure 10.4: Main driving forces involved in membrane recovery processes. (A) Pressure gradient and (B) concentration gradient (diffusion), highly involved in perstraction systems. Biomass (●) and product (●).

of ABE and determine that perstraction was the best strategy in terms to energy requirements, obtaining a 175 % more profit compared to the normal batch process [21].

Membrane bioreactors have been successfully applied for wastewater treatment, the production of glucose syrup, lactate, biofuels, and organic acids. This technology presents advantages like a high surface area per unit of volume, process is performed at low temperatures/pressures, and a high product stability by the non-addition of chemicals to the culture medium.

10.2.2 Adsorption

Adsorption is a physicochemical process used routinely in bioprocess for the concentration and separation of bioproducts. Based on their simplicity and low labor intensity, adsorbents are probably the most used ISPR. Dafoe & Daugulis found in their review that this system represents most of the recent studies in ISPR [7].

The technique is based on the adsorption of the product on the surface of the adsorbents. It occurs due to the presence of forces (i.e., hydrophobic, or ionic interaction) that maintain trapped certain molecules from its surrounding matrix. After the capture of the product, a desorption process is carried out, usually involving a change in pressure, temperature, or the addition of a displacer. The desorbing agent should be capable of releasing the molecule completely without affecting the functionality of the adsorbent and, if needed, ensure the stability of the product. An extra step of regeneration may be added after the desorption process to avoid adsorbents loss of capacity and malfunction, this process is specially reported at ion-exchange resins.

Compared to other separation systems, adsorption is an energy-effective strategy and is categorized as a mild technique, being suitable for the recovery of molecules with both negative or positive charge and low or high molecular weight. Products that have been recovered by adsorption methods include organic acids [22], pigments [23] and biofuels [24]. Also, adsorbents can be applied to a packed or fluidized bed, demonstrating a high versatility for process design.

Furthermore, they can be used either inside the bioreactor or in external packed columns where media is recirculated through. For the first configuration, the adsorbent can be dispersed freely in the medium or trapped in a compartment. Internal configuration often presents advantages compared to the external system. Environmental conditions like temperature, pressure, pH, and aeration rate are better for the cells during the extraction process. Also, better overall productivity and mass transfer rates can be achieved, while maintaining a simpler operation compared to its counterpart. External systems, in contrast, have the advantage of less abrasive impact to the resins within the reactor, no resin wear by the shear force of the impeller (specially in aerobic fermentations) and the possibility of online regeneration of the external system [6].

Despite the simplicity and versatility previously mentioned by adsorption technologies, the adsorbents present as main disadvantage the surface nonselective fouling, decreasing the resin capacity and even leading to column clogging, severely mining the separation efficiency of the process. Likewise, the desorption process usually involves the use of large pH shifts or high concentrate solution, increasing the water consumption and the waste generation.

Resin selection is equally a significant concern since it represents one of the top decisions at the successful implementation of adsorption techniques. Resin is selected depending on the molecule to be extracted and the system configuration that will be used. Typical screening methods are based on trial-and-error methods, where key factors such as resin selectivity and capacity are evaluated to determine the best resin. However, the wide number of resins and the screening process method makes the selection of the resin an intensive task. Even so, little research has been done on predicting which adsorbent will be best from the first principles, changing the selection process from a heuristic method to a rational one [7].

There are various kinds of adsorbents that can be used in extractive fermentations. Natural adsorbents (as activated charcoal) and synthetic ion-exchange resins were the first type of adsorbent materials reported. Activated carbon is commonly used due to its high adsorption capacity and low cost but presents some problems in terms of its regeneration and the lack of mechanical strength to resist agitation [25]. In the other side, ion exchange resins are often used for the removing of charged inhibitory products or by-products, like lactic acid and ammonia. They can be either cation exchange resins or anion exchange resins, showing affinity for the opposite charges in a solution. However, their use has been limited due to non-selective adsorption and resin fouling by cells and other polar compounds different from the target molecule (that decreases the binding capacity) [26].

Non-ionic resins appeared as a more selective adsorbent and are most widely used currently. Zeolites and polymeric resins are the major examples. These resins bind to specific molecules through hydrophobic interactions and the process is governed by the pore size and chemistry of the resin [7, 25]. Normally they can be divided in two groups. (i) Non-polar resins, which are typically made of hydrophobic polymers, like polystyrene or polyvinyl chloride, and are used to adsorb non-polar or hydrophobic molecules, such as lipids or hydrocarbons. (ii) And polar resins, which consist of hydrophilic polymers, such as polyacrylamide or polyethylene glycol, and adsorb polar or hydrophilic molecules. Magnetic particles are a relatively recent development in adsorbent technology used for ISPR, being its main advantage is the facility to recover it from the culture broth.

To determine the best resin for a given process, several factors should be considered. According to Jatoi *et al.*, the adsorption capacity, the adsorption speed, the cost of the adsorbent, the selectivity towards the desired product and the ease of the desorption are the most relevant parameters [25]. This process should be carried out from an integrative perspective. The adsorption speed is governed by the adsorption kinetics, and fast kinetics are preferable because it means that less contact-time is required between broth and adsorbent for product recovery. However, to reach a good separation performance, the resin must present a high capacity and selectivity to the desired product apart from the fast adsorption kinetics. Also, even if the resin has an incredible separation efficiency, it must be easy to recover the captured product from the adsorbent without affecting the molecules or the adsorbent properties. And last, but not least, the adsorbent cost has to be economically competitive to aid process viability and scalability. From this point of view, a balance between the factors must be searched in order to establish a competitive process base in product recovery using adsorption techniques. It is important to remember that all these parameters should be analyzed under the fermentation conditions, ensuring the stability of the adsorbent when is exposed to specific temperature and pH conditions.

10.3 Liquid–liquid ISPR

10.3.1 Aqueous two-phase systems (ATPS)

ATPS are ternary systems conformed by a mixture of two hydrophilic solutes above critical concentrations in aqueous media. At these concentrations, the compounds turn incompatible, and the system separates into two immiscible phases due to thermodynamic equilibrium. One of the newly formed phases acts as the extractant for the desired molecules, favoring its migration due to a higher affinity. The main advantage of ATPS is that water represents the major compound in both phases (around 70 %), creating a milder environment for biomolecules recovery [27].

This technique is especially suitable for those molecules with labile structures that are sensible to recovery with organic solvents, like antibiotics or proteins. Despite the increasing number of studies using ATPS, the physicochemical mechanisms involved in product partitioning are not yet clearly understood in a single extraction process, not to mention more complex systems like extractive fermentation. However, mass transfer between the two phases can be described by the partition coefficient and is principally governed by molecular weight, affinity, surface area, and hydrophobic and electrochemical interactions [7].

Advantages such as eco-friendly, easy in scaling-up, short processing time, suitable for continuous operation and a high efficiency at recovery and purification of biomolecules increases the attention received by this strategy. Some bioprocesses studied for its improvement using ATPS are clavulanic acid [28], lipase [29], B-carotene [14] and prodigiosin production [12]. Nevertheless, the relative high concentration of solutes (required for phase formation i.e., polymers or salts) and the risk of undesired interactions between solutes and culture broth affecting biocompatibility are some of the major concerns. Also, cost of some reagents like polymers may be prohibited for some ATPS types in terms of economic viability. Added to this, there exists a lack of information for scaling up this technique, hampering its adoption.

From the previously stated problems, it is evident that the selection of the phase forming compounds is a decisive step in ATPS successful implementation. The screening process should be done considering the biocompatibility, separation efficiency and economical aspects. ATPS-forming compounds can contain polymers, salts, alcohols, surfactants, and ionic liquids and the ATPS types can be classified into five categories:

(i) polymer–polymer system, (ii) polymer–salt, (iii) micelles, (iv) alcohol–salt and (v) ionic-liquid based ATPS.

Polymer-based ATPS are the most studied systems for extractive fermentation. They are normally composed of PEG/dextran mixtures and PEG/phosphate, sulfate, or citrate. These systems have been extensively reported for the recovery of proteins, low molecular weight molecules and compounds sensitive to ionic environments (only polymer–polymer systems) such as organelles and cells. However, special attention must be paid to the polymer concentration and molecular weight. In a study done by Ooi *et al.*, to extract lipases from a *Burkholderia pseudomallei* culture using a PEG/dextran system, they concluded that high concentrations/molecular weight of polymers can cause an increase at medium viscosity, limiting oxygen mass transfer and cell growth. In addition, ionic interactions between salts and enzymes or products have been reported to decrease viability of some extractive fermentations when testing polymer–salt systems [30].

Alcohol–salt ATPS resolves the viscosity problem and represents an economical approach compared to polymer-based systems. However, there is little research addressing this technique as ISPR due to incompatibility between the organic solvent phase and the biomolecules, threatening to denaturalize or inactivate the product [6]. In contrast, studies using micelles ATPS for extractive fermentation have increased during the last years. These systems, also known as cloud point, use non-ionic surfactants (i.e., Triton X-100 and X-114) to create a more hydrophobic aqueous-phase, reason why micelles ATPS are especially useful for the recovery of hydrophobic and amphiphilic products, like pigments and butanol, and bioparticles that may be sensitive to ionic environments too. The phase formation is triggered by an increase in the system temperature above the cloud point, causing micelles to aggregate and leading system separation into a surfactant rich phase (coacervate) and a surfactant diluted phase [31]. An important feature of cloud point systems is their ability of "milking the cells", which refers to the permeabilization of the cell membranes promoted by the surfactants, facilitating the extraction of intracellular components [32].

The most recent ATPS is the one using ionic liquids (IL), which are salts liquid at room temperature and are composed of an anion and a cation. IL-based ATPS use hydrophilic IL combined with kosmotropic salts to generate a gentler and biocompatible extraction phase for molecule recovery. The partition behavior is shaped by the chemical nature of the salt and its concentration. Electrostatic interactions between the proteins and the cations of the IL are the major driving force involved in the extraction. Also, the selectivity of the process towards a specific protein can be improved by modifying the pH [33, 34]. Penicillin, erythromycin, amoxicillin, and lactic acid are some of the examples of recovered products using ionic liquid two-phase systems.

IL systems have been proposed as an alternative to organic solvent extraction, and present advantages like high thermal stability, low viscosity, high selectivity, and salts-recyclability. However, one major problem with this technique is the potential ILs toxicity, affecting biocompatibility. Several studies have addressed this topic, and have found that IL based on fluorinated anions such as [BF$_4$]$^-$ and [PF$_6$]$^-$ usually present higher toxicity problems compared to those IL based on [Tf$_2$N]$^-$ [3].

To end this section, there are some important concepts that are necessary to understand and describe ATPS, they are summarized in Figure 10.5. The first concept is the binodal curve, a diagram that describes the formation of the phases at the system and represents the border between the monophasic and biphasic regions. Meaning that polymer–salt/polymer combinations below the curve will form a single homogeneous phase, while those above the binodal curve will form a biphasic system. The second concept is the tie-line length (TLL), which expresses the composition of the two phases and has units of % w/w. All the systems that are on the same TLL have identical compositions in their upper and bottom phases. However, these systems differ in the last concept, their volume ratio (V_R). The V_R is the ratio of the volume of the upper phase to the volume of the lower phase of the system. Literature regarding ATPS generally evaluates the influence of molecular weight, TLL and V_R over the system separation efficiency [35].

10.3.2 Organic solvent extraction

ISPR based on organic solvent addition should be the most classic and used liquid:liquid method for extractive fermentation. It uses extracts that are immiscible in water to create the biphasic systems, being products transferred to the solvents due to greater affinity. Although the principle behind this technique is simple, the selection of an ideal solvent is a challenging task when ISPR is pretended. Normally, biocompatibility, non-biodisponibility and affinity are the three most relevant criteria for a successful extractive fermentation. However, recent studies have highlighted the importance of other desired criteria like environmental impact, non-emulsion formation, mass transfer rate and the suitability of the extractant to be part of the final product.

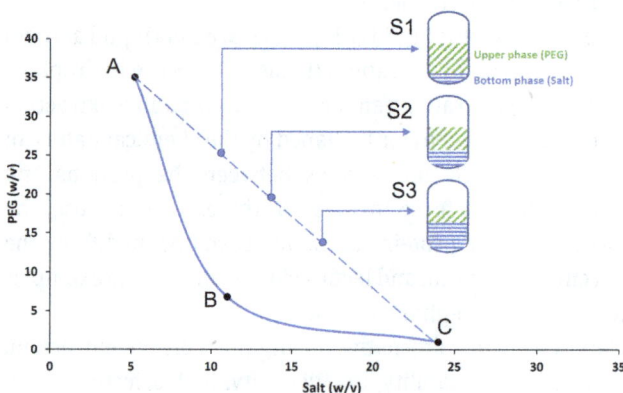

Figure 10.5: Binodal curve for a ATPS, composed from points A, B and C. S1, S2 and S3 represents systems with different V_R that are all across the same TLL.

Biocompatibility may be the major concern for organic solvent extraction. Several solvents have been previously reported to be toxic for specific microorganisms. According to Santos *et al.*, there exists a positive correlation between the toxicity for a given solvent towards microbial cells and the logarithm of the partition coefficient of the solvent (log *P*) in a mixture of equal parts of 1-octanol and water. As the log *PI* increases, also the hydrophobicity of the solvent and its affinity towards the cell membrane turns greater, becoming more toxic. But very hydrophobic solvents, with a log *P* higher than four, will not follow this rule due to the low concentration they can reach in an aqueous solution like fermentation broth, reducing their interaction with the cells. Even though this reference may be useful, the solvent restrictions can vary depending on each microorganism and its characteristics, being of extremely important to perform biocompatibility tests with the desired microorganism prior to selecting a solvent. The selected extractive phase should not be metabolized by the microbial culture. It is preferable for the solvent to be non-bioavailable to the microbes, decreasing the probability of compromising extractant stability or even the performance of the fermentation, by competition with the main substrate [3].

Recent trends are focused on the research for greener extractants, trying to find extractive phases with high affinity and partition coefficients, but with lower impact to the environment and, usually, less toxicity compared to traditional organic solvents. Screening of vegetable oils has been interestingly proposed as an alternative from pure solvents, representing a more biocompatible and economic approach [36]. Liu *et al.*, reported the use of soybean oil as an extractive phase for the improvement of prodigiosin production during a *Serratia marcescens* culture [37]. Similar approach was proposed by Salas *et al.*, using minerals and finding a significant increase in both pigment and biomass concentration [13]. However, concerns related to the increase in viscosity of the medium, yielding lower transfer rates due to rheological properties, and low partition coefficients of specific products in oleic phases should be considered. Lemos group compared the toxicity of seven vegetable oils and five pure solvents over a *Saccharomyces cerevisiae* bioethanol producing culture and obtained that only one solvent (oleic acid) was not toxic for the cells, while no vegetable oil presented toxicity. Nonetheless, among the oils, only castor oil presented a significant partition of the ethanol [36].

Based on the toxicity problem previously discussed, it is desired to minimize the contact time between the solvent extractive phase and the culture broth. A solution may be to use an external unit for the separation process, avoiding long contact between phases and cell damage or even the combination of techniques, like perstraction, using a membrane to physically separate the phases. Another interesting approach that has been recently proposed is the use of extractive phases that can take part of the final product, reducing the required downstream processing, costs, and waste generation. Biodiesel has been successfully used for extractive fermentation of butanol, obtaining a high biocompatibility with *Clostridium acetobutylicum* and a significant increase in butanol titers [38].

10.4 Gas–liquid system (stripping)

The last technique to be discussed in this chapter is gas stripping. This approach has been recognized as an attractive technique for large-scale production by several researchers due to the not interaction of the gas with the medium components, the cell's safety, and its operational simplicity and scalability. Based on these advantages, the method has become a ISPR top choice for volatile products recovery, such as ethanol and ABE bioprocesses [8, 39].

Its mechanism consists in sparging an inert gas, like carbon dioxide or nitrogen, to the broth, to separate and isolate the volatile products. The solvents present in the gas stream are later subject to condensation using cold traps for its recovery, while the gas can be recirculated to the bioreactor for a new stripping process. The efficiency of the stripping process can be changed by modifying the gas, flow rate, temperature, and pressure, giving room to process optimization [6].

Many studies have addressed gas stripping mathematical modeling, making easier the behavior and yield prediction of the process, and becoming a useful tool for operational parameters selection. Stripping macroscopic phenomena can be described by a first-order kinetic where the separation or striping rate for a given product is proportional to the concentration of the product in the medium. Also, the striping rate constant ($K_S a$), which is the linear constant in the first-order kinetic, will be directly proportional to the flow rate of the stripped gas and can be affected by the bubble size, yielding a higher $K_S a$ the processes with a bubble diameter greater than 0.13 cm [8]. However, under high flow rates foam formation may appear and negatively affect fermentation stability.

Traditional recovery techniques for volatile products, like distillation and fractionation, are used to separate products directly from the media (where they are very diluted), causing considerable operational costs due to the energy demand, especially at large scales. Gas stripping represents a potential solution to decrease operating costs, by partially removing and concentrating the volatile products in the outflow.

However, even though stripping presents some attractive features, it usually represents a high energy demand for the process compared to other ISPR options. Outram's group performed a comparison between the energy requirements of different ISPR techniques for ABE fermentation and determined that gas stripping was the most energy intensive technique for downstream, while liquid:liquid extraction represented the less energy intensive method for the same part [21]. This phenomenon is explained due to the low selectivity of gas stripping towards the products, recovering a significant amount of water and/or other impurities, diluting the gas stream and representing a burden during downstream processing. Furthermore, since gas stripping presents a low selectivity and that the separation efficiency during fermentation will be proportional to the energy injected to the system (flow rate), this technique can be a good example of a high energy intensity method at upstream and downstream. To overcome these limitations, combined methodologies have been developed, demonstrating the flexibility of gas stripping to be coupled to other primary recovery techniques. Membrane separation and solvent extraction are the two most used techniques to create hybrid processes. Combination of membrane technology with stripping is known as pervaporation, in this process the broth passes through a membrane and permeate is subjected to gas stripping. Using this approach, it is possible to either concentrate the product before the stripping process and reduce the water volume and other impurities in the vapor stream, thus, reducing the required energy of the process.

10.5 Conclusions and future trends

In situ downstream processing presents an opportunity to enhance fermentation yields by mitigating product inhibition caused by cytotoxic compounds, catabolic regulation, or the potential loss of target metabolites that may be metabolized by the cells, resulting in low yields.

However, each extraction technique comes with its own drawbacks and limitations. A recent trend is to combine different extraction techniques in hybrid modes to leverage their respective strengths in the recovery process. Nevertheless, this can increase the complexity of the procedure. Additionally, the introduction of new components to the fermentation medium may have undesired effects on the process, which must be carefully considered. Therefore, the selection and pre-testing of ISPR components are crucial to avoid negatively impacting fermentation yields through unintended interactions with the medium or microorganisms.

Tools such as High-Throughput Screening (HTS) and mathematical models can be employed to enhance the current process of selecting and optimizing an ISPR, making it more efficient and feasible. Mathematical models offer the capability to describe and predict behavior, serving as a powerful aid for process optimization.

Finally, it's crucial to highlight the close relationship between ISPR and process intensification. *In situ* product removal plays a pivotal role in process intensification strategies, as it allows for a more efficient and streamlined production process. By reducing inhibitory factors and optimizing product recovery, ISPR contributes significantly to enhancing process efficiency and overall productivity. Thus, the integration of ISPR techniques into process intensification approaches holds great promise for advancing the field of industrial bioprocessing.

Acknowledgments: The authors would like to thank the editor Dirk Holtmann for his guidance and review of this article before its publication.

References

1. Fan J, Budarin VL, Macquarrie DJ, Gomez LD, Simister R, Farmer TJ, et al. A new perspective in bio-refining: levoglucosenone and cleaner lignin from waste biorefinery hydrolysis lignin by selective conversion of residual saccharides. Energy Environ Sci 2016;9:2571–4.
2. D´Souza R, Azevedo A, Aires-Barros MR, Krajnc NL, Kramberger P, Carbajal ML, et al. Emerging technologies for the integration and intensification of downstream bioprocesses. Pharm Bioprocess 2016;1:423–40.
3. Santos AG, de Albuquerque TL, Ribeiro BD, Coelho MAZ. In situ product recovery techniques aiming to obtain biotechnological products: a glance to current knowledge. Biotechnol Appl Biochem 2021;68:1044–57.
4. Maiorella B, Blanch HW, Wilke CR. By-product inhibition effects on ethanolic fermentation by *Saccharomyces cerevisiae*. Biotechnol Bioeng 1983;25:103–21.
5. Van Hecke W, Kaur G, De Wever H. Advances in in-situ product recovery (ISPR) in whole cell biotechnology during the last decade. Biotechnol Adv 2014;32:1245–55.
6. Salas-Villalobos UA, Gómez-Acata RV, Castillo-Reyna J, Aguilar O. In situ product recovery as a strategy for bioprocess integration and depletion of inhibitory products. J Chem Technol Biotechnol 2021;96:2735–43.
7. Dafoe JT, Daugulis AJ. In situ product removal in fermentation systems: improved process performance and rational extractant selection. Biotechnol Lett 2014;36:443–60.
8. Li SY, Chiang CJ, Tseng I, He CR, Chao YP. Bio reactors and in situ product recovery techniques for acetone–butanol–ethanol fermentation. FEMS Microbiol Lett 2016;363:fnw107.
9. Chen H, Cai D, Chen C, Zhang C, Wang J, Qin P. Techno-economic analysis of acetone–butanol–ethanol distillation sequences feeding the biphasic condensate after in situ gas stripping separation. Sep Purif Technol 2019;219:241–8.
10. Ranjan A, Moholkar VS. Biobutanol: a viable gasoline substitute through ABE fermentation. Proc World Acad Sci Eng Technol 2009;51:497–503.
11. Magalhães AI, de Carvalho JC, Medina JDC, Soccol CR. Downstream process development in biotechnological itaconic acid manufacturing. Appl Microbiol Biotechnol 2017;101:1–12.
12. Soto A, Arce A, Khoshkbarchi MK. Partitioning of antibiotics in a two-liquid phase system formed by water and a room temperature ionic liquid. Sep Purif Technol 2005;44:242–6.
13. Salas-Villalobos UA, Santacruz A, Castillo-Reyna J, Aguilar O. An in-situ approach based in mineral oil to decrease end-product inhibition in prodigiosin production by *Serratia marcescens*. Food Bioprod Process 2022;135:217–26.
14. Dreyer S, Salim P, Kragl U. Driving forces of protein partitioning in an ionic liquid-based aqueous two-phase system. Biochem Eng J 2009;46:176–85.
15. Ma Y, Liu N, Greisen P, Li J, Qiao K, Huang S, et al. Removal of lycopene substrate inhibition enables high carotenoid productivity in *Yarrowia lipolytica*. Nat Commun 2022;13:572.
16. Stark D, von Stockar U. In situ product removal (ISPR) in whole cell biotechnology during the last twenty years. In: Process integration in biochemical engineering. Advances in biochemical engineering/biotechnology. Berlin, Heidelberg: Springer; 2003, vol. 80:149–75 pp.
17. Buque-Taboada EM, Straathof AJ, Heijnen JJ, Van Der Wielen LA. In situ product recovery (ISPR) by crystallization: basic principles, design, and potential applications in whole-cell biocatalysis. Appl Microbiol Biotechnol 2006;71:1–12.
18. Carstensen F, Apel A, Wessling M. In situ product recovery: submerged membranes vs. external loop membranes. J Membr Sci 2012;394:1–36.
19. Heerema L, Roelands M, Goetheer E, Verdoes D, Keurentjes J. In-situ product removal from fermentations by membrane extraction: conceptual process design and economics. Ind Eng Chem Res 2011;50:9197–208.
20. Jiménez-Bonilla P, Wang Y. In situ biobutanol recovery from clostridial fermentations: a critical review. Crit Rev Biotechnol 2012;38:469–82.
21. Outram V, Lalander CA, Lee JG, Davis ET, Harvey AP. A comparison of the energy use of in situ product recovery techniques for the acetone butanol ethanol fermentation. Bioresour Technol 2016;220:590–600.

22. Li C, Dong Y, Wu D, Peng L, Kong H. Surfactant modified zeolite as adsorbent for removal of humic acid from water. Appl Clay Sci 2011;52:353–7.
23. Wang X, Tao J, Wei D, Shen Y, Tong W. Development of an adsorption procedure for the direct separation and purification of prodigiosin from culture broth. Biotechnol Appl Biochem 2004;40:277–80.
24. Nielsen DR, Amarasiriwardena GS, Prather KL. Predicting the adsorption of second generation biofuels by polymeric resins with applications for in situ product recovery (ISPR). Bioresour Technol 2010;101:2762–9.
25. Jatoi AS, Baloch HA, Mazari SA, Mubarak NM, Sabzoi N, Aziz S, et al. A review on extractive fermentation via ion exchange adsorption resins opportunities, challenges, and future prospects. Biomass Convers Biorefin 2021;13:3543–54.
26. Phillips T, Chase M, Wagner S, Renzi C, Powell M, DeAngelo J, et al. Use of in situ solid-phase adsorption in microbial natural product fermentation development. J Ind Microbiol Biotechnol 2013;40:411–25.
27. Rito-Palomares M, Benavides J, editors. Aqueous two-phase systems for bioprocess development for the recovery of biological products. Cham: Springer International Publishing; 2017.
28. Santos VC, Hasmann FA, Converti A, Pessoa A Jr. Liquid–liquid extraction by mixed micellar systems: a new approach for clavulanic acid recovery from fermented broth. Biochem Eng J 2011;56:75–83.
29. Show PL, Tan CP, Anuar MS, Ariff A, Yusof YA, Chen SK, et al. Extractive fermentation for improved production and recovery of lipase derived from *Burkholderia cepacia* using a thermoseparating polymer in aqueous two-phase systems. Bioresour Technol 2012;116:226–33.
30. Ooi CW, Hii SL, Kamal SMM, Ariff A, Ling TC. Extractive fermentation using aqueous two-phase systems for integrated production and purification of extracellular lipase derived from *Burkholderia pseudomallei*. Process Biochem 2011;46:68–73.
31. Chávez-Castilla LR, Aguilar O. An integrated process for the in situ recovery of prodigiosin using micellar ATPS from a culture of *Serratia marcescens*. J Chem Technol Biotechnol 2016;91:2896–903.
32. Hu Z, Zhang X, Wu Z, Qi H, Wang Z. Export of intracellular *Monascus* pigments by two-stage microbial fermentation in nonionic surfactant micelle aqueous solution. J Biotechnol 2012;162:202–9.
33. Oppermann S, Stein F, Kragl U. Ionic liquids for two-phase systems and their application for purification, extraction and biocatalysis. Appl Microbiol Biotechnol 2011;89:493–9.
34. Dreyer S, Kragl U. Ionic liquids for aqueous two-phase extraction and stabilization of enzymes. Biotechnol Bioeng 2008;99:1416–24.
35. Raja S, Murty VR, Thivaharan V, Rajasekar V, Ramesh V. Aqueous two phase systems for the recovery of biomolecules – a review. Sci Technol 2011;1:7–16.
36. Lemos DA, Sonego JLS, Boschiero MV, Araujo ECC, Cruz AJG, Badino AC. Selection and application of nontoxic solvents in extractive ethanol fermentation. Biochem Eng J 2017;127:128–35.
37. Liu W, Yang J, Tian Y, Zhou X, Wang S, Zhu J, et al. An in situ extractive fermentation strategy for enhancing prodigiosin production from *Serratia marcescens* BWL1001 and its application to inhibiting the growth of *Microcystis aeruginosa*. Biochem Eng J 2021;166:107836.
38. Yen HW, Wang YC. The enhancement of butanol production by in situ butanol removal using biodiesel extraction in the fermentation of ABE (acetone–butanol–ethanol). Bioresour Technol 2013;145:224–8.
39. Silva CR, Esperança MN, Cruz AJG, Moura LF, Badino AC. Stripping of ethanol with CO_2 in bubble columns: effects of operating conditions and modeling. Chem Eng Res Des 2015;102:150–60.

Anna Dinius, Zuzanna J. Kozanecka, Kevin P. Hoffmann and
Rainer Krull*

11 Intensification of bioprocesses with filamentous microorganisms

Abstract: Many industrial biotechnological processes use filamentous microorganisms to produce platform chemicals, proteins, enzymes and natural products. Product formation is directly linked to their cellular morphology ranging from dispersed mycelia over loose clumps to compact pellets. Therefore, the adjustment and control of the filamentous cellular morphology pose major challenges for bioprocess engineering. Depending on the filamentous strain and desired product, optimal morphological shapes for achieving high product concentrations vary. However, there are currently no overarching strain- or product-related correlations to improve process understanding of filamentous production systems. The present book chapter summarizes the extensive work conducted in recent years in the field of improving product formation and thus intensifying biotechnological processes with filamentous microorganisms. The goal is to provide prospective scientists with an extensive overview of this scientifically diverse, highly interesting field of study. In the course of this, multiple examples and ideas shall facilitate the combination of their acquired expertise with promising areas of future research. Therefore, this overview describes the interdependence between filamentous cellular morphology and product formation. Moreover, the currently most frequently used experimental techniques for morphological structure elucidation will be discussed in detail. Developed strategies of *morphology engineering* to increase product formation by tailoring and controlling cellular morphology and thus to intensify processes with filamentous microorganisms will be comprehensively presented and discussed.

Keywords: filamentous microorganisms; morphological characterization; morphology engineering; mycelial pellet; product formation; surface-active agents.

Anna Dinius, Zuzanna J. Kozanecka, and Kevin P. Hoffmann contributed equally to this work.

*Corresponding author: Rainer Krull, Institute of Biochemical Engineering, Technische Universität Braunschweig, Rebenring 56, 38106 Braunschweig, Germany; and Center of Pharmaceutical Engineering, Technische Universität Braunschweig, Franz-Liszt-Str. 35a, 38106 Braunschweig, Germany, E-mail: r.krull@tu-braunschweig.de. https://orcid.org/0000-0003-2821-8610
Anna Dinius, Zuzanna J. Kozanecka and Kevin P. Hoffmann, Institute of Biochemical Engineering, Technische Universität Braunschweig, Rebenring 56, 38106 Braunschweig, Germany; and Center of Pharmaceutical Engineering, Technische Universität Braunschweig, Franz-Liszt-Str. 35a, 38106 Braunschweig, Germany

As per De Gruyter's policy this article has previously been published in the journal Physical Sciences Reviews. Please cite as: A. Dinius, Z. J. Kozanecka, K. P. Hoffmann and R. Krull "Intensification of bioprocesses with filamentous microorganisms" *Physical Sciences Reviews* [Online] 2023. DOI: 10.1515/psr-2022-0112 | https://doi.org/10.1515/9783110760330-011

List of abbreviations and symbols

API	active pharmaceutical ingredient
CLSM	confocal laser scanning microscopy
CMC	critical micelle concentration
FC	flow cytometry
GFP	green fluorescent protein
HBU	hyphal branch unit
HGU	hyphal growth unit
HLB	hydrophilic-lipophilic balance
k	growth constant of the cubic root law of pellet growth
k_{bran}	branching constant
$k_L a$	oxygen transfer coefficient
L_{hyphae}	total hyphal length
LLD	laser light diffraction
MN	Morphology number
MPEC	microparticle-enhanced cultivation
n	number of hyphal tips
q_{tip}	tip growth rate
Δr_{active}	thickness of the active pellet-layer
r_{crit}	critical pellet radius
SEC	salt-enhanced cultivation
STR	stirred tank reactor
SV	secretory vesicles
TEOA	triethanolamine
X_{pellet}	pellet-biomass
µCT	X-ray microtomography
μ_{hyphae}	total hyphal growth rate

11.1 Introduction

–   This overview describes the interdependence between filamentous cellular morphology and product formation.
–   The currently most frequently used experimental techniques for morphological structure elucidation will be discussed in detail.
–   Developed strategies of *morphology engineering* to increase product formation by tailoring and controlling cellular morphology and thus to intensify processes with filamentous microorganisms will be comprehensively presented and discussed.

Filamentous microorganisms are widely used as production systems in industrial biotechnology and have been successfully employed for the pharmaceutical and food industries, e.g., for the production of organic acids or antibiotics [1]. They are of particular importance, with their products generating annual sales of several billion dollars [1].

The first industrial biotechnological process using the filamentous fungus *Aspergillus niger* was established 1919 for the production of citric acid even without an understanding of basic molecular principles [2, 3] and only with the empirical approach of bubbling air through vessels to grow microorganisms. With the development of engineering principles by Gaden in the late 1940s for optimal supply of oxygen as critical growth-limiting substrate, as well as with the design of biochemical reactor equipment, growth of filamentous microorganisms and their production of antibiotics can proceed more efficiently. This research formed the fundamentals for large scale production of a wide range of antibiotics, such as penicillin [4] and marked a shift away from empiricism towards more knowledge-based biochemical research. With further research progress, other filamentous fungal expression systems were developed for the production of, e.g., pharmaceutical compounds, proteins and enzymes [5, 6].

Among prokaryotic (bacterial) filamentous microorganisms, actinomycetes are of primary industrial interest due to the production of mainly pharmaceutically relevant secondary metabolites [7, 8]. Overall, about 60% of the known biologically active secondary metabolites were isolated from actinomycetes in 2000 [9]. Despite some obvious differences, the morphological characteristics of actinomycetes and their implications for cultivation are similar to those of fungi. Therefore, their growth kinetics can be described using the same models [10].

A number of review articles address in detail the multitude of challenges regarding molecular biology as well as biochemical engineering in the matter of filamentous microorganisms. The first review of the investigations of microscopic hyphal growth and macroscopic pellet growth was written by Metz and Kossen [11]. In the mid-2000s, Papagianni [12] and Grimm et al. [13] summarized the kinetics of fungal morphology as well as its impact on product formation with classical engineering biochemical parameters in their widely cited review articles. A comprehensive review on cellular morphology, rheology and productivity was published by Wucherpfennig et al. [14]. Several recent reviews have focused on filamentous fungi, covering cellular morphology and productivity [14–17], process design [18] and molecular characteristics [19–21]. In 2015, a methodological compendium from the viewpoint of biochemical engineers for filaments in bioprocesses was presented [22] focussing on new monitoring methods and to some extent on their combination, with respect to the characterization of the morphological structure of the filaments. The generation of mathematical multi-scale models and the application of appropriate control strategies for exploiting the high potential of these interesting and challenging biological systems are described in detail by the corresponding reviews in the issue (see [23–27]).

The progress and trends in tailoring filamentous cellular morphology in order to increase product formation by morphology engineering techniques have been discussed by Walisko et al. [28] with the main focus on the impact of microparticles on biomass growth, morphology and product formation. Böl et al. [29] extended the challenges of influencing cellular morphology by macroparticle-enhanced cultivation, the alteration of osmolality of the culture medium by addition of inorganic salts, the salt-enhanced cultivation (SEC), and mechanically induced stress on filamentous pellet systems.

Furthermore, the effects of surface chemistry on particles in the microparticle-enhanced cultivation (MPEC) were reviewed by Laible et al. [30]. Here, the authors linked morphological, physical, and chemical properties of microparticles with effects on culture broth, filamentous morphology and molecular biology showing that microparticles interfere with gene regulation, metabolism and enzyme activity. In a recently published primer [31] the future trends of novel genomic, genetic, metabolic, imaging and modelling tools were discussed which will provide fundamental new insights into filamentous fungal growth and with some respect also into filamentous bacterial growth associated with product formation. The authors highlighted these tools and their impact on rational morphology engineering.

The present chapter summarizes the extensive work conducted in recent years in the field of improving product formation and thus intensifying biotechnological processes with filamentous microorganisms. The goal of this chapter is to provide prospective scientists with an extensive overview of this scientifically diverse, highly interesting field of study. In the course of this, multiple explicit given examples and ideas shall facilitate the combination of their acquired expertise with worthwhile areas of future research.

Subsequently, chapter 2 will give an overview about the filamentous cellular morphology and its impact on product formation. In chapter 3 the currently most frequently used experimental techniques for morphological structure elucidation will be discussed in detail. Chapter 4 presents developed strategies to increase product formation by altering and controlling morphology to intensify processes with filamentous microorganisms. Additionally, in this chapter, morphology engineering techniques to enhance product formation are defined more comprehensively than before. At present, all additives that lead to a change in filamentous cellular morphology and to an increased product formation including the supplementation of surface-active agents are subsumed here. Finally, chapter 5 will summarize the developments in intensification of biotechnological filamentous processes.

11.2 Cellular morphology and its influence on productivity

-   Chapter 2 gives an overview about the filamentous cellular morphology and its impact on product formation.
-   The mathematical derivations for characterizing micro- and macroscopic filamentous growth are discussed:
    -   for micro-morphology: through geometrical parameters obtained by digital image analysis,
    -   for macro-morphology: through the cubic root law.
-   Important morphological description parameters are presented and explained.
-   The impact of description parameters on productivity is discussed.

One of the most important characteristics of filamentous microorganisms is their diverse cellular morphology. It not only influences bioprocesses at microscale by affecting diffusibility and cell physiology, but also at macroscale, as certain morphologies can entail mixing problems [12, 14, 32]. Furthermore, the cellular morphology of filamentous fungi as well as bacteria can have a significant effect on achievable product yields [16, 27, 33, 34]. Thus, its investigation contributes to the characterization and optimization of existing bioprocesses. In this chapter the micro- and macro-morphology of filamentous microorganisms as well as important morphological description parameters are presented and explained. Additionally, its influence on the productivity is shortly discussed.

11.2.1 Micro-morphology

The lifecycle of filamentous microorganisms begins as mature spores swell and germinate, initiating the subsequent growth of hyphae. The morphological development on microscopic level includes the following steps: tip extension, septum formation and branching. In general, the hyphal length increases by polarized growth of each tip, whereas the extension rate at the hyphal apex is linear [12, 13, 16]. In the process, various precursors involved in the cell wall synthesis are enclosed in secretory vesicles (SV) and transported lengthwise the hyphae, accumulating at the tips as the so-called *Spitzenkörper*. As the growth continues and the hyphal length increases, the distance for the transport vesicles to reach the hyphal tips becomes longer. Consequently, they accumulate in different hyphal compartments before they can reach the apex. This effect results in branching: a new hyphal element growing laterally to the parental hypha. Accordingly, the formation of branches causes an increase in the number of tips [12, 13, 16] (Figure 11.2.1).

Kunz et al. [35] and Kunz and King [36] recently investigated the growth of spores or germ tubes of the filamentous fungus A. *niger* by the site-dependent and time-varying transport of SV, which contain the building blocks for the synthesis and assembly of new cell wall material. The investigations were conducted using confocal laser scanning microscopy (CLSM) in a suitable optically transparent flow cell. The first challenge to overcome was the formulation of a mathematical model for intracellular SV transport, while the studies were statistically ensured and performed without fluid dynamic loading on the microscale. Furthermore, the characterization of SV transport and product formation under cultivation-like conditions were investigated in a flow chamber with a backward-facing stage. This flow channel was developed and the fluid was mechanically characterized in order to microscopically follow polarized growth and product formation in the A. *niger* spore fixed at the bottom of the flow chamber. Subsequently, the authors plausibly examined the influence of fluid dynamic stress on SV. Moreover, a novel growth chamber for mapping of the polarized growth behavior from the spore to the hyphal tube under the influence of a time-variable wall shear stress was successfully developed and characterized. The effect of different volume flows of the incoming cultivation medium

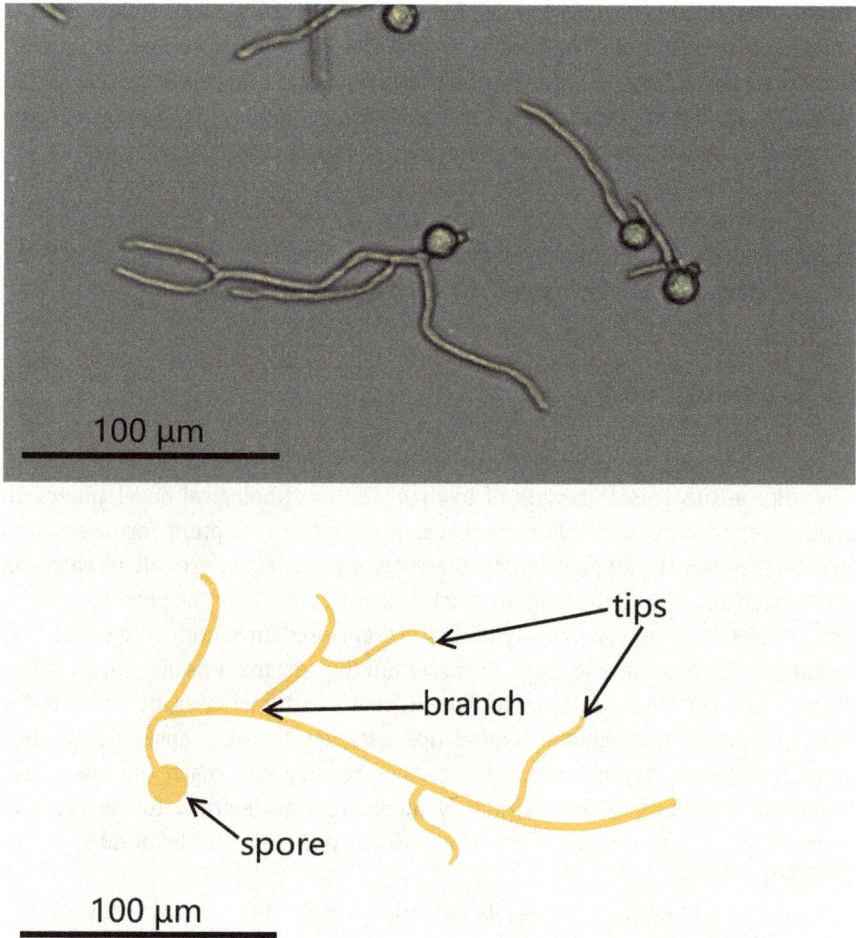

Figure 11.2.1: Microscopic hyphal growth: microscopic image (top) and schematic illustration (bottom).

on the morphology, as well as the local distribution and the dynamic development of biomass, SV and glucoamylase as a growth-associated product were microscopically recorded and further evaluated by image analysis. The wall shear stress at the chamber bottom behind the backward-facing stage was predicted by computational fluid dynamics (CFD) simulations and compared with the microscopic images for different volume flows. Here, the authors generated results, where for both very low and very high fluid dynamic stress, glucoamylase production decreased compared to the control. In conclusion, the results of this work represent an important building block for the establishment of a holistic growth model for the morphogenesis of the fungus *A. niger*.

Overall, the described mycelial growth on microscopic level can be characterized by different parameters obtained by digital image analysis. The two most fundamental parameters, which are especially relevant during the exponential growth phase of cultivation, are the total hyphal length L_{hyphae} [μm] and the number of tips n [–]. Assuming the water content, density and the diameter of the hyphal element to remain constant, while no fragmentation occurs and the growth process is unlimited, the increase in hyphal length over cultivation time can be characterized by the proportionality factor μ_{hyphae} [h^{-1}], known as the specific length growth rate (11.2.1) [13, 37]. Consequentially, the length increases exponentially under the condition of the specific growth rate μ_{hyphae} remaining constant. Taking the branching process into account, the length growth can also be defined as the product of the tip growth rate q_{tip} [μm h^{-1}] and the number of tips n (11.2.2).

$$\frac{dL_{hyphae}}{dt} = \mu_{hyphae} \cdot L_{hyphae} \tag{11.2.1}$$

$$\frac{dL_{hyphae}}{dt} = q_{tip} \cdot n \tag{11.2.2}$$

Moreover, the branching process itself in regard to the increasing number of tips has been characterized by correlation of the hyphal length L_{hyphae} and the branching constant k_{bran} [μm^{-1}h^{-1}] (11.2.3). Thus, both the hyphal length and the number of tips increase over the cultivation time with a constant specific rate (cf. (11.2.1) and (11.2.3)).

$$\frac{dn}{dt} = k_{bran} \cdot L_{hyphae} \tag{11.2.3}$$

Caldwell and Trinci [38] introduced another important fundamental micro-morphological parameter, the so-called hyphal growth unit (*HGU*), which describes the relationship between the hyphal length growth and the branching process (11.2.4). Consequently, *HGU* can also be construed as the average length of each hyphal tip. The parameter is characterized by increasing values in the beginning of mycelial growth. However, over progressing cultivation time, it gradually reaches a stagnant state [13, 37, 39]. Bergter [40] established an important mathematical correlation under these, in regard to the *HGU*, steady state conditions ($d(HGU)/dt = 0$) (11.2.5), whereas the specific length growth rate μ_{hyphae} is consequently dependent from q_{tip} and k_{bran} (11.2.6). Furthermore, a novel parameter, the hyphal branch unit (*HBU*), which is defined as the share of total hyphal length to total number of tips n_{bran} [–] was recently introduced by Schmieder [41] (11.2.7).

$$HGU = \frac{L_{hyphae}(t)}{n(t)} \tag{11.2.4}$$

$$HGU = \frac{L_{hyphae}}{n} = \sqrt{\frac{q_{tip}}{k_{bran}}} \tag{11.2.5}$$

$$\mu_{hyphae} = \sqrt{q_{tip} \cdot k_{bran}} \tag{11.2.6}$$

$$HBU = \frac{L_{hyphae}(t)}{n_{bran}(t)}$$

(11.2.7)

11.2.2 Macro-morphology

As described in the previous chapter, the mycelial growth proceeds by length growth and branching of hyphae. The subsequent complex entanglement of the hyphal elements results in three fundamental macroscopic forms of the so-called macro-morphology, depicted in Figure 11.2.2 [13, 29, 42].

Subject to the aggregation type (coagulative or disperse) of spores formed at the beginning of the mycelial growth, two principal macroscopic structures can be distinguished. Hyphae which do not agglomerate are known under the term *dispersed mycelium*. This loose macro-morphological structure can also further evolve to pellets which are nearly spherical particles composed of tightly intertwined hyphal elements. In regard to the pellet formation, so far three mechanisms can be distinguished: spore coagulative, hyphae coagulative and non-coagulative [43]. The first mechanism is defined by spores coagulating before or during germination, which results in growth and intertwining of hyphae developing as filamentous pellets thereafter. The second mechanism describes the process of spores germinating first, followed by agglomeration of hyphal developed elements leading to the formation of a hyphal clump and lastly the pellet structure. The

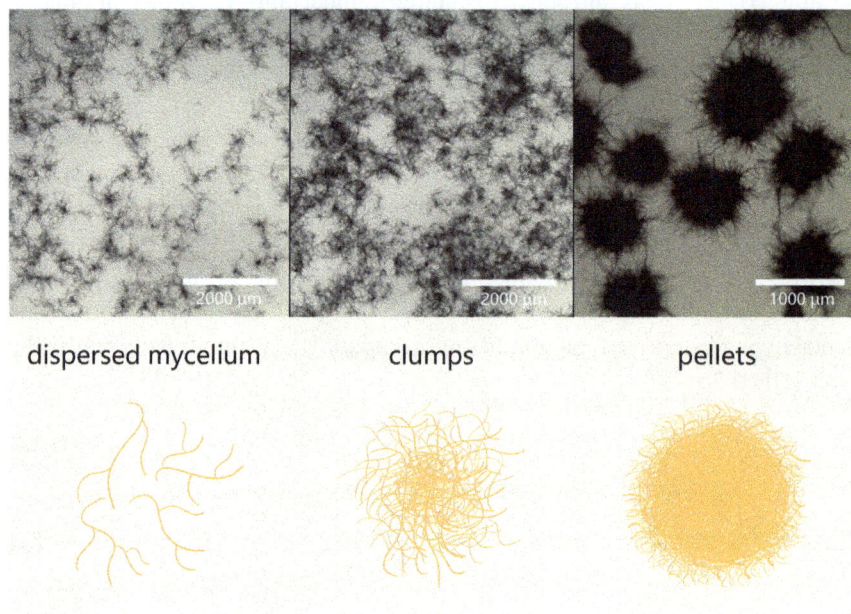

dispersed mycelium   clumps   pellets

Figure 11.2.2: The three main macro-morphological structures: microscopic images (upper row) and schematic illustrations (bottom row).

non-coagulative type describes pellet formation originating from one individual spore [12, 27, 43–48]. However, it has to be noted that pellet formation mechanisms are highly complex and dependent on cultivation conditions, so it is difficult to set a simple overall classification for every filamentous strain [49–51]. Moreover, lying between the two extremes of dispersed mycelium and pellets, the formation of rather loosely structured hyphal agglomerates termed *clumps* is also possible (Figure 11.2.2).

Studies show that so far, the macro-morphology of filamentous microorganisms has in every case proven to be strain-specific and strongly dependent on the prevailing conditions during cultivation [29, 32]. As bioprocesses involving filamentous microorganisms are highly complex and sensitive biological systems, multiple factors like the composition of the cultivation medium, pH, temperature, concentration of the inoculum, volumetric power input, hydrodynamic stress (maximum local power input) and shear stress distinctively influence the cellular morphology of a submerged culture [14, 16, 27, 29, 32, 52]. Thus, the research of filamentous cultivation systems focuses on the variation of different process parameters and on the application of additives, which has recently proven to be a highly promising approach for the modulation of cellular morphology and the increase in product formation. Therefore, the bioprocess intensification strategies involving the addition of different supplements to filamentous submerged cultures will be thoroughly presented and discussed in the chapter 4.

Without substrate limitations and under certain strain-specific environmental conditions during cultivation, filamentous microorganisms grow exponentially as pellets, whereby the size of these nearly spherical bioagglomerates can be characterized by the pellet radius r_{pellet} (Figure 11.2.3). Provided a homogenous mixing of the cultivation broth, each pellet is evenly surrounded by the dissolved substrate. Close to the pellet's surface, the substrate concentration S begins to decrease, as it is consumed by the biomass for growth and maintenance. The mass transport of substrates through the tightly intertwined hyphal network has been found to be mostly diffusive. However, it has also been proven that fluid dynamic conditions and turbulences in submerged cultivations cannot be considered negligible factors [53, 54]. In view of the fact that the diffusive transport plays a decisive role in substrate supply of the biomass within filamentous pellets, the mass transfer is strongly dependent on the porosity or biomass density of the hyphal network [34, 53, 54]. As the pellet size reaches a critical radius r_{crit}, the nutrients are completely metabolized within an outer active layer characterized by the thickness Δr_{active}, and thus they do not reach the pellet core (Figure 11.2.3). The inner pellet region within the radius r_{crit} becomes substrate-limited and its growing behavior deviates from exponential growth [14, 37, 53]. Hence, the pellet radius increases proportionally only with the thickness of the active layer Δr_{active} (11.2.8).

$$\frac{dr_{pellet}}{dt} = \mu \cdot \Delta r_{active} \qquad (11.2.8)$$

By integrating the term (11.2.8) and taking the growth of pellet biomass X_{pellet} into account, the cubic root law of pellet growth (11.2.9) has been established [14, 37, 55].

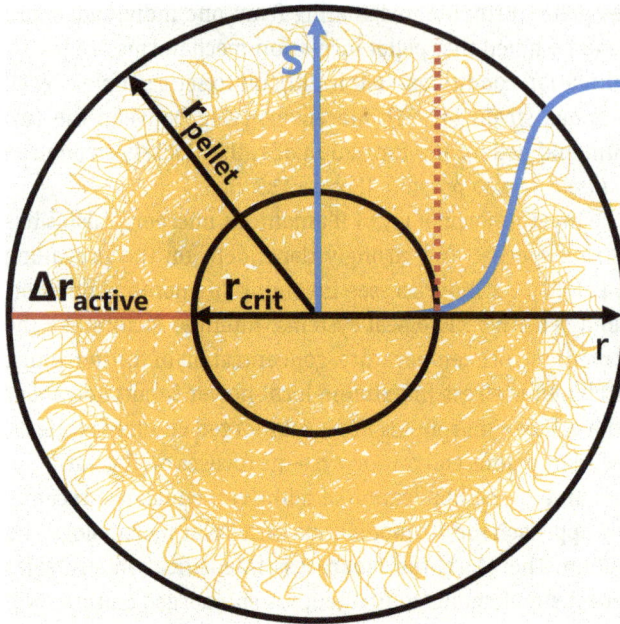

Figure 11.2.3: Macroscopic pellet growth and substrate limitation.

$$X_{pellet}^{1/3} \approx k \cdot t + X_{pellet,0}^{1/3} \tag{11.2.9}$$

The biomass concentration at $t = 0$ is described by $X_{pellet,0}$ and the strain-specific growth constant k [kg$^{1/3}$ m^{-1} h^{-1}] can be approximated as follows:

$$k \approx \left(\frac{4 \cdot \pi \cdot \rho_{pellet} \cdot c_{pellet}}{3} \right)^{1/3} \cdot (\mu \cdot \Delta r_{active}) \tag{11.2.10}$$

with the pellet density ρ_{pellet} and the pellet concentration c_{pellet}. This briefly presented correlation is only valid under the assumption of the pellet's structure being an ideal spherical particle, whereas the density of the hyphal network is homogenously distributed along the entire radius. Additionally, the correlation bases on the assumption of unlimited exponential growth only of the outer pellet active layer Δr_{active} and it does not consider pellets growing to different sizes within one culture. Thus, the cubic root law is to be viewed as a rather simple model; however, it poses a suitable approximation of the macroscopic pellet growth [14, 37].

Similar to the micro-morphology, parameters describing the filamentous morphology on the macroscopic level can be obtained mostly by microscopic investigations of the grown pellets and subsequent, often automated or semi-automated image analysis [17, 56]. The acquired images are typically processed to binary or grayscale pictures, which is followed by measurements and calculations in order to obtain typical parameters for the successive particle characterization [56]. The pellet's shape can be described by the roughness or circularity. While the latter is defined as the deviation of the particle from a true circle, roughness describes the irregularity of the pellet's perimeter based on measuring its boundary [56]. Image analysis can be also used as a tool to characterize the size of pellets in a culture and thus it poses an important method for the acquisition of particle size distribution. In reality pellets do show a deviation from the ideal circle. Therefore, popular descriptors are projected area, Feret diameter or area equivalent spherical diameter. All these characteristic values are obtained by pixel count and further calculation of the distance based on the given scale of an image [56]. However, even while the description of pellet shape and size remains a highly relevant and elementary tool of bioagglomerate characterization, in view of the very complex and variable cellular morphology of filamentous microorganisms, it cannot be considered an entirely sufficient method to describe the macro-morphological structures in filamentous submerged cultures. Moreover, obtaining a general overview of the dominating morphological structure is rather complicated due to the high number of different descriptors [57].

Overall, pellets can be distinguished from clumps by a dense and compact inner core structure surrounded by a less dense layer, referred to as the "hairy region" [58, 59]. Nonetheless, as mentioned above, the cellular morphology of filamentous microorganisms can strongly vary depending on the cultivated strain and numerous variable conditions of the bioprocess, and thus a range of intermediate structures lying between the three extremes (Figure 11.2.3) can be observed [57]. Hence, Wucherpfennig et al. [57] introduced a non-dimensional morphology descriptor, the Morphology number (MN) for *A. niger* strains as a parameter which combines the following shape and size describing values obtained by preceding image analysis: projected area, solidity (share of area to convex area), Feret diameter and aspect ratio (ratio of the largest to the smallest diameter). Pellets appearing as perfect circles result in the MN value of 1, while small mycelial fragments appearing as one-dimensional lines are assigned with MN of 0. All structures lying in between these two extremes consequently result in values between 0 and 1 [57]. In case of the glucoamylase producing strain *A. niger* AB1.13, the developed non-dimensional MN was successfully coupled with the changing of the fungus' productivity by varying the osmolality of the cultivation broth. Nevertheless, further investigations of the fructofuranosidase producing strain *A. niger* SKAn1015 (a strain genetically derived from *A. niger* AB1.13) proved the necessity to adjust the calculation formula and thus to develop and validate new specific non-dimensional morphology descriptors depending on the strain as well as new adapted shape parameters influenced by the changing cultivation conditions [60].

The up to this point generally introduced morphology descriptors pose a fundamental tool for characterizing the shape and size of hyphal agglomerates on macroscopic level. Still, due to the complexity and irregularity of filamentous morphology, a more detailed depiction of these structures remains very challenging. While attempting to characterize filamentous microorganisms by image analysis, it is necessary to evaluate their space-filling capacity, as a matter of fact, to study the fractal nature of mycelia [61, 62]. The mathematical concept of fractals, which was firstly introduced by Mandelbrot, describes geometric shapes having a non-integer, thus a broken or fractal dimension [63]. Hence, the corresponding structures are neither Euclidean lines and surfaces, nor solids but are characterized by self-similarity throughout all magnifications and scales [62]. Simultaneously, fractal geometric objects are marked by high complexity and irregularity. Barry et al. [64] successfully developed an algorithm and applied it as automatic image analysis by means of a detailed characterization of the fungal growth on a solid substrate, especially during the early microscopic hyphal development, which is known to be rather challenging to quantify. In submerged cultures, the description of macroscopic morphological structures by fractal dimensions was made by image analysis using box-counting methods [62, 65]. The processed binary images were analyzed by counting the number of boxes with a defined side length, which overlap with the mycelium. The resulting areas were calculated based on the number and area represented by each black pixel [61, 62]. In this manner, two different box-counting methods were used in order to determine two different fractal dimensions. In the first case, only boxes overlapping with the boundary of the mycelial particle were counted, leading to the box surface dimension, while the second analysis consisted of box counting applied to the mass of the mycelium, resulting in the box mass dimension [61, 62]. Consequently, characteristic ranges of the subsequently calculated dimension values were assigned to the three typical morphological structures: disperse mycelia, clumps or pellets, as the applied fractal analysis proved to be a very sensitive method for the effective quantification of fungal cellular morphology. In this manner, changes in the morphological properties of *A. niger* cultures due to a variation of the inoculum concentration as well as different compositions of the cultivation broth were successfully distinguished [62]. Another successful and innovative image analysis attempt was recently introduced by Tesche and Krull [8], where a pixel intensity thresholding method was applied in order to quantify the heterogenous morphology of the filamentous bacterial strain *Actinomadura namibiensis*. Thereby, the developed preprocessing steps of the recorded images overcame common challenges of light microscopy such as varying image brightness and contrast as well as various lens aberration effects. In this manner, the accuracy of the subsequent image analysis' results was significantly improved [8].

In conclusion, an immense amount of research with the aim of characterizing the filamentous cellular morphology in the most exact and accurate way possible has so far been conducted. Nonetheless, the complexity of these structures and their specificity linked to each individual fungal and bacterial strain cause the development of new image processing and analysis techniques to remain highly relevant and necessary for further understanding of filamentous growth.

11.2.3 Product formation

As described beforehand, the cellular morphology of filamentous microorganisms can have a significant effect on achievable product titers. As for now there are no universal and strain-independent principles for the relationship between cellular morphology and productivity. The optimal morphology rather depends not only on the used strain but also on the chosen cultivation procedure [27]. In fact, there is a large amount of process parameters which affect cellular morphology, physiology and productivity of filamentous microorganisms [14]. Hence, to understand the interrelationships between them, one must consider all these parameters collectively under non-limiting oxygen substrate conditions. On the one hand, the supply with oxygen and carbon substrates is better in dispersed mycelium and clumps than in pellets due to their looser structure [66, 67]. On the other hand, growth as dispersed mycelium and hairy clumps can significantly increase the viscosity of the culture media and can therefore result in mixing problems [68–70]. In contrast, cultivation broths containing pellets are significantly less viscous and exhibit Newtonian flow behavior [14, 51, 52]. The downside of pellet growth is that diffusion limitations can occur internally once the pellet size exceeds a critical diameter [10] resulting in oxygen and carbon substrate scarcity in the pellet core [54, 71, 72] (compare Figure 11.2.3).

For the reasons described above, the cellular morphology of small, loosely structured pellets might seem as the most favorable for optimal productivity, but this does not apply to all cases. Table 11.2.1 shows several examples of filamentous microorganisms, their products and the assumed optimal cellular morphology for improved productivity. Also depicted is the product formation type for the respective product based on the information given in the associated study. According to Gaden [73], the product formation in microorganisms can be roughly classified into three groups: I) growth-associated product formation, II) mixed product formation and III) growth-decoupled product formation. For type I, the product is formed directly as a result of primary energy metabolism, e.g., the production of ethanol or enzymes in general. Type II product formation refers to products which arise from side reactions of primary metabolism at the end of as well as after the growth phase. An example is the formation of amino acids. In contrast, Type III products are formed completely independently of primary energy metabolism and thus arise after biomass accumulation took place. Antibiotic synthesis is a well-known example for this type [37, 73]. The product formation type might be an important process variable to consider for the optimization of bioprocesses, especially with regard to filamentous microorganisms. In case of growth-associated product formation, growth-supporting cellular morphologies also lead to higher productivity [52]. But strategies to manipulate the morphology towards optimal biomass growth do not always optimize product yields in equal measure [7, 74]. However, achieving a high biomass concentration is likewise desirable to eventually attain more producing cells.

Table 11.2.1: Overview of several filamentous microorganisms that have been investigated in regard to product formation type according to Gaden [73]: (I) growth-associated product formation, II) mixed product formation, and III) growth-decoupled product formation). Also listed are the applied morphology engineering technique as well as the achieved beneficial macro-morphology.

Organism	Product	Formation type	Cultivation system(s)	Method of morphology engineering	Achieved beneficial macro-morphology	Reference
Actinomadura namibiensis	Labyrinthopeptin	III	Shaking flask	Salt concentration	Smaller pellets	[7]
Aspergillus niger	Fructofuranosidase	I	STR (fed-batch)	Macroparticles	Dispersed mycelium	[74]
Aspergillus niger	Fructofuranosidase	I	Shaking flask	Salt concentration	Pellets to dispersed mycelium	[57]
Aspergillus niger	Glucoamylase	I	STR (batch)	Aeration rate	Smaller pellets	[78]
Aspergillus niger	Glucoamylase	I	Shaking flask	Salt concentration	Pellets to dispersed mycelium	[57]
Aspergillus niger	Glucose oxidase	I	STR (batch)	Agitation speed	Dispersed mycelium	[87]
Aspergillus niger	Glucosamine	I	STR (batch)	Inoculum concentration	Pellets	[79]
Aspergillus sojae	Polygalacturonase	I	STR (batch)	pH, agitation speed	Smaller pellets	[52]
Aspergillus terreus	Lovastatin	I	Shaking flask	Inoculum concentration	Smaller pellets	[34]
Aspergillus terreus	Lovastatin	I	STR (batch)	Agitation speed, aeration gas composition	Larger pellets	[77]
Caldariomyces fumago	Chloroperoxidase	I	STR (fed-batch)	Macroparticles	Smaller pellets, single hyphae	[88]
Lentzea aerocolonigenes	Rebeccamycin	III	Shaking flask	Micro- and macroparticles	Smaller pellets	[29, 89]
Mortierella isabellina	Lipids	I	Shaking flask	Macroparticles	Dispersed mycelium	[90]
Penicillium chrysogenum	Penicillin	III	STR (fed-batch)	Agitation speed, impeller type	Dispersed mycelium	[91]
Rhizopus arrhizus	Lactic acid	I	Bubble column reactor	Medium composition, pH, aeration rate, sparger design	Smaller pellets	[80]
Rhizopus oryzae	Fumaric acid, L-malic acid	I	Shaking flask	Surfactants	Smaller pellets	[82]

Table 11.2.1: (continued)

Organism	Product	Formation type	Cultivation system(s)	Method of morphology engineering	Achieved beneficial macro-morphology	Reference
Rhizopus oryzae	Fumaric acid	I	Shaking flask	pH, temperature, agitation speed, inoculum concentration	Smaller pellets	[81]
Rhizopus chinensis	Lipase	I	Shaking flask, STR (batch)	Inoculum concentration, agitation speed	Aggregated mycelium, pellets	[83]
Streptomyces albus	Pamamycin	III	Shaking flask	Macroparticles	Smaller pellets	[75]
Streptomyces coelicolor	Actinorhodin	III	Shaking flask	Nanoparticles	Smaller pellets	[92]
Streptomyces flocculus	Lavendamycin methyl ester	II	Shaking flask, STR (batch)	Agitation speed	Dispersed mycelium	[85]
Streptomyces hygroscopicus	Geldanamycin	III	Shaking flask	Macroparticles, inoculum concentration, surfactants	Smaller pellets	[33]
Streptomyces sp. M-Z18	ε-Poly-L-lysine	II	Shaking flask, STR (batch, fed-batch)	Macroparticles	Smaller pellets	[93]

A state-of-the-art method to modify and tailor the cellular morphology to increase product formation in a given biotechnological process are morphology engineering techniques, e.g., the addition of micro- or macroparticles (see chapters 4.1 and 4.2). Hereby, Driouch et al. [74] were able to cause a morphologic shift in cultivations of *A. niger* from pellets to dispersed mycelium, which increased the productivity of fructofuranosidase eminently. Green fluorescent protein (GFP) analyses revealed that the product fructofuranosidase was produced over the whole mycelium, whereas production only occurred in a thin outer layer of the pellets in the pellet-based process, thus exhibiting the challenges of pellet growth [74]. Kuhl et al. investigated the effect of microparticles on the cultivation of *Streptomyces albus*, which produces the antibiotic pamamycin [75]. A drastic reduction of the pellet size as well as a loosened inner structure was observed, which were associated with the up to sixfold improved final product concentration. By transcriptome analysis, it was also shown that the morphological changes due to microparticles greatly influenced the gene expression. Thus, morphology is not only indirectly affecting productivity, e.g., by limiting mass transfer (see above), but likely also in a direct manner by triggering the product formation at a genetic level [75, 76]. The effect of micro- as well as macroparticles on cultivations of filamentous microorganisms is discussed further in chapter 4.

In the study of Bizukojc and Ledakowicz of *Aspergillus terreus* cultivations in shaking flasks, (+)-geodin formation seemed to be independent of the present cellular morphology, whereas lovastatin formation correlated inversely with pellet sizes [34]. In contrast, Casas López et al. found that lovastatin formation was highest in large but less dense pellets in cultivations in a stirred tank reactor (STR) [77]. Despite the fact that oxygen transfer was significantly better in the oxygen-enriched and highly agitated cultures, which resulted in small but dense pellets, lovastatin titers remained low [77]. It is reasonable to assume this would not have been the case with less compact pellets [75, 78]. Thus, pellet size alone is not sufficient to predict productivity of filamentous microorganisms. Therefore, Bizukojc and Ledakowicz also investigated the differentiation of hyphae. It was shown that the presence of active, growing zones favored lovastatin formation and adversely affected (+)-geodin formation, interrelating with the respective product formation type [34].

Besides the actual physiology of cells, the additional investigation of the compactness of pellets seems helpful to improve the prediction of final product concentration [78]. In that regard, Casas López et al. developed a ratio of pellet filament to core zones to explain the differences in product generation [77]. Beyond that, other process parameters like the specific pellet density, the hyphal or pellet fraction, the hyphal gradient [53, 54] or hyphal network spacing [7, 8] were derived to characterize the structure of pellets. In most of the presented studies (see Table 11.2.1), a looser mycelium structure was beneficial for achieving high product yields. Although this is especially true for growth as dispersed mycelium in multiple cases, it was the pellet morphology that leads to higher product generation [7, 79–82]. This might be due to the macroscopic mass transfer limitations caused by viscosity-increasing dispersed mycelium outweighing the microscopic mass transfer limitations of pellet morphology. For this, apart from the characterization of cellular morphology, parallel investigation of rheology of the culture broth seems to be a useful tool to optimize bioprocesses with filamentous microorganisms as implied in multiple studies [52, 70, 77, 83–85].

An issue regarding studies conducted on this topic is the often differing terminology to describe filamentous morphology. Nevertheless, this becomes less of an issue with the rising use of shape descriptors as well as more complex parameters (see chapter 2.2), which provide profound data for characterization [8, 86].

11.3 Methods for structure elucidation

– Chapter 3 discusses the currently most frequently used experimental techniques for morphological structure elucidation of filamentous microorganisms
   – image analysis,
   – particle size analysis by laser light diffraction,
   – pellet slicing and confocal laser scanning microscopy,
   – flow cytometry,
   – X-ray microtomography,
   – pellet sedimentation,
   – pellet mechanics and
   – oxygen microelectrode technique.

As the cellular morphology and product formation of filamentous microorganisms are interdependent, the development of methods which enable the investigation of morphological structures has become an indispensable step for an efficient bioprocess design [14, 27, 29]. Numerous studies show that the optimal morphological form in regard to highest product yields in submerged cultures strongly varies within different fungal and bacterial filamentous strains [29, 32]. Furthermore, depending on the dominating morphological structure, cultures show different properties concerning the physical traits of the cultivation broth, having either a negative or positive impact on mass and energy transfer rates within the liquid phase [14, 37]. However, at this point, attention will be drawn to pellet structure elucidating methods which allow drawing conclusions mostly about the transport processes between the bulk phase and the biomass, as well as the subsequent mass transfer within the hyphal network itself.

11.3.1 Image analysis

A well-established method to analyze the morphological structure of filamentous microorganisms is the analysis of images, most commonly acquired via microscopy. The description of the micro- and macroscopic morphology of filamentous microorganisms originates from systematic investigations of the ascomycete fungus *Neurospora* by Emerson [55], which resulted in development of the cube root law characterizing filamentous pellet growth (see chapter 2.2) [27]. While being one of the oldest methods available to characterize cellular morphology, the analysis of images provides precise information about the micro- as well as macro-morphology of filamentous microorganisms [27]. Furthermore, it enables the application of advanced imaging techniques such as scanning electron microscopy [87] or CLSM (compare chapter 3.4) [88]. However, an advantage of this method is the abundance of imaging devices that can be used, starting with simple and inexpensive imaging devices like flatbed scanners [89, 90]. For the usage of microscopes, they are usually combined with digital cameras to digitalize the images. With the development of high performance imaging and computer hardware it has become possible to process immense amounts of image information [86, 91]. Ongoing advancements like automated robots for loading and preparing samples allow the execution of high-throughput studies [92].

Once an image of a filamentous microorganism is captured, it typically undergoes several steps of processing algorithms to calculate useful parameters regarding apical growth, branching of mycelium or pellet morphology, for instance [16]. In order to analyze the objects of interest they must be distinguished from the rest of the image. The general image processing procedure usually starts with the extraction of one color channel of interest, most commonly resulting in a grayscale image. The next step is the binarization of the image leaving only black and white pixels. A simple binarization method is thresholding, which converts a pixel to black or white based on the chosen threshold value. After binarization a region detection algorithm is used to identify individual objects, which then can be further analyzed to calculate various parameters [86]. Common morphological descriptors are the projected area, Feret diameter, solidity or aspect ratio. These shape descriptors were used by Wucherpfennig et al. [57] to develop the non-dimensional *MN*, which enabled the distinct morphological characterization of *A. niger* [57]. But there are also less conventional fractal parameters like lacunarity which describes the degree of structural variance within an object, therefore being suitable for characterization of clump morphology [84]. Commonly used morphological descriptors are further explained in chapter 2.2.

One of the most popular tools for image analysis in life sciences is *ImageJ* (also known as *NIH Image* or *Fiji*). This is due to it being a freeware as well as being used for the longest period of time [93–95]. Another widespread open source image analysis tool is *CellProfiler* [96]. Many image analysis problems can be solved with both platforms. Nevertheless, ImageJ's strength lies in single-image processing while *CellProfiler* is better equipped for creating large-scale processing workflows [97]. For special purposes, universal programming languages like Java, Python, C or MATLAB are used directly [86, 98].

A major challenge of image analysis can be the preparation of samples. Capturing a representative spectrum of all available morphologies in the culture broth can be complicated, especially in case of heterogeneous suspensions with a broad range of different morphologies [8, 99]. Also, certain selectivity can hardly be prevented if samples are prepared manually, and thus a large amount of images might be necessary which can make this method very laborious. Nevertheless, image analysis is still the method of choice for the investigation of cellular morphology of filamentous microorganisms, as can be seen by the countless number of studies in which it has been used [7, 8, 54, 59, 69, 74, 79, 84, 100]. Furthermore, according to a study of the bibliographic records related to image analysis in biochemical engineering on *Google Scholar* and *Science Direct* the use of image analysis is still on the rise [86].

11.3.2 Particle size analysis by laser light diffraction

Particle size analysis in general is the measurement of size distributions of individual particles in a suspension [101]. A robust and reliable method to analyze particle sizes is the laser light diffraction (LLD) technique. The major advantage of LLD is that a vast amount of particles are measured in a short amount of time giving it high statistical certainty [16]. While this method is used predominantly for non-living particles, e.g., dry powders and liquid dispersions in pharmaceutical industry [102], its applicability for evaluating the complex cellular morphology of filamentous microorganisms has already been proven [57, 78, 99, 103, 104].

The working principle of LLD is based on the assumption, that the overall scattered light pattern captured at the detector is formed by interference of light patterns produced by each sampled particle [105]. An LLD instrument consists of a light source, an optical system to process the light beam, a sample cell and a detection system [102, 105]. The incoming light interacts with particles in the sample cell and generates a light pattern of varying intensities, which is captured by the position-sensitive detectors at the end of the instrument. The light beam interacts with a particle in multiple different ways, including basic scattering, diffraction and refraction. Multiple scattering, which occurs when already scattered light interacts again with another particle, has to remain low. Therefore, the particle concentration may not exceed a certain limit for the scattering pattern to be analyzable [105]. By mathematical deconvolution and inversion, information about the particle sizes can be derived from the resulting light pattern from which the particle size distribution can be calculated. For the interpretation of the light pattern there are mainly two different theories applied: The Fraunhofer and the Mie theory. Using the Fraunhofer model, particle sizes are calculated solely based on the spatial separation between the maxima and minima in the diffraction pattern of varying light intensities. Hence, there is no information about the refractive index of the particle needed. On the other hand, for the Mie model to be used, knowledge of complex refractive indices is required which can be difficult to determine. Nevertheless, in most modern instruments the Mie theory is employed, not least because it is applicable for a wider range of particle sizes [102, 105].

Because of its rapid and robust generation of results, LLD was first introduced by Petersen et al. [103] for the quantification of fungal morphology and the prediction of rheological characteristics of filamentous culture broths from their particle size distribution. Thus, manual sample preparation was eliminated [103]. Also, the LLD technique was successfully used in a study of Lin et al. [78] to investigate the effect of volumetric power input by aeration and agitation on the pellet size of *A. niger*. A higher fraction of power input by aeration resulted in smaller pellets with an irregular, fragmented surface [78]. In another study by Wucherpfennig et al. [57], spore germination of *A. niger* has been investigated by permanently pumping culture broth from the bioreactor to the LLD device. The duration of the lag phase and germination was determined successfully based on the sudden rise of the mean pellet diameter of the size distribution [57]. Additionally, Eslahpazir Esfandabadi et al. [104] investigated pellet fragmentation of *A. niger* by the use of LLD in a STR at different values of volumetric power input. A decline in pellet size could be observed with rising volumetric power input [104]. Rønnest et al. compared LLD with image analysis using the example of *Streptomyces coelicolor* [99]. The morphology of samples taken over the course of the whole cultivation was analyzed via both methods. By the overall similarity of both measurement results, the validity of the LLD technique could be confirmed. Although standard deviations with image analysis were lower, this might have been mainly due to the calculation procedure to convert the results to a volume-weighted distribution paired with the low number of measured particles [99]. This once more emphasizes the major advantage of LLD being the uncomplicated measurement of large amounts of particles. A disadvantage of LLD exhibits when it comes to the measurement of less spherical morphologies or particles in general. This has been shown by Kelly and Kazanjian, who analyzed monosized fiber-analog shape standards using LLD [106]. Elongated particles resulted in measurement of bimodal particle size distributions, which approximately extended from their shortest to their longest dimension resulting in higher variances [106]. Hence, LLD might be best suited for the analysis of pellets and clumps of spherical appearance.

11.3.3 Pellet slicing and confocal laser scanning microscopy

As pellets represent a hyphal network closely entangled and agglomerated to shape of nearly spherical particles, the elucidation of their inner morphological structure is of equal importance as the analysis of general macro-morphological shape and size parameters (see chapter 2.2). Moreover, the highly complex entanglement of hyphae in the pellet core requires a minimally invasive procedure of exposing the inner structure followed by high resolution optical imaging.

The pellet slicing technique as a preparation method for further microscopic imaging has made this kind of investigations possible. At first, single pellets are placed in a frozen section medium and then rapidly frozen in a cryostat microtome. Thereafter, thin slices are generated by cutting the frozen sample with a sharp blade installed in the cryochamber [78, 107]. The slices can subsequently be transferred onto a microscope slide for further investigations. In case of filamentous microorganisms, the morphological structure inside pellets can be very complex due to highly branched and closely entangled hyphae. By reason of high biomass density and thin hyphae, the resolution of single filaments within pellet slices can be very challenging while using common light microscopy. Moreover, studies show that the pellet structure can significantly vary along the radial coordinate in all spatial directions, making scans through multiple layers of one pellet slice particularly relevant for a higher reliability of the recorded data [53, 108].

CLSM is characterized by high-resolution optical imaging of desired sample layers and advanced image processing, making it a suitable method for the three-dimensional visualization of complex biological structures like hyphal networks [109–111]. Depending on the research target, there are multiple possible methods of fluorescence staining and the subsequent image analysis. In the context of filamentous microorganisms, Driouch et al. [112] carried out investigations of the microparticle addition to *A. niger* submerged cultures by cultivating a genetically modified strain of the fungus. The generated mutant [112] co-expresses the enzyme glucoamylase with the green GFP under the control of an inducible promoter. Hence, the fluorescence intensity within a pellet slice can be correlated with the activity of the biomass in regard to the product formation. Moreover, viability-staining techniques can be applied in order to label the active viable and dead regions within a pellet slice, as done by Schrinner [107], while investigating particle addition to submerged cultures of the actinomycete *Lentzea aerocolonigenes*. Veiter and Herwig used the same technique to visualize viable pellet layers as well as the inactive core of the filamentous fungus *Penicillum chrysogenum* [113]. As a more general approach for structure elucidation, hyphae can be stained with suitable fluorescent dyes in order to visualize three-dimensional pellet structures [29, 53]. In an effort to quantify hyphal density along pellet radius, Hille et al. applied a negative staining technique, whereby *A. niger* pellet slices were successfully stained with fluorescein isothiocyanate dextran [53]. Because of the inert polymer conjugated to the fluorescent molecule, it diffuses through the pores between the branched hyphae, but does not interact with the biomass: consequently, fluorescence intensity can be correlated with the porosity of the hyphal network [53].

In conclusion, the technique of pellet slicing is of crucial importance for the detailed elucidation of complex cellular morphology on the surface and eminently inside filamentous pellets. The high-resolution optical imaging accomplished by CLSM opens up numerous possibilities for subsequent investigations, not only by advanced visualizing of the morphological structure itself, but also by applying different fluorescent staining approaches in order to examine the metabolic activity, porosity and product formation within pellets.

11.3.4 Flow cytometry

Using rather conventional methods for structural pellet investigation, such as viability checks, researchers struggle to have the possibility of analyzing pellets with statistical reliability. It is, for example, in the nature of the method, that the slicing, staining and scanning of pellets with the help of CLSM can only provide a limited sample size, covering only one or a few pellets per sample, as seen in a study of Schrinner et al. [114]. Besides, the enormous workload goes hand in hand with the disadvantages of minor intervention into the pellet structure, as at-line methods, like fluorescent staining or physical techniques, do. Consequentially, flow cytometry (FC) fills the gap to unite advantages of several analysis approaches.

FC is a broadly established method, e.g., for size classification of biomass and to investigate cell viability by fluorescent staining [41, 113, 114]. As an automated high throughput method, it provides rapid analysis of chemical and physical characteristics down to single cells within a population of microorganisms. Individual cells flow in a stream in front of one or multiple light sources (mostly lasers), and the generated optical signals, first and foremost fluorescence and scattered lights, are subsequently detected and recorded [115, 116]. Initially introduced for the biological research in yeast cells, the interest for filamentous research arose recently [113, 115].

Ehgartner et al. [117] developed a screening method for rapid activity determination of spore activity in seed cultures of *P. chrysogenum* in complex media based on FC technology. By combining FC and viability assays, it was possible to distinguish between metabolically active, dormant and dead sub-populations [117]. In further work, Ehgartner et al. compared the fast FC method with the time- and personnel-intensive microscopy-based image analysis and evaluated it based on *Penicillum crysogenum* cultivations using different morphological parameters [118]. Among other things, the shape descriptors of pellet roughness and compactness were determined as well as size orders of different morphological classes (hyphae, small and large agglomerate clumps as well as pellets) were defined [118].

Discussing the potential of FC for filamentous microorganisms, Bleichrodt and Read elucidated the applicability to conidia of the fungus *Aspergillus fumigatus* [115]. Moreover, Veiter and Herwig presented FC as a quantitative method to combine morphological analysis and viability assessment of filamentous microorganisms [113], which was so far paid little attention to. In comparison to conventional methods, FC poses the potential to either be performed at-line or on-line. Additionally, as a high throughput method for various filamentous morphologies, it guarantees to obtain a statistically reliable result set, down to individual particles [113]. Apart from that, information about cellular morphology and cellular physiology are provided via light scattering [114]. The method also uses fluorescent staining to detect the overall viability, the viability of morphological classes and the viable layer of single elements, as well as information about pellet or element compactness [113]. Thus, individual pellets can be analyzed via spatially resolved signal profiles, representing approximations of the pellet's cross section [113, 114]. Moreover, Schrinner et al. confirmed good accordance of their FC results with conventional methods on the basis of cultivating the actinomycete *L. aerocolonigenes* [114]. In this context it was shown that the addition of glass beads (see chapter 4.2) resulted in decreased pellet size on the one hand and additionally in increased overall pellet viability on the other hand leading to enhanced overall metabolic activity. This study was the first to verify the increased viability by a statistically sound analysis of at least 50 pellets per sample, whereas slicing and optical analysis only provides results based on the investigation of one pellet [114]. As a result, FC is a powerful technique to elevate morphological analysis of filamentous microorganisms onto a level, which can be considered as high throughput.

11.3.5 X-ray microtomography

As the demand for non-destructive pellet analysis is unmet, several modeling approaches have been developed in order to help predicting filamentous pellet structures. For a long time, such model approaches could not be validated due to the lack of knowledge of three-dimensional (3D) pellet morphologies. However, an important key tool for the investigation of 3D morphology was the establishment of X-ray microcomputed tomography (μCT) for filamentous fungi [41, 114].

The X-ray μCT technique was established by Schmideder et al. [41, 119] and involves freeze-dried pellets obtained from submerged cultures resulting in 3D images taken by an X-ray μCT system. These images are subsequently processed by binarization and skeletonization of the hyphae and post-processed to analyze the hyphal material regarding tips and branches. Eventually, information about morphological properties like pellet diameter, total hyphal length, numbers of tips and branches, *HGU*, *HBU*, porosity and average hyphal diameter can be provided. Thus, this technique offers a lot of information that is mandatory for morphological characterization found in detail in chapter 2. Furthermore, the authors suggest that the now accessible information about hyphal fractions can give insights into symmetry of pellets. Subsequently, information about pellet evolution can be generated, which again is of interest in terms of rationally engineered aggregation events via process control or genetic modification to target a better cellular productivity [41, 119].

Equipped with such a powerful tool, Schmideder et al. were able to deliver detailed insights into the pellet structures of *A. niger* and *P. chrysogenum* [41]. Within their study, they confirmed that X-ray μCT results are in accordance with earlier general pellet investigations and furthermore, that it is advantageous to overcome the former rather superficial possibilities to gain structural pellet information [41]. Also, X-ray μCT and the subsequent mass transport computations are widely used methods to determine effective diffusivity of porous or fibrous materials in material sciences, but holistic approaches concerning the diffusivity of filamentous pellets were rather scarce, e.g., those of oxygen or other growth limiting carbon sources [119]. Thus, with the generated 3D images of fungal pellets, the authors were able to compute the effective diffusivity of *A. niger* within single pellets based on the previous micro-structural characterization by X-ray μCT and image processing. Additionally, they demonstrated the correlation between hyphal fraction and effective diffusion factor as well as tortuosity inside pellets [119].

However, the X-ray μCT technique seems to have some flaws. Apart from the high costs for measurements, there is an elaborate workflow for data processing and profound computational knowledge necessary. Furthermore, typical inaccuracies of image analysis remain unsolved so far, the voxel resolution is still low [41]. Moreover, the image processing is limited to a minimal hyphal diameter of 3 μm, which excludes most filamentous bacteria (e.g., actinomycetes) [114]. However, in the course of time progress is made steadily as the X-ray μCT technique is still in its infancy. Thus, the same authors expanded their studies on various filamentous fungi and based on the X-ray μCT-investigated pellets, they postulated a universal law for diffusivity in pellets [120].

11.3.6 Pellet sedimentation

Sedimentation is defined as the process of particles in suspension settling due to gravity force or centrifugal acceleration. According to the Stokes' law, the sedimentation velocity v_s depends on the density ρ_p and radius r of the settleable round particles, as well as the dynamic viscosity η and density of the fluid ρ_l they are suspended in:

$$v_s = \frac{2}{9} \frac{r^2 g (\rho_p - \rho_l)}{\eta} \tag{11.5.1}$$

On account of this correlation, particles can be classified by their size based on different sedimentation velocities, on the condition of all the other parameters remaining constant. In the context of filamentous microorganisms, the investigation of settling velocities arouses interest as a method to further characterize size and density of the nearly spherical pellets in submerged cultivations. The initial examinations of pellet sedimentation velocities were performed with *A. niger* pellets suspended in a sedimentation column, whereas different outer and inner pellet structures were achieved through variation of the volumetric power input by aeration and agitation [78]. The resulting velocities were not only associated with the corresponding morphological structures captured by images of pellet slices (see chapter 3.4), but also correlated with theoretically calculated sedimentation velocities. This approach is crucial in order to evaluate the shape as well as the compactness and roughness of the outer pellet structure, since the theoretically calculated values correspond to ideally smooth spherical particles of the same size and density. Compact and dense outer structure was found to be associated with impediment of mass transport to inner pellet regions on account of the larger diffusion barrier for substrates [53, 54, 78]. In this manner, prognosis concerning the optimal outer pellet structures in regard to substrate supply and productivity can be established upon the investigation of pellet settling velocities coupled with image analysis of the corresponding morphological structures.

11.3.7 Pellet mechanics

Filamentous microorganisms are particularly sensitive to shear stress, as it has direct impact on their complex cellular morphology, which in turn influences the substrate uptake rate as well as the productivity of the biomass. In submerged cultures, mechanical stress can originate from agitating elements or aeration units, depending on the reactor type and its geometry. Hence, numerous methods including different measurement strategies and numerical simulations have been developed and established in order to characterize the mechanical stress generated on different filamentous strains in various bioreactors [104, 121–126]. However, besides the shear forces generated by stirring or aeration units, the mycelial biomass experiences apparent normal forces, as pellets also dynamically collide with each other and with bioreactor walls. Consequently, investigations of the micromechanical behavior of single filamentous pellets are necessary, as such examinations allow to elucidate the plastic deformation of pellets due to defined compressive forces. Thus, they contribute to the comprehension of the precise impact of mechanical stress on the pellet structure [127].

The micromechanical behavior was for the first time successfully investigated by plate-plate compression experiments, whereas single pellets of the fungus *A. niger* and the actinomycete *A. namibiensis* were fixed by suction on a plate glued to a pipette tip and moved against a force sensor with a defined deformation rate [127]. The pellet diameter-dependent initial plate distance as well as the displacement of the pipette tip is subsequently used to calculate the compressive strain ε which can be correlated with the measured force resulting in typical force-strain curves. Further analysis of the curves during loading and unloading cycles allows the computation of the dissipated energy per unit volume. Moreover, the identified plastic deformations should be further discussed by comparing the micromechanical experiment results with the morphological pellet structures documented, e.g., via CLSM. In this manner, changes in morphological and mechanical properties caused by different bio-intensification strategies in filamentous microorganisms can be identified and characterized. In the overall context of bioprocess engineering, the preferred pellet structure lies between pellets stable to mechanical stress and simultaneously porous enough to allow unhindered oxygen supply of the entire hyphal network [127].

11.3.8 Oxygen microelectrode technique

An unhindered supply of nutrients and oxygen is crucial for an efficient growth and productivity of the biomass. Assuming a homogenous mixture of the bulk phase, the substrates are taken up through the outer surface of a pellet and are subsequently metabolized by the biomass [37]. The density of the hyphal network within outer pellet layers is therefore decisive for the diffusion rates of substrates to inner pellet regions. Mass transport limitations due to high hyphal density result in an inhomogeneous substrate distribution within pellets [16, 53]. After reaching a critical radius, pellets can be divided into two sections: the outer active substrate-consuming layer and an inner substrate-limited region (see Figure 11.2.3). As the inner core of a pellet remains inactive, cell lysis and loss of the inner pellet structure occur [16, 54, 64]. The decrease of the mean pellet size in filamentous cultures is therefore considered one of the essential effects of different bio-intensification strategies, as described in the following chapter 4. The application of microsensors to determine the local oxygen concentrations along the pellet radius has thus become an important tool for the characterization of oxygen diffusivity through the hyphal network as well as the metabolic activity of the biomass.

The oxygen sensors so far applied for micro oxygen profiling measurements are amperometric Clark-type electrodes with a tip size of only 10 μm. The working principle is based on oxygen diffusing through a membrane to the reducing cathode which in turn is polarized against an internal Ag/AgCl anode. Simultaneously, oxygen is being removed from the electrolyte by a guard cathode which contributes to minimalization of the zero current. The microsensor measures the oxygen partial pressure in the picoampere range and transfers the signal to a picoamperemeter [128]. Moreover, the microelectrode is mounted on a micromanipulator which allows highly delicate stepwise motion of the sensor along the vertical axis. In addition, the small size of the electrode tip is crucial for a minimally invasive penetration of a pellet, ensuring the preservation of the fragile and complex hyphal structure [53, 129].

The micro oxygen profiling measurements cannot be performed online and therefore the pellets need to be separated from the culture. Since the metabolic activity of microorganisms strongly depends on environmental conditions, it is of great importance to either perform the measurements as quickly as possible or provide for conditions which mimic the surroundings of the pellets in submerged culture. So far, two types of experimental setups have been developed and successfully used for micro oxygen profiling measurements in pellets of the filamentous fungi *A. niger* and *A. terreus*, respectively [108, 130]. Hille performed numerous measurements in pellets fixed on a gel-loading tip and placed in the middle of a flow cell filled with the diluted permanently aerated cultivation medium at optimum growth temperature and pH [108]. Moreover, stable and precise adjustment of the flow conditions around the pellet was ensured by a recirculation magnetic gear pump connected to the flow cell. With this experimental setup, Hille et al. [53] successfully investigated local oxygen concentrations in pellets showing different morphological structures and hyphal densities which were separated from *A. niger* cultivations at different times after inoculation. The series of experiments also included the investigation of the impact the Reynolds number has on the resulting oxygen profiles in the same pellet. Combined with digital image analysis (see chapter 3.1) and data acquired by CLSM (see chapter 3.4) the results showed that the morphological structure is strongly linked to the mass transport in fungal pellets, whereby the distribution of biomass density inside the pellet and consequently the shape of the corresponding oxygen profiles significantly vary over the cultivation time. Furthermore, the possibility to alter the Reynolds number during the profiling experiments opened up new insights into the high impact of hydrodynamic conditions on the mass transfer at the boundary layer between the pellet and the bulk phase as well as inside the pellet itself [53].

Gonciarz and Bizukojc performed oxygen profiling experiments in *A. terreus* pellets with the aim of characterizing the lovastatin biosynthesis with regard to the impact of oxygen supply [130]. Moreover, the effects of adding talc microparticles on the productivity of fungal cultures were investigated [130]. While performing the measurements, the pellets were placed on a petri dish with blue-stained agar flooded with the cultivation broth. The necessity to investigate the pellets immediately after removal out of shaking flasks over a very short period of time was clearly pointed out and critically discussed by the authors, as the measurements were performed under quiescent liquid-phase conditions and without active aeration of the cultivation broth. Nevertheless, clear differences in regard of the shape of oxygen profiles and effective diffusivities subject to the pellet size were successfully distinguished [130, 131].

In conclusion, the oxygen microelectrode technique has proven to be a highly sensitive and precise method to investigate the oxygen supply of hyphal networks in fungal pellets. Further analysis of the profiles as well as alternative application of the sensors, e.g., in combination with microkinetic cells, allows the calculation of effective diffusivities, mass fluxes and oxygen uptake kinetics [54, 130, 131]. Therefore, microelectrode techniques represent highly important tools for the elucidation of the precise impact of oxygen transfer along the pellet radius on the growth and productivity of pellet populations in filamentous cultures. So far, the results of such investigations imply a strong dependence of oxygen diffusivity on biomass density and thus morphological structure of the hyphal network, as well as the pellet size in general, as it is directly linked to the distance of the mass transfer along the pellet radius.

11.4 Bio-intensification strategies in filamentous microorganisms

- Chapter 4 presents developed strategies to increase product formation by altering and controlling morphology to intensify processes with filamentous microorganisms.
- *Morphology engineering* techniques to enhance product formation are defined more comprehensively than before.
- All additives that lead to a change in filamentous cellular morphology and to an increased product formation are subsumed here
  - microparticle supplementation,
  - macroparticle supplementation,
  - salt supplementation,
  - supplementation of surface-active agents and
  - the combination of multiple *morphology engineering* approaches.

Undeniably, a clear understanding of the detailed processes within the micro- as well as the macro- scale is of great importance to achieve satisfying cultivation performances, especially due to the direct interrelation between cellular morphology and optimal production [16]. It strongly depends on the filamentous microorganism of interest, whether the quantitative defined pellet-like or the dispersed mycelial appearance is preferable in submerged cultures. Thus, various strategies have been developed in order to implement specific targeted alterations of the morphology and to subsequently enhance the outcome of filamentous cultivations [132], especially regarding high product yields to meet industrial production requirements.

As extensively described by Krull et al. [16], rather classical approaches address the cultivation control on process level and include versatile aspects of process control. For example, the choice of spore concentration for inoculation, pH and pH shifts or the induction of mechanical stress due to agitation and aeration in STRs play key roles for cultivation performances [16]. Moreover, age, genetic factors, medium composition, mass transport, rheological properties, temperature and the geometry of reactors or stirrers are of substantial importance [27, 100]. However, latest strategies detach from fundamental process engineering, and aim for rather simple supplementation of certain substances which influence the cultivation system profoundly. In general, multiple approaches are known so far and are summarized under the term *morphology engineering* [29].

11.4.1 Microparticle supplementation

An upcoming strategy to influence the morphology and -in many cases- improve the production of new antibiotics and active pharmaceutical ingredients (API) is the addition of particles of various sizes, shapes and materials [29, 30], commonly known as microparticle-enhanced cultivation (MPEC) [133]. As MPEC was extensively discussed in several reviews [16, 27, 29, 30], only basic features and exemplary research achievements are presented in the following.

As such a terminology, MPEC saw the light of the day in a study by Kaup et al. [133], which was the first to present the addition of aluminum oxide and talc powder of varying concentrations to submerged cultivations of several filamentous microorganisms. Exemplarily, the investigation of the fungal strain *Caldariomyces fumago* showed the particle supplementation to induce positive effects on both, growth characteristics and the maximum specific productivity of the target enzyme chloroperoxidase [133]. From this moment on, MPEC became an integral part in the field of morphology engineering with the aim to enhance the cultivation outcome of numerous filamentous species. Thereby, it is of minor significance, whether mycelial pellets or dispersed mycelium is the desirable form of morphology, nor whether the cultivated strain is a filamentous fungi or bacterium, as MPEC can be customized for each application.

The beneficial effect of microparticles on filamentous fungi is widely shown by several authors (a detailed summary is presented in Ref. [30]). For instance, Driouch et al. extensively investigated the impact of MPEC on *A. niger* [74, 112, 132]. Talc powder, alumina and titanate particles were used in various screening approaches. It was shown that the particle supplementation led to distinct morphological forms – varying from dense pellets of different sizes to dispersed mycelium or short hyphal fragments, which led, inter alia, to manifold increased product titers and cellular activity [74, 112, 132]. Accordingly, Yatmaz et al. were able to enhance the production performance of *Aspergillus sojae* with regard to its severalfold higher β-mannanase activity by adding aluminum oxide and talc microparticles [134].

Filamentous bacteria, like actinomycetes, are gaining more and more attention to be employed for secondary metabolite production like API [135]. Consequently, MPEC was successfully applied to enhance and control the performances of actinomycetes in various cultivation systems. The combination of *L. aerocolonicenes* with talc of various surface properties was investigated in a study by Walisko et al. [100]. Furthermore, Kuhl et al. studied the MPEC-enhanced production of pamamycin by *S. albus* with talc microparticles [75]. In both studies, not only the product synthesis was affected and significantly enhanced, but also the cellular morphology, which is in accordance with findings in fungal MPEC experiments. Uniformly, the original pellet sizes were significantly reduced and microparticles were identified to be entangled in the hyphal pellet structure [75, 100].

Elucidating the exact mechanisms of action caused by microparticles is rather challenging. However, the whole picture is being put together piece by piece. Regarding cultures inoculated with spore suspensions, there is evidence that microparticles influence the morphology in early stages of the process by disruption of conidia aggregates and consequently may hinder spore-spore interaction and aggregation [112, 134] which are key factors for further filamentous growth (see chapter 2.2). Such results were especially obtained for coagulative spore behavior in the initial cultivation phase, while in non-coagulative species other mechanisms lead to morphological changes. Overall, the concentration of supplemented microparticles was identified to be crucial. For distinct species (spore coagulating, non-coagulating, hyphae agglomerating) a further increase in microparticle concentration beyond the optimal concentration for small pellets can even induce mycelial growth [112]. Microparticles could also be located in the pellet center, forming particle aggregates and resulting in core shell pellets, which was postulated to potentially serve as a solid scaffold for hyphal growth [132]. Moreover, microparticles were evenly embedded across entire pellets [75, 100, 132], leading to the prevailing assumption of structural changes within the pellet. Due to the particles, the former dense hyphal network may be loosened up, which would be beneficial for mass transfer. Thus, the supply of essential nutrients and oxygen would be enhanced and the substrates could reach mycelial regions further inside the pellets [75, 132]. This hypothesis, again, is supported by the evidence of overall enhanced viability in particle-supplemented pellets [112].

Nevertheless, the affection of spore interaction and structural changes due to particle incorporation do not appear to be the only factors to consider. It is imperative to investigate the intercorrelation between morphological, physical, and chemical properties within the cultivation process [30]. For instance, particle surface effects such as surface charge, adsorption of molecules, chemical and catalytic reactions, ion leaching or the potential impact of reactive oxygen species must be considered and further investigated [30], since Kuhl et al. for example, demonstrated a vast impact of the microparticles on the organism's metabolism thanks to transcriptome analysis [75].

In general, MPEC seems to be a comfortable and attractive method to modulate cellular morphology in filamentous cultivations as desired for enhanced product synthesis. However, as the optimal particle set up is highly strain specific, it has to be redefined for every cultivation system.

11.4.2 Macroparticle supplementation

Another approach of bio-intensification poses the addition of rather large sized macroparticles to cultivations of filamentous microorganisms. Pursuing the goal to add further mechanical stress, particles, such as glass beads with diameters of up to several millimeters, are added to the culture broth to modulate the bioprocess [33, 100, 136, 137]. Collisions between glass beads in the liquid phase and pellets or hyphae lead to their fragmentation and thus, an alteration of the cellular morphology by pellet breakup and dispersion is the consequence [33]. These collisions can be the result of either a collision between two macroparticles or of macroparticles colliding with a reactor wall [136]. Such effects on cellular morphology can be found in accordance with the effects of shear stress due to high power input in STRs by agitation and/or aeration. As a consequence, high shear stress conditions can cause smaller, smoother, and more compact pellets in comparison to larger and hairy pellets with low power input [33, 138].

Even though not the first choice within the modulation approaches, the supplementation of macroparticles was not just recently developed. Hotop et al. published a study dealing with the addition of glass beads (Ø 4 mm) to shake flask cultivations of the fungus *P. chrysogenum* influencing the cultivation outcome significantly [139]. Macroparticles also found their way into *A. niger* cultivations, in which millimeter sized inert glass particles increased the performance of the fungal leaching process including uranium extraction from ore [140].

Not only fungi, but also filamentous bacteria were studied in combination with macroparticles. *Streptomyces hygroscopicus* var. *geldanus* was investigated under glass bead supplementation in experiments conducted by Dobson et al. [33]. Varying bead concentrations led to higher pellet concentrations and decreased pellet sizes [33]. Furthermore, both large scales and cultivations in microtiter plate range were investigated by Sohoni et al. to avoid the formation of large pellets [141]. Herein, it was reported that macroparticles of various sizes and materials were uniformly capable to significantly reduce pellet size, enhance productivity and to keep the deviation among single experiments narrow [141]. Even for the strain NJES-13T from the newly established genus *Aptenodytes*, which originates from the gut of the Antarctic emperor penguin, a macroparticle screening of varying sizes showed that particles of 0.5 mm in diameter could increase production of the biologically active polyketide angucycline [142].

Taking into consideration the actinomycete *L. aerocolonigenes*, Walisko et al. were able to show that glass macroparticles indeed alter the amount and type of mechanical stress input into filamentous cultivations [100]. Moreover, the importance of the right mechanical stress level was indicated by a variation of the macroparticle size. While a small amount of mechanical stress input led to improved nutrient supply and thus, increased productivity of the antibiotic rebeccamycin due to less dense and more porous pellets, a very high intensity of mechanically induced stress, power input and stress frequency damaged the cells and lowered the productivity. Eventually, glass beads of 0.5 mm in diameter led to a 22-fold increase of the product titer compared to the unsupplemented control culture. However, to this point the exact mechanisms leading to the alteration still remain unclear [100]. Consequentially, a study of Schrader et al. with the same cultivation system showed a positive correlation between product concentration and addition of glass macroparticles of varying size [136]. It was further assumed, that this correlation was linked to the mechanical stress resulting from the presence of glass beads, whereby the induced stress energies increased with increasing bead diameter. This increasing stress energy loosens the pellet structure up to an optimal point of high porosity. Beyond that optimum, pellet sizes again decrease due to disintegration of the hyphal structures. Thus, within the study of Schrader et al. a deeper view into the occurring energy types, levels and origins etc. in macroparticle-supplemented shake flask cultivations was given and also the beneficial potential of concomitant CFD-DEM simulations was presented [136]. In addition to the variation in particle size (varying stress energy and stress frequency), Schrinner et al. showed several ways to modulate the cultivation productivity with macroparticle supplementation in *L. aerocolonigenes* [137]. By keeping the particle number constant while changing the size, the stress frequency remains constant, whereas the stress energy varies. Furthermore, the shaking frequency and the bead density are of particular interest, as they decisively influence the introduced stress energy and stress frequency [137].

All in all, an adequate well-defined level of macroparticle-induced stress can be beneficial for product synthesis in several filamentous microorganisms. In future studies, however, more light could be shed on macroparticle-supplementation in order to further contribute to a universal understanding of this quite promising approach of bio-intensification.

11.4.3 Salt supplementation

A further sophisticated approach for tailor-made morphology engineering is known under the term salt-enhanced cultivation (SEC) and includes the addition of inorganic salts instead of insoluble matter [7, 8, 57]. As Wucherpfennig et al. summarized [57], a number of researches dealing with the rather unspecific addition of various salt species to filamentous cultivations were published so far without an approach to systematically classify the new method. It became clear that salt addition poses the ability to influence cultivation performances significantly by altering the osmotic properties of the liquid phase. Even though osmolality effects were known for long in the field of mammalian cells, studies for filamentous applications were scarce up to this point [57].

Osmolarity or osmolality defines the quantity of osmotically active ions or particles serving as solute per liter or kilogram, respectively, in the overall solution [29]. As process parameters, both terms were identified to affect filamentous growth and productivity and represent a cheap and reliable method for morphology engineering. Hereby, common osmolalities in cultivation media can be found in the range between 0.28 and 0.32 osmol kg^{-1}. Since osmolality is a product of the media composition in a culture broth, changes during cultivation are always expected, e.g., by accumulation of metabolic products, pH control, acid or base regulation and nutrient consumption [29, 57]. So far, reported salts for SEC are particularly sodium chloride, potassium chloride and ammonium sulfate. Further sporadically tested salts with minor effects on the cultivation performance are sodium bicarbonate, sodium sulfate, calcium chloride, rubidium chloride, magnesium chloride and potassium sulfate in varying concentrations or osmolalities depending on the investigated salts and microorganisms [29]. Overall, salt supplementations of up to 2800 mM or 2550 mosmol kg^{-1}, respectively, can be found in various studies. Besides, the type of the supplemented ions is of equal importance. Polycations, for example, are beneficial to induce pellet-like growth, whereas polyanions were rather detrimental [29, 143].

As osmolality increases with an increase in salt concentrations in media, water fluxes are triggered along the osmotic gradient. The subsequent hydrostatic pressure has to be compensated as a response to changed environmental conditions by swelling or dehydration of the cells. Such mechanisms to maintain a positive cell turgor are a key factor for apical cell expansion and hyphal growth. Details for the subcellular mechanisms are summarized in Csonka and Hanson [144], Kempf and Bremer [143] and Böl et al. [29]. Acting as a modulating tool in cultivation processes, the cellular regulation by increasing osmolality has an impact on the cellular physiology and thus, on morphology and productivity due to the potential additional stress of the osmotic pressure [57].

Exemplarily, Wucherpfennig et al. showed that the enhancement of osmolality by sodium chloride or potassium chloride addition decreased the overall cell dry weight in studies with *A. niger* in bioreactor scale, whereas productivity of the enzyme fructo-furanosidase was remarkably increased 18-fold [57]. In case of the actinomycete

A. namibiensis, Tesche et al. used SEC to enhance production of the lantibiotic laby-rinthopeptin [7]. For this purpose, the addition of 50 mM ammonium sulfate turned to be the most effective supplement. Herein, the metabolization of the carbon source glycerol was significantly increased, the cellular morphology was shifted towards smaller, more circular and less frayed pellets which showed a lower level of disintegration during production phase [7].

Eventually, salt supplementation into filamentous cultivation systems seems to be a powerful tool to easily modulate the cultivation process significantly on varying scale and thus, to enhance productivity, as the morphological requirements for an individual species can be precisely adjusted. Moreover, in comparison to other morphology engineering strategies, vast disadvantages in downstream processing (e.g., elimination of solid particles) can be circumvented.

11.4.4 Supplementation of surface-active agents

Adding specific single molecules to impact biotechnological cultivation outcomes is known for a long time. For instance, polymers of various kind and properties (e.g., polyethylene glycol, agar, Carbopol or Junlon PW 110) are just a few examples among others [145]. A rather little common morphology engineering approach is the addition of surface-active agents (**surf**ace-**act**ive **agent**) [146]. In contrast to the salt and allegedly inert particle supplementation to filamentous cultivations, herewith a profound alteration of the medium properties in several aspects can be provoked already at first sight.

On the molecular level, surfactants are organic molecules consisting of a polar or ionic hydrophilic group and a nonpolar or hydrophilic chain. In aqueous solution they can either be present as single molecules (low concentrations) or can be organized as micelles (high concentrations) which markedly effects surface tension of the liquid phase. The latter is the key for their application as antifoam agents within bioprocesses to modulate emulsifying, foaming, dispersing and to serve as a detergent [33, 147]. In this context, the critical micelle concentration (CMC) is of high relevance for the broth properties, since concentrations above the CMC were reported to potentially have a toxic effect on microorganisms leading to their degradation due to the disruption of cell membranes by interacting with structural lipid components [147]. In general, a distinction between surfactants is of importance, as their application provokes various effects in the cultivation broth. For instance, non-ionic surfactants are associated with a good compatibility with microbial growth, while anionic surfactants tend to inhibit growth and cationic surfactants act antibiotically [147].

However, in case of compatible surfactants, an alteration of the surface tension poses the potential to influence aggregation processes of spores and vegetative biomass, which again may affect morphological characteristics [33]. The surface-active properties can lower the interfacial tension between hyphae and liquid medium, decreasing the potential of microorganism to form mycelial aggregates. Furthermore, the thermodynamic potential for such aggregations may decrease, attributed to a lower surface tension in the medium [148]. It was further suggested that surfactants with low hydrophilic–lipophilic balance (HLB) values, which are highly lipophilic, tend to reduce the exposure of hydrophobic spores to the liquid phase and with this, may disturb the spore aggregation which is based on hydrophobic interactions. This phenomenon results in decreased pellet sizes [149, 150].

Moreover, cell wall permeability of hyphae was further reported to potentially be increased by surfactants leading to a better migration of intracellular compounds into the media or vice versa [147]. Furthermore, in contrast to suggestions of Liu and Wu [148], Vecht-Lifshitz et al. [151] proposed that surfactants may adsorb onto cell walls, leading to increased interactions between the cells and thus, might evoke aggregations of mycelium due to more hydrophobic cell wall properties [148, 151]. In context of altering medium properties, significant changes of the $k_L a$ and thus oxygen supply within the culture broth are of great relevance [152]. Besides, certain non-ionic surfactants, e.g., Pluronic F-68, were reported to cushion fluid cell damages induced by fluid dynamic stress in STR [153, 154]. The reasons for the effectiveness of Pluronic F-68 in particular have not been fully elucidated yet and there are probably several possible explanations [152], while the coverage of bubble surfaces with the surfactant instead of cells will be the main reason. This mechanism also partially explains why the effectiveness of Pluronic F-68 significantly decreases with a decreasing Pluronic to cell concentration ratio.

Even though the exact mechanisms for the vast beneficial effects of surfactants are not yet clarified down to the smallest detail [147], several studies revealed the potential as a powerful tool in the field of morphology engineering. Hereby, the mechanisms of action are under ongoing investigation and it is perpetually elucidated how the addition of surfactants can be used for reliable enhancement of bioprocesses with filamentous microorganisms.

11.4.4.1 The impact of surfactants in filamentous cultivations

The diverse the chemical category of surfactants is, the multifaceted is the impact of each on filamentous microorganisms, as can be considered below in form of selected examples and found summarized in Table 11.4.1.

11.4.4.1.1 Antifoam
A study of Dobson et al. [33] investigating the actinomycete *S. hygroscopicus* var. *geldanus* indicates that the addition of silicone antifoam (AF) comes along with a significant increase (of up to 60%) in the product titer of geldanamycin as a result of -or in correlation with- altered morphological properties. Herein, an increase in the dispersion of pellets (36% increased pellet count) and a greater distribution of smaller pellets (50% decreased mean pellet size) could be observed [33].

Table 11.4.1: Overview of surfactant-supplemented investigations.

Surfactant	HLB [-]	Organism	Impact on cell morphology	Impact on productivity	Reference
Silicon antifoam		*Streptomyces hygroscopicus* var. *geldanus*	+	+	[33]
Tween 20	16.7	*Aspergillus oryzae* KB	+	+	[155]
		Cordyceps sinensis	+	–	[153]
		Rhizopus oryzae	o	o	[82]
Tween 40	15.6	*Aspergillus oryzae* KB	+	+	[155]
Tween 80	15	*Streptomyces hygroscopicus* var. *geldanus*	+	o/–	[33]
		Aspergillus oryzae KB	+	+	[155]
		Cordyceps sinensis	+	+	[153]
		Rhizopus oryzae	o	o	[82]
		Schizophyllum commune	+	+	[162]
		Botryosphaeria rhodina MAMB-05	o	+	[161]
Triton X-100	13.5	*Streptomyces hygroscopicus* var. *geldanus*	+	–	[33]
		Streptomyces tendae	+	*	[156]
		Rhizopus oryzae	o	o	[82]
Pluronic F68	29	*Cordyceps sinensis*	o	+/o	[153]
		Streptomyces tendae	+	*	[156]
Brij 58	15.7	*Streptomyces tendae*	+	*	[156]
Triethanolamin		*Rhizopus oryzae*	+	+	[82]
Diethanolamin		*Rhizopus oryzae*	+	–	[82]
Ethanolamine		*Rhizopus oryzae*	+	–	[82]
Lecithin	8 (soy)	*Penicillium chrysogenum*	+	*	[162]
		Anthracophylum discolor	+	+	[152]
		Lentzea aerocolonigenes	+	+	[142]

(+) positive effect; (o) no effect; (–) negative effect; (*) not investigated.

11.4.4.1.2 Tween

As a response to the addition of the non-ionic surfactant Tween in a study of Kurakake et al., the common mycelial appearance of *Aspergillus oryzae* resulted in a more pellet-like growth [150]. Correlating with the HLB value of certain Tween species (T20, T40, T60), the pellet sizes decreased with decreasing HLB. Thus, the authors suggested the main reason for such a reduction of the sizes to lie in the inhibition of the so far ungerminated spores by Tween addition [150]. Furthermore, they assumed that the alteration in morphology as well as the increased biomass formation and higher productivity in surfactant-supplemented cultivations are the result of enhanced supply of oxygen and nutritions in the culture broth. An increased membrane permeability may play a role for the increased mass transfer as well [150, 155]. Also, the addition of Tween 80 to *S. hygroscopicus* var. *geldanus* affected the bacteria's cellular morphology significantly. Increasing Tween amounts led to higher pellet concentrations and a decrease in pellet sizes, both accompanied by a correlating product concentration [33]. Fungal growth of *Cordyceps sinensis* was shifted from mycelial to pelleted growth as well as the product synthesis remarkably enhanced [148], which is in great accordance with several other studies on fungal exopolysaccharide production [148, 156, 157].

11.4.4.1.3 Triton X-100

Varying concentrations of the non-ionic surfactant Triton X-100 posed the possibility to influence pellet formation as well as the growth rate and cellular yield in cultures of *Streptomyces tendae* [151]. Furthermore, the cultivation outcome of *S. hygroscopicus* var. *geldanus* could be modified by its application during cultivation in dependence on the surfactant concentration. The mean pellet diameter increased, whereas the pellet concentration and product synthesis decreased, indicating that Triton-X-100 might exert a rather toxic than beneficial effect on this specific microorganism [33].

11.4.4.1.4 Pluronic F68

Using Pluronic F-68 in *S. tendae* led to a shift of the pellet size towards larger mean pellet diameter, correlating with the initial concentration of the surfactant and accompanied by an increase in the cell wall hydrophobicity. Once a certain concentration was exceeded, the addition even resulted in a total inhibition of pellet formation going hand in hand with a sudden recurrence of a hydrophilic cell wall character [151].

11.4.4.1.5 Brij 58

Supplementing *S. tendae* cultivations with various concentrations of Brij 58, conducted by Vecht-Lifshitz et al. [151], led to altered morphologies, which were shifted to an improved pellet formation in dependence on the amount of the added surfactant. Also, changes of the cell wall characteristics could be observed, similar to those measured during experiments with the addition of Pluronic F68 [151].

11.4.4.1.6 Triethanolamine

Wu et al. were able to show that amongst other surfactants (Triton X-100, Tween 20, Tween 80, diethanolamine, ethanolamine) the surfactant precursor triethanolamine (TEOA) has the highest impact on *Rhizopus oryzae* morphology leading to growth in pellet shape [82]. Additionally, significantly increased product formations up to a factor of 2.9 in comparison to an unsupplemented control culture could be observed. It was further shown that the origin of such effects lies in changes of the transcriptomic profile addressing carbohydrate-active enzymes regarding both, synthesis and restructuration of the cell wall. Moreover, a total number of 1094 genes were identified to being upregulated and 1841 genes faced a downregulation in comparison to the unsupplemented control cultivation [82].

11.4.4.2 Lecithin

An upcoming approach, which was paid little attention to so far, is the addition of lecithin, which acts as an emulsifier, lubricant, and surfactant due to its amphiphilic and surface-active character [158, 159]. Lecithin is a generic term for yellow-brown fatty substances containing mainly phosphoric acids, choline, fatty acids, glycerol, glycolipids, triglycerides and phospholipids. Thus, it is, inter alia, commonly used for human food, animal nutrition as well as pharmaceuticals [158, 160]. Unlike other surfactants or additives, such as Tween, Triton X-100 and polyethylene glycol, lecithin has the advantages of biodegradability, low toxicity and low cost [159], whereas on the other hand it is a quite robust agent [158]. Even though the exact mechanisms are not clarified so far, a few studies revealed its vast potential to enhance product synthesis in bioprocesses.

The supplementation of varying soya lecithin concentrations (0–10 g L^{-1}) provoked an increase in the absolute mycelial growth of the white rot fungus *Anthracophyllum discolor* [147]. However, even though Bustamante et al. summarized that soya lecithin was reported to have a beneficial effect on biotechnological enzyme production, they did observe an increased enzyme activity of a specific enzyme rather than of all considered enzymes [147]. Employing both, lecithin from a vegetable and animal source, on cultivations of *Streptomyces filipinensis* to produce the antibiotic filipin Brock et al. induced a 10–18-fold increase in the product yield [161]. Hofer et al. investigated the effect of lecithin supplementation on fungal growth of *P. chrysogenum* [162]. Hereby, it was shown that lecithin indeed enhanced biomass formation and also the product synthesis (during the cultivation process when supplemented as raw material or in form of soy bean oil). Significant effects were also observed in studies of Schrinner et al., in which the supplementation of lecithin caused an up to three-fold increase of the cell dry weight in cultivations of *L. aerocolonigenes* for the synthesis of the pharmaceutical drug rebeccamycin [137]. Regarding the product titer, almost seven times higher concentrations could be observed due to lecithin supplementation, as well as an enhanced average yield coefficient, indicating that the increased product titer is not simply the outcome of a higher biomass concentration [137]. The authors further proposed that the effect may be attributed to the antifoam properties of lecithin, such as an enhanced oxygen supply in the culture medium. Furthermore, it was suggested that decreased cell damage would be beneficial [162], but also that lecithin has the potential to be an additional nutrition source [137, 162].

11.4.5 Combination of multiple morphology engineering approaches

It should be taken into consideration that combining two or more established morphology engineering approaches can be even more beneficial than applying a single strategy. Schrinner et al., for instance, combined the sole addition of lecithin with the addition of macroparticles (see chapter 4.2) in a cultivation of *L. aerocolonigenes* [137]. Hereby, the product titer and the yield coefficient of the antibiotic rebeccamycin were almost eight-fold increased compared to the control cultivation, whereas lecithin or macroparticles alone did not result in such high titers [137].

In an experiment to investigate the fungal leaching of uranium ore, Li et al. added macroparticles of various sizes to *A. niger* cultivations containing 4 g L^{-1} uranium ore microparticles [140]. Apart from altered biomass concentrations, increased pellet sizes and a stimulation of metabolite production, the uranium extraction was significantly improved. Analysis of the bio-ore pellets' 3D-structure showed that a higher amount of ore microparticles was embedded in the hyphal pellet network. Adding stress energy by macroparticle supplementation was assumed to be the reason for the increased incorporation [140], leading to the positive effects outlined in chapter 4.1 and 4.2.

In conclusion, further investigations are necessary to discover and elucidate beneficial combinations of morphology engineering approaches. As a result, already favorable effects might be enhanced or intensified by combinatory mechanisms in a single culture.

11.5 Conclusions

Filamentous microorganisms are of major biotechnological importance. In the past, the technical challenges regarding their cultivation to increase product formation provided fundamental impulses for the development of new morphology techniques and paved the way for their process intensification.

This article provides a broad overview regarding the cellular micro- and macro-morphology of filamentous microorganisms and its effects on different product formation kinetics. To this end, a brief description of the currently most common experimental techniques for elucidating the morphological structure have been discussed and reviewed, including newly introduced techniques of flow cytometry and X-ray microtomography. Furthermore, the strategies developed to increase product formation by changing/controlling morphology and intensifying processes with filamentous microorganisms were presented. In this context, the presented morphology engineering techniques to enhance product formation were also extended to surfactants, certain additives that change the surface tension in aerobic cultivations. Promising future developments for the intensification of biotechnological processes with filamentous microorganisms lie in the combination of these already well-characterized morphology engineering techniques.

There are several challenges that need to be addressed in future studies:
- The great efforts in the morphological characterization of filamentous systems show a close relationship between cultivation conditions, cellular morphology and moreover, metabolic properties of individual cells. Closing the gap between metabolic processes and the engineering level is an important building block for future research. The underlying metabolic and regulatory mechanisms that lead to the

formation of a highly productive agglomerate form are far from providing a pervasive understanding. Experimental and *in silico* techniques provide powerful tools to gain a better understanding of the complex biological and technical challenges regarding filamentous systems.

– To date, there is no consensus on how to correlate cellular morphology and productivity of filamentous microorganisms in submerged cultivations with different product formation kinetics. Unfortunately, most published model correlations are strain-dependent and not transferable from one biological system to another. Here, at least overarching genus correlations or overarching product kinetic correlations are mandatory, so that steps towards strain independence can be made and the models become more generally applicable.

– Due to the complexity and high non-linearity of biological and physical process related parameters, simple mathematical models cannot describe the interrelationship between cellular morphology and productivity, not even for one defined scale. Dynamic models have to be developed which include morphological shape descriptors and biokinetics to estimate the evolution of substrate concentration and morphology as well as the product formation of filamentous microorganisms. These models can be used to generate a dynamical multi-objective control approach to characterize morphological shape and product yield.

– The reproducibility of cultivations with filamentous microorganisms still has much room for improvement, as cellular morphologies are often formed very heterogeneously. In order to identify a production strain as a suitable production candidate for industrial use, a heterogeneous filamentous system must be analyzed in its detailed subpopulations. The categorization of heterogeneous populations into mycelial and pellet fractions has to be considered. Therefore, especially mycelium/pellet fractions that contribute to product formation should be included in the modelling in a time-resolved manner. Therefore, the further development of advanced, powerful and automated online image analysis techniques and modelling tools is desirable for high quality prediction of culture productivity with low deviations.

– Last but not least, the key to understanding and rationally improving eukaryotic and prokaryotic filamentous systems as a common platform technology depends on overcoming the existing collaboration barriers in the fields of genetic engineering, molecular biotechnology and bioprocess engineering.

Acknowlegments: The authors acknowledge the financial support provided by the German Research Foundation (DFG) in the Priority Programme 1934 *DiSPBiotech – Dispersity, structural and phase modifications of proteins and biological agglomerates in biotechnological processes* (SPP 1934 DiSPBiotech, project number 315457657) and in the DFG project *Impact of strain-induced morphology changes on the productivity of filamentous pellet systems using Actinomadura namibiensis* (project number 463178687).

References

1. Meyer V, Andersen MR, Brakhage AA, Braus GH, Caddick MX, Cairns TC, et al. Current challenges of research on filamentous fungi in relation to human welfare and a sustainable bio-economy: a white paper. Fungal Biol Biotechnol 2016;3:6.
2. Papagianni M. Advances in citric acid fermentation by *Aspergillus niger*: biochemical aspects, membrane transport and modeling. Biotechnol Adv 2007;25:244–63.
3. Sauer M, Porro D, Mattanovich D, Branduardi P. Microbial production of organic acids: expanding the markets. Trends Biotechnol 2008;26:100–8.
4. Humphrey AE. Elmer L. Gaden, Jr., father of biochemical engineering. Biotechnol Bioeng 1991;37:995–7.
5. Ward M, Lin C, Victoria DC, Fox BP, Fox JA, Wong DL, et al. Characterization of humanized antibodies secreted by *Aspergillus niger*. Appl Environ Microbiol 2004;70:2567–76.
6. Lubertozzi D, Keasling JD. Developing *Aspergillus* as a host for heterologous expression. Biotechnol Adv 2009;27:53–75.
7. Tesche S, Rösemeier-Scheumann R, Lohr J, Hanke R, Büchs J, Krull R. Salt-enhanced cultivation as a morphology engineering tool for filamentous actinomycetes: increased production of labyrinthopeptin A1 in *Actinomadura namibiensis*. Eng Life Sci 2019;19:781–94.
8. Tesche S, Krull R. An image analysis method to quantify heterogeneous filamentous biomass based on pixel intensity values – interrelation of macro- and micro-morphology in *Actinomadura namibiensis*. Biochem Eng J 2021;166:107865.
9. Kieser T, Bibb MJ, Buttner MJ, Chater KF, Hopwood D. Practical *Streptomyces* genetics. Norwich: John Innes Foundation; 2000.
10. Nielsen J. Modelling the morphology of filamentous microorganisms. Trends Biotechnol 1996;14:438–43.
11. Metz B, Kossen NWF. The growth of molds in the form of pellets – a literature review. Biotechnol Bioeng 1977;19:781–99.
12. Papagianni M. Fungal morphology and metabolite production in submerged mycelial processes. Biotechnol Adv 2004;22:189–259.
13. Grimm LH, Kelly S, Krull R, Hempel DC. Morphology and productivity of filamentous fungi. Appl Microbiol Biotechnol 2005;69:375–84.
14. Wucherpfennig T, Kiep KA, Driouch H, Wittmann C, Krull R. Morphology and rheology in filamentous cultivations. In: Advances in Applied Microbiology, 1st ed. Elsevier; 2010, 72:89–136 pp.
15. Krull R, Cordes C, Horn H, Kampen I, Kwade A, Neu TR, et al. Morphology of filamentous fungi: linking cellular biology to process engineering using. In: Biosystems engineering II. Berlin, Heidelberg: Springer; 2010:1–21 pp.
16. Krull R, Wucherpfennig T, Esfandabadi ME, Walisko R, Melzer G, Hempel DC, et al. Characterization and control of fungal morphology for improved production performance in biotechnology. J Biotechnol 2013; 163:112–23.
17. Veiter L, Rajamanickam V, Herwig C. The filamentous fungal pellet-relationship between morphology and productivity. Appl Microbiol Biotechnol 2018;102:2997–3006.
18. Posch AE, Herwig C, Spadiut O. Science-based bioprocess design for filamentous fungi. Trends Biotechnol 2013;31:37–44.
19. Meyer V. Genetic engineering of filamentous fungi-progress, obstacles and future trends. Biotechnol Adv 2008;26:177–85.
20. Meyer V, Wu B, Ram AFJ. *Aspergillus* as a multi-purpose cell factory: current status and perspectives. Biotechnol Lett 2011;33:469–76.
21. Workman M, Andersen MR, Thykaer J. Integrated approaches for assessment of cellular performance in industrially relevant filamentous fungi. Ind Biotechnol 2013;9:337–44.

22. Krull R, Bley T. Filaments in bioprocesses. Cham, Heidelberg, New York, Dordrecht, London: Springer International Publishing; 2015.
23. Bizukojc M, Ledakowicz S. Bioprocess engineering aspects of the cultivation of a lovastatin producer *Aspergillus terreus*. Adv Biochem Eng/Biotechnol 2015;149:133–70.
24. Meyer V, Fiedler M, Nitsche B, King R. The cell factory *Aspergillus* enters the big data era: opportunities and challenges for optimising product formation. Adv Biochem Eng/Biotechnol 2015;149:91–132.
25. Quintanilla D, Hagemann T, Hansen K, Gernaey KV. Fungal morphology in industrial enzyme production-modelling and monitoring. Adv Biochem Eng/Biotechnol 2015;149:29–54.
26. Serrano-Carreón L, Galindo E, Rocha-Valadéz JA, Holguín-Salas A, Corkidi G. Hydrodynamics, fungal physiology, and morphology. Adv Biochem Eng/Biotechnol 2015;149:55–90.
27. Walisko R, Moench-Tegeder J, Blotenberg J, Wucherpfennig T, Krull R. The taming of the shrew-controlling the morphology of filamentous eukaryotic and prokaryotic microorganisms. Adv Biochem Eng/Biotechnol 2015;149:1–27.
28. Walisko R, Krull R, Schrader J, Wittmann C. Microparticle based morphology engineering of filamentous microorganisms for industrial bio-production. Biotechnol Lett 2012;34:1975–82.
29. Böl M, Schrinner K, Tesche S, Krull R. Challenges of influencing cellular morphology by morphology engineering techniques and mechanical induced stress on filamentous pellet systems – a critical review. Eng Life Sci 2021;21:51–67.
30. Laible AR, Dinius A, Schrader M, Krull R, Kwade A, Briesen H, et al. Effects and interactions of metal oxides in microparticle-enhanced cultivation of filamentous microorganisms. Eng Life Sci 2021;7:491.
31. Meyer V, Cairns T, Barthel L, King R, Kunz P, Schmideder S, et al. Understanding and controlling filamentous growth of fungal cell factories: novel tools and opportunities for targeted morphology engineering. Fungal Biol Biotechnol 2021;8:8.
32. Gibbs PA, Seviour RJ, Schmid F. Growth of filamentous fungi in submerged culture: problems and possible solutions. Crit Rev Biotechnol 2000;20:17–48.
33. Dobson LF, O'Cleirigh CC, O'Shea DG. The influence of morphology on geldanamycin production in submerged fermentations of *Streptomyces hygroscopicus* var. *geldanus*. Appl Microbiol Biotechnol 2008;79: 859–66.
34. Bizukojc M, Ledakowicz S. The morphological and physiological evolution of *Aspergillus terreus* mycelium in the submerged culture and its relation to the formation of secondary metabolites. World J Microbiol Biotechnol 2010;26:41–54.
35. Kunz PJ, Barthel L, Meyer V, King R. Vesicle transport and growth dynamics in *Aspergillus niger*: microscale modeling of secretory vesicle flow and centerline extraction from confocal fluorescent data. Biotechnol Bioeng 2020;117:2875–86.
36. Kunz P, King R. Secretory vesicle and glucoamylase distribution in *Aspergillus niger* and macromorphology in regions of varying shear stress. Front Microbiol 2022;13:842249.
37. Krull R, Wucherpfennig T, Hempel DC. Bioverfahrenstechnik. In: Bender B, Göhlich D, editors. Dubbel: Taschenbuch für den Maschinenbau 3: Maschinen und Systeme, 26th ed. Berlin, Heidelberg: Springer Vieweg; 2020.
38. Caldwell IY, Trinci AP. The growth unit of the mould *Geotrichum candidum*. Arch Mikrobiol 1973;88:1–10.
39. Prosser JI, Trinci AP. A model for hyphal growth and branching. J Gen Microbiol 1979;111:153–64.
40. Bergter F. Kinetic model of mycelial growth. Z Allg Mikrobiol 1978;18:143–5.
41. Schmideder S, Barthel L, Friedrich T, Thalhammer M, Kovačević T, Niessen L, et al. An X-ray microtomography-based method for detailed analysis of the three-dimensional morphology of fungal pellets. Biotechnol Bioeng 2019;116:1355–65.
42. Pamboukian CRD, Facciotti MCR. Production of antitumoral retamycin during fed-batch fermentations of *Streptomyces olindensis*. Appl Biochem Biotechnol 2004;112:111–22.
43. Takahashi J, Yamada K. Studies on the effect of some physical conditions on the submerged mold culture: Part II. On the two types of pellet formation in the shaking culture. J Agric Chem Soc Jpn 1959;33:707–9.

44. Kowalska A, Boruta T, Bizukojc M. Kinetic model to describe the morphological evolution of filamentous fungi during their early stages of growth in the standard submerged and microparticle-enhanced cultivations. Eng Life Sci 2019;19:557–74.
45. Kowalska A, Boruta T, Bizukojc M. Performance of fungal microparticle-enhanced cultivations in stirred tank bioreactors depends on species and number of process stages. Biochem Eng J 2020;161:107696.
46. Nielsen J. Modelling the growth of filamentous fungi. Adv Biochem Eng/Biotechnol 1992;46:187–223.
47. Metz B, Kossen NWF. The growth of molds in the form of pellets – a literature review. Biotechnol Bioeng 1977;19:781–99.
48. El Enshasy HA. Fungal morphology: a challenge in bioprocess engineering industries for product development. Curr Opin Chem Eng 2022;35:100729.
49. Zhang J, Zhang J. The filamentous fungal pellet and forces driving its formation. Crit Rev Biotechnol 2016; 36:1066–77.
50. Vecht-Lifshitz SE, Magdassi S, Braun S. Pellet formation and cellular aggregation in *Streptomyces tendae*. Biotechnol Bioeng 1990;35:890–6.
51. Pazouki M, Panda T. Understanding the morphology of fungi. Bioprocess Eng 2000;22:127–43.
52. Oncu S, Tari C, Unluturk S. Effect of various process parameters on morphology, rheology, and polygalacturonase production by *Aspergillus sojae* in a batch bioreactor. Biotechnol Prog 2007;23:836–45.
53. Hille A, Neu TR, Hempel DC, Horn H. Oxygen profiles and biomass distribution in biopellets of *Aspergillus niger*. Biotechnol Bioeng 2005;92:614–23.
54. Hille A, Neu TR, Hempel DC, Horn H. Effective diffusivities and mass fluxes in fungal biopellets. Biotechnol Bioeng 2009;103:1202–13.
55. Emerson S. The growth phase in *Neurospora* corresponding to the logarithmic phase in unicellular organisms. J Bacteriol 1950;60:221–3.
56. Paul GC, Thomas CR. Characterisation of mycelial morphology using image analysis. Adv Biochem Eng/ Biotechnol 1998;60:1–59.
57. Wucherpfennig T, Hestler T, Krull R. Morphology engineering-osmolality and its effect on *Aspergillus niger* morphology and productivity. Microb Cell Factories 2011;10:58.
58. Cox PW, Thomas CR. Classification and measurement of fungal pellets by automated image analysis. Biotechnol Bioeng 1992;39:945–52.
59. Cox PW, Paul GC, Thomas CR. Image analysis of the morphology of filamentous micro-organisms. Microbiology (Reading, Engl) 1998;144:817–27.
60. Lakowitz A. Charakterisierung der Morphologie von *Aspergillus niger* mittels automatischer Bildanalyse. Braunschweig, Germany: Technische Universität Braunschweig; 2011.
61. Obert M, Pfeifer P, Sernetz M. Microbial growth patterns described by fractal geometry. J Bacteriol 1990; 172:1180–5.
62. Papagianni M. Quantification of the fractal nature of mycelial aggregation in *Aspergillus niger* submerged cultures. Microb Cell Factories 2006;5:5.
63. Mandelbrot BB. The fractal geometry of nature. New York: W. H. Freeman and Company; 1983.
64. Barry DJ, Chan C, Williams GA. Morphological quantification of filamentous fungal development using membrane immobilization and automatic image analysis. J Ind Microbiol Biotechnol 2009;36:787–800.
65. Wucherpfennig T. Cellular morphology – a novel process parameter for the cultivation of eukaryotic cells. Braunschweig, Germany: Technische Universität Braunschweig; 2013.
66. Phillips DH. Oxygen transfer into mycelial pellets. Biotechnol Bioeng 1966;8:456–60.
67. Wittler R, Baumgartl H, Lübbers DW, Schügerl K. Investigations of oxygen transfer into *Penicillium chrysogenum* pellets by microprobe measurements. Biotechnol Bioeng 1986;28:1024–36.
68. Riley GL, Tucker KG, Paul GC, Thomas CR. Effect of biomass concentration and mycelial morphology on fermentation broth rheology. Biotechnol Bioeng 2000;68:160–72.
69. Rodríguez Porcel EM, Casas López JL, Sánchez Pérez JA, Fernández Sevilla JM, Chisti Y. Effects of pellet morphology on broth rheology in fermentations of *Aspergillus terreus*. Biochem Eng J 2005;26:139–44.

70. Bliatsiou C, Schrinner K, Waldherr P, Tesche S, Böhm L, Kraume M, et al. Rheological characteristics of filamentous cultivation broths and suitable model fluids. Biochem Eng J 2020;163:107746.
71. Olsvik E, Tucker KG, Thomas CR, Kristiansen B. Correlation of *Aspergillus niger* broth rheological properties with biomass concentration and the shape of mycelial aggregates. Biotechnol Bioeng 1993;42:1046–52.
72. Hille A, Neu TR, Hempel DC, Horn H. Einfluss der Morphologie auf Stofftransport und -umsatz in *Aspergillus niger*-Pellets. Chem Ing Tech 2006;78:627–32.
73. Gaden EL. Fermentation process kinetics. Biotechnol Bioeng 1959;1:413–29.
74. Driouch H, Roth A, Dersch P, Wittmann C. Optimized bioprocess for production of fructofuranosidase by recombinant *Aspergillus niger*. Appl Microbiol Biotechnol 2010;87:2011–24.
75. Kuhl M, Gläser L, Rebets Y, Rückert C, Sarkar N, Hartsch T, et al. Microparticles globally reprogram *Streptomyces albus* toward accelerated morphogenesis, streamlined carbon core metabolism, and enhanced production of the antituberculosis polyketide pamamycin. Biotechnol Bioeng 2020;117: 3858–75.
76. van Wezel GP, McDowall KJ. The regulation of the secondary metabolism of *Streptomyces*: new links and experimental advances. Nat Prod Rep 2011;28:1311–33.
77. Casas López JL, Sánchez Pérez JA, Fernández Sevilla JM, Rodríguez Porcel EM, Chisti Y. Pellet morphology, culture rheology and lovastatin production in cultures of *Aspergillus terreus*. J Biotechnol 2005;116:61–77.
78. Lin P-J, Scholz A, Krull R. Effect of volumetric power input by aeration and agitation on pellet morphology and product formation of *Aspergillus niger*. Biochem Eng J 2010;49:213–20.
79. Papagianni M, Mattey M. Morphological development of *Aspergillus niger* in submerged citric acid fermentation as a function of the spore inoculum level. Application of neural network and cluster analysis for characterization of mycelial morphology. Microb Cell Factories 2006;5:3.
80. Zhang ZY, Jin B, Kelly JM. Effects of cultivation parameters on the morphology of *Rhizopus arrhizus* and the lactic acid production in a bubble column reactor. Eng Life Sci 2007;7:490–6.
81. Das RK, Brar SK. Enhanced fumaric acid production from brewery wastewater and insight into the morphology of *Rhizopus oryzae* 1526. Appl Biochem Biotechnol 2014;172:2974–88.
82. Wu N, Zhang J, Ou W, Chen Y, Wang R, Li K, et al. Transcriptome analysis of *Rhizopus oryzae* seed pellet formation using triethanolamine. Biotechnol Biofuels 2021;14:230.
83. Teng Y, Xu Y, Wang D. Changes in morphology of Rhizopus chinensis in submerged fermentation and their effect on production of mycelium-bound lipase. Bioproc Biosyst Eng 2009;32:397–405.
84. Wucherpfennig T, Lakowitz A, Krull R. Comprehension of viscous morphology-evaluation of fractal and conventional parameters for rheological characterization of *Aspergillus niger* culture broth. J Biotechnol 2013;163:124–32.
85. Xia X, Lin S, Xia X-X, Cong F-S, Zhong J-J. Significance of agitation-induced shear stress on mycelium morphology and lavendamycin production by engineered *Streptomyces flocculus*. Appl Microbiol Biotechnol 2014;98:4399–407.
86. Jung S-K. A review of image analysis in biochemical engineering. Biotechnol Bioproc Eng 2019;24:65–75.
87. Collins SP, Pope RK, Scheetz RW, Ray RI, Wagner PA, Little BJ. Advantages of environmental scanning electron microscopy in studies of microorganisms. Microsc Res Tech 1993;25:398–405.
88. Park Y, Tamura S, Koike Y, Toriyama M, Okabe M. Mycelial pellet intrastructure visualization and viability prediction in a culture of *Streptomyces fradiae* using confocal scanning laser microscopy. J Ferment Bioeng 1997;84:483–6.
89. O'Cleirigh C, Walsh PK, O'Shea DG. Morphological quantification of pellets in *Streptomyces hygroscopicus* var. *geldanus* fermentation broths using a flatbed scanner. Biotechnol Lett 2003;25:1677–83.
90. Sasamoto H, Azumi Y, Shimizu M, Hachinohe Y, Suzuki S. In vitro bioassay of allelopathy of *Arabidopsis thaliana* by sandwich method and protoplast co-culture method with digital image analysis. Plant Biotechnol 2017;34:199–202.
91. Lee D, Mehta N, Shearer A, Kastner R. A hardware accelerated system for high throughput cellular image analysis. J Parallel Distr Comput 2018;113:167–78.

92. Churgin MA, Jung S-K, Yu C-C, Chen X, Raizen DM, Fang-Yen C. Longitudinal imaging of *Caenorhabditis elegans* in a microfabricated device reveals variation in behavioral decline during aging. eLife 2017;6: e26652.
93. Collins TJ. ImageJ for microscopy. BioTechniques 2007;43:25–30.
94. Eliceiri KW, Berthold MR, Goldberg IG, Ibáñez L, Manjunath BS, Martone ME, et al. Biological imaging software tools. Nat Methods 2012;9:697–710.
95. Gallagher SR. Digital image processing and analysis with ImageJ. Curr Protoc Essent Lab Tech 2014;9: A.3C.1–A.3C.29.
96. Carpenter AE, Jones TR, Lamprecht MR, Clarke C, Kang IH, Friman O, et al. CellProfiler: image analysis software for identifying and quantifying cell phenotypes. Genome Biol 2006;7:R100.
97. Dobson ETA, Cimini B, Klemm AH, Wählby C, Carpenter AE, Eliceiri KW. ImageJ and CellProfiler: complements in open-source bioimage analysis. Curr Protoc 2021;1:e89.
98. Ljungqvist MG, Nielsen ME, Ersbøll BK, Frosch S. Image analysis of pellet size for a control system in industrial feed production. PLoS One 2011;6:e26492.
99. Rønnest NP, Stocks SM, Lantz AE, Gernaey KV. Comparison of laser diffraction and image analysis for measurement of *Streptomyces coelicolor* cell clumps and pellets. Biotechnol Lett 2012;34:1465–73.
100. Walisko J, Vernen F, Pommerehne K, Richter G, Terfehr J, Kaden D, et al. Particle-based production of antibiotic rebeccamycin with *Lechevalieria aerocolonigenes*. Process Biochem 2017;53:1–9.
101. Dane JH, Topp GC, Campbell GS, editors. Methods of soil analysis: Part 4: physical methods. Madison, WI: Soil Science Society of America; 2002.
102. Shekunov BY, Chattopadhyay P, Tong HHY, Chow AHL. Particle size analysis in pharmaceutics: principles, methods and applications. Pharm Res 2007;24:203–27.
103. Petersen N, Stocks S, Gernaey KV. Multivariate models for prediction of rheological characteristics of filamentous fermentation broth from the size distribution. Biotechnol Bioeng 2008;100:61–71.
104. Eslahpazir Esfandabadi M, Wucherpfennig T, Krull R. Agitation induced mechanical stress in stirred tank bioreactors-linking CFD simulations to fungal morphology. J Chem Eng Jpn 2012;45:742–8.
105. Jillavenkatesa A, Dapkunas SJ, Lum L. Particle size characterization. Washington: National Institute of Standards and Technology; 2001.
106. Kelly RN, Kazanjian J. Commercial reference shape standards use in the study of particle shape effect on laser diffraction particle size analysis. AAPS PharmSciTech 2006;7:E49.
107. Schrinner K. Micro- and macroparticle enhanced cultivation of filamentous *Lentzea aerocolonigenes* for increased rebeccamycin production. Braunschweig, Germany: Technische Universität Braunschweig; 2021.
108. Hille A. Stofftransport- und Stoffumsatzprozesse in filamentösen Pilzpellets. Braunschweig, Germany: Technische Universität Braunschweig; 2008.
109. Jonkman J, Brown CM. Any way you slice it – a comparison of confocal microscopy techniques. J Biomol Tech 2015;26:54–65.
110. Walla PJ. Modern biophysical chemistry: detection and analysis of biomolecules, 2nd ed. Weinheim, Germany: Wiley-VCH; 2014.
111. Teng X, Li F, Lu C. Visualization of materials using the confocal laser scanning microscopy technique. Chem Soc Rev 2020;49:2408–25.
112. Driouch H, Sommer B, Wittmann C. Morphology engineering of *Aspergillus niger* for improved enzyme production. Biotechnol Bioeng 2010;105:1058–68.
113. Veiter L, Herwig C. The filamentous fungus *Penicillium chrysogenum* analysed via flow cytometry – a fast and statistically sound insight into morphology and viability. Appl Microbiol Biotechnol 2019;103:6725–35.
114. Schrinner K, Veiter L, Schmideder S, Doppler P, Schrader M, Münch N, et al. Morphological and physiological characterization of filamentous *Lentzea aerocolonigenes*: comparison of biopellets by microscopy and flow cytometry. PLoS One 2020;15:e0234125.

115. Bleichrodt R-J, Read ND. Flow cytometry and FACS applied to filamentous fungi. Fungal Biol Rev 2019;33: 1–15.
116. Kumar A, Galaev IY, Mattiasson B. Cell separation. Berlin, Heidelberg: Springer Berlin Heidelberg; 2007.
117. Ehgartner D, Herwig C, Neutsch L. At-line determination of spore inoculum quality in *Penicillium chrysogenum* bioprocesses. Appl Microbiol Biotechnol 2016;100:5363–73.
118. Ehgartner D, Herwig C, Fricke J. Morphological analysis of the filamentous fungus *Penicillium chrysogenum* using flow cytometry-the fast alternative to microscopic image analysis. Appl Microbiol Biotechnol 2017; 101:7675–88.
119. Schmideder S, Barthel L, Müller H, Meyer V, Briesen H. From three-dimensional morphology to effective diffusivity in filamentous fungal pellets. Biotechnol Bioeng 2019;116:3360–71.
120. Schmideder S, Müller H, Barthel L, Friedrich T, Niessen L, Meyer V, et al. Universal law for diffusive mass transport through mycelial networks. Biotechnol Bioeng 2021;118:930–43.
121. Papagianni M, Mattey M, Kristiansen B. Citric acid production and morphology of *Aspergillus niger* as functions of the mixing intensity in a stirred tank and a tubular loop bioreactor. Biochem Eng J 1998;2: 197–205.
122. Wang L, Ridgway D, Gu T, Moo-Young M. Effects of process parameters on heterologous protein production in *Aspergillus niger* fermentation. J Chem Technol Biotechnol 2003;78:1259–66.
123. Kelly S, Grimm LH, Hengstler J, Schultheis E, Krull R, Hempel DC. Agitation effects on submerged growth and product formation of *Aspergillus niger*. Bioproc Biosyst Eng 2004;26:315–23.
124. Peter CP, Suzuki Y, Büchs J. Hydromechanical stress in shake flasks: correlation for the maximum local energy dissipation rate. Biotechnol Bioeng 2006;93:1164–76.
125. Tan R-K, Eberhard W, Büchs J. Measurement and characterization of mixing time in shake flasks. Chem Eng Sci 2011;66:440–7.
126. Eslahpazir M, Krull R, Krühne U. Computational fluid dynamics. In: Moo-Young M, editor. Comprehensive biotechnology. Amsterdam: Pergamon; ScienceDirect; 2019:95–107 pp.
127. Dittmann J, Tesche S, Krull R, Böl M. The influence of salt-enhanced cultivation on the micromechanical behaviour of filamentous pellets. Biochem Eng J 2019;148:65–76.
128. Unisense A/S. Oxygen microsensor; 2020. https://www.unisense.com/O2/ [Accessed 19 Apr 2022].
129. Revsbech NP. An oxygen microsensor with a guard cathode. Limnol Oceanogr 1989;34:474–8.
130. Gonciarz J, Bizukojc M. Adding talc microparticles to *Aspergillus terreus* ATCC 20542 preculture decreases fungal pellet size and improves lovastatin production. Eng Life Sci 2014;14:190–200.
131. Bizukojc M, Gonciarz J. Influence of oxygen on lovastatin biosynthesis by *Aspergillus terreus* ATCC 20542 quantitatively studied on the level of individual pellets. Bioproc Biosyst Eng 2015;38:1251–66.
132. Driouch H, Hänsch R, Wucherpfennig T, Krull R, Wittmann C. Improved enzyme production by bio-pellets of *Aspergillus niger*: targeted morphology engineering using titanate microparticles. Biotechnol Bioeng 2012;109:462–71.
133. Kaup B-A, Ehrich K, Pescheck M, Schrader J. Microparticle-enhanced cultivation of filamentous microorganisms: increased chloroperoxidase formation by *Caldariomyces fumago* as an example. Biotechnol Bioeng 2008;99:491–8.
134. Yatmaz E, Germec M, Karahalil E, Turhan I. Enhancing β-mannanase production by controlling fungal morphology in the bioreactor with microparticle addition. Food Bioprod Process 2020;121:123–30.
135. Pommerehne K, Walisko J, Ebersbach A, Krull R. The antitumor antibiotic rebeccamycin-challenges and advanced approaches in production processes. Appl Microbiol Biotechnol 2019;103:3627–36.
136. Schrader M, Pommerehne K, Wolf S, Finke B, Schilde C, Kampen I, et al. Design of a CFD-DEM-based method for mechanical stress calculation and its application to glass bead-enhanced cultivations of filamentous *Lentzea aerocolonigenes*. Biochem Eng J 2019;148:116–30.
137. Schrinner K, Schrader M, Niebusch J, Althof K, Schwarzer FA, Nowka P-F, et al. Macroparticle-enhanced cultivation of *Lentzea aerocolonigenes*: variation of mechanical stress and combination with lecithin

supplementation for a significantly increased rebeccamycin production. Biotechnol Bioeng 2021;118: 3984–95.

138. Bellgardt KH. Process models for production of beta-lactam antibiotics. Adv Biochem Eng/Biotechnol 1998;60:153–94.

139. Hotop S, Möller J, Niehoff J, Schügerl K. Influence of the preculture conditions on the pellet size distribution of *Penicillium chrysogenum* cultivations. Process Biochem 1993;28:99–104.

140. Li G, Sun J, Li F, Wang Y, Li Q. Macroparticle-enhanced bioleaching of uranium using *Aspergillus niger*. Miner Eng 2022;180:107493.

141. Sohoni SV, Bapat PM, Lantz AE. Robust, small-scale cultivation platform for *Streptomyces coelicolor*. Microb Cell Factories 2012;11:9.

142. Zhu W-Z, Wang S-H, Gao H-M, Ge Y-M, Dai J, Zhang X-L, et al. Characterization of bioactivities and biosynthesis of angucycline/angucyclinone derivatives derived from *Gephyromycinifex aptenodytis* gen. nov, sp. nov. Mar Drugs 2021;20:34.

143. Kempf B, Bremer E. Uptake and synthesis of compatible solutes as microbial stress responses to high-osmolality environments. Arch Microbiol 1998;170:319–30.

144. Csonka LN, Hanson AD. Prokaryotic osmoregulation: genetics and physiology. Annu Rev Microbiol 1991; 45:569–606.

145. Hobbs G, Frazer C, Gardner D, Cullum J, Oliver S. Dispersed growth of *Streptomyces* in liquid culture. Appl Microbiol Biotechnol 1989;31:272–7.

146. Rosen MJ. Surfactants and interfacial phenomena, 4th ed. Hoboken: John Wiley & Sons Incorporated; 2012.

147. Bustamante M, González ME, Cartes A, Diez MC. Effect of soya lecithin on the enzymatic system of the white-rot fungi *Anthracophyllum discolor*. J Ind Microbiol Biotechnol 2011;38:189–97.

148. Liu Y-S, Wu J-Y. Effects of Tween 80 and pH on mycelial pellets and exopolysaccharide production in liquid culture of a medicinal fungus. J Ind Microbiol Biotechnol 2012;39:623–8.

149. Dynesen J, Nielsen J. Surface hydrophobicity of *Aspergillus nidulans* conidiospores and its role in pellet formation. Biotechnol Prog 2003;19:1049–52.

150. Kurakake M, Hirotsu S, Shibata M, Takenaka Y, Kamioka T, Sakamoto T. Effects of nonionic surfactants on pellet formation and the production of β-fructofuranosidases from *Aspergillus oryzae* KB. Food Chem 2017; 224:139–43.

151. Vecht-Lifshitz SE, Magdassi S, Braun S. Effects of surface active agents on pellet formation in submerged fermentations of *Streptomyces tendae*. J Dispersion Sci Technol 1989;10:265–75.

152. Nienow AW. Reactor engineering in large scale animal cell culture. Cytotechnology 2006;50:9–33.

153. Oh S, Nienow AW, Al-Rubeai M, Emery AN. The effects of agitation intensity with and without continuous sparging on the growth and antibody production of hybridoma cells. J Biotechnol 1989;12:45–61.

154. Nienow AW, Langheinrich C, Stevenson NC, Emery AN, Clayton TM, Slater NK. Homogenisation and oxygen transfer rates in large agitated and sparged animal cell bioreactors: some implications for growth and production. Cytotechnology 1996;22:87–94.

155. Žnidaršič P, Komel R, Pavko A. Influence of some environmental factors on *Rhizopus nigricans* submerged growth in the form of pellets. World J Microbiol Biotechnol 2000;16:589–93.

156. Silva CC, Dekker RF, Silva RSS, Silva MLC, Barbosa AM. Effect of soybean oil and Tween 80 on the production of botryosphaeran by *Botryosphaeria rhodina* MAMB-05. Process Biochem 2007;42:1254–8.

157. Hao L, Xing X, Li Z, Zhang J-C, Sun J-X, Jia S-R, et al. Optimization of effect factors for mycelial growth and exopolysaccharide production by *Schizophyllum commune*. Appl Biochem Biotechnol 2010;160:621–31.

158. Li C, Li Q, Wang Z, Ji G, Zhao H, Gao F, et al. Environmental fungi and bacteria facilitate lecithin decomposition and the transformation of phosphorus to apatite. Sci Rep 2019;9:15291.

159. Gao X, Li Y, Zhang J. Soybean lecithin enhanced cellulase production by *Penicillium oxalicum* JG in a scaled-up bioconversion process. Cellul Chem Technol 2020;54:705–12.

160. Jolly M, Vidal R, Marchand PA. Lecithins: a food additive valuable for antifungal crop protection. Int J Econ Plants 2018;5:104–7.
161. Brock TD. The effect of oils and fatty acids on the production of filipin. Appl Microbiol 1956;4:131–3.
162. Hofer A, Herwig C, Spadiut O. Lecithin is the key material attribute in soy bean oil affecting filamentous bioprocesses. AMB Express 2018;8:90.

Jochen Schaub*, Andreas Ankenbauer, Tobias Habicher, Michael Löffler,
Nicolas Maguire, Dominique Monteil, Sebastian Püngel, Lisa Stepper,
Fabian Stiefel, Judith Thoma, Andreas Unsöld, Julia Walther,
Christopher Wayne and Thomas Wucherpfennig

12 Process intensification in biopharmaceutical process development and production – an industrial perspective

Abstract: Process intensification aims to increase productivity in biologics manufacturing. Significant progress has been made in academia, the biopharmaceutical industry, and by the regulatory guidance since the 2000s. Process intensification can include all unit operations of a drug substance manufacturing process. The applied upstream concepts have consequences on the downstream process (DSP). The DSP process must manage larger product amounts while ensuring the required quality and impurity profiles, and cope with the available time frame as per scheduling requirements in a facility. Further, intensification in DSP is not based on a single technology only but rather on various technologies. This contribution provides an industry perspective on process intensification, describing basic concepts, technical and engineering aspects as well as the impact on the manufacturing process given existing facilities and a product portfolio to be manufactured. It also covers scientific approaches that support understanding and design of intensified bioprocesses. From an implementation perspective, the technologies used for intensification must be robust, scalable, and suitable for commercial manufacturing. Specific examples for a high seeding density fed batch (using N-1 perfusion) and a continuous process are provided for Chinese hamster ovary (CHO) cells producing therapeutic antibodies. Economic and sustainability aspects are addressed as well. Process intensification in an industrial environment is complex and many factors need to be considered, ranging from characteristics of a specific molecule to its commercial manufacturing at internal or external sites for global or regional markets.

---

*Corresponding author: Jochen Schaub, Bioprocess Development Biologicals, Boehringer Ingelheim Pharma GmbH & Co. KG, Biberach a.d. Riß, Germany, E-mail: jochen.schaub@boehringer-ingelheim.com
Andreas Ankenbauer and Christopher Wayne, DevOps Biologics DS Germany, HP BioP Launch&Innovation, Boehringer Ingelheim Pharma GmbH & Co. KG, Biberach a.d. Riß, Germany
Tobias Habicher, Michael Löffler, Nicolas Maguire, Sebastian Püngel, Lisa Stepper, Fabian Stiefel, Judith Thoma, Andreas Unsöld, Julia Walther and Thomas Wucherpfennig, Bioprocess Development Biologicals, Boehringer Ingelheim Pharma GmbH & Co. KG, Biberach a.d. Riß, Germany
Dominique Monteil, Cell Culture, HP BioP Mammalian, Boehringer Ingelheim US Biopharma, Fremont, CA, USA

As per De Gruyter's policy this article has previously been published in the journal Physical Sciences Reviews. Please cite as: J. Schaub, A. Ankenbauer, T. Habicher, M. Löffler, N. Maguire, D. Monteil, S. Püngel, L Stepper, F. Stiefel, J. Thoma, A. Unsöld, J. Walther, C. Wayne and T. Wucherpfennig "Process intensification in biopharmaceutical process development and production – an industrial perspective" *Physical Sciences Reviews* [Online] 2023. DOI: 10.1515/psr-2022-0113 | https://doi.org/10.1515/9783110760330-012

Keywords: N-1 perfusion; downstream process; monoclonal antibody; cell culture medium; modeling; sustainability

12.1 Introduction

The first pioneering antibody was approved by the Food and Drug Administration (FDA) in 1986. Only recently, in 2021, the 100th antibody product has been approved by the FDA and it took only 6 years since the 50th antibody was approved in 2015 [1]. This group of biologics make up almost 20 % of new drug approvals per year and, based on recent years, an approval rate of about 10 per year is currently considered to be a baseline estimation [1]. The late commercial pipeline of antibody therapeutics indeed grew by about 30 % in the past year [2] and so did the number of targets for antibody therapeutics that was approved in the United States or the European Union in the past years. In 2021, eight medicines were approved and eight were in the review process, in 2016–2020 approvals added up to 28 new targets altogether, in 2011–2015 to 16 new targets, and in 2006–2010 to seven new targets [2]. At the same time, competition for promising targets is very intense. The majority of approvals (45 %) were treatments against cancer, followed by immune-mediated disorders (27 %) [2]. Important from both medicinal but also Chemistry and Manufacturing Control (CMC) aspects are the increasing number of antibody formats, with the first antibody fragment (including antigen-binding fragments (Fabs), single-chain variable region constructs (scFv) or nanobodies) receiving FDA approval in 1994, the first antibody-drug conjugate (ADC) in 2000, and the first bispecific antibody in 2014 [1]. In the field of antibody therapeutics, various multifunctional formats such as immunocytokines are in development [3]. Also commercially, antibody therapeutics are quite successful with six biologics in addition to three mRNA-based Covid-19 vaccines or antibody-based medicines against Covid-19 being in the Top 20 of worldwide sales in 2021 [4]. The global market for biopharmaceutical products experienced a significant growth within recent years being valued at $237.2 billion in 2018 [5]. From 129 biopharmaceutical products approved by the FDA for human use (5-year period between January 2014 and December 2019) 66 (≈50 %) are monoclonal antibodies or conjugates [5]. By the year 2023 the global monoclonal antibody market alone is projected to reach $200 billion. To serve this growing and competitive market, biopharmaceutical companies are under constant pressure to reduce production costs and to increase facility output.

Advanced therapy medicinal products (ATMPs) comprising cellular therapy, gene therapy as well as tissue-based products are another area of active research, development, and product commercialization with specific CMC requirements. Only recently, the impressively fast and successful development of mRNA-based vaccines against Covid-19 can be considered as a certain breakthrough for this technology, with a promising potential (including financial resources) to expand into further areas such as oncology.

Despite these great achievements, costs of treatments using these medicines are high and health care systems generally are struggling globally. Obviously, pricing of medicines is a complex topic and costs for CMC development and commercial manufacturing of biologics is only one (minor) part, both with respect to selling prices but also within a (bio) pharmaceutical company [6] that requires high research and development invests in order to secure a profitable future portfolio and business due to e.g. attrition rates or the competitive landscape. On the other side, competition for biotherapeutics after patent expiry (typically 20 years after the filing date of the relevant patent application) can be expected to increase significantly as has been the case for small molecules. As of today, 36 biosimilars are approved by the FDA [7] targeting mainly the current (and future) biologics blockbusters, for example Humira/Adalimumab (seven FDA approved biosimilars, ten approved by the European Medicines Agency (EMA)). Patent litigations and legal settlements could partly extend market exclusivity for major products, but the next years will show how prices will actually develop, providing a broader picture. Price is a relevant differentiator for biosimilars and development costs are in general lower due to reduced clinical efforts. Yet, efficient bioprocesses and competitive manufacturing costs will be more important for these medicines. This might generally stimulate innovations in process development and manufacturing.

Overall, this variety of medicines likely will also require more, innovative and/or (more) flexible CMC platforms in order to develop and produce these medicinal products. On the other hand, well established, robust development and manufacturing platforms with a proven track record, detailed understanding of scale-up/scale-down as well as process transfer are of high value for CMC functions. Especially processes intended for multi-product facilities within internal production networks or at external partners benefit from platform experience and comprehensive data. In the pursuit to bring medicines to the market and ensure a robust supply for patients worldwide, any CMC issue needs to be avoided.

When looking into more detail into the Top 6 selling antibody-based medicines 2021 [4], which are Humira/Adalimumab (Abbvie), Keytruda/Pembrolizumab (Merck Sharp & Dohme/MSD), Stelara/Ustekinumab (Janssen-Cilag), Eylea/Aflibercept (Bayer/Regeneron), Opdivo/Nivolumab (Bristol-Myers Squibb (BMS)), and Dupixent/Dupilumab (Sanofi), only Stelara is produced in a continuous perfusion process (using SP2/0 cells) whereas all others according to publicly available EMA information are CHO based fed batch processes (note that above listings provide the product name, the international nonproprietary name, and the primary sponsoring company of the first approval process). The first EU approval for these products was between 2003 and 2017. Three of these products (Humira/arthritis, Keytruda/cancer, Opdivo/cancer) have four manufacturing sites for the biological active substance, two (Stelara, Dupixent) have three sites. Having more (than one) manufacturing site(s) certainly is an advantage (e.g. as general risk mitigation or to efficiently supply regional markets), yet management of several sites to meet high demands is complex and associated efforts and costs for transfers, ramp-up and manufacturing are high, in particular when (several) external partners (CMOs) are part of such a network as well. This can become dispensable or greatly reduced if high performing (intensified) processes are developed from the beginning. Especially N-1 perfusion coupled to HSD fed batch production is of increasing interest for originator companies and is also offered up to large scale volumes by CMO companies (e.g. Boehringer Ingelheim, Samsung or Fuji).

Unfortunately, recent information on products in development is publicly not available that allow a more across-the-board view on the biopharmaceutical industry with respect to where process intensification is heading to from a technology perspective and, more importantly, when. Referring to granted antibody therapeutics in the EU or the US in 2021 (until 1st November) [2], eight antibody therapeutics were granted, not including orphan drugs and Covid-19 related medicines. With some uncertainty due to limited information, only one product was developed as HSD fed batch process including N-1 perfusion, none as continuous. The N-1 perfusion process was Aduhelm/Aducanumab (Biogen) with manufacturing sites in North Carolina and a newly built facility in Solothurn, Switzerland. The project Aduhelm is also well known since it was the first new medicine against Alzheimer disease for almost two decades receiving FDA approval. There were controversies on the pharmaceutical agent and Biogen withdrawed the marketing authorisation for the EU. Regarding manufacturing technology, this translates into about 10–15 % of products intended to use N-1 process intensification. Clearly, data from one year (2021) are not sufficient for more general conclusions; however data from 2022 to date do not provide further insight currently. Share of continuous processes for approved biotherapeutics using cell culture is estimated to be fewer than 10 % [8], referring to publications issued in 2015 [9]. In an overview from 2018 [6], most approvals for perfusion processes date to the time frame from 1994 to 2009. Interestingly, the 100th approved therapeutic antibody [1] was developed as a fed batch process and the product is Jemperli/Dostarlimab (GlaxoSmithKline, GSK). The Indication is cancer targeting PD-1, which is currently the most competitive target with four approved medicines and another five in review by FDA or EMA.

Without being a focus of this contribution, the role of single-use technology and the innovations by the various suppliers (often based on initial academic research and university spin-offs), making this technology commercially available and their usage common, have to be acknowledged. An example for this kind of innovation from a downstream perspective is multi-column chromatography (see e.g. an overview in [10]) and from an upstream perspective the miniaturized and highly automated Ambr® system with perfusion options.

Today, 2000 L single-use bioreactors are well established, CMOs (e.g. Wuxi) offer scales up to 4000 L, suppliers (e.g. Thermo Fisher) offer up to 5000 L, and even 6000 L working volume can be obtained for customized solutions (ABEC). The footprint of single-use solutions is in general lower and other principles apply in the facility layout (e.g. scale-out, ballroom concept) compared to stainless-steel manufacturing plants. Single-use facilities claim to be more flexible, can be constructed in less time with lower initial investment [11]. Thus, it (also) facilitates regional or temporary manufacturing. Given the broad range of biologics in development and on the market, today single-use technology complements large scale stainless steel production in many companies. On the other side, single-use equipment itself comes at high costs and dependency on functional supply chains is higher. Midterm, sustainability topics might become more important and flexibility is not only relevant with respect to equipment but even more with respect to well-trained employees and associated costs, favoring balanced facility utilization rates to ensure high overall productivity in CMC functions within a company. Single-use technology certainly has brought innovation and flexibility. It also brought additional impetus for those biopharmaceutical companies that have an existing (stainless steel) infrastructure in place to develop strategies to further increase productivity of their current plants in order to stay competitive.

In the past about 30 years, big progress was achieved in research, development and manufacturing of antibody therapeutics. Titers in commercial manufacturing increased from tens to hundreds of mg/L with about 0.3–0.5 g/L of monoclonal antibody (mAb) titers in commercial processes in the mid-1990s [6]. About 3–5 g/L were reached in the mid-2000s [6], at which titers at the lower end of this span are reported as well with about 3 g/L in the 2010s [12]. Obviously, harvest titers also need to be put into context with the product portfolio as well as the cultivation time, which varies between 10 and 15 days for typical fed batch processes. In the past years, CHO fed batch platforms for mAbs were developed yielding product titers approaching 10 g/L [13] or even beyond [14].

A high performing CHO bioprocess development platform consists of a stable, monoclonal cell line with high cell specific productivity, a strong media platform supporting cell growth, high cell viabilities and optimal production, and finally a cell culture process format enabling high product titers in upstream, high yields in downstream while meeting desired product quality profiles. These three elements (cell line, media, and process) mutually depend on each other. Hence, they need to be well balanced and designed to achieve an overall optimum that, ideally, supports a wide range of molecules.

Briefly, cell line development includes activities such as vector design, transfection technology, screening and selection of cell pools and cell clones that are predictive for commercial scale process development. Cell line engineering strategies, which are summarized in Section 12.8, are applied to e.g. modulate product quality attributes, improve performance, ensure stability or inhibit early apoptosis. An important area from a platform perspective is the identification of gene integration sites with high transcriptional activity resulting in high cell specific productivity, preferably with few integration sites in order to minimize variability of cell clones. This is especially important for difficult to express molecules and/or complex novel antibody formats. In a representative example specific productivities of about 50 pg/cell/d can be achieved for a monoclonal antibody, with single clones reaching above 75 pg/cell/d [15].

Chemically defined media platforms are well established in biopharmaceutical companies for some years now [5, 16, 17]. Several companies use proprietary media which are stoichiometrically balanced (e.g. for amino acids), functionally understood and optimized (e.g. trace elements or vitamins) and well characterized (e.g. with respect to raw material variability or potential impurities). Ideally, such media platforms are an integral part of a CHO development and manufacturing platform and are applied end-to-end from cell line development to commercial manufacturing. With respect to process intensification, it is desirable that those media platforms are adjustable to support intensified process formats (e.g. when switching from fed batch to HSD fed batch applying an N-1 perfusion step). Supporting high cell densities and high titers, as well as the challenge of managing high media volumes (in N-1 perfusion pre-stage or in continuous processes) are important factors. To address the latter, efforts were taken to maximize media concentration using different approaches [18]. Cell culture media aspects are described in Section 12.4.

In the 2000s, the concept of continuous processing for the production of therapeutic antibodies by mammalian cell culture was actively addressed in joint efforts by academia and industry, including close interactions with regulatory agencies (resulting e.g. in the FDA guidance document on continuous manufacturing of drug substances and drug products available in current draft version ICH Q13 as issued in June 2021). In 2013, the Engineering Conferences International (ECI) established a dedicated, biennial conference series on integrated continuous biomanufacturing, and in 2014 Konstantin Konstantinov (Genzyme/Sanofi, now Codiak BioSciences) published a white paper on continuous bioprocessing, stating a transformative potential of this technology and a potential for a paradigm shift in biologics manufacturing [19]. He also mentioned the important role of the equipment vendors for implementation. Depending on molecule specifics, volumetric productivities of up to 2 g/L/d will be common, technically values of 4–6 g/L/d are reported, provided high performing cell clones and media enabling cell densities of $> 100 \times 10^6$ cells/mL (e.g. [20]).

The promises of continuous manufacturing are manifold. These are high volumetric productivity, reduced spatial footprints including smaller equipment size and consequently lower capital expenditures (especially for newly built single-use facilities), increased flexibility (including regional or temporary manufacturing), more streamlined and automated process flows (due to more steady-state like operation mode enabling (semi-) continuous harvest and purification options), lower process cycle times and, in particular, higher facility utilization (by relatively shorter downtimes between runs, especially when long run durations are possible). In the past (1990s), perfusion processes were mainly developed for sensitive products that are prone to degradation, such as blood factors (e.g. human coagulation factor VIII) and enzymes (e.g. imiglucerase), or have adverse effects on the cell culture production process itself (e.g. growth factors). In the 2000s, this changed to some extent and also antibodies were produced in perfusion processes (e.g. Alemtuzumab/MabCampath by Genzyme (now Sanofi) approved in 2001 or Golimumab/Simponi and Ustekinumab/Stelara by Janssen Biotech (now part of Johnson & Johnson) approved in 2009). Yet, in the past about 10+ years, to our knowledge and observation, a significant rise in commercial N-stage perfusion processes has not occurred and the estimated share of approved continuous processes is estimated to be below 10 % as described above.

One explanation for this can be that technological development for (existing) fed batch platforms continued in parallel and significant progress and productivity gains were achieved for cell lines, cell culture media and fed batch bioprocessing (e.g. by optimized feeding regimes). The substantial technological progress within the fed batch framework and, for many companies, the fact that they have a fed batch-based manufacturing infrastructure in place to serve clinical and commercial demands, questions the need to (completely) change process development and biomanufacturing. Robustness, predictable scale-up, proven transferability, highly productive modern fed batch processes, operational simplicity, a vast amount of process knowledge in addition to a track record of successful approvals and agency interactions in many cases are further, supportive arguments to prefer the fed batch mode. Provided a high-performing fed batch platform, Kelley et al. [8], for example, questions whether further titer improvements balance additional investments or potential process risks. Specifically, he mentions diminishing returns for titers above 8 g/L (e.g. when downstream bottlenecks are not resolved) and a rather low number of products that require such high titers to supply patients. Further technical concerns or scientific challenges are associated with the retention devices (e.g. fouling) as needed in any perfusion mode or challenges in media design and supply (e.g. media concentration to keep media volumes economically) as well as downstream feed streams (e.g. buffer volumes).

Another reason to build on fed batch might be associated with the molecule portfolio of an originator company. As outlined above, the diversity of molecule formats in research pipelines has significantly increased. Some of those formats have impact on CMC and require adaptation, optimization or even innovation of one or more unit operations in a (fed batch) platform process, in upstream and/or downstream. Further changes might induce additional development challenges, costs and/or impact timelines.

For completeness, it has to be noted that several intensified process variants are described in the spectrum from fed batch to N-stage perfusion, often based on a fed batch and referred to as "concentrated fed batch". Even hybrid approaches (short-term perfusion directly coupled to fed batch N-stage) exist [21]. In most cases, a "concentration" in such a fed batch occurs by achieving higher viable cell densities in the bioreactor, either from the cultivation start or during the process by cell accumulation (i.e. cell retention). In few cases not only the cells are retained/recirculated but also the product by application of filters with further reduced pore sizes. One prominent example was the PER.C6 antibody process developed by Perciva (a joint venture of Crucell and DSM) in 2009 with a product concentration of up to 27 g/L and use of an alternating tangential flow (ATF) system from Repligen. Along with the concentrated product come high concentrations of, for example, cellular DNA as well as host cell proteins in addition to higher cell counts/more cell debris which pose challenges for efficient clarification and downstream processing. More importantly, product quality can be impacted as well. Also, the risk for product degradation increases in such a "concentrated" environment. Technically, challenges due to membrane fouling and/or product retention are well known from N-stage perfusion processes. In principle, these challenges also need to be considered for concentrated fed batch modes, in particular when very high cell concentrations and product titers are achieved in the bioreactor. Here, we will focus on N-1 perfusion coupled to a fed batch process using high seeding cell densities. The aim is to increase productivity but stay within a fed batch and process design space where such additional challenges (technical, robustness, impact on product quality) do not arise.

In summary, as the need for efficiency in bioprocess manufacturing can be expected to increase, intensified process formats are likely to become more common in the future. The overall biologics pipeline is growing as is the biotherapeutics market overall. Apparently, market demands for specific products are difficult to estimate. In the best case, these demands can increase substantially (e.g. due to approvals in additional indications and/or countries, or also due to market exclusivity) and development of a second generation process with increased productivities and yields can be one option. In such a scenario, the question is how much further improvement is achievable based on an existing (fed batch) platform, considering the big improvements as described above, and assuming that these have been already implemented by most companies in the past years. Or, in other words, is the progress of the past a reasonable prediction for the future, namely when looking into more recent antibody formats or dosage forms (e.g. highly concentrated pharmaceutical formulations)? This kind of productivity increase is a prerequisite to realize more output in a given manufacturing network if nothing else is changed. In contrast, integration of process intensification options such as N-1 perfusion would increase overall capacity, flexibility, and adaptability to overall product demands in many cases with moderate adaptations in a given facility.

For now, the decision taking and balancing of pros and cons seem to be case dependent (e.g. [22]). Currently, one can also observe varying assessments of companies within the biopharmaceutical industrial community but also continuous high interest in current developments (e.g. in conferences or industry fora including publications such as the Biomanufacturing Technology Roadmap of the BioPhorum Operations Group [11]). Additional aspects that have not yet been extensively discussed to date can be expected to become more important in the future such as reliable supply chains and availability of skilled personnel.

In this contribution, we will focus on process intensification for antibody therapeutics using CHO cells and implementation efforts from small scale development to large scale (>10,000 L) manufacturing. At this, a relevant aspect is the consideration of an existing production network that, in our case, to a significant extent is based on fed batch technology. Process intensification via N-1 perfusion is a technology that enables increase in productivity and feasible implementation efforts in existing facilities. Continuous processing options will also be covered. However, from a manufacturing perspective, this technology differs significantly from large scale fed batch facilities and typically are established in newly planned, dedicated facilities (e.g. with significantly reduced bioreactor volumes in manufacturing stage, large extent to which single-use technology is applied and, as a result, a different optimal facility layout). We will describe the basic concepts for process intensification in upstream and downstream processing as well as the relevant technical and engineering aspects for implementation. This will be complemented by cell line and clone selection aspects as well as cell culture media requirements. Modern scientific approaches (e.g. systems biotechnology tools) and process analytical technology (PAT) will also be covered. Specifically, we will present examples for intensified CHO cell culture processes producing therapeutic antibodies. In dedicated chapters economic as well as sustainability aspects will be addressed. Finally, this contribution provides a summary and a future perspective.

12.2 Basic concepts for process intensification

A key goal of process intensification is to increase the "output" of a production facility. However, production facilities vary in their scales (e.g. 2000 L vs. 12,000 L bioreactor size), their purpose (i.e. launch vs commercial production vs mixed launch/commercial facilities), and their biologic production portfolio. Therefore, the precise definition of facility "output" for different facilities is sensitive to this heterogeneity in facility scale and facility purpose.

Most straightforwardly, for commercial production, the output of a facility may be measured in terms of kilograms of drug substance produced per annum. Process intensification in this context traditionally involves the pursuit of methods to increase upstream harvest titer without necessitating any substantial decrease in the harvest frequency (defined as production runs per week). For the downstream, this translates to the purification of increased product amounts in the given harvest frequency.

More subtly, for non-commercial production facilities (such as e.g. product-launch facilities), facility output might appropriately be measured in terms of production batches per year. Non-commercial production activities (e.g. generation of material for clinical trials, transfer runs or process performance qualification (PPQ) runs) generally require relatively smaller quantities of manufactured drug substance. To achieve the optimal utilization of manufacturing facilities, thereby leveraging manufacturing capacity for development pipeline throughput, increasing the number of available production slots – and thus increasing plant utilization – is of great importance.

To optimize output across all scenarios, the ability to increase both the harvest frequency as well as the harvest titer would be ideal. However, there is often an optimization challenge between increasing titer and realizing a high harvest frequency. Many methods that increase either the harvest titer or frequency can lead to decreased performance in the corresponding variable. Minimizing, or even avoiding, this trade-off, is a key challenge for upstream process development and manufacturing. Similarly, the accomplished titer and the available time for purification define the purification challenge in downstream process development and manufacturing. However, product quality and process related impurities represent further crucial factors with regards to downstream process intensification opportunities.

The appropriate definition of a facility's output is therefore dependent upon the commercial status of the biopharmaceutical and the specialization of the facility (commercial or launch). For upstream, the task of process intensification is to try to maximize

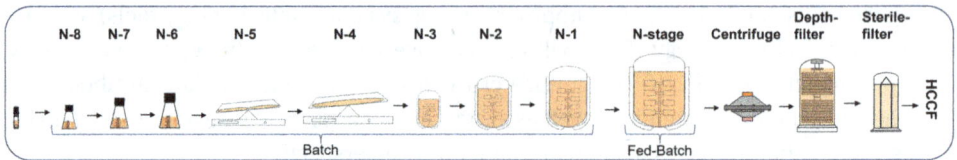

Figure 12.1: Schematic overview of a representative, commercial upstream process using stainless steel and controlled bioreactors. The individual cell cultivation from N-8 to N-6 is generally done in shake flasks of increasing size. N-5 and N-4 represent wave bioreactors with single-use bags. N-3, N-2, N-1, and N-stage represent stainless steel stirred tank reactors of increasing size with pH and dissolved oxygen (DO) feedback control (SS STR). Centrifugation is generally performed with disk centrifuges. Filtration steps include a dead-end depth filter and sterile filter.

harvest titer whilst simultaneously maintaining a high harvest-frequency capacity. Further, optimizing product quality lightens the burden for downstream processing. The task of downstream process intensification is to purify higher product amounts in shorter time, while meeting drug substance release criteria. Obviously, from a commercial production perspective, optimization of routine manufacturing has highest priority due to the high number of runs to be performed. In the following, the focus is on the upstream process as it triggers process intensification in the first steps. Technical aspects for both upstream and downstream are described in Section 12.3.

A typical upstream process starts with thaw of a frozen cell vial and ends with centrifugation and filtration steps (Figure 12.1). At the end of the upstream process the cell free fluid containing the product of interest, which is named harvested cell culture fluid (HCCF), enters the downstream process. The individual steps in an upstream process are generally grouped as following:

I. Seedtrain: thawing, N-8, N-7, N-6, N-5, N-4
II. Pre-stages: N-3, N-2 and N-1
III. Production stage: N-stage
IV. Harvest: centrifugation and filtration (depth and sterile filter)

Different strategies to intensify such an upstream process were described. Intensification strategies targeting the seed train mainly focus on a reduction of the overall process duration. Decreasing the duration of a seed train can be achieved with high-density cell banks (HDCB). A perfusion process capable of reaching high cell densities combined with high cell viabilities allows establishing required high-density cell banks. These cell banks are then used to inoculate bioreactors of increased size instead of starting the seed train with shake flasks. However, a perfusion process can also be directly integrated into the seed train. Thus, the number of passages during the seed train can be reduced and the production bioreactor can be inoculated with cells from a bioreactor of smaller size compared to a standard pre-stage bioreactor operated in batch mode. The application of high-density cell banks and seed-train perfusion processes were reviewed recently [22, 23].

Table 12.1: Overview of the investigated operational modes at N-1 and N-stage for process intensification.

Intensification strategy	N-1 operational mode	N-stage operational mode	Characteristic	Reference
Non-perfusion HSD Fed batch	Fed batch	Fed batch	High seeding density in N-stage achieved with N-1 Fed batch	[12]
HSD Fed batch	Perfusion	Fed batch	High seeding density in N-stage	[12, 24–32]
Perfusion	Batch	Perfusion	With cell bleed (stable operation)	[33–41]
Dynamic Perfusion	Batch	Perfusion	Minimal/without cell bleed	[42]
Intensified dynamic perfusion	Perfusion	Perfusion	N-1 perfusion followed by dynamic perfusion	[43]
Cascade	Perfusion/ Chemostat	Chemostat	N-1 process coupled to N-stage	[44]
Concentrated Fed batch	Batch	Perfusion	N-stage perfusion with product retention	[45]
Hybrid	Perfusion	Perfusion/Fed batch	N-stage perfusion with switch to fed batch	[21]

The focus of upstream process intensification is to maximize productivity (respectively, harvest titer for a specific cell culture duration) of antibody therapeutics by optimizing the production process. Currently, fed batch production processes are the standard in the biopharmaceutical industry. In a fed batch process nutrient limitations are prevented by adding highly concentrated feeds, thereby increasing culture longevity and finally productivity. In the last decades biopharmaceutical companies have constantly optimized these fed batch processes in combination with cell lines and media, thereby increasing titers to 3–8 g/L [8] and even above 10 g/L [14].

In the last decade, intensification strategies targeted the operational mode of pre-stage (N-1) and production stage (N-stage). Various combinations of operation modes at the N-1 and N-stage were investigated highlighting the potential to outperform fed batch processes. Table 12.1 shows a summary of the published strategies.

The overview in Table 12.1 highlights the current trend of investigating continuous process options – mainly perfusion – to intensify upstream processes. This trend is also reflected in a series of reviews from academia and industry related to process intensification and continuous bioprocessing [5, 8, 22, 23, 46–50]. Continuous bioprocesses are characterized by constantly adding media while removing spent medium at the same rate. If cells are removed with the spent medium the process is termed chemostat. If cells are retained in the bioreactor the process is called perfusion. In the following, challenges and possibilities related to perfusion processes are listed in addition to detailed description in the text for many items.

Challenges [48, 51].
– Technical challenges for scale-up and scale-down.
– Technical challenges regarding long-term operability and maintenance of sterility.
– Less flexibility to handle multiple products, due to long run times.
– Establishment of new upstream production platform.
– Integration into an existing (fed batch) development and manufacturing network.
– Higher cost and longer time required for process development at lab scale.
– Genetic instability of cells.

Opportunities [8, 22, 40, 49].
– High volumetric productivity (g/L/day) due to high cell densities.
– Smaller manufacturing facilities coupled with decreased capital investment.
– Reduced residence time for instable molecules.
– More uniform product quality profile.
– Efficient bioprocesses for new modalities (e.g. cell therapy, viral vectors).

Importantly, most of the publications as listed above related to different upstream intensification strategies show lab scale results (Table 12.1). Developing and scaling such processes to manufacturing scale poses additional challenges that are only partially addressed in these publications. Most of the biopharmaceutical companies have made significant investments in the construction of large scale (multi-product) fed batch facilities. Thus, many of the described intensification strategies aiming for N-stage perfusion are not transferable without massive reconstruction of these facilities (upstream as well as downstream), which is highly questionable to occur for business-management reasons but also from a capacity perspective provided that a running portfolio needs to be supplied. Due to the high capital expenditures for new production facilities – despite potential for smaller footprints – combined with cost pressure to constantly reduce operational expenditures, the transferability of processes into existing large scale production facilities plays an essential role.

In contrast, upstream process intensification using N-1 fed batch or N-1 perfusion followed by an HSD fed batch process offers the possibility to use existing production facilities [12, 23, 52]. Further, harvest and subsequent downstream processing can be operated batch-wise using the existing downstream manufacturing infrastructure. Despite (typically) minor modifications of the N-1 bioreactor being necessary (cell retention device and media tank), the fed batch production bioreactor does not need technical modifications. Further advantages are that process knowledge, control- and scale-up strategies can be leveraged from the existing fed batch platform. With the reported increase in titer or reductions in process run times combined with stable product quality attributes (Table 12.1; N-1 perfusion followed by HSD fed batch), the yearly facility output and overall facility utilization can be enormously enhanced (see detailed analysis in Section 12.9).

To summarize, the described possibilities for upstream process intensification highlight the potential to further increase productivity, thereby decreasing the production costs of therapeutic antibodies. While for some of the intensification strategies promising lab scale results were shown (e.g. concentrated fed batch [45]), other strategies are already brought into the production environment (e.g. N-1 perfusion [31, 53]). As described in the introduction, several aspects need to be considered for the development and commercialization of (intensified) bioprocesses, with the molecule portfolio and the (existing or newly to be constructed) manufacturing network being important drivers at the beginning and the end of the CMC value chain. From the perspective of a biopharmaceutical company having fed batch production platforms and facilities in place and aiming to increase the output of these facilities, it appears as a logical next step to bring N-1 perfusion processes and HSD fed batch processes into production. This allows available knowledge on how to develop robust, scalable, and transferable processes to be leveraged before embarking on a fully continuous bioprocessing future. In the case of specific project dependent requirements (e.g. driven by molecule needs such as product instability) or in case of investments in novel (single-use) facilities, the answer might be different.

12.3 Technical aspects and engineering principles for process intensification

To pursue process intensification, one may start with a fundamental technical question: "What limits facility output?".

In the context of non-continuous upstream processes in both existing as well as newly built facilities, the following four points serve as high-level answers to this question:
- Product specific – which range of molecule specifics/formats should be covered by an intensified technical platform
- Upstream harvest titer, when isolated from its influence on limiting downstream process scheduling, limits facility output.
- Upstream facility scheduling constraints – which are generally a function of upstream processing times (e.g. run times in N-1 and N-stage) as well as available upstream technical infrastructure (e.g. number of bioreactors in N-1 and N-stage) as well as organizational aspects (e.g. work schedules) – limit facility output.
- Downstream facility scheduling and capacity constraints – which are highly dependent on harvest titer. Purification of higher product masses requires an increased capacity of the individual unit operations or accelerated processing of the unit operations in a serial or parallelized approach whereby different cycling strategies can be applied. Also, organizational aspects (e.g. work schedules) apply.

All facilities are limited by at least one, if not all four, of the above dimensions depending on the characteristics of a certain production portfolio in a certain facility. In this section, an introduction to the technical and engineering factors of the listed limitations is given, and various strategies to overcome these are outlined.

12.3.1 Upstream

A bioreactor capable of running a continuous process requires additional hardware components compared to conventional batch or fed batch bioreactors. A continuous process is characterized by a continuous in-flow of fresh medium while removing culture liquid at the same rate thereby keeping the bioreactor volume constant. The control of the in- and out-flow rates is achieved by feedback control loops. At lab scale bioreactors weighing scales are generally applied while in larger scale bioreactors flow meters and filling probes are used for process control. In this chapter, technical aspects and engineering principles related to continuous bioprocessing are presented. The focus lies on equipment (cell retention device and pump) as well as scale-up considerations.

Perfusion processes are the industry standard for continuous bioprocessing of mammalian cells. In contrast to a chemostat, a perfusion bioreactor is equipped with a cell retention device to retain the cells within the bioreactor and to only withdraw spent medium. The principle of cell retention is based on either cell size, density, or aggregation under ultrasonic waves [46]. In the past, different methods to retain cells within the bioreactor have been described. They include membrane filters (static, dynamic or tangential flow), centrifuges, hydroclones, gravitational settlers and acoustic wave separators [46, 54–56]. Nowadays, the most widely used systems for cell retention are hollow fiber filters [56]. Various manufacturers exist that offer single-use and multi-use hollow fiber filters with surface areas suitable for lab to production scale. Other hollow fiber characteristics that generally can be chosen are the membrane material, pore size (micro and ultrafiltration membranes), lumen diameter and length of the hollow fibers. Manufacturer of hollow fiber filters for bioprocessing applications offer lab, pilot and production scale filters with constant lumen diameter and fiber length. Thus, the surface area is increased only by increasing the number of individual fibers. Furthermore, this allows keeping constant hydrodynamic conditions within the hollow fibers between scales when using the flow per fiber or wall shear rate as a scale-up parameter.

For perfusion processes external pumps are used to establish the tangential flow required for hollow fiber filters. Rotary lobe pumps are widely used in the biopharmaceutical industry, including perfusion applications [53]. However, it was shown that the small gaps and clearances in these pumps can cause high shear stress [57–59]. Since mammalian cells are sensitive towards shear stress, it is essential that for perfusion applications pumps with low-shear forces are used. Diaphragm pumps (e.g. ATF from Repligen) cause low shear due to the absence of narrow gaps and fast-moving components while they can create an alternating tangential flow (ATF) [33]. However, these pumps are available in limited size, which poses challenges during scale-up. Magnetically levitating pumps were also shown to have low shear forces [60, 61]. Apart from this, magnetically levitating pumps are available from lab to production scale and can be operated with single-use pump heads with no bearings to wear out or seals to breakdown. It was shown that the use of such pumps leads to a similar process performance when compared to an ATF diaphragm pump [62]. It was recently shown that also centrifugal pumps can be arranged to create an alternating tangential flow [63, 64].

Selecting adequate perfusion equipment is crucial for a robust and reliable scale-up of perfusion processes. Besides pumps and cell retention devices, the media requirements should be carefully considered. Depending on the process characteristics, the reactor volume is exchanged several times throughout the cultivation. Thus, adequately sized media hold-up tanks are required in large scale facilities. Downsizing of media hold-up tanks can be achieved by using media concentrates in combination with in-line dilution [18]. Recently, a study was presented that demonstrates the feasibility of recycling permeate in perfusion cultures [65]. Another characteristic of perfusion processes is that cell densities are generally higher compared to batch and fed batch processes. Processes with higher cell densities require more oxygen. Thus, it is essential to characterize the used large scale perfusion vessels in terms of the volumetric mass-transfer coefficient (kLa) and to determine oxygen uptake rates [66, 67]. The same applies to the mixing time. In case insufficient oxygen transfer or mixing capabilities are identified during scale-up preparations, either the process is adapted to the given limitations or hardware components (e.g. stirrer, sparger and mass flow controllers) are adapted to cope with increased cell densities. If characteristics of the produced molecule are further included, parameters such as the maximum allowable residence time in the bioreactor must be considered, which in turn limits the ratio between the reactor volume and the perfusion rate [22, 68].

To conclude, perfusion processes require robust and scalable equipment, which generally does not find application in batch and fed batch processes. Thus, knowledge about the equipment is key for a successful process development and scale-up. The published data also highlight that different equipment from different manufacturer led to robust processes thus fulfilling their requirements. But it is crucial to weigh pros and cons for optimal use. This means that a 4 to 6-day N-1 perfusion process certainly has different requirements to e.g. a hollow fiber filter than a 30-day perfusion process. The same applies to the scalability of the equipment. A perfusion process running in a 500 L vessel certainly requires differently sized pumps than a process running in 5000 L. However, tremendous progress has been made in terms of the availability of perfusion equipment for different scales by the vendors. This paves the way for developing robust and high yielding perfusion processes based on biochemical engineering principles and transferring them to manufacturing facilities.

12.3.2 Downstream

Downstream processes for antibody therapeutics comprise several unit operations. These usually include a protein A capture step, virus inactivation, polishing chromatography and filtration steps. The downstream process depletes product and process related impurities. The impurity profile and the product quality of the material supplied by the upstream process have significant impact on the design and yield of the downstream process. For instance, aggregated or fragmented product in the harvest titer decrease the total product yield. Further, clearance of undesired impurities, such as host cell proteins, is usually associated with an impact on process yield [69–71]. Separation technologies make use of physico-chemical principles. If impurities and the product have similar characteristics with regards to charge, size and hydrophobicity, the available separation technologies compromise yield to achieve high impurity clearance factors.

The downstream process design forms the basis for the dimensioning and layout during scale-up. The process layout sets the framework and determines the amount of product that can be purified. Multiple factors influence the process layout in downstream manufacturing: the harvest titer and volume, the harvest frequency and shift models, continuous or batch processing, and constraints of the downstream facility. Facility constraints for intensification of the downstream process can be diverse. Factors potentially limiting intensification in existing facilities include the shop floor area and the allowable floor load. Both are physical constraints for filter areas, allowable column dimensions, the number and size of buffer and product pool tanks and skids. Eventually, the available shop floor area might also limit parallelization opportunities. The water for injection (WFI) supply and the wastewater treatment can pose further constraints in existing facilities.

The harvest titer and volume, the harvest frequency and the mode of operation are related to the upstream process. These factors define the amount of product to be purified and the available time period, aiming for maximal facility output in kg per hour. With the introduction of N-1 perfusion to an existing batch facility, typically the titer rises. Titers of up to 10 g/L or higher can be expected. However, the harvest frequency does not necessarily increase. On the downstream side, the facility output can be either maximized by increasing the capacity of individual unit operations, or by decreasing the required processing times. Decreased processing times in the downstream operations can be achieved with serial or parallel processing strategies, process designs using fewer unit operations, or by faster processing, i.e. higher flow rates. Scale-up of unit operations or higher load capacities on the other hand increase the product output. A combination of these measures is possible, either between different unit operations or even within a unit operation. This higher degree of freedom allows greater flexibility for the implementation of intensified processes and there is a good chance that conventional separation technologies are capable to purify batch sizes of up to 100 kg [72]. However, the flexibility also creates a higher complexity, as intensification in downstream is not based on one core technology. Different approaches apply to the individual unit operations.

In context of chromatography columns, there are further technical constraints. In practice, 2.0 m is the upper limit of the inner diameter of an axial flow chromatography column. At 25 cm bed height, this translates into a maximum column volume of 785 L. Construction and manufacturing of larger inner diameter columns becomes difficult with regards to stability. Based on our experience, the capex for large columns grows exponentially. For pre-packed columns, available column volumes are far smaller. Radial flow columns promise to have lower space and weight demand at large-scale. Saxena and Weil have already reported about the potential of radial flow chromatography for scale-up in the 1980s [73]. However, the flow properties in radial columns are different from axial columns. Small-scale development solutions and scale-down models for radial flow chromatography columns are less intuitive, compared to axial columns. Ideally, small-scale experiments use traditional axial flow columns and are still representative for large-scale in radial flow columns. Besselink et al. demonstrated comparability between the breakthrough performance of axial and a radial flow-based affinity chromatography [74]. As more comparative data becomes available, radial flow chromatography might represent a good opportunity to further increase column volumes.

The preferred option to intensify column chromatography is to increase column loads. In recent years, significant progress has been achieved. Various vendors have commercialized high-capacity resins that demonstrate also appropriate selectivity for the purification challenges of antibody therapeutics [75–77]. Especially high-capacity protein A resins are important in context of process intensification. Resin suppliers have advanced load capacities of protein A approximately two-fold, for instance by oligomerization of the protein A binding motifs [78, 79]. Capturing of the product from HCCF at high load capacities and elution at high product concentrations affects the pool volumes of subsequent steps. The application of a high-capacity protein A step further allows to either maximize the amount of captured product per cycle. Alternatively, the cost of goods can be reduced in early clinical supply, as protein A column volumes can be reduced. The cost of goods is comparatively smaller for typical polishing resins. Thus, serial operation is less attractive and operation of polishing columns in a single cycle per product batch is favorable in many cases. High-capacity ion exchange resins enable polishing of large product amounts in bind-and-elute operation [80]. Flow-through chromatography steps are usually not a technical constraint for process intensification. Flow-through operation might be a smart choice for intensification of hydrophobic interaction chromatography. Typical capacities for hydrophobic interaction chromatography resins are not competitive to modern ion exchange resins [81].

Reducing residence times is a third option for intensification of column chromatography. To purify increasing product amounts, this approach needs to be combined with a cycling strategy. Further, optimization of the residence time requires careful balancing, as the residence time correlates with the dynamic binding capacity. Multi flow rate loading strategies address this correlation. The loading capacity of a protein A affinity step could be increased by 20 %, using such a multi flow rate loading strategy [82]. In existing facilities, the maximum flow rates accomplished by the pumps pose a further limit to chromatography and filtration unit operations. These flow rates must further be applicable to the fixed piping, at least in context of a stainless-steel facility.

Intensification of the filtration steps can also contribute to process intensification. This includes depth filtration, virus filtration and tangential flow filtration. The use of modern filter products can significantly reduce the required time for these process steps [83]. Chemically defined filters may be considered to reduce the buffer and WFI consumption. Optimization of filter loadings and pool concentrations can further increase the capacity of the downstream process [84, 85]. In case of tangential flow filtration, the formulation of the drug substance plays a role. Highly concentrated formulations with comparatively low viscosity are beneficial with regards to process intensification. Low viscosities even at high product concentrations enable robust and fast processing. The development of tangential flow processes that use only one diafiltration buffer can further decrease the buffer/WFI consumption. In this case, appropriate control over the Donnan effect is key. For instance, control can be accomplished by *in silico* modelling [86].

Various publications investigate the potential of non-conventional separation technologies with regards to process intensification. Investigated approaches include aqueous two-phase systems, multi-column chromatography, single-pass tangential flow filtration (SP-TFF). Madsen et al. provide a comprehensive review on SP-TFF [87]. The technologies address different pain points related to the intensification of downstream processes. SP-TFF concentrates large (intermediate) pool volumes in one single pass. The serial configuration allows fully continuous operation. SP-TFF can reduce pool volumes to levels that can be handled in existing facilities. This effect is especially useful if SP-TFF is applied early in the downstream process. A sufficient product stability at high concentrations in intermediate pools is a pre-requisite. Further, the subsequent purification steps must robustly perform at high (intermediate) pool concentrations. Protein A chromatography is a major driver for the Cost of Goods (see Section 12.9.2) in the downstream process. However, to our best knowledge, selectivity of protein A is yet unmet. Alternative process trains without protein A capturing suffer from lower impurity removal, lower yield. For instance, aqueous two-phase systems can be applied for capturing [88]. Other examples use a chelating agent and/or a series of filtrations. Corresponding data either indicates lower impurity removal compared to protein A or is not published [89, 90].

Continuous process options for the downstream promise a reduced footprint, less buffer consumption, and higher productivities [91]. Continuous operation of chromatography columns is accomplished by multi-column chromatography. Two or more chromatography columns are operated with switching valves. In general, the breakthrough of the first column during loading phase is adsorbed by the next column, which allows for full utilization of the resin capacity. Continuous operation of TFF can be accomplished using an SP-TFF system [87]. Continuous approaches for virus inactivation include multiple incubation chambers [92] and the use of a coiled flow inversion reactor coupled to a protein A capturing step [93].

The previously described, alternative technologies come with a higher risk for scale-up, process transfer and process validation [72, 91]. This additional risk must not lead CMC to the critical path, especially if conventional separations technologies are capable to solve the challenges of process intensification. Further implications apply to continuous processes. Variations in the upstream process need to be accommodated in a continuous downstream process. Integrated process models (see Section 12.5.2) and PAT (see Section 12.6.2) are supportive tools for the implementation of continuous downstream processes [94].

12.4 Chemically defined cell culture media for intensified CHO processes

12.4.1 General requirements

Chemically defined (CD) media are routinely used in the production of biologics in CHO cell culture and provide enhanced raw material control. Nutrient optimized CD media is an important path to increase cell growth and mAb productivity in recombinant CHO cell lines [95].

Cell culture media design is a significant part of overall process development since cells depend on the nutrient sources in the media to grow and survive. Cell growth, product accumulation, and product quality have a direct correlation to the composition of media [96, 97]. Cell culture media formulations are optimized to meet cell demand for nutrients and to diminish formation of inhibitory byproducts (e.g. lactate, ammonium) which then, impact process performance. Three important considerations when developing media formulations for mammalian cell culture need to be ensured. They need to be stable, promote high-density growth, and enable high therapeutic protein production of appropriate quality [98].

12.4.2 Media for pre-stage perfusion

The high cell densities to be achieved in pre-stage perfusion to inoculate an HSD fed batch culture require an adequate supply of substrates, which is necessary for optimal cell growth and productivity. In addition, perfusion allows continuous removal of undesirable metabolites produced by mammalian cells. Therefore, an ideal perfusion culture medium needs to support a perfusion rate that provides the right balance between the addition of substrates and the removal of toxic metabolites [95]. Running an HSD fed batch process needs higher volumes of prepared media due to the N-1 perfusion step to enable high seeding cell densities of up to 5–15×10^6 cells/mL for the intensified production stage. Therefore, more batches have to be prepared and stored before entering the process. This requires higher amounts/volumes of raw materials such as complex base powders and consumables (e.g. membrane filters for final medium filtration/sterilization). Obviously, these media aspects result in additional costs and efforts. However, both with respect to process economics (Section 12.9) as well as sustainability (Section 12.10), these investments pay off for an overall CHO process.

12.4.3 Media for HSD fed batch processes

In order to optimize the N-stage production process it is well known that the concentration of metabolites in the medium needs to be carefully tuned. A "poor" medium may not provide enough nutrients to the HSD fed batch culture, whereas a too "rich" medium may be disadvantageous because either the cells do not use all of the available nutrients, or worse, they "over-consume" those and turn them into toxic byproducts [99].

For an HSD fed batch medium optimization it is the primary goal to increase the cells longevity and specific productivity. Peak cell densities are no primary focus since they can have a negative effect on process performance. However, the integral of the cell density is more important than the peak cell density and can be regulated by seeding cell density. Therefore, studies mainly focus on the investigation of metabolic bottlenecks, inhibitory metabolites, or/and antioxidant capacity.

In the very early days of cell culture medium development, various substances were added to the cell culture process to prevent the cells from running into metabolite bottlenecks [96]. Today, finetuning based on the specific product and the associated cell specific demands becomes more important. In particular for intensified processes with increased cell numbers balancing the cellular pathways and also considering the generation of inhibitory metabolites by media adaptations is one of the big challenges. The analysis of potentially inhibiting metabolites revealed an early accumulation of these substances in HSD cultures. Future work might address the reduction of these metabolites by media adaptions targeting their metabolic precursors, similar to the work from Mulukutla et al. [100]. The prolonged productivity phase of an HSD fed batch poses new challenges for the production cells. For example, Brunner et al. revealed distinct metabolic phases with high cell-specific productivities and high formation of reactive oxygen species (ROS) in an HSD process [32]. Consequently, an optimized feeding strategy was applied. Chevallier et al. and Henry et al. also described the reduction of ROS species via media optimization of antioxidant compounds such as vitamins, thiols, or α-ketoacids as well as genetic engineering approaches that can further improve culture longevity in HSD processes [101, 102]. There are different tools available that can be used, also in combination, to identify optimization possibilities. RNA data in combination with an FBA (Flux Balance Analysis) was used by Brunner et al. to identify metabolic phases with high formation of ROS and high cell-specific productivities in the HSD cultures [32]. Quantitative analysis of metabolites that are present inside and outside the cell provides evidence which pathways and reactions are active in the cell under given culture conditions [103].

Such data can be integrated into metabolic models. Brunner et al. used metabolite data and metabolic modeling to reveal the accumulation of inhibiting metabolites early in the process and FBA led to the assumption that ROS were contributing to the fast decrease in cell viability [32]. Based on this data a titer increase of up to 47 % was achieved by adapting the feeding strategy and balancing the lactate/cysteine ratio, and even increased by 97 % when compared to the fed batch process [32] Johnson et al. observed an increase in product titer for a fed batch process by almost 30 % (from about 7 g/L to 9 g/L for a 14 d CHO process [13]) or by about 67 % for an N-1 perfusion pre-stage followed by an HSD fed batch process (from about 6 g/L to > 10 g/L [104]). Salim et al. demonstrated in their study an increased cell growth and mAb productivity when using a medium that was optimized based on coupling multivariate data analytics with amino acid stoichiometric balances [105].

12.5 Scientific approaches supporting process intensification

As for the development of CHO based fed batch processes, also intensified process formats such as HSD or continuous processes benefit from advancements in mammalian cell culture and process science. Some relevant areas such as cell line development (Section 12.8), media development (Section 12.4) and process analytical technologies (PAT, Section 12.6) are described in more detail in the referred sections. The importance of single-use technology in various areas also has been mentioned, some of them enabling significantly higher throughput and higher degree of automation (such as the Ambr® system, to provide one specific example for cell culture). Yet, many more exist for the development of cell lines, cell culture processes, media, downstream processes, analytical methods as well as in pharmaceutical development. In general, these aim to implement high throughput-screenings (HTS) which are integrated into automated systems and digital environments including e.g. laboratory information management systems (LIMS), experimental planning approaches (e.g. design of experiments), comprehensive databases, data visualization and analysis tools and reporting functions. These are not described in detail here but must be mentioned as they are important enablers for scientific progress. Given the extent and depth of a biopharmaceutical value chain in development (and manufacturing) as well as that e.g. non-GMP and GMP regulations apply, digitalization efforts are complex and rely on commercial tools but also require profound internal IT know-how to achieve end-to-end solutions. On the one hand side this is challenging. On the other hand, it is a prerequisite in an industrial setting where time constraints and resource pressure are common boundary conditions.

In the following, two further areas will be shortly mentioned. One is the field of systems biotechnology, the other is process modeling. Modeling with focus on facility utilization and cost of goods can be found below in Section 12.9.

12.5.1 Systems Biotechnology

The spectrum of systems biotechnology is broad. From an experimental and data perspective it covers genomics, transcriptomics, proteomics, and metabolomics but can include more areas such as epigenomics or lipidomics. From a data analysis perspective, computational approaches are necessary to cope with the huge amount of data generated using these tools. These are described in Section 12.5.2. Although big data sets from systems biotechnology approaches are often challenging to interpret and also are not required or necessary in many cases (e.g. when the focus is on detailed biochemical mechanisms versus large-scale system analyses), they provide such a wealth of information that their usage will continue to expand. A widely applied technology is next-generation sequencing (NGS) to support cell line development, engineering as well as characterization. Examples from our development comprise genomics [106, 107], epigenomics [108], media development [32, 109] as well as process development [28]. Media development aspects and the use of systems biotechnology tools for the development of HSD media are also briefly discussed in Section 12.4.3. Many publications are available from academia but also industry. A condensed overview and introduction was published by Stolfa et al. [110]. A broader overview on CHO systems biotechnology was provided in two dedicated Biotechnology Journal issues by Borth and Hu [111].

12.5.2 Advanced data analysis and modeling

The availability of large-scale data sets from research and development (e.g. from systems biotechnology approaches) as well as from manufacturing require advanced data analysis and modeling tools. In addition, many research based data are publicly accessible, also data tools and model repositories exist. In the field of systems biotechnology, in short, statistical approaches, stochiometric models (e.g. metabolic flux analysis (MFA) or flux balance analysis (FBA)), kinetic models (mass balances and kinetics captured in differential equations) and combinations thereof are applied. Whereas metabolic models are particularly useful to increase understanding on the cellular level or for media design [32], kinetic models, for example, support the design of optimized process regimes in N-1 perfused systems [28]. Modeling of cell culture processes faces the challenge that mammalian cells are inherently complex (e.g. metabolic networks underlie regulatory control) in addition to the biochemical process model and the manifold interactions with the cell culture medium. This is one reason why statistical or hybrid approaches appear promising.

Statistical and hybrid models can also be established for the downstream process and at the interface between upstream and downstream. Design of experiments (DoE) approaches are commonly used [112]. The optimization of cell culture media with regards to monoclonal antibody titer and the host cell protein profile is one published exampled [113]. Deeper physical understanding of the downstream process is provided by mechanistic models. Significant progress has been achieved in recent years employing mechanistic modeling in the development and characterization of downstream processes for therapeutic antibodies and antibody formats [114]. Recent improvements include a standardization of model calibration workflows [115] and predictions across different molecules based on protein descriptors [116]. In addition to novel modelling approaches for chromatography [117], we have progressed the development of mechanistic models for ultrafiltration and diafiltration (UF/DF) [86] as well as pH adjustment steps required for the integrated modeling of multiple DSP unit operations [118]. Coupling of models for individual unit operations forms digital twins of the entire downstream process. Digital twins enable real-time process decisions and increase process understanding [94, 119]. The impact of critical process parameters on a certain critical quality attribute can be determined across the whole downstream process. This additional process knowledge can be used to improve process efficiency. Further, we invested in the qualification of unit operation models to facilitate the application of mechanistic models for *in silico* process characterization studies [120, 121].

Other approaches, such as machine learning or artificial neural networks, have been described in [122]. Generic and specific recurrent neural network (RNN) models can be used to predict the influence of critical parameter variations on key process indicators such as metabolites, cell growth and product titer [123]. Eventually, specific unit operation models in upstream and downstream need to be coupled to allow for a comprehensive understanding along the process steps in a manufacturing process. Obviously, this task is even more demanding. However, steps are being taken towards such holistic process models which in our example use the Bayesian inference and numerical Markov chain Monte Carlo calculations [124].

To support the scale-up and transfer of intensified processes an in-depth knowledge of production equipment properties is essential. Experimental characterization of equipment such as stirred tanks is usually supported by computational fluid dynamics (CFD), which can be applied over multiple steps of the biopharmaceutical production chain, using dedicated simulation approaches for the determination of power numbers, flow fields, energy dissipation, mixing times and shear rates [125–127]. Newly popular lattice-Boltzmann (LB) simulations can accelerate classical CFD modeling and have been recently validated [128]. An application example is given by Kuschel et al., where the scale up from a 2 L glass perfusion bioreactor to 100 L and 500 L disposable perfusion pilot scale systems is investigated [129].

12.6 PAT tools

The concept of process analytical technology (PAT) is based on the idea of designing, analyzing, and controlling pharmaceutical manufacturing by measuring critical process parameters (CPPs) that affect critical quality attributes (CQAs) with the goal of delivering a product with consistent quality and performance [130]. To achieve this, it is necessary to understand the effects of process inputs such as process parameters, metabolites, and raw materials on the CQAs in a given design space. To control or influence a CQA, having the process inputs available as real-time information is advantageous [94]. A variety of sensors are available that can be used as PAT tools for real-time monitoring of process parameters in biopharmaceutical processes. The utilization of standard equipment such as probes for temperature, pH and dissolved oxygen in bioreactors, $UV_{280 \text{ nm}}$, pH and conductivity for chromatographies or pressure and flow sensors in filtrations already enable monitoring and automated control of certain process parameters in mAb production processes. Recent advances in sensor technology as well as instrumentations with smaller footprints and high-throughput capabilities expanded the portfolio of in-line and at-line technologies for real-time monitoring of biopharmaceutical production [131].

12.6.1 Upstream

In the following, three PAT tools are highlighted which are becoming highly relevant in-line and at-line analytical methods supporting process understanding, control and knowledge in common and intensified upstream processes. Each technology is already being used in the industrial environment in both the development and the manufacture of CHO bioprocesses for therapeutic antibody production.

12.6.1.1 Raman spectroscopy

Raman spectroscopy is a method of vibrational spectroscopy that measures scattering events in which photons lose energy and excite molecular vibrations from the ground state to the first excited state (known as Stokes scattering). Weaker anti-Stokes scattering events (photons gain energy by returning molecules from an excited state to the ground state), are often neglected in biopharmaceutical applications [132]. Raman spectroscopy with autoclavable probes has been used in a variety of cultivations with CHO cells for online prediction of substrates [133–135], metabolites [133, 134], cell count [136, 137], titer [136] and, in recent years mAb product quality attributes [38, 138, 139]. An extended overview of monitoring targets in CHO bioprocesses using in-line Raman spectroscopy is described by Park et al. [140]. Online real-time data from Raman spectroscopy utilized in a two-substrate (glucose and arginine) feedback control loop, demonstrated a significant reduction in the cumulative volume of multicomponent feeding compared to a dynamically fed glucose-based culture without influencing the culture performance [134]. In a different example, Raman spectroscopy was used to control both glucose and lactate at predetermined concentration over the duration of fed batch CHO cultivation. This feeding strategy resulted in lower ammonium concentrations and, more importantly, an increase in mAb galactosylation levels of about 50 %, demonstrating the capability of real-time feedback-control of multiple substrates to extend the possibilities of influencing mAb product quality [141].

Besides using real-time monitoring data for direct control of substrates like glucose, lactate or various amino acids by adjusting the nutrient feeding rate, high temporal resolution data provide additional value in the development and application of *in silico* cell culture models. Several industrial experts agreed that in line Raman Spectroscopy as

PAT tool is of high business value in both development and production bioreactors [142]. Already used for glucose control in cGMP applications, the technology enhances the development of robust and intensified bioprocesses. As such, the technology is maturing into a standard sensor in industrial upstream bioprocesses.

12.6.1.2 Capacitance probes

The use of capacitance probes for in-line measurement of biomass is already an established method in CHO cultivations. The capacitance measurement for viable cell count prediction has been described for several scales (including GMP) and has been used, for example, for automation of seed trains, automatic feeding based on calculated number of viable cells or control of cell-specific perfusion rate (CSPR) [143–146].

In the exponential growth phase, the capacitance correlates linearly with the number of viable cells, provided that the cells have a constant diameter. However, the correlation may no longer be linear due to physiological changes, such as change in cell diameter or signal interference from non-viable cells [143, 147, 148]. By using multi-frequency capacitance signals and multivariate data analysis (MVDA) such as PLS models, the prediction of viable cell count can deal with changing cell diameters and signal interference [143, 149]. Furthermore, additional information and predictive capabilities such as early detection of apoptosis can be extracted from the frequency spectra [148, 150].

In N-1 perfusion, the CSPR is a critical process parameter that describes the volumetric rate at which the system is perfused relative to the number of viable cells. It affects both the quality of cells generated for the following fed batch process and the consumption of expensive cell culture medium [25, 144]. Capacitance measurement is ideally suited for accurate control of CSPR due to the typically high viabilities in perfusion with near constant cell diameter [151].

In a study by Schulze et al., two different CSPRs were established using capacitance probes to investigate the effect of CSPR on cell performance, metabolic profiles, and apoptotic parameters in N-1 perfusion as well as performance in the subsequent N-stage fed batch [25]. In a study by Rittershaus et al., capacitance controlled CSPR was implemented in a platform process. Compared to the optimized volume-specific offline controlled exchange rates, media consumption could be reduced by 24 % with capacitance controlled CSPR without affecting viability [144].

Capacitance measurement is a mature PAT technology for biomass prediction with GMP applications in both N-1 perfusion and N-stage. For N-stage cultures, the incorporation of multiple frequencies allows accurate prediction of the number of viable cells despite changes in cell diameter. Inline capacitance probes are essential for the development of N-1 perfusion platforms. By using capacitance measurement as part of a PAT strategy, advantages are gained in controlling the critical process parameter CSPR and monitoring number of viable cells, resulting in improved process robustness and contributing to process intensification.

12.6.1.3 Multi-dimensional fluorescence spectroscopy

Another analytical application with non-invasive inline measurement probes as well as a wide use in off-line analytics is multi-dimensional fluorescence spectroscopy, also called 2D fluorescence, 2DF or 2D-FL, in which excitation-emission matrices (EEM) are collected over varying excitation and emission wavelengths. In addition to applications for in-line prediction of total cell count, viable cell count, viability and titer as well as the development of batch evolution models in CHO processes [152–154], 2D-FL using in-line probes shows great potential for cell culture medium quality assessment [155] and monitoring of the cell culture medium preparation process [17].

Although not (yet) routinely used in industrial applications, several applications in media preparation have already been described, such as evaluating the effects of light-induced changes in media composition [156], quantitative monitoring of photo-degradation in cell culture media [157] and predicting the efficiency of fed batch media after storage-induced changes [158]. This demonstrates the added value of 2D-FL in broadening the understanding of media degradation and its influence on cultivations.

In a study investigating the effect of media preparation parameters on media quality, 2D-FL was able to distinguish between different introduced variations caused by varying trace element concentration, lot, and supplier based on a PCA model fingerprint [17]. By monitoring the entire media preparation process with 2D-FL, Brunner et al. were able to understand which step in the media preparation process caused the variation of fluorescence, providing a valuable basis for root cause analysis [17]. In a design of experiments (DoE) approach, the authors investigated the effect of critical scale-up media preparation parameters such as temperature, specific power input, and preparation time. Their data showed that different degrees of media degradation occurred depending on the preparation conditions which could be detected with 2D FL and to some extent even quantified using PLS regression models.

Competing with other inline PAT tools such as Raman spectroscopy and biocapacity for monitoring of cell culture processes, 2D-FL shows unparalleled potential for monitoring media degradation and preparation processes, providing cross-scale insight into media preparation and thus improving cell culture cultivations through media robustness.

12.6.1.4 Summary

The three highlighted PAT tools all apply to the concept of measuring critical process parameters that affect critical quality attributes (Table 12.2). Generating enormous amounts of data with high temporal resolution, PAT challenges scientists and engineers with the need for advanced data analysis, while also paving the way for advanced process development. The application of PAT-based technologies, automation, and advanced data analytics enables a comprehensive understanding of biopharmaceutical processes to define a robust QbD strategy for manufacturing [131].

Table 12.2: Comprehensive overview of current examples of highly relevant in-line and at-line PAT tools that support process understanding, control, and knowledge in both common and intensified processes.

Technology		Examples for application	Reference
Raman Spectroscopy	Process monitoring	Amino acids and antibody N-glycosylation in perfusion culture	[38]
		Glycation and glycosylation at lab- and manufacturing scale in fed batch culture	[138]
		Size variants and high mannose species in perfusion culture	[139]
		Amino acids in fed batch culture	[159]
		Antibody glycosylation site occupancy in batch culture	[160]
	Feed control	Glucose and arginine feed control in fed batch culture	[134]
		Glucose and phenylalanine in fed batch culture	[161]
		Glucose and lactate in fed batch culture	[141]
		Glucose in presence of high autofluorescence	[162]
Capacitance Probes	Detection of apoptosis	Earlier detection compared to traditional viability tests in an at-line setup	[150]
		Detecting shifts in cell physiology and utilize information to partial or full recovery from early apoptosis	[148]
	Process control	Impact of N-1 CSPR on N-stage performance	[25]
		Automated feed strategy based on capacitance in N-stage	[146]
		Automated dilution of seed train cultures and prediction of glucose demand in GMP manufacturing	[145]
		Implementation of CSPR control in a platform process and utilization of capacitance data in N-1 scale-up	[144]
		Multi-frequency capacitance measurements to compensate for varying cell diameters	[143]
Multi-dimensional fluorescence spectroscopy	Process monitoring	Prediction of total and viable cell count as well as insight into different cell statuses inside the bioreactor	[152]
		Prediction capability of process relevant fluorophores and evaluation of a batch evolution model	[153]
		Quantitative prediction of glycoprotein yield	[154]
	Medium quality assessment	Impact of ambient light induced media photo-degradation on media quality, cell physiology, and titer	[156]
		Rapid quantitative analysis of media degradation caused by ambient light exposure.	[157]
		Prediction of fed batch media efficiency after storage-induced changes	[158]
		Assessment of the impact of preparation parameters on media quality	[17]

12.6.2 Downstream

PAT tools can also contribute to intensification of the downstream process. Real-time data for critical quality attributes and process indicators can reduce potential technical stops during processing. The time required for off-line analyses of in-process samples can sum up to several hours for a typical monoclonal antibody downstream process. For instance, real-time concentration measurements during TFF promise process time savings, replacing one or several offline measurements. Published approaches include mid-infrared spectroscopy [163], near-infrared spectroscopy [164], and Raman spectroscopy [165]. Near-infrared spectroscopy and Raman spectroscopy further promise to provide excipient data, replacing other off-line high pressure liquid chromatography analyses.

Benefits of PAT further extend to a modified control strategy. The breakthrough of a monoclonal antibody loading onto a protein A column can be monitored in real-time. Applied methodologies are partial least square regression modelling on UV/Vis absorption spectra [166] or near-infrared spectroscopy [167]. The real-time monitoring allows increasing protein A load capacities, accounting also for changes over the resin life time. Higher utilization of the protein A capacity is of special interest, considering protein A resins can account for a major part of the cost derived from long-lead items in the downstream process. Similarly, near-infrared sensors can also be used to monitor titer variations from 1-7 g/L in continuous or perfusion upstream processes [168]. A dynamic control strategy can be established, accommodating variations in the upstream by adaptations in the downstream process.

The referenced PAT tools for downstream processing do not represent a complete list. Several other technologies are being pursued, also for other quality attributes and process indicators. To-date, most of the solutions are still in development stages. Especially the quantification of very low impurity levels towards the end of the downstream process is challenging, yet necessary to ensure patient safety and to fulfill regulatory requirements [169]. Further work needs to demonstrate accuracy, robustness and scaleability of PAT suitable for the downstream process.

12.7 Examples for intensified CHO cell culture processes producing therapeutic antibodies

12.7.1 HSD fed batch process in stainless steel bioreactor

HSD processes shift the cell mass accumulation to the expansion bioreactor phase, allowing for high cell concentration in the N-1 stage by use of perfusion bioreactors. This enables 5–15 times higher seeding densities for the production stage [170] (see also grey line in Figure 12.2 for fed batch reference). Thus, the unproductive-growth phase in the production stage is skipped and product formation starts immediately after seeding. An

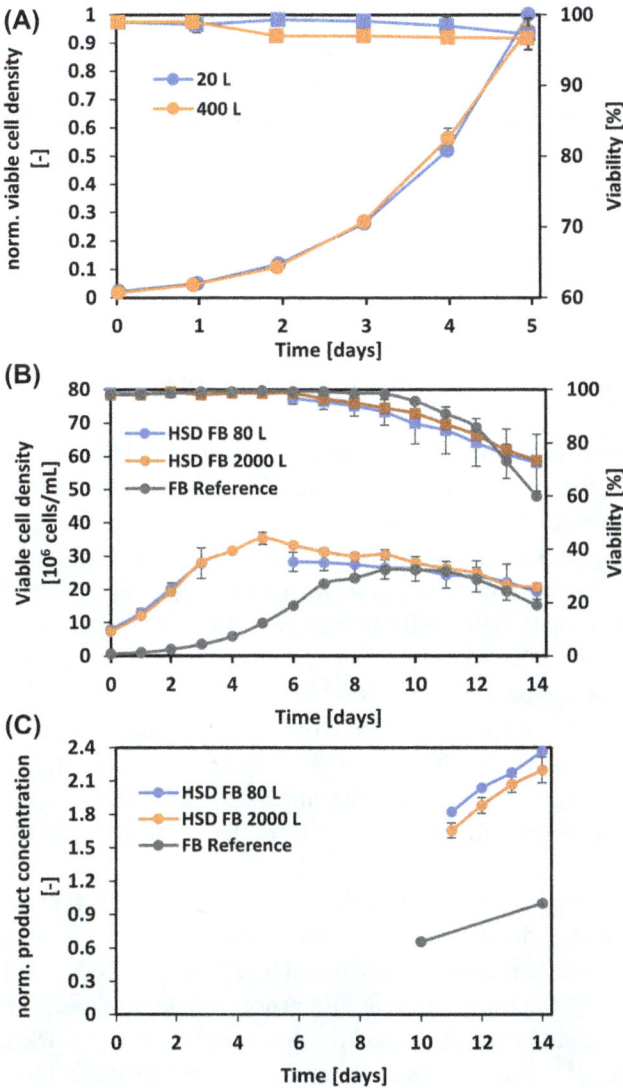

Figure 12.2: Scale-up of a perfusion process from 20 L (*n* = 3) to 400 L (*n* = 3) stainless steel tank (A), scale-up of the HSD fed batch from 80 L (*n* = 2) to 2000 L (*n* = 3) stainless steel tank (B) and corresponding normalized product concentration during the N-stage fed batch process (C). As reference a fed batch in 2 L (*n* = 1) scale is included. Viable cell density (circles) and viability (squares).

exemplary HSD process (Figure 12.2) was developed in lab- and scaled-up to pilot scale. In numbers this means, the N-1 perfusion step was scaled-up from 20 L to 400 L and the N-stage from 80 L to 2000 L for a highly complex, multispecific antibody. A 5-day perfusion process enabled high seeding densities of 7×10^6 cells/mL in the fed batch N-stage by achieving high viable cell densities (VCDs) and a viability of about 97 % at transfer. The production reactors were harvested after 14 days with a VCD of around 21×10^6 cells/mL and a viability ranging between about 68 and 74 %. Besides VCD and viability, product concentration proved to be comparable over both scales demonstrating robustness and scalability of the HSD process applied for this molecule. Overall, process output was improved by 120 % compared to the fed-batch. Additionally, product quality for the scale-up of the process from lab (80 L, 200 L) to pilot scale (2000 L) was comparable.

12.7.2 Continuous perfusion process in single-use bioreactor

As discussed in previous sections, different process intensifications can yield more efficient processes depending on the market need and facility fit as well as the facility availability. Initially, as discussed in the introduction, continuous processing was performed at steady state or fixed low cell densities due to limitations in technology. In the recent years, advances in cell retention devices and media concentrates have enabled N-stage perfusion cultivation to push to higher cell densities. In addition, newer dynamic N-stage perfusion processes typically have no cell bleed and thus are more productive. Figure 12.3 shows a dynamic N-stage perfusion in a 100 L single use bioreactor in comparison to a fed-batch. The N-stage which was seeded from an N-1 perfusion, demonstrated high cell densities reaching $>120 \times 10^6$ cells/mL and an 8-fold increase in cell density compared to the fed batch process. Similarly, the cumulative productivity of the continuous process exhibited an 8-fold increase. Product from the production N-stage perfusion process was continuously harvested and purified.

Currently, most newly developed continuous perfusion processes are performed in flexible SUB facilities. When considering the scale of operation of these processes, a 500 L SUB continuous process would produce more than double that of a 2000 L fed batch process in such an example. Such a continuous SUB process could thus easily be scaled-out allowing for an efficient and flexible facility that could meet the market demand for multiple products. However, for existing larger stainless-steel facilities, continuous perfusion processes require numerous modifications in order to implement as described.

12.8 Cell line and clone selection aspects

CHO cell lines are the main expression host for complex biopharmaceuticals such as monoclonal antibodies, enzymes, and hormones due to the ability of proper protein folding, post-translational modifications, and the ability to grow in suspension in large

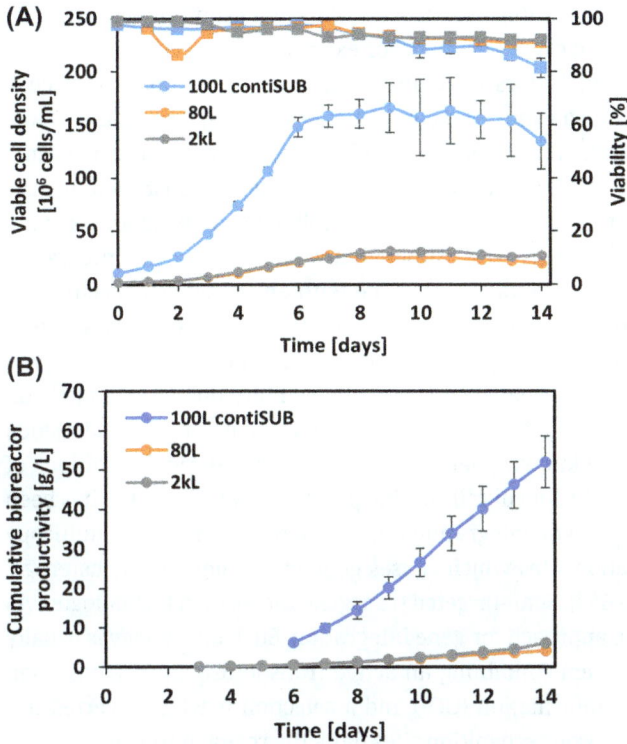

Figure 12.3: Comparison of intensified N-stage perfusion process (blue lines, n = 3) in a 100 L SUB to standard large scale fed batch process in 80 L (grey lines) and 2000 L (black lines) stainless steel bioreactors. (A) Viable cell density (circles) and viability (squares). (B) Cumulative productivity per bioreactor volume.

bioreactors [171]. Typically, hundreds to thousands of stable single cell clones are screened in a labor-intensive and time-consuming cell line development process to identify the optimal manufacturing cell line [172]. Beside upstream process performance, cell line stability and product quality are evaluated. As mentioned previously, the trend of process intensification is striving towards increased process outputs, reduced facility footprints and improved facility utilization [11]. Among others, Chun et al. [48] summarized that intensification in an upstream process can be achieved considering essentially the following three variables impacting the volumetric productivity (per definition the mass of product per bioreactor volume and time unit). These are improved cell specific productivity, increased total viable cell mass, and enhanced product formation phase (vs. cell growth). In all intensification approaches, clonal cell lines, being the production factories in biopharmaceutical processes, play a crucial role. Hence, cell line development activities are integral part of an intensified development and manufacturing platform.

Depending on the detailed intensification strategy, different requirements might be set during cell line generation and clone selection. For example, cell specific productivity is highly dependent on different factors including complexity of the target molecule, vector design and transgene integration technology. A state-of-the-art cell line development workflow is based on random transgene integration whereby a linearized plasmid encoding for a gene of interest (GOI) as well as a selection marker is transfected into the host cell nucleus to be stably integrated into the genome following a metabolic selection process to recover transgenic mutants [173]. Although a random transgene integration technology might lead to a final manufacturing clone with enhanced cell specific productivity, obtained clones are not generally suitable for intensified processes due to phenotypic and/or genetic instabilities which does not allow extended run times. In case of continuous manufacturing a key challenge is cell line stability due to extended run times over several weeks in intensified processes [174–176]. Though not fully understood yet, cell line instability is a well-known phenomenon during long-term cultivations of CHO cell lines that can lead to a decline in cell specific productivity [177–179]. It has been demonstrated that random transgene integration is prone to the formation of multicopy concatemers at single integration sites which causes genetic and phenotypic heterogeneity and thus instability [180–183]. Semi-targeted transgene integration technologies via transposase are an alternative approach for gene integration. Such technology is usually based on a two-component system containing an active transposase and a transposon vector comprising the genetic information (GOI) and a selection marker. Inverted terminal repeats (ITRs) and transposon recognition sites flank the region harboring the GOI and selection marker. The transposase enzyme recognizes the ITRs and transfers the integrated transposon element into the host cell genome via a "cut-and-paste" mechanism into transcriptionally active and/or accessible genomic regions as mostly single intact copy integration at multiple sites across the host cell genome avoiding concatemer-induced recombination and genetic instability [184, 185]. Due to a reduced likelihood of phenotypic instability via transgene silencing or loss of transgene copies, semi-targeted transgene integration enables process intensification with regard to extended run times.

From a cell line perspective, the cell production itself and the harvestability play an important role for the overall manufacturability, i.e. not only productivity but also the cellular growth rate, harvest viability, or the level of process-related impurities might be ranked of higher importance during clone selection compared to a fed batch. For the clone selection process and cell line stability assessment ideally, an appropriate small scale model is used that is suitable for application of the intended process mode [22]. Classically, the clone selection process from an upstream process perspective evaluates growth performance, titer, and metabolic profiling. However, it might be necessary to expand selection criteria dependent on the selected process mode. According to our experience and a recent study conducted by Müller et al. [24] clone ranking is dependent on the selected process mode and can differ significantly between intensified and non-intensified approaches. In case of perfusion processes, the need for mimicking cell retention, media exchange and longer run times imposes a high technical complexity.

Since fed batch platforms currently prevail in the biopharmaceutical industry, the number of high-throughput perfusion models is still limited. In literature models in deep well [18], spin tubes [186–188] or microbioreactors [24, 189–192] are discussed leading to an appropriate clone selection process.

As described above, cell line stability is highly dependent on the transgene integration technology and technologies are available to overcome those challenges. Therefore, and because the harvest titer might no longer be one of the highest ranked clone selection criteria as the process output has been debottlenecked, product quality aspects can move even more in the focus. Potential CQAs including but not limited to aggregation, charge profile, glycosylation, HCP content as well as binding kinetics and effector functions are important parameters to decide for the final manufacturing clone for fed batch but also for intensified processes where platform knowledge is not readily available yet to a comparable extent.

Genetic engineering of CHO cells might be a further approach to improve cell line phenotypes and characteristics. In particular, overexpression, knockout or down-regulation of certain genes corresponding to oxidative stress, osmolality or shear stress are of interest. Classically, those approaches are focusing on the enhancement of cell specific productivity, the control of (critical) quality attributes or improvement of cell line stability [193–195]. Due to the availability of the first CHO–K1 genomic sequence since 2011 target identification and editing has been highly simplified to perform e.g. targeted gene KOs [196]. Finally, also omics approaches (see Section 12.5.1) may and will be used to further optimize CHO host and production cells.

12.9 Economic considerations

The average expenses for research and development of a new biopharmaceutical entity until market entry significantly increased over the last decade from $1.8 billion in 2010 [197] to $2.8 billion in 2016 [198]. Consequently, to target the CMC cost share, process options to diminish costs need to be identified. This circumstance puts a special emphasis on the economics of the commercial production of biopharmaceuticals. That applies in general for products like mAbs, but even to a greater extent if, for example, low indication molecules or new markets in developing countries are targeted. To improve economic production, the capacity of the production facility and the cost of goods (COGs) during manufacturing are important optimization goals.

In this regard, process intensification is an effective method to maximize facility output by increasing the yield per batch and concomitantly reducing the COGs per gram of mAb. Assuming a large-scale facility with eight 12,000 L bioreactors, the COGs can be reduced drastically following a power law like function when increasing the product yield per batch and even more by increasing the number of batches per year [199]. The impact of implementing N-1 perfusion with HSD fed batch on the capacity of such a facility and on the costs was evaluated using (a) facility modeling (see Section 12.9.1) and (b) cost of goods modeling (see Section 12.9.2).

12.9.1 Facility modeling

The impact of process intensification on manufacturing output can be analyzed in greater detail using facility models. In the biopharmaceutical industry, products are ultimately anchored in a facility, which is typically described as a system where multiple processes are operating in parallel, to enable the main manufacturing process for the product [200]. A facility model should consider the production facility (vessels and capacities), production recipes (processing times/rates, resource requirements), material and resource availability, processing time, targets/due dates and finally production costs [201]. The aim is to find an optimal schedule which meets production targets without limitation by resource constrains [201]. Different problems can be solved with such models, for instance minimizing cost, maximizing throughput, or even maximizing sustainability [202]. These goals can be facilitated by the selection and sizing of batches to be carried out, the assignment of tasks to processing units such as tanks or bioreactors and the sequencing and timing of tasks on each unit [201]. Models should be appropriately detailed to present a plant virtually at both the unit operation and process level. Once available, such a digital twin of a production facility can be used to perform a "what if" analysis to debottleneck a process or test ideas for optimization *in silico* prior to implementation [203]. For the downstream part, cycling strategies can be optimized and evaluated *in-silico* against scale-up of individual unit operations.

A number of different software suites for this type of detailed material and process flow analysis are commercially available. Besides GAMS, INOSIM, gPROMS, SuperPro Designer, Anylogic and ExtendSim cited in [201] also Real-Time Modelling Systems (by Emerson) and Virtecs must be mentioned. A review of available software solutions with focus on the downstream is provided by Rolinger et al. [169]. The review includes a comparison between static and dynamic facility models. Simulation and scheduling tools are important throughout the life cycle of process development and product commercialization [204]. The evaluation of process alternatives or development targets is very useful during process development. Later in a project life cycle, such models can be used to support process transfer to manufacturing and process fitting analysis [204].

To create a facility model a flow sheet comprising all unit operations and recipes with their respective operation and prep times is established. Auxiliary operations like SIP/CIP, available water, and buffer preparation should be included as well. A greatly simplified flow sheet just depicting the upstream and downstream process is shown in Figure 12.4.

To better understand the capabilities of a facility model, an example for the upstream operations of a stainless-steel facility with ten 12,000 L N-stage bioreactors is given. This facility model was created using the real-time modelling systems software (Emerson Electric Co., St. Louis, USA). The downstream process was not modeled in detail in this example. Here only the complete purification time from capture to formulation was considered.

UPSTREAM PROCESS   **DOWNSTREAM PROCESS**

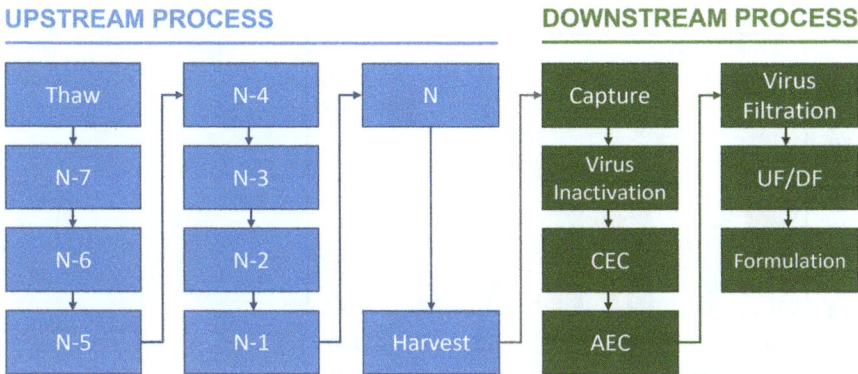

Figure 12.4: Simplified process overview for an exemplary process model. Auxiliary unit operations like CIP/ SIP are not depicted.

Using the established facility model a fed batch process was compared with an HSD fed batch utilizing N-1 perfusion. Both processes were considered in six-run campaigns in a multi-product facility. The fed batch process for a complex format yielded 0.6 g/L in 14 days, while 1.5 g/L was achieved in 14 days with the intensified process format. The model was used to investigate how yield and runs per year change if the percentage of intensified processes increases from 0 to 100 % (Figure 12.5A). With an increasing number of HSD processes in this facility, the number of runs that can be conducted per year decreases by approximately 15 % (Figure 12.5A). This decrease is due to limitations and bottlenecks within the facility, which arise with a longer N-1 run time and the amounts of perfusion media needed for the process. At the same time the annual yield of the facility increases to more than 220 %, because the yield of the HSD processes is so much higher than that of the fed batch process. The yield of the facility can be further improved by accelerating the downstream process. An acceleration of 24 h can increase the facility yield by another 20 % (Figure 12.5B). As discussed earlier, facility models can be beneficial in setting development targets on how a process can be developed that maximizes the output of a production facility. In Figure 12.5B the impact of the run time of the N-stage cultivation on facility yield is shown. In the investigated case longer N-stage run times clearly lead to higher facility yields. In summary, the facility model demonstrates that the yield of commercial production facilities can be significantly increased by intensified upstream processes. It has to be noted that the model complexity increases, the more diverse a product portfolio in a given plant is.

12.9.2 Cost of goods modeling

The production costs per gram of mAb were diminishing over the years e.g. due to improvements of processes or in manufacturing, resulting in current costs well below $100 per gram [205]. In order to substantially further decrease cost of manufacturing,

(A)

(B)

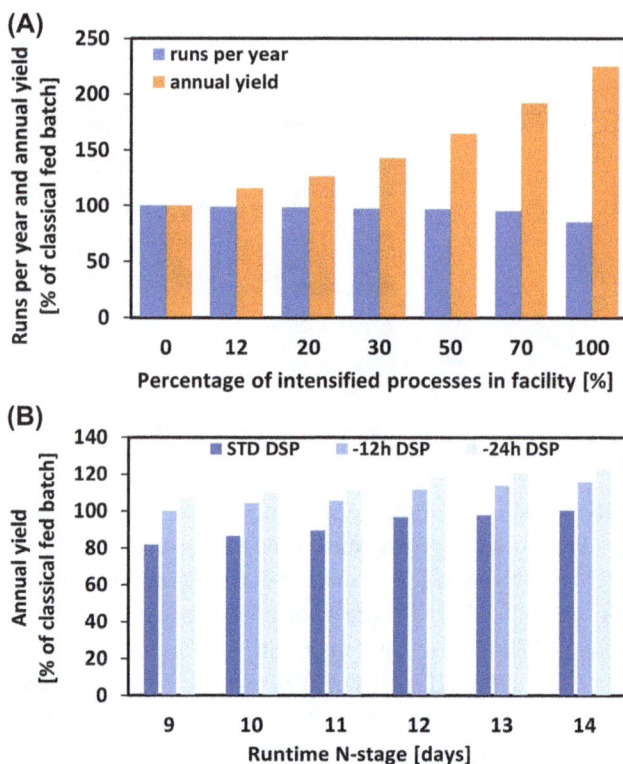

Figure 12.5: Impact of HSD processes and N-stage runtime on production efficiency and yield. (A) Shows the maximal number of runs per year (blue bars) with increasing percentage (0 % equals a fed batch) of HSD processes in the modelled multi-product facility. The number of runs per year is normalized to the maximum number of fed batch runs (=100 %). Orange bars show how the annual yield of the facility changes if percentage of HSD fed batch runs is increased. Yields are normalized to the maximum output per year of the fed batch process (=100 %). In (B) the annual yield, normalized to the process, is shown in dependence of the N-stage cultivation run time, from 9 up to standard 14 days. The effect of a faster downstream process is shown in the lighter blue colors for the respective bioreactor run times.

variable costs such as culture media, buffer, resins, and consumables, are in focus since these costs account for approximately 28 % of all COGs. The major share are fixed costs like depreciation, insurances, maintenance, and labor [199].

Process intensification through e.g. HSD fed batch with N-1 perfusion or concentrated fed batch as well as continuous perfusion, are common ways to optimize the overall yield and thus reducing the costs per gram of drug substance. However, the highest titer does not necessarily correlate to the best economic option. Even though a concentrated fed batch showed the highest volumetric productivity, the overall cost of goods per gram drug substance ($81/g) turned out to be higher whereas HSD fed batch ($63/g) and perfusion ($59/g) were lower compared to a fed batch ($71/g) as shown in a case scenario of Xu et al.

[206]. A concentrated fed batch must yield nearly 10 times higher product output compared to a fed batch to achieve similar media costs [206]. Another simulation study showed the superiority of a continuous manufacturing approach over a fed batch process with almost 50 % reduced COGs [47].

Regarding capital expenditures, N-1 perfusion constitutes a relatively straightforward way to implement process intensification into a production facility since an existing plant can be upgraded at comparable low cost (<0.5 Mio$) for cell recirculation pumps and perfusion skids with tangential flow filtration. In order to evaluate the impact on production costs, an existing mAb manufacturing process using a fed batch was compared with an intensified HSD fed batch approach in a multi-product facility. The facility comprises six 12,000 L bioreactors as production stages. This evaluation is based on calculations performed with GeminiBEAT, a tool by Gemini Bioprocessing. For simplification, we assumed similar fixed costs (production equipment, maintenance activities and labor) which leaves the main cost drivers within variable costs during upstream and downstream manufacturing (for example chemicals, consumables, and resins). As a case study, a fed batch process was assessed with a titer of 2.8 g/L mAb and with an overall process yield of 78 % after final formulation. For an intensified process we assume the exchange of 6 additional bioreactor volumes during N-1 perfusion and an HSD fed batch, that yields the double amount of titer (5.6 g/L).

Such a process intensification would increase the costs per batch by 56 %, which combines a plus of 44 % for USP and a plus of 71 % for DSP based on the legacy equipment. However, the COG per gram mAb can be decreased by 22 % in this example with a doubling of titer. This reduction is higher than reported in [206] where costs for an HSD fed batch compared to a fed batch could be decreased by about 11 % (i.e. from $71/g in FB to $63/g in HSD FB). The major cost share in upstream represents the media demand in N-1, which, in this case, is six times higher compared to a standard N-1 batch.

Another important aspect concerning the costs of an upstream process is its performance, more precisely the productivity of the cells. This raises the question what needs to be the minimum titer increase to be cost efficient through intensification. As shown in Figure 12.6A at least 30 % more mAb production will reduce the COGs through an intensified strategy, whereas low yield improvements (<30 %) even increase the COG/g mAb for this specific evaluation. The reason lies in the high costs for perfusion medium that remain equal independent of the plus of product output, whereas capture cycles for affinity chromatography increase with titer as shown in Figure 12.6A. Affinity chromatography, anion, and cation exchange steps are the main cost drivers for purification of higher product titers. To circumvent the increase of capture cycles, either higher resin volumes or resins with higher capacity are in the focus of optimization for a more efficient antibody purification. Another question is whether the reduction of N-stage run time benefits the facility in terms of costs since process intensification can yield higher titer in a shorter time compared to the fed batch as shown in Figure 12.2. Based on a facility utilization with 30 % intensified campaigns in a multi-product facility, a reduction of N-stage run time is not beneficial. As shown in Figure 12.6B, the longest N-stage process

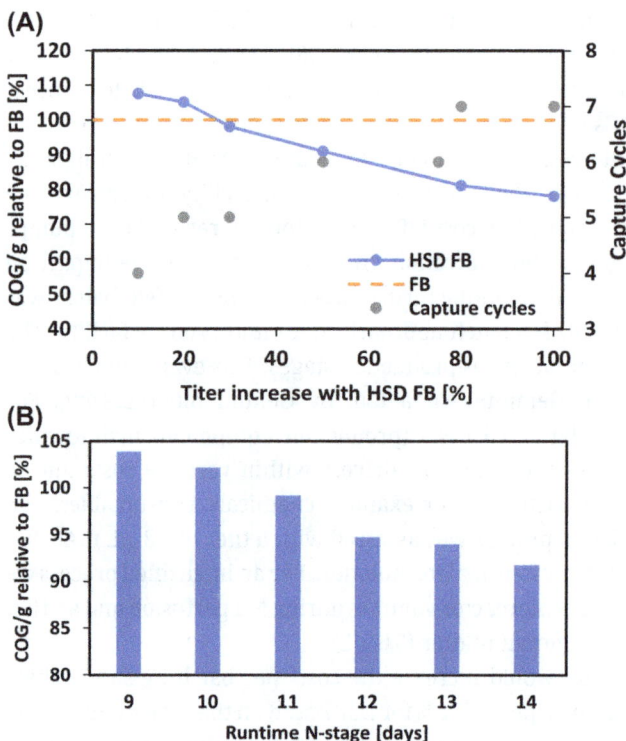

Figure 12.6: Impact of titer increase and HSD N-stage runtime on costs of goods (COG) per gram mAb. (A) Evolution of COG per gram mAb as a function of titer increase compared to a fed batch (FB). Higher titers result in higher protein A affinity capture cycles assuming the legacy setup of the facility in this example. (B) Cost of goods per gram mAb within a production year with 30 % intensified campaigns and 70 % fed batch campaigns in relation to 100 % fed batch campaign as a function of HSD N-stage run time.

length results in the lowest COG/g/year (−8%) whereas a 5-day shorter process would increase the COG/g by 4 % within one production year. In summary, process optimization through HSD fed batch is key to increase the facility output and concomitantly reduce the production costs, but all depending on the relative titer increase per batch.

Major drivers for the COG in downstream are virus filters [207] and the protein A resin [208]. A COG evaluation with Protein A suggests that product loss caused by partial breakthrough can be economically justified [209]. According to the evaluation, maximum utilization of protein A resin capacity can be reasonable, depending on the protein A resin price. A comparison of batch versus continuous downstream processing reveals that relative savings by continuous downstream processing increase at higher titers and volumes [210]. Single-use equipment further reduces the COG in this example. Another evaluation concludes that single-use continuous processes are especially beneficial at yearly productions between 100 and 500 kg [211]. Process development cost for continuous processes outweigh COG advantages for lower yearly productions. At higher yearly production rates, media cost dominates the COG and USP becomes the largest contributor to COG. Yet, all COG models heavily rely on individual assumptions. The findings published in literature need to be carefully considered. Scientifically sound conclusions require a dedicated COG model that involves specific circumstances, such as available facilities, labor cost and the market demand.

12.10 Sustainability aspects

In an historic event, the 2016 Paris Agreement within the United Nations Framework Convention on Climate Change (UNFCCC) [212] activated and elevated public attention on greenhouse gas emissions (GHGE). In the same year, the Scripps Institutions of Oceanography reported that carbon dioxide level exceeded the symbolic 400 ppm threshold for the first time in human history with even rising levels above 420 ppm in 2022 [213]. Based on comprehensive data, it is overwhelmingly clear that aggressive reduction targets, as set in the 2016 Paris Agreement, are a necessity. As until now, the pharmaceutical industry felt relatively little pressure to reduce GHGE as it provides treatments, vaccinations and medication to improve health and was not under specific scrutiny to evaluate their GHGE. Especially biopharmaceuticals, which do not require energy intensive chemical synthesis (e.g. high temperatures, pressure and/or other problematic materials), in addition with relatively little public attention, were not a focus area in the past in the context of GHGE. However, in a pioneering study [214], the overall emission intensity of the pharmaceutical industry in 2015 was estimated to be 55 % higher than that of the automotive sector. In absolute, the estimated aggregate of global emissions of the pharmaceutical sector is 11 % higher than the global automotive sector. In a life cycle impact assessment (LCA) for a 2000 L single-use process for monoclonal antibodies, it was determined that 1 kg of drug substance produces 22.7 tons of CO_2 equivalence. Electricity consumption dominates all categories, accounting for 80 % of the impact on climate change, CO_2 footprint and freshwater use [215]. In this study, most of the electricity consumption was due to the powering of the cleanroom infrastructure. Based on these initial data, it can be expected that the pharmaceutical sector will also face political and social pressure to address its environmental impact in the coming decades. One of the steps to achieve these goals is by improving the overall manufacturing efficiency. An initial and easy to apply method of assessing the impact of a manufacturing process to the environment is calculating the process mass intensity (PMI) [216]. This method calculates all the masses such as water, raw materials and consumables used to produce one kg of a medicinal product, here a therapeutic antibody. Even though PMI does not comprehensively measure the absolute environmental footprint of a process as its primary focus is on material efficiency, PMI calculations are relatively simple and fast to conduct.

Intensified processes, with its superior volumetric productivity and potentially shorter process times can be a significant contributor to the reduction of the PMI. Figure 12.7 shows the PMI calculations of upstream and downstream for two 2000 L

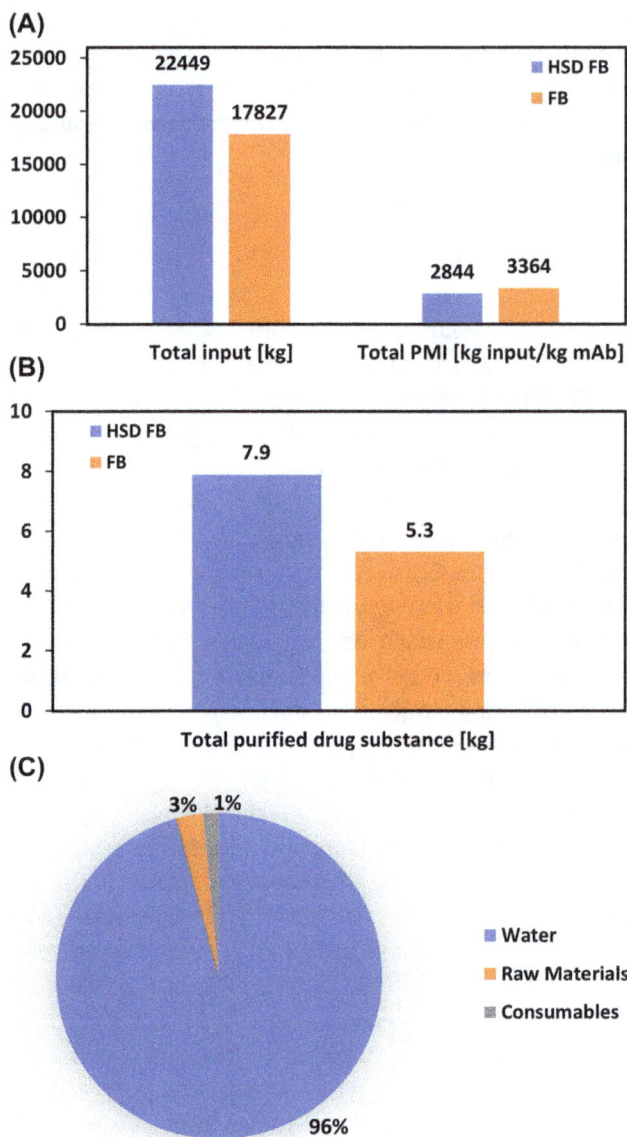

Figure 12.7: Overview of a process mass intensity (PMI) calculation based on a 2000 L process. (A) Total material input and PMI value per kg of kg antibody (mAb). (B) Total amount of purified drug substance. (C) Graphical representation of the percentage of the overall process masses of the HSD fed batch.

processes, one time as a fed batch and the other as an HSD fed batch. The differences are that the HSD fed batch consists of an N-1 perfusion and a high seeding density N-Stage. For the downstream process, 15 % more material demand was estimated based on experience. Overall, the HSD fed batch has a higher total material input (Figure 12.7B) due to the perfusion step in N-1 with elevated media demands. Furthermore, hollow fiber modules in upstream and more resin for the affinity chromatography in downstream increase the amount of consumables used. As depicted in Figure 12.7C, for the HSD fed batch process the main consumed mass is WFI for media and buffer generation. WFI is water of a very high purity grade, often achieved by energy demanding distillation and osmosis processes. However, overall higher amount of purified product (Figure 12.7B) is more than a compensation for the higher material demand. The total PMI for the overall process is 16 % lower for the HSD fed batch than for the fed batch. This clearly shows that increasing efficiency of the manufacturing process plays a pivotal role in reducing the footprint of biopharmaceutical production.

12.11 Summary and future perspectives

The overall biologics pipelines are growing as well as the biotherapeutics market in total. Compared to the first monoclonal antibodies, a variety of antibody formats is now being developed and provided to patients. Further formats and novel biologics such as advanced therapy medicinal products (ATMPs) will add to future portfolios. This increasing diversity of medicinal products directly impacts process development and manufacturing. This is good news as it triggers rethinking of the current *modus operandi* and will likely foster more innovation in CMC areas and implementations thereof. In the past years, we have seen significant progress in CHO platform development with mAb titers approaching 10 g/L or even beyond. Cell lines are stable, well characterized and enable high expression. Recent technologies, such as the use of transposase mediated transgene integration, provide additional consistency by a more targeted direction to actively transcribed genomic regions. Cell culture media nowadays are chemically defined and raw material variability is better understood due to the use of advanced analytical methods and PAT technologies. Experience from development and manufacturing of large numbers of diverse antibody therapeutics resulted in the development of upstream and downstream platforms that ensure robust processes and reliable scale-up. Multiple data analysis and modeling tools support these efforts based on big data sets along the value chain. Thus, the progress made in cell line, media, upstream and downstream development puts the industry in a good position to take next steps. One area is process intensification with the aim to increase the productivity in the manufacture of biologics. Since about the 2000s, the concept of continuous manufacturing is actively discussed in academia and industry in the field of therapeutic antibodies including regulatory agencies. Principally, the technology and regulatory framework is available and some commercial products use this technology. With respect to continuous processing, in our view, a "transformation" has not (yet) taken place in the last almost 20 years. One reason might be that, within existing platforms, which are traditionally based on fed batch processes, sufficient productivity gains could be achieved, thereby challenging the need to change established production technology. Another reason for biopharmaceutical companies that have established fed batch based manufacturing networks is that a technology switch is not feasible within an existing production network. However, N-1 perfusion coupled to HSD fed batch can be integrated with moderate efforts. In recent years, several facilities in the biopharmaceutical industry consider this technology either in technical upgrades in existing plants or when extending sites, respectively building new facilities. Using the available technology and the basic concepts as described in this contribution, our experience confirms that this technology works robust and reliable across scales implemented so far. We also have seen significant titer increases for diverse antibody formats ranging from 30 % to 150 %.

CMC functions can expect increasing and more diverse portfolios in the future. At the same time both, overall and CMC cost pressure increases. Thus, there is a need for more efficient bioprocess development and manufacturing. Available platform technologies are a solid ground to proceed and implement innovation. Looking into the various elements of a CHO platform (such as expression vectors, cell line, media, upstream and downstream platform) and how they all contributed to further enhance productivity and yields, we expect intensified processes as a next step and lever although implementation efforts are higher compared to the other elements. Also, economic considerations (e.g. facility utilization and output analysis, cost of goods) and sustainability goals clearly favor the implementation of intensified processes. The path forward for the upstream section appears relatively straightforward. For the downstream part, intensification options for the individual process steps exist and the basic technology is available as well. However, which exact elements and combinations thereof will prevail and lead to a scalable, intensified downstream platform is yet dependent from the particular facility, its limitations and the specific molecule characteristics. Eventually, but most importantly, these improvements will hopefully also support the efforts to make as many (novel) biological medicines available for patients as possible.

Acknowledgments: The authors would like to thank the editor Dirk Holtmann for their guidance and review of this article before its publication.

References

1. Mullard A. FDA approves 100th monoclonal antibody product. Nat Rev Drug Discov 2021;20:491–5.
2. Kaplon H, Chenoweth A, Crescioli S, Reichert JM. Antibodies to watch in 2022. mAbs 2022;14:2014296.
3. Jin S, Sun Y, Liang X, Gu X, Ning J, Xu Y, et al. Emerging new therapeutic antibody derivatives for cancer treatment. Signal Transduct Targeted Ther 2022;7:39. https://doi.org/10.1038/s41392-021-00868-x.
4. Dunleavy K. The top 20 drugs by worldwide sales in 2021. In: Fierce pharma; 2022. [Online]. Available: https://www.fiercepharma.com/special-reports/top-20-drugs-worldwide-sales-2021 [Accessed 31 May 2022].

5. O'Flaherty R, Bergin A, Flampouri E, Mota LM, Obaidi I, Quigley A, et al. Mammalian cell culture for production of recombinant proteins: a review of the critical steps in their biomanufacturing. Biotechnol Adv 2020;43:107552.

6. Jagschies G. Biopharmaceutical processing: development, design, and implementation of manufacturing processes. Amsterdam, Netherlands: Elsevier; 2018.

7. FDA. FDA/biosimilar product information; 2022. https://www.fda.gov/drugs/biosimilars/biosimilar-product-information [Accessed 28 Sep 2022].

8. Kelley B, Kiss R, Laird M. A different perspective: how much innovation is really needed for monoclonal antibody production using mammalian cell technology? Adv Biochem Eng Biotechnol 2018;165:443–62.

9. Le H, Chen C, Goudar CT. Biopharmaceuticals—continuous processing in upstream operations. In: Chemical engineering progress; 2015. [Online]. Available: https://www.aiche.org/resources/publications/cep/2015/december/sbe-special-section-biopharmaceuticals-continuous-processing-upstream-operations.

10. Vogg S, Müller-Späth T, Morbidelli M. Current status and future challenges in continuous biochromatography. Curr Opin Chem Eng 2018;22:138–44.

11. Sawyer D, Malmberg L, Ramasubramanyan N, Sanderson K, Lu R, Sur E, et al Biomanufacturing technology roadmap-overview. In: Biomanufacturing Technol. Roadmap; 2017. S.1–48 pp. [Online]. Available: https://www.biophorum.com/wp-content/uploads/bp_downloads/Overview-1.pdf.

12. Yongky A, Xu J, Tian J, Oliveira C, Zhao J, McFarland K, et al. Process intensification in fed-batch production bioreactors using non-perfusion seed cultures. mAbs 2019;11:1502–14.

13. Johnson AS, Casey ME, Guillen N, Lawrence SM. Medium development strategies and scale down models for a high density high productivity cell line. In: Presented at the cell culture engineering XVI. Tampa, Florida, USA: Cell culture engineering XVI; 2018. https://dc.engconfintl.org/ccexvi/223.

14. Zhang L, Shawley R, Gawlitzek M, Wong B. Development towards a high-titer fed-batch CHO platform process yielding product titers > 10 g/L. In: Presented at the cell culture engineering XVI. Tampa, Florida, USA: Cell culture engineering XVI; 2018. https://dc.engconfintl.org/ccexvi/234.

15. Schmieder V, Fieder J, Drerup R, Gutierrez EA, Guelch C, Stolzenberger J, et al. Towards maximum acceleration of monoclonal antibody development: leveraging transposase-mediated cell line generation to enable GMP manufacturing within 3 months using a stable pool. J Biotechnol 2022;349:53–64.

16. Ritacco FV, Wu Y, Khetan A. Cell culture media for recombinant protein expression in Chinese hamster ovary (CHO) cells: history, key components, and optimization strategies. Biotechnol Prog 2018;34: 1407–26.

17. Brunner M, Brosig P, Losing M, Kunzelmann M, Calvet A, Stiefel F, et al. Towards robust cell culture processes—Unraveling the impact of media preparation by spectroscopic online monitoring. Eng Life Sci 2019;19:666–80.

18. Lin H, Leighty RW, Godfrey S, Wang SB. Principles and approach to developing mammalian cell culture media for high cell density perfusion process leveraging established fed-batch media. Biotechnol Prog 2017;33:891–901.

19. Konstantinov KB, Cooney CL. White paper on continuous bioprocessing. May 20–21, 2014 continuous manufacturing Symposium. J Pharmaceut Sci 2015;104:813–20.

20. Barrett S, Franklin J, Stangl M, Cvetkovic A, He W. Intensification of a multi-product perfusion platform – managing growth characteristics at high cell density for maximized volumetric productivity. In: Presented at the cell culture engineering XVI. Tampa, Florida, USA: Cell culture engineering XVI; 2018. https://dc.engconfintl.org/ccexvi/236.

21. Hiller GW, Ovalle AM, Gagnon MP, Curran ML, Wang W. Cell-controlled hybrid perfusion fed-batch CHO cell process provides significant productivity improvement over conventional fed-batch cultures. Biotechnol Bioeng 2017;114:1438–47.

22. Bielser J-M, Wolf M, Souquet J, Broly H, Morbidelli M. Perfusion mammalian cell culture for recombinant protein manufacturing – a critical review. Biotechnol Adv 2018;36:1328–40.

23. Wong HE, Chen C, Le H, Goudar CT. From chemostats to high-density perfusion: the progression of continuous mammalian cell cultivation. J Chem Technol Biotechnol 2021. https://doi.org/10.1002/jctb. 6841.
24. Müller D, Klein L, Lemke J, Schulze M, Kruse T, Saballus M, et al. Process intensification in the biopharma industry: improving efficiency of protein manufacturing processes from development to production scale using synergistic approaches. Chem Eng Process – Process Intensif 2022;171:108727.
25. Schulze M, Lemke J, Pollard D, Wijffels RH, Matuszczyk J, Martens DE. Automation of high CHO cell density seed intensification via online control of the cell specific perfusion rate and its impact on the N-stage inoculum quality. J Biotechnol 2021;335:65–75.
26. Xu J, Xu X, Huang C, Angelo J, Oliveira CL, Xu M, et al. Biomanufacturing evolution from conventional to intensified processes for productivity improvement: a case study. mAbs 2020;12:1770669.
27. Xu J, Rehmann MS, Xu M, Zheng S, Hill C, He Q, et al. Development of an intensified fed-batch production platform with doubled titers using N-1 perfusion seed for cell culture manufacturing. Bioresour Bioprocess 2020;7:17. https://doi.org/10.1186/s40643-020-00304-y.
28. Stepper L, Filser FA, Fischer S, Schaub J, Gorr I, Voges R. Pre-stage perfusion and ultra-high seeding cell density in CHO fed-batch culture: a case study for process intensification guided by systems biotechnology. Bioproc Biosyst Eng 2020;43:1431–43.
29. Yang WC, Lu J, Kwiatkowski C, Yuan H, Kshirsagar R, Ryll T, et al. Perfusion seed cultures improve biopharmaceutical fed-batch production capacity and product quality. Biotechnol Prog 2014;30:616–25.
30. Padawer I, Ling WLW, Bai Y. Case Study: an accelerated 8-day monoclonal antibody production process based on high seeding densities. Biotechnol Prog 2013;29:829–32.
31. Pohlscheidt M, Jacobs M, Wolf S, Thiele J, Jockwer A, Gabelsberger J, et al. Optimizing capacity utilization by large scale 3000 L perfusion in seed train bioreactors. Biotechnol Prog 2013;29:222–9.
32. Brunner M, Kolb K, Keitel A, Stiefel F, Wucherpfennig T, Bechmann J, et al. Application of metabolic modeling for targeted optimization of high seeding density processes. Biotechnol Bioeng 2021;118: 1793–804.
33. Karst DJ, Serra E, Villiger TK, Soos M, Morbidelli M. Characterization and comparison of ATF and TFF in stirred bioreactors for continuous mammalian cell culture processes. Biochem Eng J 2016;110:17–26.
34. Bielser J-M, Aeby M, Caso S, Roulet A, Broly H, Souquet J. Continuous bleed recycling significantly increases recombinant protein production yield in perfusion cell cultures. Biochem Eng J 2021;169:107966.
35. Xu S, Chen H. High-density mammalian cell cultures in stirred-tank bioreactor without external pH control. J Biotechnol 2016;231:149–59.
36. Wolf MKF, Closet A, Bzowska M, Bielser JM, Souquet J, Broly H, et al. Improved performance in mammalian cell perfusion cultures by growth inhibition. Biotechnol J 2019;14:1700722.
37. Warikoo V, Godawat R, Brower K, Jain S, Cummings D, Simons E, et al. Integrated continuous production of recombinant therapeutic proteins. Biotechnol Bioeng 2012;109:3018–29.
38. Schwarz H, Mäkinen ME, Castan A, Chotteau V. Monitoring of amino acids and antibody N-glycosylation in high cell density perfusion culture based on Raman spectroscopy. Biochem Eng J 2022;182:108426.
39. Dowd JE, Jubb A, Kwok KE, Piret JM. Optimization and control of perfusion cultures using a viable cell probe and cell specific perfusion rates. Cytotechnology 2003;42:35–45.
40. Walther J, Lu J, Hollenbach M, Yu M, Hwang C, McLarty J, et al. Perfusion cell culture decreases process and product heterogeneity in a head-to-head comparison with fed-batch. Biotechnol J 2019;14:1700733.
41. Clincke M-F, Mölleryd C, Zhang Y, Lindskog E, Walsh K, Chotteau V. Study of a recombinant CHO cell line producing a monoclonal antibody by ATF or TFF external filter perfusion in a WAVE Bioreactor™. BMC Proc 2011;5(8 Suppl):105.
42. Zhou H, Fang M, Zheng X, Zhou W. Improving an intensified and integrated continuous bioprocess platform for biologics manufacturing. Biotechnol Bioeng 2021;118:3618–23.

43. Sartorius. Dynamic perfusion. https://www.sartorius.com/en/applications/biopharmaceutical-manufacturing/process-intensification/dynamic-perfusion [Accessed 11 Nov 2022].
44. Kundu AM, Hiller GW. Hydrocyclones as cell retention devices for an N-1 perfusion bioreactor linked to a continuous-flow stirred tank production bioreactor. Biotechnol Bioeng 2021;118:1973–86.
45. Yang WC, Minkler DF, Kshirsagar R, Ryll T, Huang Y-M. Concentrated fed-batch cell culture increases manufacturing capacity without additional volumetric capacity. J Biotechnol 2016;217:1–11.
46. Rathore AS, Zydney AL, Anupa A, Nikita S, Gangwar N. Enablers of continuous processing of biotherapeutic products. Trends Biotechnol 2022;40:804–15.
47. Yang O, Prabhu S, Ierapetritou M. Comparison between batch and continuous monoclonal antibody production and economic analysis. Ind Eng Chem Res 2019;58:5851–63.
48. Chun C, Edward HW, Chetan TG. Upstream process intensification and continuous manufacturing. Curr Opin Chem Eng 2018;22:191–8.
49. Karst DJ, Steinebach F, Morbidelli M. Continuous integrated manufacturing of therapeutic proteins. Curr Opin Biotechnol 2018;53:76–84.
50. Stube J, Ditz R, Kornecki M, Huter M, Schmidt A, Thiess H, et al. Process intensification in biologics manufacturing. Chem Eng Process – Process Intensif 2018;133:278–93.
51. Croughan MS, Konstantinov KB, Cooney C. The future of industrial bioprocessing: batch or continuous? Biotechnol Bioeng 2015;112:648–51.
52. Jordan M, Kinnon NM, Monchois V, Stettler M, Broly H. Intensification of large-scale cell culture processes. Curr Opin Chem Eng 2018;22:253–7.
53. Karst DJ, Ramer K, Hughes EH, Jiang C, Jacobs PJ, Mitchelson FG. Modulation of transmembrane pressure in manufacturing scale tangential flow filtration N-1 perfusion seed culture. Biotechnol Prog 2020:e3040. https://doi.org/10.1002/btpr.3040.
54. Bettinardi IW, Castan A, Medronho RA, Castilho LR. Hydrocyclones as cell retention device for CHO perfusion processes in single-use bioreactors. Biotechnol Bioeng 2020;117:1915–28.
55. Voisard D, Meuwly F, Ruffieux P -A, Baer G, Kadouri A. Potential of cell retention techniques for large-scale high-density perfusion culture of suspended mammalian cells. Biotechnol Bioeng 2003;82:751–65.
56. MacDonald MA, Noebel M, Recinos DR, Martínez VS, Schulz BJ, Howard CB, et al. Perfusion culture of Chinese Hamster Ovary cells for bioprocessing applications. Crit Rev Biotechnol 2021;1–17. https://doi.org/10.1080/07388551.2021.1998821.
57. Johnstone P, Mast E, Hughes E, Peng H. Development of a small-scale rotary lobe-pump cell culture model for examining cell damage in large-scale N-1 seed perfusion process. Biotechnol Prog 2020:e3044. https://doi.org/10.1002/btpr.3044.
58. Amer M, Vaca A, Bowden M. Evaluating shear in perfusion rotary lobe pump using nanoparticle aggregates and computational fluid dynamics. Bioproc Biosyst Eng 2022;45:1–12.
59. Kamaraju H, Wetzel K, Kelly WJ. Modeling shear-induced CHO cell damage in a rotary positive displacement pump. Biotechnol Prog 2010;26:1606–15.
60. Blaschczok K, Kaiser SC, Loeffelholz C, Imseng C, Burkart J, Boesch P, et al. Investigations on mechanical stress caused to CHO suspension cells by standard and single-use pumps. Chem-Ing-Tech 2013;85:144–52.
61. Dittler I, Kaiser SC, Blaschczok K, Loeffelholz C, Boesch P, Dornfeld W, et al. A cost-effective and reliable method to predict mechanical stress in single-use and standard pumps. Eng Life Sci 2014;14:311–7.
62. Wang S, Godfrey S, Ravikrishnan J, Lin H, Vogel J, Coffman J. Shear contributions to cell culture performance and product recovery in ATF and TFF perfusion systems. J Biotechnol 2017;246:52–60.
63. Weinberger ME, Kulozik U. On the effect of flow reversal during crossflow microfiltration of a cell and protein mixture. Food Bioprod Process 2021;129:24–33.
64. Weinberger ME, Kulozik U. Understanding the fouling mitigation mechanisms of alternating crossflow during cell-protein fractionation by microfiltration. Food Bioprod Process 2022;131:136–43.

65. Madabhushi SR, Huang CJ, Wang X, Bui A, Atieh TB, Rayfield WJ, et al. An innovative strategy to recycle permeate in biologics continuous manufacturing process to improve material efficiency and sustainability. Biotechnol Prog 2022;38:e3262.

66. Martínez-Monge I, Roman R, Comas P, Fontova A, Lecina M, Casablancas A, et al. New developments in online OUR monitoring and its application to animal cell cultures. Appl Microbiol Biotechnol 2019;103: 6903–17.

67. Seidel S, Maschke RW, Werner S, Jossen V, Eibl D. Oxygen mass transfer in biopharmaceutical processes: numerical and experimental approaches. Chem-Ing-Tech 2021;93:42–61.

68. Konstantinov K, Goudar C, Ng M, Meneses R, Thrift J, Chuppa S, et al. The "Push-to-Low" approach for optimization of high-density perfusion cultures of animal cells. Adv Biochem Eng Biotechnol 2006;101: 75–98.

69. Kornecki M, Mestmäcker F, Zobel-Roos S, de Figueiredo LH, Schlüter H, Strube J. Host cell proteins in biologics manufacturing: the good, the bad, and the ugly. Antibodies 2017;6. https://doi.org/10.3390/ antib6030013.

70. Levy NE, Valente KN, Choe LH, Lee KH, Lenhoff AM. Identification and characterization of host cell protein product-associated impurities in monoclonal antibody bioprocessing. Biotechnol Bioeng 2014;111: 904–12.

71. Kelley B. Industrialization of mAb production technology: the bioprocessing industry at a crossroads. mAbs 2009;1:443–52.

72. Kelley B. Very large scale monoclonal antibody purification: the case for conventional unit operations. Biotechnol Prog 2007;23:995–1008.

73. Saxena V, Weil A. Radial flow columns: a new approach to scaling-up biopurifications. Biochromatography 1987;2:90–7.

74. Besselink T, van der Padt A, Janssen AEM, Boom RM. Are axial and radial flow chromatography different? J Chromatogr A 2013;1271:105–14.

75. Müller E. Properties and characterization of high capacity resins for biochromatography. Chem Eng Technol 2005;28:1295–305.

76. Marina Graalfs H, Joehnck M, Jacob LR, Frech C. Cation-exchange chromatography of monoclonal antibodies: characterisation of a novel stationary phase designed for production-scale purification. mAbs 2010;4:395–404.

77. Chen J, Tetrault J, Ley A. Comparison of standard and new generation hydrophobic interaction chromatography resins in the monoclonal antibody purification process. J Chromatogr A 2008;1177:272–81.

78. Müller E, Vajda J. Routes to improve binding capacities of affinity resins demonstrated for protein A chromatography. J Chromatogr B 2016;1021:159–68.

79. Ramos-de-la-Peña AM, González-Valdez J, Aguilar O. Protein A chromatography: challenges and progress in the purification of monoclonal antibodies. J Separ Sci 2019;42:1816–27.

80. Lu W, Zhang Z, Zhang S, Zhang T, Wan Y, Li Y. Screening of six cation exchange resins for high binding capacity, monomer purity and step yield: a case study. Protein Expr Purif 2022;199:106155.

81. Ghose S, Tao Y, Conley L, Cecchini D. Purification of monoclonal antibodies by hydrophobic interaction chromatography under no-salt conditions. mAbs 2013;5:795–800.

82. Ramakrishna A, Maranholkar V, Hadpe S, Iyer J, Rathore A. Optimization of multi flow rate loading strategy for process intensification of protein A chromatography. J Chromatogr Open 2022;2:100049.

83. Brough H, Antoniou C, Carter J, Jakubik J, Xu Y, Lutz H. Performance of a novel viresolve NFR virus filter. Biotechnol Progr 2002;18:782–95.

84. Bohonak DM, Mehta U, Weiss ER, Voyta G. Adapting virus filtration to enable intensified and continuous monoclonal antibody processing. Biotechnol Prog 2020;37:e3088.

85. Goodrich EM, Bohonak DM, Genest PW, Peterson E. Recent advances in ultrafiltration and virus filtration for production of antibodies and related biotherapeutics. In: Matte A, editor. Approaches to the

purification, analysis and characterization of antibody-based therapeutics. Amsterdam: Elsevier; 2020. 137–66 pp.

86. Briskot T, Hillebrandt N, Kluters S, Wang G, Studts J, Hahn T, et al. Modeling the Gibbs–Donnan effect during ultrafiltration and diafiltration processes using the Poisson–Boltzmann theory in combination with a basic Stern model. J Membrane Sci 2022;648:120333.

87. Madsen E, Kaiser J, Krühne U, Pinelo M. Single pass tangential flow filtration: critical operational variables, fouling, and main current applications. Separ Purif Technol 2022;291:120949.

88. Kruse T, Kampmann M, Rüddel I, Greller G. An alternative downstream process based on aqueous two-phase extraction for the purification of monoclonal antibodies. Biochem Eng J 2020;161:107703.

89. Prouzeau T, Pezzini J, Mothes B. EASY: a disruptive mAb purification process to reduce cost of goods. Biopharm Int 2023;34:26–30.

90. Nadar S, Shooter G, Somasundaram B, Shave E, Baker K, Lua LHL. Intensified downstream processing of monoclonal antibodies using membrane technology. Biotechnol J 2021;16:e2000309.

91. Gerstweiler L, Bi J, Middelberg APJ. Continuous downstream bioprocessing for intensified manufacture of biopharmaceuticals and antibodies. Chem Eng Sci 2021;231:116272.

92. Gillespie C, Holstein M, Mullin L, Cotoni K, Tuccelli R, Caulmare J, et al. Continuous in-line virus inactivation for next generation bioprocessing. Biotechnol J 2019;14:1700718.

93. Kateja N, Nitika N, Fadnis RS, Rathore AS. A novel reactor configuration for continuous virus inactivation. Biochem Eng J 2021;167:107885.

94. Chopda V, Gyorgypal A, Yang O, Singh R, Ramachandran R, Zhang H, et al. Recent advances in integrated process analytical techniques, modeling, and control strategies to enable continuous biomanufacturing of monoclonal antibodies. J Chem Technol Biotechnol 2022;97:2317–35.

95. Kuiper M, Spencer C, Faeldt R, Vuillemez A, Holmes W, Samuelsson T, et al. Repurposing fed-batch media and feeds for highly productive CHO perfusion processes. Biotechnol Progr 2019;35:e2821.

96. Eagle H. Nutrition needs of mammalian cells in tissue culture. Science 1955;122:501–4.

97. Freshney RI. Culture of specific cell types. Culture of Animal Cells 2005. https://doi.org/10.1002/0471747599.cac023.

98. BioPhorum. Raw materials: Best practice guide for preparation of cell culture media solution. London: BioPhorum; 2021. https://www.biophorum.com/download/raw-materials-best-practice-guide-for-preparation-of-cell-culture-media-solution/.

99. Pérez-Fernández BA, Fernández-de-Cossio-Díaz J, Boggiano T, León K, Mulet R. In-silico media optimization for continuous cultures using genome scale metabolic networks: the case of CHO-K1. Biotechnol Bioeng 2021;118:1884–97.

100. Mulukutla BC, Kale J, Kalomeris T, Jacobs M, Hiller GW. Identification and control of novel growth inhibitors in fed-batch cultures of Chinese hamster ovary cells. Biotechnol Bioeng 2017;114:1779–90.

101. Chevallier V, Andersen MR, Malphettes L. Oxidative stress-alleviating strategies to improve recombinant protein production in CHO cells. Biotechnol Bioeng 2020;117:1172–86.

102. Henry MN, MacDonald MA, Orellana CA, Gray CA, Gillard M, Baker K, et al. Attenuating apoptosis in Chinese hamster ovary cells for improved biopharmaceutical production. Biotechnol Bioeng 2020;117: 1187–203.

103. Pereira S, Kildegaard HF, Andersen MR. Impact of CHO metabolism on cell growth and protein production: an overview of toxic and inhibiting metabolites and nutrients. Biotechnol J 2018;13:1700499.

104. Kshirsagar R, Raju R, Ali A, Kwiatkowski C, McElearney K, Gilbert A. Application of -omics knowledge yields enhanced bioprocess performance. In: Presented at the cell culture engineering XVI. Tampa, Florida, USA: Cell culture engineering XVI; 2018. https://dc.engconfintl.org/ccexvi/217.

105. Salim T, Chauhan G, Templeton N, Ling WLW. Using MVDA with stoichiometric balances to optimize amino acid concentrations in chemically defined CHO cell culture medium for improved culture performance. Biotechnol Bioeng 2022;119:452–69.

106. Koenitzer JD, Mueller MM, Leparc G, Pauers M, Bechmann J, Schulz P, et al. A global RNA-seq-driven analysis of CHO host and production cell lines reveals distinct differential expression patterns of genes contributing to recombinant antibody glycosylation. Biotechnol J 2015;10:1412–23.
107. Birzele F, Schaub J, Rust W, Clemens C, Baum P, Kaufmann H, et al. Into the unknown: expression profiling without genome sequence information in CHO by next generation sequencing. Nucleic Acids Res 2010;38: 3999–4010.
108. Wippermann A, Rupp O, Brinkrolf K, Hoffrogge R, Noll T. The DNA methylation landscape of Chinese hamster ovary (CHO) DP-12 cells. J Biotechnol 2015;199:38–46.
109. Schaub J, Clemens C, Schorn P, Hildebrandt T, Rust W, Mennerich D, et al. CHO gene expression profiling in biopharmaceutical process analysis and design. Biotechnol Bioeng 2010;105:431–8.
110. Stolfa G, Smonskey MW, Boniface R, Hachmann AB, Gulde P, Joshi AD, et al. CHO-Omics review: the impact of current and emerging technologies on Chinese hamster ovary based bioproduction. Biotechnol J 2018; 13:1700227.
111. Borth N, Hu W. Enhancing CHO by systems biotechnology. Biotechnol J 2018;13:1800488.
112. Matte A. Recent advances and future directions in downstream processing of therapeutic antibodies. Int J Mol Sci 2022;23. https://doi.org/10.3390/ijms23158663.
113. Gronemeyer P, Ditz R, Strube J. DoE based integration approach of upstream and downstream processing regarding HCP and ATPE as harvest operation. Biochem Eng J 2016;113:158–66.
114. Rischawy F, Saleh D, Hahn T, Oelmeier S, Spitz J, Kluters S. Good modeling practice for industrial chromatography: mechanistic modeling of ion exchange chromatography of a bispecific antibody. Comput Chem Eng 2019;130:106532.
115. Saleh D, Wang G, Mueller B, Rischawy F, Kluters S, Studts J, et al. Straightforward method for calibration of mechanistic cation exchange chromatography models for industrial applications. Biotechnol Progr 2020; 36:e2984.
116. Saleh D, Hess R, Ahlers-Hesse M, Rischawy F, Wang G, Grosch JH, et al. A multiscale modeling method for therapeutic antibodies in ion exchange chromatography. Biotechnol Bioeng 2022;120:125–38.
117. Briskot T, Hahn T, Huuk T, Wang G, Kluters S, Studts J, et al. Analysis of complex protein elution behavior in preparative ion exchange processes using a colloidal particle adsorption model. J Chromatogr A 2021; 1654:462439.
118. Rischawy F, Briskot T, Schimek A, Wang G, Saleh D, Kluters S, et al. Integrated process model for the prediction of biopharmaceutical manufacturing chromatography and adjustment steps. J Chromatogr A 2022;1681:463421.
119. Taylor C, Marschall L, Kunzelmann M, Richter M, Rudolph F, Vajda J, et al. Integrated process model applications linking bioprocess development to quality by design milestones. Bioengineering 2021;8. https://doi.org/10.3390/bioengineering8110156.
120. Saleh D, Wang G, Rischawy F, Kluters S, Studts J, Hubbuch J. In silico process characterization for biopharmaceutical development following the quality by design concept. Biotechnol Progr 2021;37:e3196.
121. Saleh D, Wang G, Mueller B, Rischawy F, Kluters S, Studts J, et al. Cross-scale quality assessment of a mechanistic cation exchange chromatography model. Biotechnol Progr 2021;37:e3081.
122. Smiatek J, Jung A, Bluhmki E. Towards a digital bioprocess replica: computational approaches in biopharmaceutical development and manufacturing. Trends Biotechnol 2020;38:1141–53.
123. Smiatek J, Clemens C, Herrera LM, Arnold S, Knapp B, Presser B, et al. Generic and specific recurrent neural network models: applications for large and small scale biopharmaceutical upstream processes. Biotechnology Reports 2021;31:e00640.
124. Herrera LM. Holistic process models: a Bayesian predictive ensemble method for single and coupled unit operation models. Process 2022;10:662.
125. Wutz J, Waterkotte B, Heitmann K, Wucherpfennig T. Computational fluid dynamics (CFD) as a tool for industrial UF/DF tank optimization. Biochem Eng J 2020;160:107617.

126. Wutz J, Steiner R, Assfalg K, Wucherpfennig T. Establishment of a CFD-based kLa model in microtiter plates to support CHO cell culture scale-up during clone selection. Biotechnol Progr 2018;34:1120–8.
127. Wutz J, Lapin A, Siebler F, Schaefer JE, Wucherpfennig T, Berger M, et al. Predictability of kLa in stirred tank reactors under multiple operating conditions using an Euler–Lagrange approach. Eng Life Sci 2016;16: 633–42.
128. Kuschel M, Fitschen J, Hoffmann M, von Kameke A, Schlüter M, Wucherpfennig T. Validation of Novel Lattice Boltzmann Large Eddy Simulations (LB LES) for equipment characterization in biopharma. Process 2021;9:950.
129. Kuschel M, Wutz J, Salli M, Monteil D, Wucherpfennig T. CFD supported scale up of perfusion bioreactors in biopharma. Front Chem Eng 2023;18:1076509.
130. FDA. PAT—a framework for innovative pharmaceutical development, manufacturing, and quality assurance. In: Pharmaceutical CGMPs; 2004. [Online]. Available: https://www.fda.gov/regulatory-information/search-fda-guidance-documents/pat-framework-innovative-pharmaceutical-development-manufacturing-and-quality-assurance.
131. Wasalathanthri DP, Rehmann MS, Song Y, Gu Y, Mi L, Shao C, et al. Technology outlook for real-time quality attribute and process parameter monitoring in biopharmaceutical development—a review. Biotechnol Bioeng 2020;117:3182–98.
132. Buckley K, Ryder AG. Applications of Raman spectroscopy in biopharmaceutical manufacturing: a short review. Appl Spectrosc 2017;71:1085–116.
133. Berry B, Moretto J, Matthews T, Smelko J, Wiltberger K. Cross-scale predictive modeling of CHO cell culture growth and metabolites using Raman spectroscopy and multivariate analysis. Biotechnol Progr 2015;31: 566–77.
134. Domján J, Pantea E, Gyuerkés M, Madarász L, Kozák D, Farkas A, et al. Real-time amino acid and glucose monitoring system for the automatic control of nutrient feeding in CHO cell culture using Raman spectroscopy. Biotechnol J 2022;17:2100395.
135. Romann P, Kolar J, Tobler D, Herwig C, Bielser J, Villiger TK. Advancing Raman model calibration for perfusion bioprocesses using spiked harvest libraries. Biotechnol J 2022;17:2200184.
136. Santos RM, Kessler J, Salou P, Menezes JC, Peinado A. Monitoring mAb cultivations with in-situ Raman spectroscopy: the influence of spectral selectivity on calibration models and industrial use as reliable PAT tool. Biotechnol Progr 2018;34:659–70.
137. Abu-Absi NR, Kenty BM, Cuellar ME, Borys MC, Sakhamuri S, Strachan DJ, et al. Real time monitoring of multiple parameters in mammalian cell culture bioreactors using an in-line Raman spectroscopy probe. Biotechnol Bioeng 2011;108:1215–21.
138. Gibbons L, Rafferty C, Robinson K, Abad M, Maslanka F, Le N, et al. Raman based chemometric model development for glycation and glycosylation real time monitoring in a manufacturing scale CHO cell bioreactor process. Biotechnol Progr 2021;e3223. https://doi.org/10.1002/btpr.3223.
139. Liu Z, Zhang Z, Qin Y, Chen G, Hun J, Wang Q, et al. The application of Raman spectroscopy for monitoring product quality attributes in perfusion cell culture. Biochem Eng J 2021;173:108064.
140. Park S-Y, Park C-H, Choi D-H, Hong JK, Lee D-Y. Bioprocess digital twins of mammalian cell culture for advanced biomanufacturing. Curr Opin Chem Eng 2021;33:100702.
141. Eyster T, Talwar S, Fernandez J, Foster S, Hayes J, Allen R, et al. Tuning monoclonal antibody galactosylation using Raman spectroscopy-controlled lactic acid feeding. Biotechnol Progr 2021;37. https://doi.org/10.1002/btpr.3085.
142. Gillespie C, Wasalathanthri DP, Ritz DB, Zhou G, Davis KA, Wucherpfennig T, et al. Systematic assessment of process analytical technologies for biologics. Biotechnol Bioeng 2021. https://doi.org/10.1002/bit.27990.
143. Metze S, Blioch S, Matuszczyk J, Greller G, Grimm C, Scholz J, et al. Multivariate data analysis of capacitance frequency scanning for online monitoring of viable cell concentrations in small-scale bioreactors. Anal Bioanal Chem 2020;412:2089–102.

144. Rittershaus ESC, Rehmann MS, Xu J, He Q, Hill C, Swanberg J, et al. N-1 perfusion platform development using a capacitance probe for biomanufacturing. Bioengineering 2022;9:128.
145. Moore B, Sanford R, Zhang A. Case study: the characterization and implementation of dielectric spectroscopy (biocapacitance) for process control in a commercial GMP CHO manufacturing process. Biotechnol Progr 2019;35:e2782.
146. Zhang A, Liu Tsang V, Moore B, Shen V, Huang YM, Kshrisagar R, et al. Advanced process monitoring and feedback control to enhance cell culture process production and robustness. Biotechnol Bioeng 2015;112. https://doi.org/10.1002/bit.25684.
147. Opel CF, Li J, Amanullah A. Quantitative modeling of viable cell density, cell size, intracellular conductivity, and membrane capacitance in batch and fed-batch CHO processes using dielectric spectroscopy. Biotechnol Progr 2010;26:1187–99.
148. Ma F, Zhang A, Chang D, Velev OD, Wiltberger K, Kshirsagar R. Real-time monitoring and control of CHO cell apoptosis by in situ multifrequency scanning dielectric spectroscopy. Process Biochem 2019;80: 138–45.
149. Cannizzaro C, Gügerli R, Marison I, von Stockar U. On-line biomass monitoring of CHO perfusion culture with scanning dielectric spectroscopy. Biotechnol Bioeng 2003;84:597–610.
150. Wu S, Ketcham SA, Corredor CC, Both D, Drennen JK, Anderson CA. Rapid at-line early cell death quantification using capacitance spectroscopy. Biotechnol Bioeng 2022;119:857–67.
151. Woodgate JM. Perfusion N-1 culture—opportunities for process intensification. In: Biopharmaceutical Processing - Development, Design, and Implementation of Manufacturing Processes. Amsterdam: Elsevier; 2018:755–768 pp.
152. Claßen J, Graf A, Aupert F, Solle D, Höhse M, Scheper T. A novel LED-based 2D-fluorescence spectroscopy system for in-line bioprocess monitoring of Chinese hamster ovary cell cultivations – Part II. Eng Life Sci 2019;19:341–51.
153. Graf A, Claßen J, Solle D, Hitzmann B, Rebner K, Hoehse M. A novel LED-based 2D-fluorescence spectroscopy system for in-line monitoring of Chinese hamster ovary cell cultivations – Part I. Eng Life Sci 2019;19:352–62.
154. Li B, Shanahan M, Calvet A, Leister KJ, Ryder AG. Comprehensive, quantitative bioprocess productivity monitoring using fluorescence EEM spectroscopy and chemometrics. Analyst 2014;139:1661–71.
155. Ryder AG. Cell culture media analysis using rapid spectroscopic methods. Curr Opin Chem Eng 2018;22: 11–7.
156. Neutsch L, Kroll P, Brunner M, Pansy A, Kovar M, Herwig C, et al. Media photo-degradation in pharmaceutical biotechnology – impact of ambient light on media quality, cell physiology, and IgG production in CHO cultures. J Chem Technol Biotechnol 2018;93:2141–51.
157. Calvet A, Li B, Ryder AG. A rapid fluorescence based method for the quantitative analysis of cell culture media photo-degradation. Anal Chim Acta 2014;807:111–9.
158. Ryan PW, Li B, Shanahan M, Leister KJ, Ryder AG. Prediction of cell culture media performance using fluorescence spectroscopy. Anal Chem 2010;82:1311–7.
159. Bhatia H, Mehdizadeh H, Drapeau D, Yoon S. In-line monitoring of amino acids in mammalian cell cultures using Raman spectroscopy and multivariate chemometrics models. Eng Life Sci 2018;18:55–61.
160. Li M, Ebel B, Paris C, Chauchard F, Guedon E, Marc A. Real-time monitoring of antibody glycosylation site occupancy by in situ Raman spectroscopy during bioreactor CHO cell cultures. Biotechnol Progr 2018;34: 486–93.
161. Webster TA, Hadley BC, Dickson M, Busa JK, Jaques C, Mason C. Feedback control of two supplemental feeds during fed-batch culture on a platform process using inline Raman models for glucose and phenylalanine concentration. Bioproc Biosyst Eng 2021;44:127–40.
162. Matthews TE, Smelko JP, Berry B, Romero-Torres S, Hill D, Kshirsagar R, et al. Glucose monitoring and adaptive feeding of mammalian cell culture in the presence of strong autofluorescence by near infrared Raman spectroscopy. Biotechnol Progr 2018;34:1574–80.

163. Milewska A, Baekelandt G, Boutaieb S, Mozin V, Falconbridge A. In-line monitoring of protein concentration with MIR spectroscopy during UFDF. Eng Life Sci 2023;23:e2200050.

164. Thakur G, Hebbi V, Rathore AS. Near Infrared Spectroscopy as a PAT tool for monitoring and control of protein and excipient concentration in ultrafiltration of highly concentrated antibody formulations. Int J Pharm 2021;600:120456.

165. Rolinger L, Hubbuch J, Rüdt M. Monitoring of ultra- and diafiltration processes by Kalman-filtered Raman measurements. Anal Bioanal Chem 2023;415:841–54.

166. Rüdt M, Brestrich N, Rolinger L, Hubbuch J. Real-time monitoring and control of the load phase of a protein A capture step. Biotechnol Bioeng 2017;114:368–73.

167. Thakur G, Hebbi V, Rathore AS. An NIR-based PAT approach for real-time control of loading in protein A chromatography in continuous manufacturing of monoclonal antibodies. Biotechnol Bioeng 2020;117: 673–86.

168. Thakur G, Bansode V, Rathore AS. Continuous manufacturing of monoclonal antibodies: automated downstream control strategy for dynamic handling of titer variations. J Chromatogr A 2022;1682:463496.

169. Rolinger L, Rüdt M, Hubbuch J. A critical review of recent trends, and a future perspective of optical spectroscopy as PAT in biopharmaceutical downstream processing. Anal Bioanal Chem 2020;412: 2047–2064. https://doi.org/10.1007/s00216-020-02407-z.

170. Schulze M, Niemann J, Wijffels RH, Matuszczyk J, Martens DE. Rapid intensification of an established CHO cell fed-batch process. Biotechnol Progr 2022;38:e3213.

171. Walsh G. Biopharmaceutical benchmarks 2018. Nat Biotechnol 2018;36:1136–45.

172. Hunter M, Yuan P, Vavilala D, Fox M. Optimization of protein expression in mammalian cells. Curr Protoc Protein Sci 2019;95:e77.

173. Wurm FM. Production of recombinant protein therapeutics in cultivated mammalian cells. Nat Biotechnol 2004;22:1393–8.

174. Bertrand V, Karst DJ, Bachmann A, Cantalupo K, Soos M, Morbidelli M. Transcriptome and proteome analysis of steady-state in a perfusion CHO cell culture process. Biotechnol Bioeng 2019;116:1959–72.

175. Wolf MKF, Pechlaner A, Lorenz V, Karst DJ, Souquet J, Broly H, et al. A two-step procedure for the design of perfusion bioreactors. Biochem Eng J 2019;151:107295.

176. Karst DJ, Steinebach F, Soos M, Morbidelli M. Process performance and product quality in an integrated continuous antibody production process. Biotechnol Bioeng 2017;114:298–307.

177. Kim M, O'Callaghan PM, Droms KA, James DC. A mechanistic understanding of production instability in CHO cell lines expressing recombinant monoclonal antibodies. Biotechnol Bioeng 2011;108:2434–46.

178. Veith N, Ziehr H, MacLeod RAF, Reamon-Buettner SM. Mechanisms underlying epigenetic and transcriptional heterogeneity in Chinese hamster ovary (CHO) cell lines. BMC Biotechnol 2016;16:6.

179. Jamnikar U, Nikolic P, Belic A, Blas M, Gaser D, Francky A, et al. Transcriptome study and identification of potential marker genes related to the stable expression of recombinant proteins in CHO clones. BMC Biotechnol 2015;15:98.

180. Yusufi FNK, Lakshmanan M, Ho YS, Loo BLW, Ariyaratne P, Yang Y, et al. Mammalian systems biotechnology reveals global cellular adaptations in a recombinant CHO cell line. Cell Syst 2017;4: 530–42.e6.

181. Lee JS, Park JH, Ha TK, Samoudi M, Lewis NE, Palsson BO, et al. Revealing key determinants of clonal variation in transgene expression in recombinant CHO cells using targeted genome editing. ACS Synth Biol 2018;7:2867–78.

182. Kostyrko K, Neuenschwander S, Junier T, Regamey A, Iseli C, Schmid-Siegert E, et al. MAR-Mediated transgene integration into permissive chromatin and increased expression by recombination pathway engineering. Biotechnol Bioeng 2017;114:384–96.

183. Dahodwala H, Lee KH. The fickle CHO: a review of the causes, implications, and potential alleviation of the CHO cell line instability problem. Curr Opin Biotech 2019;60:128–37.

184. Balasubramanian S, Rajendra Y, Baldi L, Hacker DL, Wurm FM. Comparison of three transposons for the generation of highly productive recombinant CHO cell pools and cell lines. Biotechnol Bioeng 2016;113: 1234–43.

185. Muñoz-López M, García-Pérez JL. DNA transposons: nature and applications in genomics. Curr Genomics 2010;11:115–28.

186. Villiger-Oberbek A, Yang Y, Zhou W, Yang J. Development and application of a high-throughput platform for perfusion-based cell culture processes. J Biotechnol 2015;212:21–9.

187. Gomez N, Ambhaikar M, Zhang L, Huang C, Barkhordarian H, Lull J, et al. Analysis of Tubespins as a suitable scale-down model of bioreactors for high cell density CHO cell culture. Biotechnol Progr 2017;33: 490–9.

188. Wolf MKF, Lorenz V, Karst DJ, Souquet J, Broly H, Morbidelli M. Development of a shake tube-based scale-down model for perfusion cultures. Biotechnol Bioeng 2018;115:2703–13.

189. Kreye S, Stahn R, Nawrath K, Goralczyk V, Zoro B, Goletz S. A novel scale-down mimic of perfusion cell culture using sedimentation in an automated microbioreactor (SAM). Biotechnol Progr 2019;35:e2832.

190. Yin L, Au WY, Yu CC, Kwon T, Lai Z, Shang M, et al. Miniature auto-perfusion bioreactor system with spiral microfluidic cell retention device. Biotechnol Bioeng 2021;118:1951–61.

191. Jin L, Wang Z-S, Cao Y, Sun R-Q, Zhou H, Cao R-Y. Establishment and optimization of a high-throughput mimic perfusion model in ambr® 15. Biotechnol Lett 2021;43:423–33.

192. Mozdzierz NJ, Love KR, Lee KS, Lee HLT, Shah KA, Ram RJ, et al. A perfusion-capable microfluidic bioreactor for assessing microbial heterologous protein production. Lab Chip 2015;15:2918–22.

193. Hong JK, Lakshmanan M, Goudar C, Lee D-Y. Towards next generation CHO cell line development and engineering by systems approaches. Curr Opin Chem Eng 2018;22:1–10.

194. Cui Y, Cui P, Chen B, Li S, Guan H. Monoclonal antibodies: formulations of marketed products and recent advances in novel delivery system. Drug Dev Ind Pharm 2017;43:1–39.

195. Fischer S, Handrick R, Otte K. The art of CHO cell engineering: a comprehensive retrospect and future perspectives. Biotechnol Adv 2015;33:1878–96.

196. Xu X, Nagarajan H, Lewis NE, Pan S, Cai Z, Liu X, et al. The genomic sequence of the Chinese hamster ovary (CHO)-K1 cell line. Nat Biotechnol 2011;29:735–41.

197. DiMasi JA, Grabowski HG, Hansen RW. Innovation in the pharmaceutical industry: new estimates of R&D costs. J Health Econ 2016;47:20–33.

198. Paul SM, Mytelka DS, Dunwiddie CT, Persinger CC, Munos BH, Lindborg SR, et al. How to improve R&D productivity: the pharmaceutical industry's grand challenge. Nat Rev Drug Discov 2010;9:203–14.

199. Jagschies G. Management of Process Economy—Case Studies. In: Biopharmaceutical processing. Amsterdam: Elsevier; 2018:1191–1223 pp.

200. Roush D, Asthagiri D, Babi AK, Benner S, Bilodeau C, Carta G, et al. Toward in silico CMC: an industrial collaborative approach to model-based process development. Biotechnol Bioeng 2020;117:3986–4000.

201. Maracelias CT. Chemical production scheduling mixed integer programming models and methods | Chemical engineering | Cambridge University Press. In: Cambridge series in chemical engineering. Cambridge: Cambridge University Press; 2021.

202. Babi DK, Griesbach J, Hunt S, Insaidoo F, Roush D, Todd R, et al. Opportunities and challenges for model utilization in the biopharmaceutical industry: current versus future state. Curr Opin Chem Eng 2022;36: 100813.

203. Pistikopoulos EN, Tian Y, Bindlish R. Operability and control in process intensification and modular design: challenges and opportunities. AIChE J 2021;67. https://doi.org/10.1002/aic.17204.

204. Petrides D, Carmichael D, Siletti C, Koulouris A. Biopharmaceutical process optimization with simulation and scheduling tools. Bioengineering 2014;1:154–87.

205. Farid SS, Baron M, Stamatis C, Nie W, Coffman J. Benchmarking biopharmaceutical process development and manufacturing cost contributions to R&D. mAbs 2020;12:1754999.

206. Xu S, Gavin J, Jiang R, Chen H. Bioreactor productivity and media cost comparison for different intensified cell culture processes. Biotechnol Progr 2017;33:867–78.
207. Souza JD, Scott K, Genest P. Virus-filtration process development optimization: the key to a more efficient and cost-effective step. BioProcess Int 2016;41:62–74.
208. Franzreb M, Müller E, Vajda J. Cost estimation for protein a chromatography: an in silico approach to MAb purification strategy. BioProcess Int 2014;12:44–52.
209. Broly H, Costioli M, Guillemot-Potelle C, Mitchell-Logean C. Cost of goods modeling and quality by design for developing cost-effective processes. Biopharm Int 2010;23.
210. Hummel J, Pagkaliwangan M, Gjoka X, Davidovits T, Stock R, Ransohoff T, et al. Modeling the downstream processing of monoclonal antibodies reveals cost advantages for continuous methods for a broad range of manufacturing scales. Biotechnol J 2019;14:1700665.
211. Mahal H, Branton H, Farid SS. End-to-end continuous bioprocessing: impact on facility design, cost of goods, and cost of development for monoclonal antibodies. Biotechnol Bioeng 2021;118:3468–85.
212. UNFCC the Paris agreement. UNFCCC; 2016. [Online]. Available: https://unfccc.int/process-and-meetings/the-paris-agreement/the-paris-agreement. [Accessed 28 Sep 2022].
213. Carbon dioxide now more than 50 % higher than pre-industrial levels. [Online]. Available: https://www.noaa.gov/news-release/carbon-dioxide-now-more-than-50-higher-than-pre-industrial-levels?fbclid=IwAR3_PAk4AmI4czOO5ikK_CAGca94LMwQwIEfG9lo3ZWi72BeR6KaX05hHSw. [Accessed 28 Sep 2022].
214. Belkhir L, Elmeligi A. Carbon footprint of the global pharmaceutical industry and relative impact of its major players. J Clean Prod 2019;214:185–94.
215. Budzinski K, Constable D, D'Aquila D, Smith P, Madabhushi SR, Whiting A, et al. Streamlined life cycle assessment of single use technologies in biopharmaceutical manufacture. New Biotechnol 2022;68:28–36.
216. Budzinski K, Blewis M, Dahlin P, D'Aquila D, Esparza J, Gavin J, et al. Introduction of a process mass intensity metric for biologics. New Biotechnol 2019;49:37–42.

Index

https://doi.org/10.1515/9783110760330-013

www.ingramcontent.com/pod-product-compliance
Lightning Source LLC
Chambersburg PA
CBHW080707220326
41598CB00033B/5341